Advances in

GEOPHYSICS

VOLUME 50

Advances in Geophysics

Volume 50

Earth Heterogeneity and Scattering Effects on Seismic Waves

Series Editor

RENATA DMOWSKA

School of Engineering and Applied Sciences
Harvard University
Cambridge, Massachusetts, USA

Guest Editors

HARUO SATO
Department of Geophysics,
Graduate School of Science,
Tohoku University
Sendai, Japan

MICHAEL C. FEHLER
Department of Earth,
Atmospheric, and Planetary Sciences,
Massachusetts Institute of Technology
Cambridge, USA

AMSTERDAM • BOSTON • HEIDELBERG • LONDON
NEW YORK • OXFORD • PARIS • SAN DIEGO
SAN FRANCISCO • SINGAPORE • SYDNEY • TOKYO

Academic Press is an imprint of Elsevier

Academic Press is an imprint of Elsevier
Radarweg 29, PO Box 211, 1000 AE Amsterdam, The Netherlands
32 Jamestown Road, London NW1 7BY, UK
360 Park Avenue South, New York, NY 10010-1710
30 Corporate Drive, Suite 400, Burlington, MA 01803, USA
525 B Street, Suite 1900, San Diego, CA 92101-4495, USA

First edition 2008

Copyright © 2008 Elsevier Inc. All rights reserved.

No part of this publication may be reproduced, stored in a retrieval system or transmitted in any form or by any means electronic, mechanical, photocopying, recording or otherwise without the prior written permission of the publisher

Permissions may be sought directly from Elsevier's Science & Technology Rights Department in Oxford, UK: phone (+44) (0) 1865 843830; fax (+44) (0) 1865 853333; email: permissions@elsevier.com. Alternatively you can submit your request online by visiting the Elsevier web site at http://elsevier.com/locate/permissions, and selecting Obtaining permission to use Elsevier material

Notice
No responsibility is assumed by the publisher for any injury and/or damage to persons or property as a matter of products liability, negligence or otherwise, or from any use or operation of any methods, products, instructions or ideas contained in the material herein. Because of rapid advances in the medical sciences, in particular, independent verification of diagnoses and drug dosages should be made

ISBN: 978-0-12-374509-5

ISSN: 0065-2687

For information on all Academic Press publications
visit our website at books.elsevier.com

Printed and bound in Hungary

08 09 10 11 12   10 9 8 7 6 5 4 3 2 1

Working together to grow
libraries in developing countries

www.elsevier.com | www.bookaid.org | www.sabre.org

ELSEVIER   BOOK AID International   Sabre Foundation

Dedicated to Keiiti Aki (1930–2005)
Pioneer of short-period seismology

CONTENTS

Contributors xv
Preface xvii

Chapter 1

Coherent Back-Scattering and Weak Localization of Seismic Waves

Ludovic Margerin

1. Introduction 1
2. Weak Localization Effect: A Heuristic View 3
3. The Role of Source Mechanism and Wavefield Polarization 6
   3.1. Effect of Source Mechanism 6
   3.2. Review of Multiple Scattering Formalism 7
   3.3. Theoretical Results for Acoustic and Elastic Waves 10
4. Geophysical Applications 13
   4.1. Measurement of the Dispersion Relation of Surface Waves 13
   4.2. Measurement of the Diffusion Constant in Strongly
         Scattering Media 16
5. Conclusion 17
   Acknowledgments 17
   References 17

Chapter 2

Theory of Transmission Fluctuations in Random Media with a Depth-Dependent Background Velocity Structure

Yingcai Zheng and Ru-Shan Wu

1. Introduction 21
2. Acoustic Waves in Stratified Media and WKBJ Green Function 23
3. Rytov Solution to the Wave Equation in a Heterogeneous Medium . . . 25
4. Complex Phase ψ Due to a Plane Wave Incidence 26
5. Coherence Function Between Two Plane Waves 29
6. Coherence Functions Using Delta-Correlated Assumption 33
7. Coherence Functions in a Constant Background Medium 34
8. Numerical Examples 34
9. Validity of the Delta-Correlated Assumption 36
10. Discussions and Conclusions 37
    Acknowledgments 38
    References 39

Chapter 3

Synthesis of Vector-Wave Envelopes in Random Elastic Media on the Basis of the Markov Approximation

HARUO SATO AND MICHAEL KORN

1. Introduction 44
   1.1. Markov Approximation for the Wave Envelope Synthesis 44
   1.2. Analyses of Seismogram Envelopes 45
   1.3. Objectives 48
2. Vector-Wave Envelopes for the Plane Wavelet Incidence 51
   2.1. Three-Dimensional Random Elastic Media 51
   2.2. Two-Dimensional Random Elastic Media 64
3. Vector-Wave Envelopes for the Radiation from a Point Source 69
   3.1. Three-Dimensional Random Elastic Media 69
   3.2. Two-Dimensional Random Elastic Media 78
4. Discussions 83
   4.1. RTT with the Born Approximation Scattering Coefficients 83
   4.2. Realistic ACFs for Random Media 85
5. Summary 86
   Acknowledgments 87
   Appendix 87
   References 91

Chapter 4

Geometrical Optics of Acoustic Media with Anisometric Random Heterogeneities: Travel-Time Statistics of Reflected and Refracted Waves

AYSE KASLILAR, YURY A. KRAVTSOV AND SERGE A. SHAPIRO

1. Introduction 95
2. Basic Elements of the GO Method 97
   2.1. Basic Equations of the GO 97
   2.2. Model of Quasi-Homogeneous Fluctuations of Medium Parameters 99
   2.3. Travel-Time Covariance Function in a Medium with Anisometric Fluctuations 100
   2.4. Boundary of GO Applicability 101
3. Travel-Time Fluctuations in Reflection Geometry 102
   3.1. Reflection Geometry 102
   3.2. Travel-Time Covariance Function for Small Offsets 104
   3.3. Double Passage Effect 106
4. Travel-Time Fluctuations in Refraction Geometry 107
   4.1. Refracting Medium with a Constant Velocity Gradient 107
   4.2. Travel-Time Variance along a Curvilinear Ray 108
   4.3. Dependence of Travel-Time Variance on Offset 110
   4.4. Inverse Problem Solution for Refraction Geometry 112
5. Results of Numerical Simulations 115
6. Discussion and Conclusion 118
   Acknowledgements 120
   References 120

Chapter 5

Attenuation of Seismic Waves Due to Wave-Induced Flow and Scattering in Randomly Heterogeneous Poroelastic Continua

TOBIAS M. MÜLLER, BORIS GUREVICH AND SERGE A. SHAPIRO

1. Introduction 123
2. Meso- and Macroscopic Heterogeneity in the Earth and Its
   Description as a Random Medium 126
3. Attenuation and Dispersion of Seismic Waves due to
   Wave-Induced Flow 129
   3.1. Biot's Equations of Dynamic Poroelasticity and
        Associated Green's Functions 129
   3.2. The Basic Poroelastic Scattering Equation 133
   3.3. First-Order Statistical Smoothing Approximation 134
   3.4. Effective Fast Wave Number Accounting for
        Conversion Scattering into Slow P Waves 135
   3.5. Attenuation and Dispersion due to Wave-Induced Flow 138
   3.6. Asymptotic Behavior at Low and High Frequencies 145
4. Attenuation of Seismic Waves in Random Porous
   Media due to Scattering 148
   4.1. The Generalized ODA Formalism 148
   4.2. Effective Wave Number in 3-D Random Media 149
   4.3. Scattering Attenuation and Asymptotic Behavior 154
5. The Interplay Between Attenuation Due Interlayer Flow
   and Scattering 158
   5.1. 1-D Poroelastic Random Media 158
   5.2. Asymptotic Scaling of Attenuation 162
6. Concluding Remarks 163
   Acknowledgments 164
   References 164

Chapter 6

Observing and Modeling Elastic Scattering in the Deep Earth

PETER M. SHEARER AND PAUL S. EARLE

1. Introduction 167
2. Data Stacking 168
   2.1. Shallow-Versus Deep-Earthquake Teleseismic P Coda 170
   2.2. Regional Variations in Teleseismic P Coda Amplitude 170
3. Monte Carlo Methods 174
   3.1. Seismology Applications 175
   3.2. Monte Carlo Implementation 176
   3.3. The Monte Carlo Source 177
   3.4. Particle Trajectories 178
   3.5. Scattering Angles 181
   3.6. Intrinsic Attenuation 185
4. Fit to Teleseismic P Coda 187

5. Conclusions	188
Acknowledgments	190
References	190

Chapter 7

A Scattering Waveguide in the Heterogeneous Subducting Plate

TAKASHI FURUMURA AND BRIAN L.N. KENNETT

1. Introduction	196
2. Anomalous Intensity Patterns from Two Deep Events in the Subducted Philippine Sea Plate and in the Subducted Pacific Plate	198
2.1. Separation of Low-Frequency Precursors and High-Frequency Coda	201
2.2. Frequency Selective Propagation Properties in the Subducting Plate	201
3. 2D FDM Modeling of Scattering Wavefield	204
4. 2D FDM Modeling of Slab Guided Waves	208
4.1. Base Model: High-Q and High-V Subduction Zone	208
4.2. Heterogeneous Plate Model: Isotropic Heterogeneities in the Plate	210
4.3. Anisotropic Heterogeneities in the Plate	211
4.4. Effect of Plate Thickness	213
4.5. Effect of Heterogeneity Scale in the Plate	214
5. Discussion and Conclusion	215
Acknowledgments	216
References	216

Chapter 8

Laboratory Experiments of Seismic Wave Propagation in Random Heterogeneous Media

OSAMU NISHIZAWA AND YO FUKUSHIMA

1. Introduction	219
2. Laboratory Experiments	221
2.1. Statistical Description of Heterogeneity	221
2.2. Wave Fields in Random Media	223
3. Scale-Invariant Expression	227
4. Waveform Analysis	230
4.1. Travel-Time Fluctuation	231
4.2. Cross Spectra Between Waves	231
4.3. Shear-Wave Particle Velocities	236
4.4. Waveform Envelope	238
5. Key Features of Wave Fluctuation in Random Media	240
5.1. Masking Signal Waves by Small-Scale Heterogeneities	240
5.2. Boundary Between EHM and SRM	241
5.3. Diffraction of Scattered Waves	242

6. Conclusions 243
   6.1. Validity of Equivalent Homogeneous Medium Assumption 243
   6.2. Random Media Effect on Seismic Data Processing 244
   6.3. Role of Laboratory Experiments for Studying Seismic
        Wave Propagation 244
   Acknowledgments 244
   References 245

Chapter 9

Measurements of the Earth at the Scale of Logs, Crosswells, and VSPs

ARTHUR C.H. CHENG

1. Introduction 247
2. Acoustic Logging 249
   2.1. Dipole Logging 250
   2.2. Modern Array Processing 250
   2.3. Depth of Investigation 253
3. Crosswell Seismic Survey 255
   3.1. Resolution of a Crosswell Seismic Survey 256
4. Vertical Seismic Profiling 259
5. Discussions and Summary 261
   Acknowledgments 262
   References 262

Chapter 10

Coda Energy Distribution and Attenuation

KAZUO YOSHIMOTO AND ANSHU JIN

1. Introduction 265
2. Coda Energy Distribution and Measurement on $Q_{P,S}^{-1}$
   using Local Seismograms 268
   2.1. Uniformity of Coda Energy Distribution 268
   2.2. Nonuniform Coda Energy Distribution in
        Tectonically Active Regions 273
3. Temporal Decay Rate of Coda Energy: Q_C^{-1} 278
   3.1. Lapse-Time Dependence 278
   3.2. Frequency Dependence 279
   3.3. Geographic Variation 281
   3.4. Temporal Variation 284
   3.5. Models to Explain the Spatio-Temporal Correlation
        Between Q_C^{-1} and Seismicity 289
4. Closing Remarks 292
   Acknowledgments 293
   References 293

Chapter 11

Imaging Inhomogeneous Structures in the Earth by Coda Envelope Inversion and Seismic Array Observation

KIN'YA NISHIGAMI AND SATOSHI MATSUMOTO

1. Introduction 301
2. Analysis of Seismic Network Data 302
    2.1. Inversion of Coda Envelope 302
    2.2. Kirchhoff Coda Migration 306
3. Analysis of Seismic Array Data 307
    3.1. Detection of Seismic Signals by Array Observations 307
    3.2. Single-Scattering Model for Seismic Array 309
    3.3. Characteristics of Coda Waves Based on Array Observations 311
    3.4. Scatterer/Inhomogeneity Distribution Inferred from
         Seismic Array Data 312
4. Summary 315
    Acknowledgments 316
    References 316

Chapter 12

Source Effects From Broad Area Network Calibration of Regional Distance Coda Waves

WILLIAM SCOTT PHILLIPS, RICHARD JEROME STEAD, GEORGE EDWARD RANDALL, HANS EDWARD HARTSE AND KEVIN MITSUO MAYEDA

1. Introduction 319
2. Data Analysis 321
3. Coda Calibration Methodology 326
    3.1. Coda Start Time Calibration 331
    3.2. Coda Shape Calibration and Amplitude Measurement 331
    3.3. Intrastation Site Calibration 333
    3.4. 2-D Path and Interstation Site Calibration 335
    3.5. Source to Coda Transfer Function 340
4. Coda Spectral Results 342
5. Discussion 345
6. Conclusions 348
    Acknowledgments 349
    References 349

Chapter 13

Seismic Wave Scattering in Volcanoes

EDOARDO DEL PEZZO

1. Introduction 353
    1.1. Volcanic Earthquakes 353
    1.2. A Brief Review of Coda-Q^{-1} Observation on Volcanoes 354

2. Separated Estimates of Intrinsic and Scattering Attenuation 357
   2.1. The Method of Wennerberg 358
   2.2. The Energy-Flux Model 359
   2.3. 2-D Transport Theory Applied to Volcanic Tremor 360
3. Diffusion Model Applied to Shot Data 361
   3.1. Uniform Half Space 361
   3.2. Two-Layer Media 363
4. Energy-Transport Theory Applied to Earthquake Data 365
   4.1. Uniform Half Space 365
   4.2. Possible Bias Introduced by Assuming a Uniform Diffusive Layer 366
   4.3. Coda-Localization Effects 366
5. Concluding Remarks 367
   Acknowledgments 369
   References 369

Chapter 14

Monitoring Temporal Variations of Physical Properties in the Crust by Cross-Correlating the Waveforms of Seismic Doublets

GEORGES POUPINET, JEAN-LUC GOT AND FLORENT BRENGUIER

1. Introduction 374
2. Selection of Doublets 374
3. Basic Processing 375
   3.1. Time Delays Measured from Cross-Correlation
        or Cross-Spectrum 375
   3.2. Cross-Spectral Moving Window or Cross-Correlation
        Moving Window Technique 377
4. Relocating Doublets from P and S Travel-Time Delays 378
   4.1. Double-Difference Location 378
   4.2. Two Synthetic Examples with IASP91 Travel Times 379
   4.3. Possible Technical and Intrinsic Difficulties 380
5. An Example of Observed Delays: An Excellent Doublet in Japan 382
6. Slope of the Delay in the Coda and the Measurement of
   S-Velocity Temporal Variation 384
7. Possible Artifacts in $\Delta V_S/V_S$ Measurement: Arguments from
   the Coda of Spatial Doublets 385
8. Search for Temporal Variation of S-Wave Splitting 389
9. Search for Temporal Variation of Coda Attenuation 390
10. "Virtual Doublets" Computed by Cross-Correlating Seismic Noise 391
11. PKP from Teleseismic Doublets and the Rotation
    of the Inner Core 393
12. Conclusion 395
    Acknowledgements 395
    References 395

Chapter 15

Seismogram Envelope Inversion for High-Frequency Seismic Energy Radiation from Moderate-to-Large Earthquakes

HISASHI NAKAHARA

1. Introduction	402
2. Envelope Inversion Methods	403
2.1. General Framework	403
2.2. A Classification of Current Envelope Inversion Methods	404
2.3. The Method of Nakahara *et al.* (1998)	406
3. Data Analysis and the Results	411
3.1. An Example of Practical Data Analysis	411
4. Compilation of the Results	413
4.1. Frequency Dependence of High-Frequency Seismic Energy	414
4.2. Scaling of High-Frequency Seismic Energy	418
4.3. Spatial Relationship Between Asperities and High-Frequency Sources	420
5. Conclusions	423
Acknowledgments	424
References	424

Chapter 16

On the Random Nature of Earthquake Sources and Ground Motions: A Unified Theory

DANIEL LAVALLÉE

1. Introduction	427
2. Random Model of Earthquakes Slip Spatial Distribution and Consequences for the Ground Motions	429
2.1. From the Source	429
2.2. ... to the Ground Motion	433
3. The 2004 Parkfield Earthquake	435
3.1. Random Model of the Source	435
3.2. Random Model of the Ground Motion PGA	440
3.3. Random Model of the Ground Motion PGV	443
4. The 1999 Chi-Chi Earthquake	444
4.1. Random Model of the Source	444
4.2. Random Model of the Ground Motion PGA	445
5. Limitations of the Model	448
6. Conclusion: From Randomness to Invariance	454
Acknowledgments	455
References	458

GLOSSARY	463
INDEX	469

CONTRIBUTORS

Numbers in parentheses indicate the pages on which the authors' contributions begin.

BRENGUIER, F. (373) LGIT, Université Joseph Fourier & CNRS, BP53, 38041, Grenoble, France

CHENG, A.C.H. (247) Cambridge GeoSciences, 14090 Southwest Freeway, Suite 300, Sugar Land, TX 77478, USA

DEL PEZZO, E. (353) INGV - Osservatorio Vesuviano. Via Diocleziano, 328. 80124 Napoli January 21, 2008

EARLE, P.S. (167) United States Geological Survey, MS 966 DFC, Denver, CO 80225

FURUMURA, T. (219) Disaster Prevention Research Institute Kyoto University, Gokasho, Uji, Kyoto, 611-0011

GOT, J.-L. (373) LGIT, Université de Savoie & CNRS, 73000, Le Bourget du Lac, France

GUREVICH, B. (123) Curtin University of Technology and CSIRO Division of Petroleum resources, Perth, Australia

HARTSE, H.E. (319) Los Alamos National Laboratory

JIN, A. (265) 139W. Phillips Blvd. #315, Pomona, CA 91766, USA

KASLILAR, A. (95) Istanbul Technical University, Mining Faculty, Department of Geophysics, 34390 Maslak, Istanbul, Turkey

KENNETT, B.L.N. (195) Research School of Earth Sciences, The Australian National University, Canberra, ACT 0200, Australia

KORN, M. (43) Institute of Geophysics and Geology, University of Leipzig, Talstrasse 35, D-04103 Leipzig, Germany

KRAVTSOV, Y.A. (95) Institute of Physics, Maritime University of Szczecin, Szczecin, 70-500, Poland and Space Research Institute, Russian Academic Science, Moscow, 117 997, Russia

LAVALLÉE, D. (427) Institute for Crustal Studies, University of California, Santa Barbara, California, USA

MARGERIN, L. (1) Centre Européen de Recherche et d'Enseignement de Géosciences de l'Environnement, Université Aix-Marseille, CNRS, BP 80, 13545 Aix en Provence, France

MATSUMOTO, S. (301) Institute of Seismology and Volcanology, Faculty of Sciences, Kyushu University, Japan

MAYEDA, K.M. (319) Weston Geophysical Corporation

MÜLLER, T.M. (123) Geophysikalisches Institut, Universität Karlsruhe, Germany

NAKAHARA, H. (401) Department of Geophysics, Graduate School of Science, Tohoku University, Aoba-ku, Sendai 980-8578, Japan

NISHIGAMI, K. (301) Disaster Prevention Research Institute, Kyoto University, Japan

NISHIZAWA, O. (219) National Institute of Advanced Industrial Science and Technology, 1-1-1 Higashi, Tsukuba, Ibaraki, 305-8567, Japan

PHILLIPS, W.S. (319) Los Alamos National Laboratory

POUPINET, G. (373) LGIT, Université Joseph Fourier & CNRS, BP53, 38041, Grenoble, France

RANDALL, G.E. (319) Los Alamos National Laboratory

SATO, H. (43) Department of Geophysics, Graduate School of Science, Tohoku University, Aramaki-Aza-Aoba 6-3, Aoba-ku, Sendai-shi, Miyagi-ken, 980-8578, Japan

SHAPIRO, S.A. (95, 123) Fachrichtung Geophysik, Freie Universität Berlin, 74-100, Haus D, 12249 Berlin, Germany

SHEARER, P.M. (167) Institute of Geophysics and Planetary Physics, U.C. San Diego, La Jolla, CA 92093-0225

STEAD, R.J. (319) Los Alamos National Laboratory

WU R.-S. (21) Institute of Geophysics and Planetary Physics, University of California, Santa Cruz, USA, 95064

YOSHIMOTO, K. (265) Natural/Basic and Applied Sciences, International Graduate School of Arts and Sciences, Yokohama City University, 22-2 Seto, Kanazawa-ku, Yokohama 236-0027, Japan

ZHENG, Y. (21) Institute of Geophysics and Planetary Physics, University of California, Santa Cruz, USA, 95064

PREFACE

Seismic waves generated by earthquakes have been interpreted to provide us information about the Earth's structure across a variety of scales. As a scientific activity of the Commission on Seismological Observation and Interpretation of the IASPEI, focusing on the seismic wave scattering in the Earth from heterogeneities having various types and scales, we organized a task group on "Scattering and Heterogeneity of the Earth." As the first product of this task group, Wu and Maupin (2007) edited a book entitled "Advances in Wave Propagation in Heterogeneous Earth" as the 48th volume of "Advances in Geophysics" (Series Editor, R. Dmowska). That volume mainly contains introductions to and basic review of modeling methods for elastic waves in laterally heterogeneous structures that are most commonly used in contemporary seismology.

For short-period seismic waves (e.g., those having periods less than 1 s), scattering due to randomly distributed small-scale heterogeneities in the Earth significantly changes the envelope of seismograms with increasing travel distance and excites coda waves. Models of propagation through deterministic structures such as those with horizontally uniform velocity layers used in traditional seismology cannot explain these phenomena. In addition to the invention of the velocity tomography, the study of coda waves in the heterogeneous lithosphere started by Aki (1969) marked a new era in short-period seismology. The former reveals the existence of three-dimensional deterministic heterogeneity from onset readings; the latter reveals the existence of small-scale random heterogeneity. The two approaches are complementary for the construction of a unified image of the real Earth; however, here we mainly focus on the latter subject, seismic wave scattering by random small-scale heterogeneity in the Earth.

This book is edited as the second product of the task group. Topics covered are recent developments in wave theory and observation including: weak localization of seismic waves, synthesis of short-period seismic wave envelopes, laboratory investigations of ultrasonic wave propagation in rock samples, coda wave analysis for mapping medium heterogeneity and for monitoring temporal variation of physical properties in the crust, radiation of short-period seismic waves from an earthquake fault, and borehole measurements of Earth properties on a range of scales. Various types of forward modeling and inversion schemes are introduced.

As a compelling description of the value of the study of the field of seismic wave scattering in the heterogeneous Earth, we refer to words of late professor Keiiti Aki in a letter he wrote to Dr. V.I. Kelis-Borok in 2003 from his lecture note (Aki, 2003), "... To a geodynamicist, the earth's property is smoothly varying within bodies bounded by large-scale interfaces. Most seismologists also belong to this 'smooth earth club,' because once you start with an initial model of smooth earth your data usually do not require the addition of small-scale heterogeneity to your initial model. As summarized well in a recent book by Sato and Fehler (1998), the acceptance of coda waves in the data set is needed for the acceptance of small-scale seismic heterogeneity of the lithosphere. There are an increasing number of seismologists who accept it, forming the 'rough earth club.' I believe that you are also a member of the rough earth club, judging from the emphasis on the hierarchical heterogeneity of the lithosphere ..."

This book starts with theoretical approaches for modeling wave propagation and scattering in randomly inhomogeneous media. *Chapter 1* (Margerin) reviews recent theoretical developments on the weak localization of coda waves: the amplitude of coherent back-scattered waves in the vicinity of the source is larger than what predicted from the classical radiative transfer theory. For cases where the wavelength is shorter than the scale of medium inhomogeneity, the WKBJ approximation is used in *Chapter 2* (Zheng and Wu) to arrive at a new stochastic theory for the coherence function of log amplitude and phase for waves passing through random media with a depth-dependent background velocity structure. As a statistical extension of the phase screen method for the parabolic wave equation, the Markov approximation is known to be an effective method to predict wave envelopes in random media for high-frequency waves. *Chapter 3* (Sato and Korn) reviews an extension of that approximation for scalar waves to vector waves. The newly developed theory reliably predicts envelope broadening and the excitation of the orthogonal component of motion (the transverse component for P-waves) with increasing travel distance. The validity of the approach is tested by comparison with sets of wave traces generated by finite differences. *Chapter 4* (Kaslilar et al.) discusses the travel time statistics of acoustic waves in random media based on geometrical optics. They develop a method to estimate the statistical parameters characterizing the random media from travel-time fluctuations of reflected and refracted waves. *Chapter 5* (Müller et al.) presents a theory for attenuation and dispersion of compressional seismic waves in inhomogeneous, fluid-saturated porous media in the framework of wave propagation in continuous random media. The statistical smoothing method treats both intrinsic attenuation due to wave-induced flow and scattering attenuation as the redistribution of wave energy in space and time in a unified manner.

The following two chapters treat practical modeling of seismic wave propagation through the heterogeneous Earth. *Chapter 6* (Shearer and Earle) focuses on the envelopes of teleseismic P waves traveling through the heterogeneous mantle. Envelopes calculated by using a statistical synthesis based on the Born scattering amplitudes for random elastic media are fitted to the observed stacked P wave envelopes. *Chapter 7* (Furumura and Kennett) presents a scattering slab model for the Pacific plate and the Philippine Sea plate beneath Japan that explains the observed efficient wave-guide for high-frequency seismic waves in this region. The heterogeneous component of their slab model consists of an anisotropic random velocity fluctuation with a longer correlation distance in the plate down-dip direction and a much shorter correlation distance across the plate thickness. Precise numerical simulations well explain the frequency selective wave propagation effect.

The following two chapters treat laboratory experiment and scaling issues in borehole surveys. *Chapter 8* (Nishizawa and Fukushima) presents laboratory experiments of ultrasonic wave propagation through heterogeneous rock samples by using a laser Doppler vibrometer. Variations in travel times, fluctuations of amplitude, phase, and particle-motion, as well as envelope formation are examined with respect to the statistical properties of random heterogeneities of rock in the range of millimeters. *Chapter 9* (Cheng) reviews the latest technologies in down-hole seismic measurements: acoustic logging, cross well seismic and vertical seismic profiling. They cover a frequency range from about 10 kHz down to about 10 Hz, and can investigate heterogeneity in the Earth from a scale of 10 s of centimeters to 100 s of meters. This chapter contains a discussion of the scale over which the various methods can resolve heterogeneity.

The following chapters treat various types of observations and analyses of coda waves. *Chapter 10* (Yoshimoto and Jin) presents the general characteristics of coda waves of local earthquakes and theoretical models based on the radiative transfer theory. This chapter discusses the nonuniform distribution of coda energy in tectonically active regions. The measurement of coda attenuation is focused especially as a useful tool for monitoring the temporal change in physical parameters in the curst. *Chapter 11* (Nishigami and Matsumoto) presents the inversion of coda wave envelopes of local earthquakes for the spatial distribution of scattering strength in the crust. The idea is based of the assumption that the lapse-time dependent residual of individual coda envelope from a smooth master curve reflects the spatial variation of scattering strength. Applying this method to data retrieved in the San Andreas Fault system, they show a good correlation between sub-parallel active faults and relatively stronger scattering zones in the crust. This chapter also has a discussion about slant stacking of seismic array waveform data for the energy evaluation under the assumption of a single scattering model. *Chapter 12* (Phillips *et al.*) develops a calibration technique to estimate the source spectra from the spectra of Lg and Sn coda waves of local earthquakes. Applying these techniques to records registered at stations across central and eastern Asia, they determine the regional variation of coda attenuation and apparent stress of earthquakes. *Chapter 13* (Del Pezzo) reviews scattering studies in various volcanic regimes. In some cases, the frequency dependence of coda attenuation in volcanoes is found to be less than that measured in nonvolcanic areas. According to the multiple lapse time window analysis of the data, scattering loss dominates over intrinsic loss with increasing frequency because of strong heterogeneity in volcanoes. Different from the above envelope analyses, *Chapter 14* (Poupinet *et al.*) focuses on the phase information of coda waves and presents a cross-correlation (-spectrum) moving window technique of coda waves of local earthquake doublets for monitoring the temporal change in the velocity structure of the crust. This technique is tested by earthquake doublet seismograms registered by a digital seismic network with a high time precision. This chapter also presents a technique that creates "virtual doublets" from the correlation of long seismic noise sequences.

The last two chapters treat earthquake strong motions and source models. *Chapter 15* (Nakahara) presents a seismogram envelope inversion for short-period seismic energy radiation from an earthquake fault. The basic idea is to use the envelope Green function derived from the multiple isotropic scattering model for short period S-waves to invert for the spatial variation in radiation from a fault. This chapter compiles the characteristics of short-period seismic energy radiation from moderate to large earthquakes. *Chapter 16* (Lavallée) presents an earthquake source model based on the assumption that the slip distribution obeys a Lévy law. This model predicts that the sum of these amplitudes observed at a given distance from the sources will also be distributed according to a Lévy law.

The text is written for graduate students, scientists, and engineers of geophysics, physics, acoustics, civil engineering, environmental sciences, geology, and planetary sciences. A glossary of special terms relevant to the study of scattering of waves in random media is placed at the end of this book. For further understanding, there are monographs that treat medium heterogeneity and wave scattering as follows: Chandrasekhar (1960) is a classic text for radiative transfer theory in scattering media. Ishimaru (1978) and Rytov *et al.* (1987) offer advanced mathematical tools for the study of wave propagation in random media and a link between wave theory and the radiative transfer theory. Shapiro and Hubral (1999) puts special focus on wave propagation through stratified random media focusing on 1D problems. Goff and Holliger (2002) summarizes observations of crustal heterogeneity. Sato and Fehler (1998) reviews seismological observation facts and mathematical models of scattering phenomena especially focusing on short period seismic waves and small-scale heterogeneity.

We thank the following scientists for their careful reviews of different chapters: Joe Andrews, Nirenda Biswas, Daniel Burns, Arthur C.H. Cheng, Vernon F. Cormier, Edoardo Del Pezzo, Karl Ellefsen, William L. Ellsworth, Alexander A. Gusev, David Higdon, Lianjie Huang, Ludek Klimes, Michael Korn, Yury A. Kravtsov, Ludovic Margerin, Gary Mavko, Steve McNutt, Tobias Müller, Takeshi Nishimura, Masakazu Ohtake, Lenya Rhyzik, Steve Roecker, Tatsuhiko Saito, Sergei Shapiro, Roel Snieder, Anna Tramelli, Kasper Van Wijk, Ulrich Wegler, Kazuo Yoshimoto, and Yuehua Zeng.

We thank the IASPEI, in particular the ex-president E. Robert Engdahl, the Secretary-General, Peter Suhadolc, the chairman of the Commission on Seismological Observation and Interpretation, Dmitry Storchak, and the current president, Zhongliang Wu for their support during this book project.

This book also owes a lot to Renata Dmowska, the editor of the series "Advances in Geophysics." We thank her for her continuous encouragement and help for the editing work.

Haruo Sato and Michael C. Fehler
June 7, 2008

REFERENCES

Aki, K. (1969). Analysis of seismic coda of local earthquakes as scattered waves. *J. Geophys. Res.* **74,** 615–631.

Aki, K. (2003). *Seismology of Earthquake and Volcano Prediction*. pp. 1–219. NIED and YIES, Tsukuba.

Chandrasekhar, S. (1960). *Radiative Transfer*. Dover, New York.

Goff, J.A., Holliger, K. (2002). *Heterogeneity in the Crust and Upper Mantle – Nature, Scaling and Seismic Properties*. Kluwer Academic/Plenum Publishers, Dordrecht, The Netherlands.

Ishimaru, A. (1978). *Wave Propagation and Scattering in Random Media*. Academic, San Diego.

Rytov, S.M., Kravtsov, Y.A., Tatarskii, V.I. (1987). Wave propagation through random media. *In Principles of Statistical Radio Physics*, vol. 4, Springer-Verlag, Berlin.

Shapiro, S.A., Hubral, P. (1999). *Elastic Waves in Random Media—Fundamentals of Seismic Stratigraphic Filtering*. Springer-Verlag, Berlin.

Sato, H., Fehler, M.C. (1998). *Seismic Wave Propagation and Scattering in the Heterogeneous Earth*. Springer-Verlag, New York.

Wu, R.S., Maupin, V. (2007). Advances in wave propagation in heterogeneous earth. *In Advances in Geophysics* (Series Ed., R. Dmowska), vol. 48, Academic Press, San Diego.

COHERENT BACK-SCATTERING AND WEAK LOCALIZATION OF SEISMIC WAVES

Ludovic Margerin

Abstract

I present a review of the weak localization effect in seismology. To understand this multiple scattering phenomenon, I begin with an intuitive approach illustrated by experiments performed in the laboratory. The importance of reciprocity and interference in scattering media is emphasized. I then consider the role of source mechanism, again starting with experimental evidence. Important theoretical results for elastic waves are summarized, that take into account the full vectorial character of elastic waves. Applications to the characterization of heterogeneous elastic media are discussed.

Key Words: Multiple scattering, interference, reciprocity, elastic waves. © 2008 Elsevier Inc.

1. Introduction

In strongly scattering media, the propagation of multiply-scattered waves is best described by considering the transport of the energy. An elastic scattering medium is an inhomogeneous medium where the wavespeed and the density vary laterally. It can also contain embedded obstacles such as cracks or cavities. Upon propagation, an incoming plane wave with well-defined wavevector **k** will transfer energy to all possible space directions, a phenomenon known as scattering. The energy transport approach has been developed by astrophysicists at the beginning of the twentieth century and has given birth to the theory of radiative transfer (Chandrasekhar, 1960; Apresyan and Kravtsov, 1996). Phenomenologically, the transfer equation for acoustic, electromagnetic, and elastic waves can be derived from a detailed local balance of energy that neglects the possible interference between wave packets (see, e.g., Sato, 1994; Sato and Fehler, 1998; and Margerin, 2005, for seismic applications). This important assumption is justified by the fact that the phase of the wave is randomized by the scattering events. Thus at a given point, the field can be written as a sum of random phasors and on average, intensities can be added, rather than amplitudes. This seemingly convincing argument can actually be shown to be wrong and one of the goals of this paper will be to put forward the role of interferences in scattering media in specific cases. To begin with, we present the results of an experiment of ultrasound propagation in a granular material, which can be defined as a material containing many individual solid particles with arbitrary sizes. An array of 128 acoustic transducers has been placed at the surface of a box with lateral dimensions 0.15 m × 0.15 m × 0.15 m containing commercial sand for aquariums. The sand does not have a well-defined granulometry but the typical size of a grain is ∼2 mm. The central transducer emits a short pulse in the 1–1.5 MHz frequency range and the waves are recorded by the whole array. The wavespeed of the dominant ballistic pulse is about 1000 m/s, which gives a dominant wavelength of ~0.8 mm, which is of the same order as the size of one transducer of the array (~0.55 mm). The logarithm of the energy of the wavefield along the array after time averaging over four cycles is shown in Fig. 1, as a function of distance from the center of the array in millimeters (horizontal axis) and time in microseconds (vertical axis).

In this figure, one can identify direct waves that propagate along the array and decay exponentially with distance because of the energy losses due to scattering. They are followed by a diffuse coda, which can be thought of as the result of the random walk of the energy in the scattering medium. As a rule of thumb, the multiple scattering halo grows like $\sqrt{Dt}$, where $D = vl^*/3$ is the diffusion constant of the waves in the medium and v is the wave velocity. The transport mean free path l^* (see, e.g., Sheng, 1995, for a rigorous definition) represents the typical step length of the random walk of the energy in the scattering medium and is much larger than the wavelength. In the case of sand samples, the transport mean free path is roughly 10 times larger than the wavelength. According to diffusion theory (Akkermans and Montambaux, 2005), at fixed time $t = t_0$, the energy distribution in the scattering medium is approximately proportional to $e^{-3r^2/(4vt_0 l^*)}$, where r is the distance from the source. Therefore, at fixed time, the energy in the diffuse halo is expected to vary significantly on the scale of l^*. Yet, at the center of Fig. 1, the reader will notice a clear, but highly localized increase of intensity. This is not

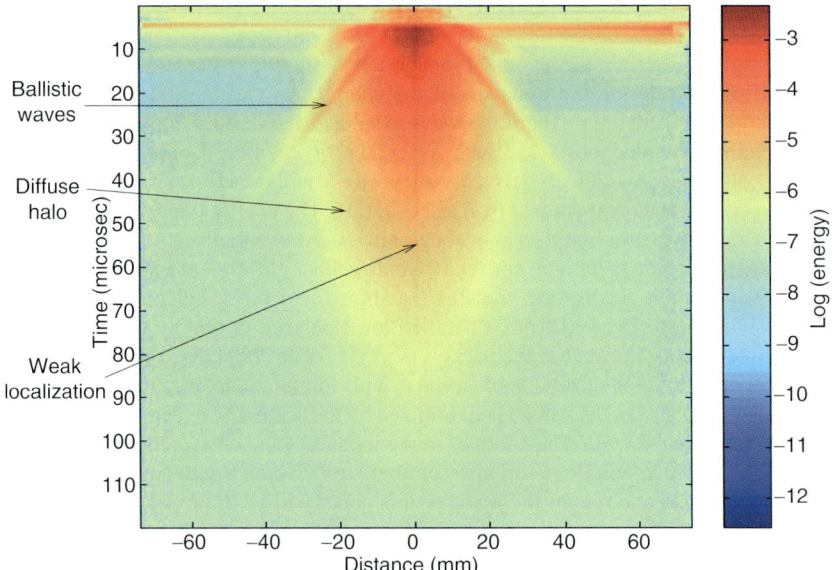

FIG. 1. Energy of the wavefield recorded at the surface of a granular material as a function of time. A short pulse with a central frequency of 1.2 MHz is shot at the central transducer. Direct waves propagating along the array rapidly attenuate. They are followed by coda waves that form a diffuse halo in the medium. Note the sharp increase of intensity at the center of the array, where energy was initially released. Experiment performed at the Laboratoire Ondes et Acoustique, Paris, by R. Hennino and A. Derode.

an artifact. In this particular experiment, the typical width of the zone of enhanced intensity is roughly equal to the size of one transducer. Other experiments, to be described later, have demonstrated that the zone of enhancement actually coincides with the central wavelength of the waves, which is the clear signature of an interference effect that takes place around the source in the multiple scattering medium: this is known as weak localization. In the next section, I provide a simple explanation for this observation.

2. Weak Localization Effect: A Heuristic View

In what follows, I represent a scalar partial wave as a complex number $\psi = Ae^{i\phi}$, where A and ϕ are real numbers denoting the amplitude and phase, respectively. Each partial wave follows an arbitrarily complicated scattering path from source to receiver in the medium. At a given point, the measured field u is a superposition of a large number of partial waves that have propagated along different scattering paths

$$u = \sum_j A_j e^{i\phi_j}, \tag{1}$$

where A_j and ϕ_j are random and uncorrelated because of the multiple scattering events, and j can be understood as a "label" for the different paths. The representation (1) is strictly valid for point scatterers and will suffice for the present purposes. Typical examples of scattering paths are shown in Fig. 2.

In Eq. (1), I now pair direct and reciprocal scattering paths as shown in Fig. 2. The direct and reciprocal paths are characterized by the fact that the same scatterers are visited, but the sequences of scattering events are opposite. To illustrate this definition, in Fig. 2, the sequence S, A, B, C, D, R (solid line) and S, D, C, B, A, R (dashed line) represent the direct and reciprocal paths, respectively. One obtains

$$u = \sum_{j'} \left(\psi_{j'}^{d} + \psi_{j'}^{r} \right), \tag{2}$$

where ψ denotes the complex partial waves, the superscripts d and r stand for "direct" and "reciprocal," and a new label j' has been introduced to emphasize the new representation of the field. The intensity I is proportional to $|u|^2$ and reads

$$I = \sum_{j',k'} \left(\psi_{j'}^{d} + \psi_{j'}^{r} \right) \overline{\left(\psi_{k'}^{d} + \psi_{k'}^{r} \right)}, \tag{3}$$

where the overbar denotes complex conjugation. In Eq. (3), it is reasonable to assume that the waves visiting *different* scatterers will have random phase differences and after averaging over scatterer positions will have no contribution. Thus, we can restrict the summation to the case $j' = k'$ to obtain

$$I = \sum_{j'} \left(\left| \psi_{j'}^{d} \right|^2 + \left| \psi_{j'}^{r} \right|^2 \right) + \sum_{j'} (\psi_{j'}^{d} \overline{\psi_{j'}^{r}} + \overline{\psi_{j'}^{d}} \psi_{j'}^{r}). \tag{4}$$

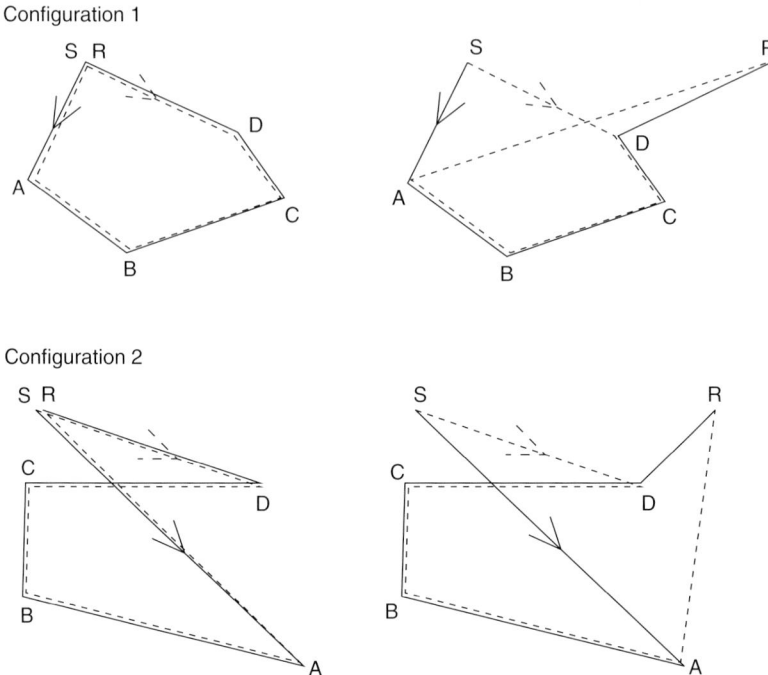

FIG. 2. Examples of multiple scattering paths from source S to receiver R. Scattering events are labeled with letters A, B, C, and D. Solid and dashed lines represent direct and reciprocal paths, respectively. The two configurations differ by the position of the scattering events. On the left: source and receiver coincide. On the right, source and receiver are typically a few wavelengths apart. Reprinted figure with permission from Margerin *et al.* (2001). Blackwell Publishing.

The first sum in Eq. (4) represents the usual incoherent contribution to the measured intensity, which can be calculated with radiative transfer theory (see Wegler et al., 2006 for recent applications). The second sum can be interpreted as the *interference* term between the direct and reciprocal paths in the scattering medium. In a reciprocal medium, that is, a medium where the reciprocity principle is verified, the amplitude and phase of the direct and reciprocal wave paths are exactly the same, that is, $A_j^d = A_j^r$ and $\phi_j^d = \phi_j^r$, provided that source and receiver are located at the same place. Therefore, the total intensity which includes the interference term is exactly double of the classical incoherent term. This is the interference term which causes the intensity to be higher in the experiment shown in Fig. 1. Reciprocity is a general property of wave equations such as the acoustic and elastic wave equation. In a simple scalar case, it means that the response measured at $\mathbf{r}_2$ due to a source at $\mathbf{r}_1$ is the same as the response measured at $\mathbf{r}_1$ due to a source at $\mathbf{r}_2$. This remarkable property can be broken when an external field acts on the system. An interesting seismic example of broken reciprocity is the effect of the Coriolis force on the seismic wave motion at long period where the effect of the rotation of the Earth is important. A generalized reciprocity relation can still be given upon exchange of source and receiver, which involves the inversion of the instantaneous rotation vector of the Earth (see Dahlen and Tromp, 1998, for a thorough discussion).

Although it had been theoretically predicted in several pioneering papers published around 1970 (Watson, 1969; de Wolf, 1971; Barabanenkov, 1973), it is only in the mid-eighties that the role of interference in multiple scattering has been appreciated with the discovery of the coherent backscattering of light (Kuga and Ishimaru, 1984; van Albada and Lagendijk, 1985; Wolf and Maret, 1985; Kaveh *et al.*, 1986). Later the coherent backscattering effect has been predicted and observed for acoustic and elastic waves in both stationary and dynamic experiments (Bayer and Niederdrank, 1993; Sakai *et al.*, 1997; Tourin *et al.*, 1997; de Rosny *et al.*, 2000). Today, coherent backscattering or weak localization is still a very active topic of research. The coherent backscattering for moving scatterers has been studied by Snieder (2006) and Lesaffre *et al.* (2006). Derode *et al.* (2005) have used the coherent backscattering effect to measure the heterogeneity of human bones. Aubry and Derode (2007) have devised an ingenious technique to measure lateral variations of the diffusion constant of strongly scattering media based upon the separation of the incoherent and coherent intensities.

Note that the term "coherent backscattering" refers to the intensity enhancement observed in a small cone of direction in the far-field of a disordered sample for plane wave sources with *fixed* incident direction $\hat{k}$. Although the basic physics of coherent backscattering and weak localization are identical, the latter term indicates that the loops of interference occur *inside* the disordered sample, and should therefore be preferred to describe the seismic experiments. These interference loops result in a deviation from the diffusive behavior (Haney and Snieder, 2003). When the wavelength is of the same order as the mean free path, the interference effects can completely block the transport of energy away from the source, a phenomenon known as strong localization (see, e.g., Sheng, 1995; Akkermans and Montambaux, 2005). Weak localization is therefore a basic phenomenon to explain the transition from the diffuse to the localized propagation regime. Note that there exists a number of other mechanisms of intensity enhancement. One of the most famous is the "opposition effect" in astrophysics, which manifests itself as an increase of the reflectance of celestial bodies such as the Moon when the light of the Sun reflected from the regolith is observed close to the backscattering direction. Hapke *et al.* (1993) have shown that the opposition effect is partly explained by the coherent backscattering of light. I refer the interested reader to the paper by Barabanenkov *et al.* (1991) for an extensive review of backscattering enhancement phenomena in optics. In particular, these authors discuss the case of backscattering enhancement by several deterministic scatterers which can also be of interest in seismology. Let me finally point out that weak localization is only one manifestation of the role of the phase in the seismic coda (see Campillo, 2006, for a review).

I have shown with a very simple argument that interference effects have to be incorporated in the usual transport theory, but for the moment, it has not been explained why the enhancement due to interference is so highly localized. In Fig. 2, I schematically represent the more usual case where source and detection are not collocated. In that case there is a phase shift between the two wavepaths which is acquired during the propagation from the source to the first scattering event and from the last scattering event to the receiver. Clearly, if the distance between source and detection is "large enough," the phase shift will fluctuate randomly from one configuration of the scatterers to the other. Therefore, the interference term is expected to vanish upon averaging when source and receiver are sufficiently far apart. One of the goals of the paper is to demonstrate and illustrate the fact that the enhancement zone is actually narrow and about the size of the wavelength. A final comment on the role of multiple scattering is in order. It is clear that the representation of the wavefield [Eq. (3)] makes sense only if one can pair the direct and reciprocal propagation paths. If there is a single scattering event, there is only one possible path from source to receiver and therefore no interference is possible. This basic observation proves that weak localization is indeed a genuine multiple scattering effect. In what follows, I pursue the experimental approach of weak localization by considering the role of the source mechanism.

3. The Role of Source Mechanism and Wavefield Polarization

Because of their vectorial nature, the weak localization of elastic waves cannot be fully explained by the simple intuitive approach presented above. As will be shown shortly, the reciprocity principle of elastic waves in its full extent has to be obeyed in order to preserve the factor of 2 enhancement at the source position. This subtle effect is first examined through a laboratory experiment with ultrasound.

3.1. Effect of Source Mechanism

The most common seismic sources, that is, explosions and earthquakes are combinations of dipoles and/or couples of forces. We must therefore consider more complex sources than simple isotropic point sources. In the case of earthquakes, the radiation is strongly anisotropic and the radiation pattern displays nodal planes with reversal of the polarity of first motions. de Rosny et al. (2001) have studied the weak localization of elastic waves propagating in a chaotic reverberant cavity. The nature of the disorder is different from the scattering medium, but until a time known as the Heisenberg time, the mechanisms of enhancement are similar and can be based on an intuitive ray description. Beyond the Heisenberg time, the eigenmodes of the system can be resolved and the statistical properties of the eigenfunctions lead to an enhancement of intensity by a factor 3 around the source, as demonstrated experimentally and theoretically by Weaver and Burkhardt (1994) and Weaver and Lobkis (2000). This result is valid in chaotic cavities only. Using an interferometer, de Rosny et al. (2001) have recorded the vertical motions of Lamb waves generated by vertical monopole and dipole sources in a thin (0.5 mm thickness) chaotic plate of total area 2335 mm^2, with the shape of a quarter stadium. The dominant frequency of the signal is 1.0 MHz and the typical wavelength is $\sim$2.5 mm. After time averaging between lapse time $t = 200$ μs and $t = 500$ ms, they have measured the intensity patterns shown in Fig. 3. The beginning of the signal is excluded in order to avoid the first reflection on the boundary of the plate and the choice of the end of the time window is dictated by the signal to noise ratio. The distribution of energy is perfectly homogeneous in the plate, except in a small area centered around the source where, in the case of a monopole, it is the double of the background intensity. An important result of this study is the confirmation of the typical wavelength size of the zone of intensity enhancement, which shows that weak localization is a near-field effect. In the 3-D case, the increase of intensity would occur inside a sphere centered at the source.

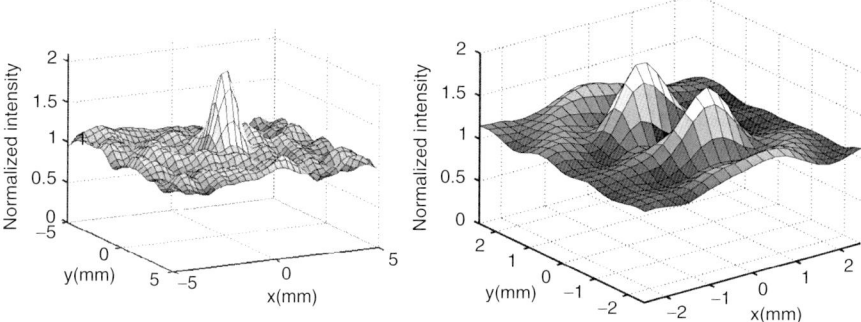

FIG. 3. Enhanced backscattering of elastic waves in a 2-D chaotic cavity. The central frequency is 1 MHz and the dominant wavelength is 2.5 mm. The integrated intensity between lapse time $t = 200$ μs and $t = 5$ ms is represented as a function of position around the source. *Left*: Monopolar source. *Right*: Dipolar source. In the dipolar case, the enhancement disappears on the line going through the source and perpendicular to the dipole axis. Along the dipole, the intensity enhancement presents two maxima located about half a wavelength away from the source. Reprinted figure with permission from de Rosny *et al.* (2001). Copyright (2001) by the American Physical Society.

In addition, Fig. 3 (right) highlights the importance of the source mechanism. In the dipolar case, there are actually two zones of enhancement with an enhancement factor of about 1.6, separated by a line of zero enhancement. To understand this puzzling observation, one must consider the full reciprocity principle for elastic waves. In the present experiment, there is a lack of symmetry between the dipolar source and the monopolar detection. To restore reciprocity one would like to measure not the field itself, but its directional derivative along the dipole axis. When this operation is performed and intensity is redefined as the square of the partial derivative, de Rosny *et al.* (2001) have demonstrated both experimentally and theoretically that the factor of 2 enhancement at the source is restored. Being motivated by this result, I now give an asymptotic but rigorous theory of weak localization for vector waves.

3.2. Review of Multiple Scattering Formalism

To obtain a satisfactory theory of weak localization, one needs to develop a transport theory of the energy that keeps track of all polarization indices at both source and receiver. As shown by Weaver (1990), the necessary information is contained in the fourth rank coherence tensor $\Gamma_{ij \to kl}$ of the elastic wavefield defined as

$$\Gamma_{ij \to kl}(t, \mathbf{r}_1, \mathbf{r}_2 \to \mathbf{r}_3, \mathbf{r}_4) = \langle G^{\alpha}_{ki}(t, \mathbf{r}_3, \mathbf{r}_1) \overline{G^{\alpha}_{lj}(t, \mathbf{r}_4, \mathbf{r}_2)} \rangle, \tag{5}$$

where $G^{\alpha}_{ki}(t, \mathbf{r}_3, \mathbf{r}_1)$ is the element of the Green matrix corresponding to a point force applied at $\mathbf{r}_1$ in direction i, and measured displacements in direction k at $\mathbf{r}_3$. The superscript α is introduced to label the realizations of the random medium. To each α there corresponds exactly one medium of the statistical ensemble. t denotes the time elapsed since energy has been released by sources with a common origin time. Note that in the analysis that follows, the signal is assumed to be band-pass filtered in a narrow frequency band with central angular frequency ω. In order to simplify notation, all tensor quantities are assumed to depend implicitly on ω. The tensor $\Gamma_{ij \to kl}(t, \mathbf{r}_1, \mathbf{r}_2 \to \mathbf{r}_3, \mathbf{r}_4)$ describes the transfer of the displacement correlation function from source (displacement indices i, j and positions $\mathbf{r}_1, \mathbf{r}_2$) to receiver (displacement indices k, l and positions $\mathbf{r}_3, \mathbf{r}_4$). The brackets denote an average over α, that is, over an ensemble of random media with prescribed statistics. In what follows we assume that the property of statistical homogeneity holds, in which case the ensemble-averaged Green tensor depends on the difference of the position vectors of the source and detector only

$$\langle G_{ki}^{\alpha}(t, \mathbf{r}_3, \mathbf{r}_1) \rangle = G_{ki}(t, \mathbf{r}_3 - \mathbf{r}_1), \tag{6}$$

where for notational simplicity G_{ki} (without superscript) denotes the ensemble averaged Green tensor.

The complete evolution of the tensor $\Gamma_{ij \to kl}$ is described by the Bethe-Salpeter equation which contains all correlations among all possible scattering paths in the medium (Sheng, 1995). This is far too detailed for the present purposes, and one usually contents oneself with the approximate calculation of two terms: the classical contribution—also termed "diffuson"—and the interference term between reciprocal paths—also termed "cooperon"—(see Akkermans and Montambaux, 2005, for the origin of this terminology). In the radiative transfer equation, information on source is usually integrated out and the cooperon term is neglected (see Margerin, 2005, for details). Apresyan and Kravtov (1996) have suggested a modification of the radiative transfer equation, which includes contribution of the cooperon and thereby is able to describe coherent phenomena such as weak localization. The cooperon and diffuson contributions are conveniently represented by Feynman diagrams which are both computationally efficient and physically meaningful. Typical diagrams are shown in Fig. 4. In the ladder diagrams, the Green function (upper line) and its complex conjugate (lower line) visit the same scatterers in the same order. In the crossed diagrams, first introduced for multiple scattering of classical waves in the pioneering papers of Barabanenkov (1973, 1975), the upper line is unchanged but in the lower line the sequence of scattering is reversed. The ladder and crossed diagrams correspond to the classical (incoherent) and interference (coherent) contributions, respectively. To make the link with the elementary treatment given in Section I, the reader can think of the ladder diagrams alone, as the result of summing the intensities of the direct path (solid line in Fig. 2) and reciprocal path (dashed line in Fig. 2). The ladder and crossed diagrams altogether are the result of first summing and then squaring the fields of the direct and reciprocal paths.

Below, I give a long-time asymptotic formula for the ladder term. To calculate the crossed term, one can make use of the following reciprocity argument: the field produced by a force in direction k at $\mathbf{r}_2$ and recorded in direction l at $\mathbf{r}_4$ after scattering at $A, B, C, D, \ldots$ is equal to the field produced by a force in direction l at $\mathbf{r}_4$ and recorded in direction k at $\mathbf{r}_2$ after scattering at $\ldots, D, C, B, A$. This is equivalent to saying that every crossed diagram can be turned into a ladder diagram after suitable exchange of the polarization indices, and positions on the lower line. We now decompose the tensor $\Gamma_{ij \to kl}$ into the fundamental diffuson and cooperon contributions and write

$$\Gamma_{ij \to kl}(t, \mathbf{r}_1, \mathbf{r}_2 \to \mathbf{r}_3, \mathbf{r}_4) = L_{ij \to kl}(t, \mathbf{r}_1, \mathbf{r}_2 \to \mathbf{r}_3, \mathbf{r}_4) + C_{ij \to kl}(t, \mathbf{r}_1, \mathbf{r}_2 \to \mathbf{r}_3, \mathbf{r}_4). \tag{7}$$

The previous discussion implies the following fundamental reciprocity relation between $C_{ij \to kl}$ and $L_{ij \to kl}$

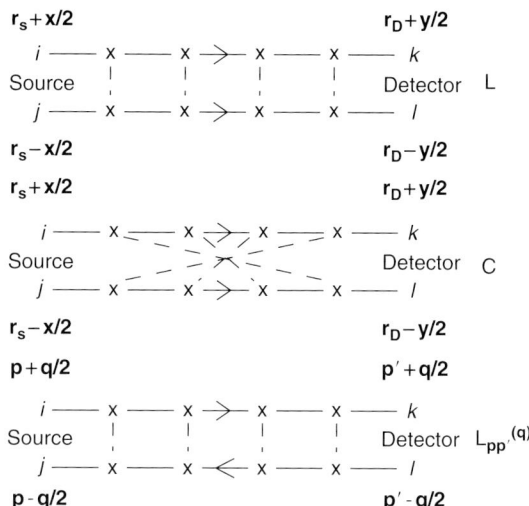

FIG. 4. Typical Feynman diagrams representing the classical (*top*) and interference (*middle*) contribution to the intensity pattern. Crosses connected by dashed lines represent the *same* scatterer. The crosses are also connected by solid lines which represent Green functions describing the propagation between different scatterers. Note that in the lower line, the Green functions are complex conjugated. The upper diagram L, often pictorially termed "ladder diagram," gives the classical contribution to the measured intensity. The middle diagram termed "crossed diagram" represents the interference term between reciprocal wavepaths. Bottom: scattering diagram in the wavenumber domain. Reprinted figure with permission from van Tiggelen *et al.* (2001). Copyright (2001) by the American Institute of Physics.

$$C_{ij \to kl}(t, \mathbf{r}_1, \mathbf{r}_2 \to \mathbf{r}_3, \mathbf{r}_4) = L_{il \to kj}(t, \mathbf{r}_1, \mathbf{r}_4 \to \mathbf{r}_3, \mathbf{r}_2). \tag{8}$$

As above, I draw the attention of the reader to the fact that relation (8) is true in multiple scattering only. Fortunately, in nonabsorbing media, the single scattering contribution vanishes exponentially (see, e.g., Sato and Fehler, 1998) while, at large lapse time, the ladder contribution can be shown to be the solution of a diffusion equation (Barabanenkov and Ozrin, 1991, 1995; Sheng, 1995; Akkermans and Montambaux, 2005) and therefore decays only algebraically. In the presence of absorption, both the single scattering and ladder contribution exhibit an algebro-exponential decay but the single scattering term still vanishes faster because it suffers from scattering losses. These facts have been illustrated in papers by Gusev and Abubakirov (1987), Hoshiba (1991), and Zeng *et al.* (1991), where solutions of the single scattering and full multiple scattering problem are presented. Therefore, relation (8) applies after one mean free time, which is the average time between two scattering events. In addition, from Eq. (8), one can easily infer that when source and detection coincide $\mathbf{r}_1 = \mathbf{r}_2 = \mathbf{r}_3 = \mathbf{r}_4$, and the polarization of source and detection are identical $i = j = k = l$, the interference term equals exactly the diffuson term. Our next task is to provide a general formula to predict the exact shape of the weak localization effect in the long time limit. This requires an asymptotic solution of the Bethe-Salpeter equation, which is presented in what follows.

3.3. Theoretical Results for Acoustic and Elastic Waves

The Bethe-Salpeter equation is an exact equation for the full coherence tensor of the wavefield. The complete solution of this equation seems out of reach in general, but asymptotic solutions for the diffuson contribution in the long time limit have been presented by Barabanenkov and Ozrin (1991, 1995), which can be applied to classical scalar and vectorial linear waves, independent of the underlying equation of motion. In the theoretical approach, one considers a narrowly band-passed signal with central period T which is much smaller than the typical duration of the coda. This is known in the literature as the slowly varying envelope approximation (Sheng, 1995). The results of Barabanenkov and Ozrin are valid in this limit and I refer the interested reader to the original publications for further technical details. Asymptotically $t \to \infty$, the ladder or diffuson or classical contribution takes the form

$$L_{ij \to kl}(t, \mathbf{r}_1, \mathbf{r}_2 \to \mathbf{r}_3, \mathbf{r}_4) = \frac{e^{-t/\tau_a}}{(Dt)^{3/2}} \mathrm{Im} G_{ij}(\mathbf{r}_1 - \mathbf{r}_2) \mathrm{Im} G_{lk}(\mathbf{r}_4 - \mathbf{r}_3), \tag{9}$$

where Im denotes the imaginary part of a complex number, D is the diffusion constant of the waves in the random medium, and τ_a denotes a phenomenological absorption term. Note that in the last equation, the tensor G_{ij} stands for the spatial part of the ensemble-averaged elastic Green tensor at angular frequency $\omega = 2\pi/T$.

Because of the underlying assumption of statistical homogeneity, the tensor G_{ij} depends on the separation vector between source and station only and is an implicit function of ω. The tensor $\Gamma_{ij \to kl}$ depends on both central frequency of the waves, and lapse time in the coda. Equation (9) is valid in the slowly varying envelope approximation. The reader is referred to Sheng (1995) and Apresyan and Kravtov (1996) for more mathematical details. According to Eq. (8), the cooperon term reads

$$C_{ij \to kl}(t, \mathbf{r}_1, \mathbf{r}_2 \to \mathbf{r}_3, \mathbf{r}_4) = \frac{e^{-t/\tau_a}}{(Dt)^{3/2}} \mathrm{Im} G_{il}(\mathbf{r}_1 - \mathbf{r}_4) \mathrm{Im} G_{jk}(\mathbf{r}_2 - \mathbf{r}_3). \tag{10}$$

Let us investigate some consequences of Eq. (10) for a given source station configuration: $\mathbf{r}_1 = \mathbf{r}_2, \mathbf{r}_3 = \mathbf{r}_4$. To illustrate the fact that interference effects are important only in a region of the size of one wavelength around the source, we consider the scalar case and introduce the enhancement factor defined as the total intensity normalized by the diffuson contribution: $E = 1 + C/L$. For scalar waves in 3-D, the enhancement profile E is given by

$$E = 1 + \left(\frac{\sin(kr)}{kr}\right)^2, \tag{11}$$

where $k = \omega/c$ is the central wavenumber of the scalar waves with velocity c and r is the source receiver distance. For scalar waves in 2-D, the corresponding formula is

$$E = 1 + J_0(kr)^2 \tag{12}$$

where J_0 denotes the Bessel function of the first kind of order 0. These results are valid in the long lapse time limit and when the mean free path is much larger than the wavelength, which is the most common situation in practice. When scattering is extremely strong, the wavelength can be of the order of the mean free path, and strong localization can set in, a regime which is very difficult to reach (Akkermans and Montambaux, 2005). The cardinal sine and Bessel function in Eqs. (11) and (12) are proportional to the imaginary part of the Green function of the Helmoltz equation in 2-D and 3-D, respectively. Equation (11) has been verified by numerical simulations (Margerin et al., 2001), while Eq. (12) has been checked experimentally by de Rosny et al. (2000) (see also the left part of Fig. 3). It is important to note that the enhancement profile is independent from absorption. This is not surprising since absorption does not affect the reciprocity principle. This is often illustrated by the following sentence (van Tiggelen and Maynard, 1997): "If you can see me, I can see you!" that applies even in the fog. This independence of the weak localization effect on absorption is used later in the paper to give estimates of the scattering properties of heterogeneous media.

I now explore theoretically the role of the source mechanism. As an example, I will give a simple formula that explains all the features of the experiments described in Fig. 3 with a dipole source. The radiation of the dipole is obtained by taking the directional partial derivative of the Green function with respect to source coordinates. For a dipole source along the *x* axis of the coordinate system, the coherence function Γ of the scalar field is given by

$$\Gamma(t, \mathbf{r}_1, \mathbf{r}_2 \to \mathbf{r}_3, \mathbf{r}_4) = \langle \partial_{x_1} G^\alpha(t, \mathbf{r}_3, \mathbf{r}_1) \partial_{x_2} \overline{G^\alpha(t, \mathbf{r}_4, \mathbf{r}_2)} \rangle. \tag{13}$$

Because the operation of taking derivatives is linear, the ∂_{x_1} and ∂_{x_2} symbols can be taken outside of the ensemble average brackets. For the diffuson and cooperon contributions, one obtains respectively

$$L = \frac{e^{-t/\tau_a}}{(Dt)^{3/2}} \mathrm{Im} G(\mathbf{0}) \partial_{x_1} \partial_{x_2} \mathrm{Im} G(\mathbf{r}_2 - \mathbf{r}_1)\vert_{\mathbf{r}_2 = \mathbf{r}_1} \tag{14}$$

$$C = \frac{e^{-t/\tau_a}}{(Dt)^{3/2}} \partial_{x_1} \mathrm{Im} G(\mathbf{r}_1 - \mathbf{r}_4)\vert_{\mathbf{r}_4 = \mathbf{r}_3} \partial_{x_2} \mathrm{Im} G(\mathbf{r}_3 - \mathbf{r}_2)\vert_{\mathbf{r}_2 = \mathbf{r}_1}. \tag{15}$$

In the 2-D case, after some algebra, one obtains the following enhancement pattern for the dipole source

$$E = 1 + (J_1(kr)\cos(\theta))^2, \tag{16}$$

where J_1 denotes the Bessel function of the first kind of order 1. This theoretical prediction matches the observations of Fig. 3 (left) very closely.

The calculation of the weak localization effect has thus been illustrated in simple cases. The calculations in the full elastic case with arbitrary moment tensor sources have been performed by van Tiggelen et al. (2001). The principle is the same as above but the calculations are much more tedious. To illustrate the effect of broken symmetry between source and receiver, I show in Fig. 5 the calculation of the weak localization profile for explosion and dislocation sources in an infinite elastic medium. As usual, an explosion

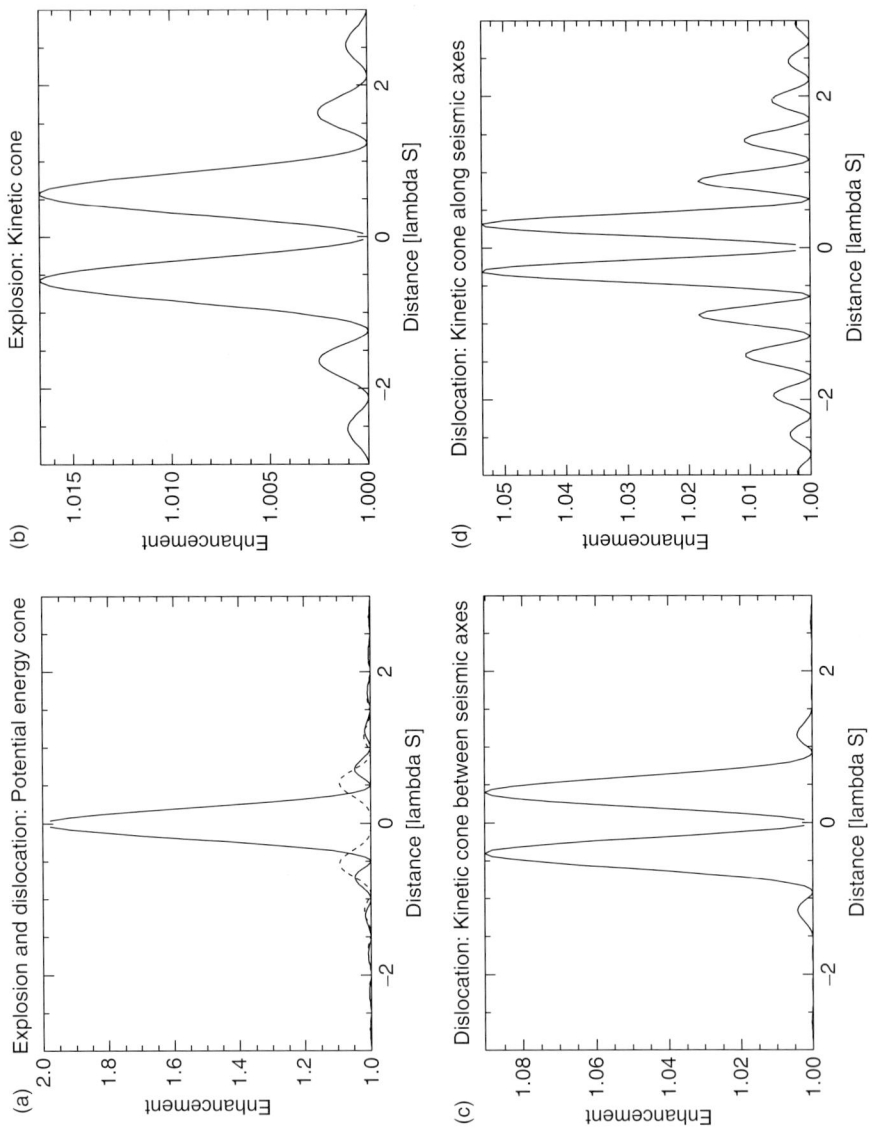

and a seismic dislocation will be represented by three mutually orthogonal dipole of forces, and by a double couple of forces, both with zero net applied linear and angular momentum, respectively. The enhancement profiles in Fig. 5 are calculated in the plane perpendicular to the axis of the double couple. In this plane, there exist two perpendicular directions, termed seismic axes, with respect to which the moment tensor is diagonal. For the simple dislocation source, the seismic axes define the direction of maximum radiation of P waves and make an angle $45°$ with the applied forces. Note that the direction of application of the forces also coincide with the maximum radiation of S waves. Figure 5 demonstrates that the backscattering enhancement can be totally destroyed if the operations carried out at the source and at the detection are different. For instance, the enhancement of kinetic energy due to a dislocation is typically less than 10% and vanishes exactly at the source position. The angular dependence of the enhancement profile in the plane containing the double couple is illustrated in Fig. 5c and d. It is seen that the enhancement is slightly higher along the direction of maximum radiation of S waves. Although I shall not prove it, it is always possible to recover the factor 2 enhancement by measuring the appropriate quantity. For the case of a dipole source, one needs to measure the derivative of the field along the dipole axis. This of course complicates the experimental setup but it can be done in practice (see, e.g., Hennino *et al.*, 2001). For an explosion, one needs to evaluate the divergence of the wavefield (i.e., the compressional energy of the waves), which is even more demanding. Although this seems discouraging, I present below some experiments with seismic waves that demonstrate the potential usefulness of weak localization.

4. Geophysical Applications

At this point, I would like to convince the reader that the weak localization effect is not only a theoretical or mathematical curiosity but a useful interference effect which can be used to probe the medium properties when multiple scattering hampers the detection of ballistic waves. I give two examples of applications of weak localization: the first one illustrates the measurement of elastic properties of a concrete slab (Larose *et al.*, 2006); the second one reports the measurement of the mean free path in a volcano (Larose *et al.*, 2004).

4.1. Measurement of the Dispersion Relation of Surface Waves

Let me first consider an application at a scale which is intermediate between the laboratory and the field. An array of vertical accelerometers with typical bandwidth 0.1–5 kHz has been set up on a steel reinforced concrete slab. Upon propagation, waves undergo strong multiple scattering and reflections. In the time domain, the ballistic signals are difficult to separate from the multiple reflections on the sides of the slab and

---

FIG. 5 Enhancement profiles as a function of the distance from the source, in units of the shear wavelength, for an infinite Poissonian medium. (a) The enhancement of the potential energy for an explosion (solid) and a dislocation (dashed). (b) Enhancement profile for the kinetic energy near an explosion. (c) and (d) Enhancement profiles for the kinetic energy between (at an angle $45°$ to) and along the seismic axes of a dislocation source. The term "cone" shown on top of each plot refers to the zone of intensity enhancement around the source. Reprinted figure with permission from van Tiggelen *et al.* (2001). Copyright (2001) by the American Institute of Physics.

are also rapidly attenuating. In such a situation, an alternative to classical signal processing techniques is welcome. The elastic waves are generated by an approximately vertical hammer strike, thus ensuring that source and detection verify the general reciprocity condition previously discussed. In the frequency range of interest (from 500 to 1500 Hz), the elastic energy is transported through the slab by the fundamental antisymmetric Lamb mode, which is known to be strongly dispersive. The phase speed of the fundamental Lamb mode at 1 kHz is ∼1000 m/s. The geometry of the experiment can thus be considered quasi 2-D and the shape of the weak localization can be accurately predicted using the theoretical expression (12). I refer the reader to Larose et al. (2006) for further details.

To illustrate the practical measurement of the weak localization effect, I show in Fig. 6 (top) a schematic view of the spatial distribution of the field intensity at a given instant of time. The upper plot in Fig. 6 is not the outcome of the experiment but serves as an

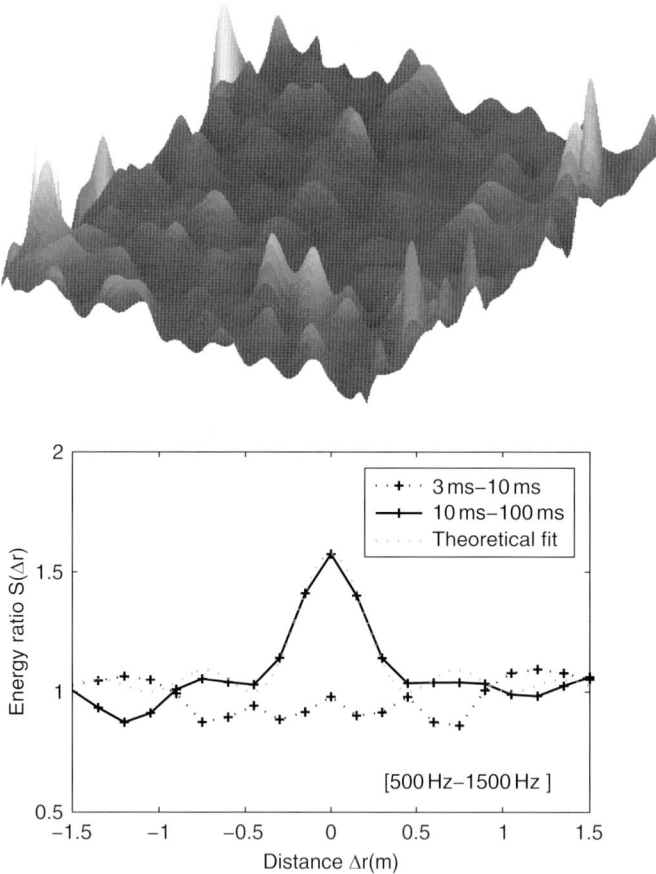

FIG. 6. (*Top*) Schematic snapshot of random spatial intensity fluctuations measured in the coda. One can imagine that the typical side length is ∼2 m. The weak localization effect is hidden in this speckle pattern. (*Bottom*) Reprinted figure with permission from Larose et al. (2006). Copyright (2006) by the American Physical Society: Typical profile of intensity enhancement around the source obtained after averaging over a large number of speckle patterns.

illustration of the experimental process involved in the measurement of the weak localization effect. This complex interference pattern shows rapid spatial fluctuations and is known in optics as a speckle pattern. In the speckle, the maxima can be viewed as random constructive interferences between multiply-scattered waves that mask the weak localization spot (at the center of Fig. 6 (top)). At another time instant, the speckle pattern will have completely changed, except for the deterministic constructive interference between reciprocal waves around the source. By averaging speckle patterns over sufficiently long time windows, one suppresses unwanted fluctuations thus revealing the weak localization intensity pattern as shown in Fig. 6 (bottom). In the present example, the averaging is performed in a lapse time window ranging from 10 to 100 ms. This simple procedure makes the link between averaging in theory and in practice. The reader will notice that the enhancement factor is lower than 2 in this experiment. Presumably, this is caused by the application of a force with a significant horizontal component, thus breaking the symmetry between source and detection. Despite the imperfection of the reconstruction, it is still possible to determine with reasonable accuracy the width of the weak localization zone as a function of frequency using the theoretical relation (12), thus providing the dispersion of the fundamental antisymmetric Lamb mode. In Fig. 7, the results of the dispersion measurements are plotted together with a theoretical fit. Indeed, the phase velocity c of the fundamental Lamb mode obeys the following rather complicated dispersion relation (see, e.g., Royer and Dieulesaint, 2000)

$$c = \sqrt{\frac{2\pi}{\sqrt{3}} f h \beta \sqrt{1 - \frac{\beta^2}{\alpha^2}}}, \qquad (17)$$

where f is the frequency, h is the slab thickness, and α and β denote the P and S wave speeds, respectively. In the present experiment, h is known and the values of the wavespeeds can be adjusted to fit the experimental data. This technique agrees with independent measurements to within a few percents, which is highly satisfactory.

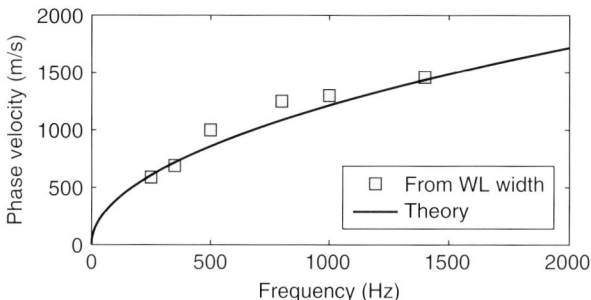

FIG. 7. Dispersion relation of the fundamental antisymmetric Lamb mode in a concrete slab. (Squares) Experimental data obtained from the width of the weak localization zone. The theoretical relation between the wavelength and the width of the weak localization zone is given by Eq. (12). (Solid line) Theoretical fit of the experimental data with Eq. (17). Reprinted figure with permission from Larose *et al.* (2006). Copyright (2006) by the American Physical Society.

4.2. Measurement of the Diffusion Constant in Strongly Scattering Media

Recent active experiments have shown the extreme character of the propagation in volcanic regions (see, e.g., Wegler and Lühr, 2001; Wegler, 2004). In the short period band, it is hardly possible and even sometimes impossible to extract the coherent ballistic waves from the complex recorded waveform. In such extreme cases, a meaningful measurement of the medium heterogeneity is the diffusion constant of the waves. However, in practice, its measurement can be affected by a number of factors, such as anelastic dissipation or local boundary conditions (see Friedrich and Wegler, 2005 for details). As demonstrated above, absorption does not influence the measurement of the weak localization effect. Unfortunately, the weak localization intensity profile does not depend on the scattering properties of the medium and therefore does not offer direct access to the scattering properties of the medium. However, we know that weak localization is a multiple scattering process which sets in only after the single scattering intensity has become negligible. Such a time dependence has been confirmed in a numerical study by Margerin *et al.* (2001), which concludes that after a time of the order of the mean free time, the weak localization pattern is measurable. This has also been verified experimentally by Larose *et al.* (2004). They measured the weak localization effect on a volcano in the 10–20 Hz frequency band and found that there exists a characteristic rise time of the intensity enhancement. Their result is shown in Fig. 8. In the case of the Puy des Goules, one can infer a mean free time of the order of 1 s which gives a mean free path of a few hundred meters. Note that in their experiment, the width of the weak localization depends in a nonlinear way on frequency. This is explained by the fact that according to the equipartition principle (Weaver, 1990; Hennino *et al.*, 2001), the energy at the surface is largely dominated by the fundamental mode Rayleigh wave, whose dispersion is caused by the complex layering at the surface. Although the

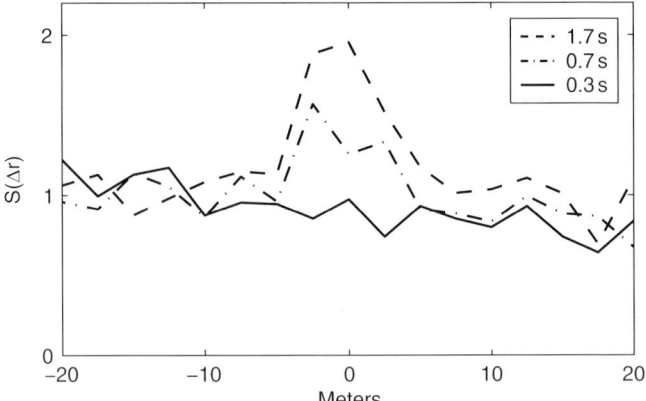

FIG. 8. Emergence of the weak localization spot in the coda. The normalized intensity is represented as a function of distance around the source. The dominant frequency of the waves is 15 Hz and the dominant signal is the Rayleigh wave with a wavespeed $\sim$250 m/s. The wavelength is of the order of 15 m. The different curves correspond to the averaged intensity profiles obtained at different lapse time in the coda (as indicated in the inset). Reprinted figure with permission from Larose *et al.* (2004). Copyright (2004) by the American Physical Society.

observation of weak localization on a volcano requires active sources, it is free of the effect of absorption and therefore offers direct access to the scattering properties of the medium.

5. Conclusion

In this paper, I have given a brief introduction to the weak localization effect in seismology. I have summarized general formulas that enable the calculation of the weak localization effect for a wide range of practical cases. The fundamental role of reciprocity between source and detection has been emphasized and illustrated with experimental results. In practice, the control of the source mechanism is crucial. The simplest solution is to measure the vertical displacements generated by vertical forces. Applications to the characterization of scattering or bulk elastic properties have been presented. In particular, I have shown that the emergence time of weak localization yields an estimate of the scattering mean free path, independent of absorption effects. Thus, weak localization combined with other measurements such as time and space dependence of the coda could be used to discriminate anelastic and scattering attenuation.

Acknowledgments

I would like to thank E. Larose, A Derode, J. de Rosny, and R. Hennino for their invaluable help with the figures and for many discussions on the weak localization effect. I would like to thank M. Campillo and B. van Tiggelen for their constant input. Comments by H. Sato, Yu. Kravtsov, and an anonymous reviewer greatly helped to improve the presentation and the readability of this chapter.

References

Akkermans, E., Montambaux, G. (2005). Physique mésoscopique des électrons et des photons CNRS Editions, Paris: EDP Sciences, Paris.
Apresyan, L.A., Kravtsov, Yu. A. (1996). Radiation Transfer: Statistical and Wave Aspects. Amsterdam: Gordon and Breach, Amsterdam.
Aubry, A., Derode, A. (2007). Ultrasonic imaging of highly scattering media from local measurements of the diffusion constant: Separation of coherent and incoherent intensities. *Phys. Rev. E* **75**, 026602.
Barabanenkov, Yu.N. (1973). Wave corrections to the transfer equation for "back" scattering. *Radiophysics and Quantum Electronics* **16**, 65–71.
Barabanenkov, Yu.N. (1975). Multiple scattering of waves by ensembles of particles and the theory of radiation transport. *Sov. Phys. Usp* **18**, 673–689.
Barabanenkov, Yu.N., Ozrin, V.D. (1991). Asymptotic solution of the Bethe-Salpeter equation and the Green-Kubo formula for the diffusion constant for wave propagation in random media. *Phys. Lett. A* **154**, 38–42.
Barabanenkov, Yu.N., Ozrin, V.D. (1995). Diffusion asymptotics of the Bethe-Salpeter equation for electromagnetic waves in discrete random media. *Phys. Lett. A* **206**, 116–122.
Barabanenkov, Yu.N., Kravtsov, Yu.A., Ozrin, V.D., Saichev, A.I. (1991). Enhanced backscattering in optics. *Prog. Opt.* **29**, 65–197.
Bayer, G., Niederdrank, T. (1993). Weak localization of acoustic waves in strongly scattering media. *Phys. Rev. Lett.* **70**, 3884–3887.

Campillo, M. (2006). Phase and correlation in 'random' seismic fields and the reconstruction of the Green function. *Pure Appl. Geophys.* **163**, 475–502.

Chandrasekhar, S. (1960). Radiative Transfer. New York: Dover, New York.

Dahlen, F.A., Tromp, J. (1998). Theoretical Global Seismology. Princeton, New Jersey: Princeton University Press, Princeton, New Jersey.

Derode, A., Mamou, V., Padilla, F., Jenson, F., Laugier, P. (2005). Dynamic coherent backscattering in an absorbing heterogeneous medium: Application to human trabecular bone characterization. *Appl. Phys. Lett.* **87**, 114101.

de Rosny, J., Tourin, A., Fink, M. (2000). Coherent backscattering of elastic wave in a chaotic cavity. *Phys. Rev. Lett.* **84**, 1693–1695.

de Rosny, J., Tourin, A., Fink, M. (2001). Observation of a coherent backscattering effect with a dipolar source for elastic waves: Highlight of the role played by the source. *Phys. Rev. E* **64**, 066604, doi:10.1103/PhysRevE.64.066604.

de Wolf, D.A. (1971). Electromagnetic reflection from an extended turbulent medium: Cumulative forward-scatter single-backscatter approximation. *IEEE Trans Antennas Propagation* **19**, 254–262.

Friedrich, C., Wegler, U. (2005). Localization of seismic coda at Merapi volcano (Indonesia). *Geophys. Res. Lett.* **32**, L14312, doi: 10.1029/2005GL023111.

Gusev, A., Abubakirov, I.R. (1987). Monte-Carlo simulation of record envelope of a near earthquake. *Phys. Earth Planet. Inter.* **49**, 30–36.

Haney, M., Snieder, R. (2003). Breakdown of wave diffusion in 2D due to loops. *Phys. Rev. Lett.* **91**, 093902.

Hapke, B.W., Nelson, R.M., Smythe, W.D. (1993). The opposition effect of the moon—The contribution of coherent backscatter. *Science* **260**, 509–511.

Hennino, R., Tregoures, N., Shapiro, N.M., Margerin, L., Campillo, M., van Tiggelen, B.A., Weaver, R.L. (2001). Observation of equipartition of seismic waves. *Phys. Rev. Lett.* **86**, 3447–3450.

Hoshiba, M. (1991). Simulation of multiple-scattered coda wave excitation based on the energy conservation law. *Phys. Earth Planet. Inter.* **67**, 123–136.

Kaveh, M., Rosenbluh, M., Edrei, I., Freund, I. (1986). Weak localization and light scattering from disordered solids. *Phys. Rev. Lett.* **57**, 2049–2052.

Kuga, Y., Ishimaru, A. (1984). Retroreflection from a dense distribution of spherical particles. *J. Opt. Soc. Am. A* **8**, 831–835.

Larose, E., Margerin, L., Campillo, M., van Tiggelen, B.A. (2004). Weak localization of seismic waves. *Phys. Rev. Lett.* **93**, 048501.

Larose, E., de Rosny, J., Margerin, L., Anache, D., Gouedard, P., Campillo, M., van Tiggelen, B. (2006). Observation of multiple scattering of kHz vibrations in a concrete structure and application to monitoring weak changes. *Phys. Rev. E* **73**, 016609.

Lesaffre, M., Atlan, M., Gross, M. (2006). Effect of the photon's Brownian Doppler shift on the weak-localization coherent-backscattering cone. *Phys. Rev. Lett.* **97**, 033901.

Margerin, L. (2005). Introduction to radiative transfer of seismic waves. In: Levander, A., Nolet, G., (Eds.), *Seismic Earth: Array Analysis of Broadband Seismograms: Geophysical Monograph Series*, Vol. 157, Chapter 14. AGU, Washington, pp. 229–252.

Margerin, L., Campillo, M., van Tiggelen, B.A. (2001). Coherent backscattering of acoustic waves in the near field. *Geophys. J. Int.* **145**, 593–603.

Royer, D., Dieulesaint, E. (2000). Elastic Waves in Solids, Part I. Free and Guided Propagation. Berlin Heidleberg: Springer-Verlag, Berlin Heidleberg.

Sakai, K., Yamamoto, K., Takagi, K. (1997). Observation of acoustic coherent backscattering. *Phys. Rev. B* **56**, 10930–10933.

Sato, H. (1994). Multiple isotropic scattering model including P-S conversions for the seismogram envelope formation. *Geophys. J. Int.* **117**, 487–494.

Sato, H., Fehler, M. (1998). Seismic Wave Propagation in the Heterogeneous Earth. Heidelberg: Springer, Heidelberg.

Sheng, P. (1995). Introduction to Wave Scattering, Localization and Mesoscopic Phenomena San Diego: Academic Press, San Diego.

Snieder, R. (2006). The coherent backscattering effect for moving scatterers. *Europhys. Lett.* **74**, 630–636.

Tourin, A., Roux, P., Derode, A., van Tiggelen, B.A., Fink, M. (1997). Time dependent coherent backscattering of acoustic waves. *Phys. Rev. Lett.* **79**, 3637–3639.

van Albada, M.P., Lagendijk, A. (1985). Observation of weak localization of light in a random medium. *Phys. Rev. Lett.* **55**, 2692–2695.

van Tiggelen, B.A., Maynard, R. (1997). Reciprocity and coherent backscattering of light. *In:* Papanicolaou, G.R., (Ed.), *Wave Propagation in Complex Media (IMA Volumes in Mathematics and Its Applications, Vol. 96)*. Springer, New York, pp. 247–271.

van Tiggelen, B.A., Margerin, L., Campillo, M. (2001). Coherent backscattering of elastic waves: Specific role of source, polarization, and near field. *J. Acoust. Soc. Am.* **110**, 1291–1298.

Watson, K. (1969). Multiple scattering of electromagnetic waves in an underdense plasma. *J. Math. Phys.* **10**, 688–702.

Weaver, R.L. (1990). Diffusivity of ultrasound in polycrystals. *J. Mech. Phys. Solids* **38**, 55–86.

Weaver, R., Burkhardt, J. (1994). Weak Anderson Localization and enhanced backscatter in reverberation rooms and quantum dots. *J. Acoust. Soc. Am.* **96**, 3186–3190.

Weaver, R.L., Lobkis, O.I. (2000). Enhanced backscattering and modal echo of reverberant elastic waves. *Phys. Rev. Lett.* **84**, 4942–4945.

Wegler, U. (2004). Diffusion of seismic waves in a thick layer: Theory and application to Vesuvius volcano. *J. Geophys. Res.* **109**, doi: 10.1029/2004JB003048.

Wegler, U., Lühr, B.G. (2001). Scattering behavior at Merapi volcano (Java) revealed from an active seismic experiment. *Geophys. J. Int.* **145**, 579–592.

Wegler, U., Korn, M., Przybilla, J. (2006). Modelling full seismogram envelopes using radiative transfer theory with Born scattering coefficients. *Pure Appl. Geophys.* **163**, 503–531.

Wolf, P.E., Maret, G. (1985). Weak localization and coherent backscattering of photon. *Phys. Rev. Lett.* **55**, 2696–2699.

Zeng, Y., Su, F., Aki, K. (1991). Scattering wave energy propagation in a random isotropic scattering medium. 1. Theory. *J. Geophys. Res.* **96**, 607–619.

THEORY OF TRANSMISSION FLUCTUATIONS IN RANDOM MEDIA WITH A DEPTH-DEPENDENT BACKGROUND VELOCITY STRUCTURE

YINGCAI ZHENG AND RU-SHAN WU

ABSTRACT

An extended theory on the coherence function of log amplitude and phase for waves passing through random media is developed for a depth-dependent background medium using the WKBJ-approximated Green's function, the Rytov approximation, and the stochastic theory of the random velocity field. The new theory overcomes the limitation of the existing theory that can only deal with constant background media. Our extended coherence functions depend jointly on the angle separation between two incident plane waves and the spatial lag between receivers. The theory is verified through numerical simulations using the iasp91 background velocity model with two layers of random media. The current theory has the potential to be used to invert for the depth-dependent spectrum of heterogeneities in the Earth.

KEY WORDS: Transmission fluctuation, coherence function, random media, heterogeneity spectrum, Phase, amplitude, WKBJ, Rytov © 2008 Elsevier Inc.

1. INTRODUCTION

The statistical approach, complementary to the deterministic method such as seismic tomography, is indispensable in probing the small-scale inhomogeneities in the Earth using scattered seismic waves. Its characterization of heterogeneities using statistical parameters has yielded many important physical constraints and interpretations for the lithosphere to the core-mantle boundary region, and led to key implications regarding the Earth's compositional constituents, thermal state, dynamic mixing, and others.

In 1970s, many attempts (Capon, 1974; Capon and Berteussen, 1974; Berteussen, 1975; Berteussen *et al.*, 1975, 1977) had been made to infer the spatial spectrum of the velocity heterogeneity using observed fluctuations of the logarithmic amplitude ($logA$) and the phase ϕ across a seismic array since the pioneering work by Aki (1973). Aki essentially employed the Chernov theory (Chernov, 1960) to study the transverse coherence function (TCF) using the data from the Large Aperture Seismic Array (LASA), Montana. He found that a 60-km random layer in the lithosphere with ~10 km scale length for the heterogeneities and 4% root-mean-square (rms) P-wave velocity fluctuation could explain the data. The Chernov theory (1960) studies wave propagation in a single layer of stationary random velocity heterogeneities with a Gaussian correlation function and with a constant background velocity and as such, the TCF has a nice closed mathematical form. By the word stationary, we mean the spatially translational invariance of the statistic (e.g., correlation function). The TCF for random media of general spectral type can be found in Tatarskii (1971) and Ishimaru (1978) and it does not possess a simple analytical expression in general. It is interesting to note that this stochastic approach predated the deterministic tomography method (Aki and Lee, 1976; Aki et al., 1976, 1977).

The heterogeneity spectrum of the Earth is fractal based on well-logging data (Wu et al., 1994b; Jones et al., 1997; Goff and Holliger, 1999) and it may not be Gaussian. Capon and Berteussen (1974) found that the Chernov theory was not applicable to fluctuations of log$A$ and phase data under the Norwegian Seismic Array (NORSAR); however, they attributed this to the validity of the Born approximation at high frequencies involved in the theory. The Earth's large-scale structures are largely stratified in depth and the herteogeneity spectrum can be slowly varying with depth. Flatté and Wu (1988) introduced a new kind of coherence function, called the angular coherence function (ACF), to resolve the depth-dependent spectra under the NORSAR. A two-layer model with power-law type medium was favored over a one-layer stationary Gaussian medium. Wu and Flatté (1990) also formulated the joint transverse and angular coherence function (JTACF) in which both the spatial lag between two seismic stations and the angular separation between two plane waves were taken into account. Their formulation was based on wave perturbation theory and assumed the Rytov and the parabolic approximations for the wave equation. Chen and Aki (1991) independently obtained the same JTACF result using the Born approximation. Parametric (Flatté et al., 1991a) and nonparametric (Wu and Xie, 1991) inversions have been carried out to process real data or to study the depth resolution in the spectral inversion. Wu and Xie (1991) called such inversion "stochastic tomography" and they found that the JTACF has the best depth resolution, that the ACF has limited depth resolution close to the surface, and that the TCF has no depth resolution at all. To invert for the spectrum of a single-layer stationary random heterogeneity, Zheng et al. (2007) proposed a new scheme using only the TCF data for log$A$ and phase, in which a Fourier transform was established between the heterogeneity spectrum and the combination of the log$A$ and phase TCF data. There was some concern on the discrepancy on the phase coherence function between the theory and the numerical simulation (Line et al., 1998a; Hong et al., 2005). However, recent investigation has shown that this discrepancy was caused by the incorrect phase picking method used in the numerical and field experiments. Application of the correct phase measuring method has resulted in excellent agreement between numerical tests and the theory (Zheng and Wu, 2005). The coherence function formation using combined data from arrays with different apertures was investigated by Flatté et al. (1991b) and Flatté and Xie (1992). The theory of transmission fluctuation was also applied to seismic reflection data to obtain heterogeneities in the upper crust (Line et al., 1998b). The theory of JTACF has been applied to the NORSAR data (Wu et al., 1994a) and the Southern California Seismic Network data (Liu et al., 1994; Wu et al., 1995). For earlier reviews on the subject, see Wu (2002) and Sato et al. (2002).

Besides the coherence function study, many other seismological methods have been devised to characterize the small-scale heterogeneities. The seismic coda envelope analysis (Sato and Fehler, 1998) has been widely used to investigate the scattering strength in the Earth. Through studying the radiated power carried by precursors to the PKIKP phase (Cleary and Haddon, 1972), the spectrum for P-wave volumetric scatters in the D'' region, or for the core-mantle boundary topographical relief (Bataille and Flatté, 1988; Bataille et al., 1990), or for the whole mantle (Hedlin et al., 1997) and mid-mantle (Hedlin and Shearer, 2002) has been constrained.

Despite significant progress that has been made using the coherence function to characterize the Earth's small-scale heterogeneities, all these theoretical treatment of the problem is based on wave propagation in a homogeneous background medium. This is inadequate for the real Earth, which has a depth-dependent velocity profile to first order. In this chapter, we show how we can generalize the theory of coherence functions (log$A$ and phase ϕ) to arbitrary random medium superimposed on general depth-dependent background medium. This generalized theory can be directly utilized for the real Earth.

2. Acoustic Waves in Stratified Media and WKBJ Green Function

The linearized wave equation for pressure in the frequency domain is

$$\rho \nabla \cdot \left(\frac{1}{\rho} \nabla p \right) + \frac{\omega^2}{c^2} p = 0. \tag{1}$$

Let ω be the angular frequency, $p = p(x, y, z, \omega)$ the pressure field, and ∇ the spatial gradient operator. Define z as depth variable. ρ and c are density and wave propagation speed, respectively. For a stratified medium, $c = c(z), \rho = \rho(z)$, the 2-D Fourier transform can be performed to Eq. (1) with respect to spatial coordinates, x and y, to obtain

$$k_z^2(z) p + \frac{\partial^2 p}{\partial z^2} - \rho^{-1} \frac{\partial \rho}{\partial z} \frac{\partial p}{\partial z} = 0, \tag{2}$$

where k_z is the vertical wave number defined as

$$k_z(z) = \pm \sqrt{\frac{\omega^2}{c^2}(z) - k_x^2 - k_y^2}. \tag{3}$$

k_x and k_y are horizontal wave numbers corresponding to x and y, respectively. The plus sign in Eq. (3) corresponds to the downgoing wave and the minus sign the upgoing wave. We use the same symbol p to denote the pressure before and after the Fourier transform and this should not cause confusion. Plugging the trial solution in the form of $p = A(z) \exp[i\phi(z)]$ into Eq. (2) and the real part is

$$k_z^2 A + \left(A'' - \phi'^2 A \right) - \rho^{-1} \frac{\partial \rho}{\partial z} A' = 0, \tag{4}$$

and the imaginary part reads

$$2\phi' A' + A\phi'' - \rho^{-1} \frac{\partial \rho}{\partial z} A\phi' = 0. \tag{5}$$

The prime represents the partial derivative with respect to depth z. If the amplitude is slowly varying with depth, we can use the WKBJ approximation (i.e., get rid of all terms that involve derivatives of A) in Eq. (4) and it reduces to

$$k_z^2 - \phi'^2 = 0 \Rightarrow \phi' = k_z. \tag{6}$$

Therefore, the phase function is solved as

$$\phi(z) = \int_{z_s}^{z} k_z(z')dz', \tag{7}$$

where z_s is a reference depth, commonly taken as the source depth. The validity of the WKBJ approximation is assured if the wavelength is shorter than the characteristic scale of the background velocity model. The WBKJ also implies energy conservation for the transmitted wave, which means that no reflected waves are produced (Wu and Cao, 2005). Substituting Eq. (6) in Eq. (5), we have

$$2\frac{d \ln A}{dz} + \frac{d \ln k_z}{dz} - \frac{d \ln \rho}{dz} = 0, \tag{8}$$

Rearranging Eq. (8), we get

$$d \ln A = d \ln \sqrt{\frac{\rho(z)}{k_z(z)}}, \tag{9}$$

So the solution for the amplitude in Eq. (9) is

$$A(z) = C\sqrt{\frac{\rho(z)}{k_z(z)}}, \tag{10}$$

where C is a constant. The final solution to Eq. (2) is

$$p(k_x, k_y, z, \omega) = C\sqrt{\frac{\rho(z)}{k_z(z)}} \exp\left[i \int_{z_s}^{z} k_z(z')dz'\right]. \tag{11}$$

Once we have obtained Eq. (11), the solution for Eq. (1) is just the inverse Fourier transform

$$p(x, y, z, \omega) = (2\pi)^{-2} \int_{-\infty}^{+\infty} \int_{-\infty}^{+\infty} p(k_x, k_y, z, \omega)\exp\left[i(k_x x + k_y y)\right]dk_x dk_y. \tag{12}$$

Next, let us investigate the solution for a point source. Under this case, the Eq. (1) has an additional source term

$$\rho \nabla \cdot \left(\frac{1}{\rho}\nabla p\right) + \frac{\omega^2}{c^2} p = -\delta(x - x_s)\delta(y - y_s)\delta(z - z_s). \tag{13}$$

x_s, y_s and z_s are source position coordinates. For simplicity, the source is placed at the origin. In the homogeneous case, the solution in frequency domain is

$$p(x, y, z, \omega) = \frac{1}{4\pi r}\exp\left(\frac{i\omega r}{c}\right), \quad r = \sqrt{x^2 + y^2 + z^2}. \tag{14}$$

When a receiver is sufficiently close to the source, the solution (12) should coincide with (14). In order to compare both solutions in the wave number domain, we first expand the point source solution (14) into plane wave components using the Weyl integral (e.g., Aki and Richards, 2002 p. 190). The coefficient to component (k_x, k_y) is $1/2ik_z(z_s)$ and the corresponding coefficient in Eq. (12) is $C/\sqrt{\rho(z_s)/k_z(z_s)}$. Clearly, these two should be equal. Thus, this constant C is

$$C = \frac{1}{2i\sqrt{\rho(z_s)k_z(z_s)}}. \tag{15}$$

Therefore, the complete Green's function for both 3-D and 2-D cases can be obtained. Let us rewrite the Green's function in 3-D case

$$G(x,y,z,\omega) = (2\pi)^{-2} \int_{-\infty}^{+\infty}\int_{-\infty}^{+\infty} \frac{1}{2i\sqrt{k_z(z_s)k_z(z)}}\sqrt{\frac{\rho(z)}{\rho(z_s)}}$$
$$\exp\left\{i\left[k_x x + k_y y + \int_{z_s}^{z} k_z(z')\mathrm{d}z'\right]\right\}\mathrm{d}k_x\,\mathrm{d}k_y. \tag{16}$$

In 2-D case, just drop all terms pertaining to k_y and the inverse Fourier transform constant is $(2\pi)^{-1}$. Also remember that the 2-D Fourier transform becomes a 1-D transform.

3. Rytov Solution to the Wave Equation in a Heterogeneous Medium

For convenience, we rewrite the monochromatic wave equation

$$\rho\nabla\cdot(\rho^{-1}\nabla p) + \frac{\omega^2}{c^2}p = 0. \tag{17}$$

The solution for the pressure wavefield is sought in the form of

$$p = p_0 e^{\psi}. \tag{18}$$

$\psi = \psi(\vec{x}',\omega)$ is the complex phase and $p_0 = p_0(\omega,\vec{x}')$ is the background incident wavefield. In what follows we suppress the explicit dependence on ω for the wavefield for simplicity in notation. Assuming no lateral density perturbation and substituting (18) into (17), we have

$$(\rho\nabla\cdot\nabla\rho^{-1} + k^2)(p_0\psi) = -2k^2\gamma p_0 - p_0\nabla\psi\cdot\nabla\psi, \tag{19}$$

where $\gamma = (c_0^2/c^2 - 1)/2 \approx -\delta c/c_0$ is the scattering potential, $\delta c = c - c_0$. The background wave number k is defined as $k = \omega/c_0(z')$. The Rytov approximation assumes $(\nabla\psi)^2 \ll 2k^2\gamma$. The right-hand side of Eq. (19) is a spatially distributed source term. The complex phase can be solved as

$$\psi(\vec{x}) = \frac{2}{p_0(\vec{x})}\int_V k^2(\vec{x}')G(\vec{x},\vec{x}',\omega)\gamma(\vec{x}')p_0(\vec{x}')\mathrm{d}^3\vec{x}'. \tag{20}$$

From Eq. (20), it can be seen that contributions of scattering potentials γ at different locations to the complex phase are independent from each other. However, this linear relationship is to the complex phase, not to the field itself, which is different from the Born approximation. Note that this complex phase is naturally "unwrapped." It can be directly used to obtain the phase velocity. We see that the Rytov approximation is a single-scattering approximation in terms of complex phase. A local Rytov approximation has been developed (Huang et al., 1999) to calculate wavefields, in which case the multiple forward scattering is included.

4. Complex Phase ψ Due to a Plane Wave Incidence

To obtain the complex phase at depth z, we can rewrite Eq. (20) in an explicit form

$$\psi(\vec{x}) = \frac{2}{p_0(\vec{x})} \int_z^L k^2(z')dz' \int\int G(\vec{x}_T - \vec{x}'_T, z, z', \omega) \gamma(\vec{x}'_T, z') \\ \times p_0(\vec{x}') d^2\vec{x}'_T, \qquad (21)$$

where $\vec{x} = (\vec{x}_T, z)$ is the receiver location and $\vec{x}' = (\vec{x}'_T, z')$ is the location for the heterogeneity. In the case of an upgoing plane wave (negative vertical wave number), incidence with ray parameter $\vec{q}$,

$$\frac{p_0(\vec{x}')}{p_0(\vec{x})} = \sqrt{\frac{\rho(z')}{\rho(z)}} \sqrt{\frac{k_z(q,z)}{k_z(q,z')}} \exp\left[i\omega\vec{q} \cdot (\vec{x}'_T - \vec{x}_T) - i\int_z^{z'} k_z(q,s)ds\right], \qquad (22)$$

where $q = |\vec{q}|$ and $k_z(\vec{q}, z) = \omega\sqrt{1/c^2(z) - q^2}$. We always choose a nonnegative $\omega \geq 0$ because of the complex conjugate symmetry $p(\omega, \vec{x}) = p^*(-\omega, \vec{x})$. Substituting Eqs. (16) and (22) into Eq. (21) and recognizing that it represents a convolution, we can easily put down the complex phase in the wave number domain:

$$\psi(\vec{q}, \vec{x}) = 2(2\pi)^{-2} \int_z^L dz' \int\int_\kappa \frac{k^2(z')}{2i\sqrt{k_z(\vec{\kappa} + \omega\vec{q}, z')k_z(\vec{\kappa} + \omega\vec{q}, z)}} \sqrt{\frac{k_z(q,z)}{k_z(q,z')}} \\ \times \exp\left\{i\int_z^L [k_z(\vec{\kappa} + \omega\vec{q}, z'') - k_z(q, z'')]dz'' + i\vec{\kappa} \cdot \vec{x}\right\} dv)(\vec{\kappa}, z'), \qquad (23)$$

where $\vec{\kappa}$ is the horizontal wave number of the heterogeneity spectrum and $dv(\vec{\kappa}, z')$ is the Fourier-Stieltjes spectral density (Yaglom, 1962). Such a spectral representation is very technical and it seems harmless to replace $dv(\vec{\kappa}, z')$ by $v(\vec{\kappa}, z')d^2\vec{\kappa}$, where $v(\vec{\kappa}, z')$ is the Fourier spectrum of the velocity perturbation field $\gamma(\vec{x}_T, z')$ in the transverse plane at depth z'. If one wants to pursue the mathematical exactness of the Green function (16), all wave numbers should be integrated, including both propagating (real k_z) and evanescent components (imaginary k_z). However, the evanescent components are not used in our case. For high-frequency wave propagation in random media, most scattered energy is in forward direction within an angle that spans $\sim 1/(ka)$, with a being the scale length of the heterogeneity. The forward direction is understood as the incoming direction of the incident wave. If we consider high-frequency wave propagation, Eq. (23) can be simplified as

$$\psi\left(\vec{q},\vec{x}\right) \approx (2\pi)^{-2} \int_0^L dz' \iint_\kappa \frac{k^2(z')}{ik_z(q,z')} e^{i\vec{\kappa}\cdot\vec{x} + i\int_0^{z'}[k_z(\vec{\kappa}+\omega\vec{q},z) - k_z(q,z)]dz} dv\left(\vec{\kappa},z'\right), \quad (24)$$

where L is the depth of the lower boundary of the heterogeneous layer and receivers are placed at zero depth (Fig. 1). Here we define

$$D\left(\vec{\kappa},z'\right) = \frac{1}{\omega}\int_0^{z'}\left[k_z\left(\vec{\kappa}+\omega\vec{q},z\right) - k_z(q,z)\right]dz. \quad (25)$$

Under the forward scattering approximation ($ka > 1$, k is the wave number and a is the characteristic scale length of the heterogeneity), only wave numbers around $\vec{q}$ are integrated. We can expand the D function [Eq. (25)] around $\vec{\kappa} = 0$ using the Taylor expansion,

$$D\left(\vec{\kappa},z'\right) \approx -\vec{\kappa}\cdot\vec{r}\left(\vec{q},z'\right)\frac{1}{\omega} + \theta\left(\vec{\kappa},\vec{q},\omega,z'\right) + \cdots. \quad (26)$$

$\vec{r}\left(\vec{q},z'\right)$ is the transverse vector between the piercing points at depths 0 and z' for a ray incidence from below with slowness vector $\vec{q}$. The quadratic term of κ is

$$\theta\left(\vec{\kappa},\vec{q},\omega,z'\right) = -\frac{1}{2}\frac{I_1(q,z')\kappa^2}{\omega^2} - \frac{1}{2}\frac{I_2(q,z')(\vec{\kappa}\cdot\vec{q})^2}{\omega^2}, \quad (27)$$

where

$$I_1(q,z') = \int_0^{z'}\left[c^{-2}(z) - q^2\right]^{-1/2}dz, \quad (28)$$

and

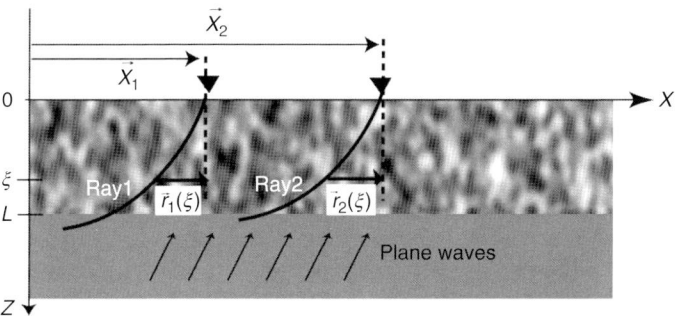

FIG. 1. Schematic geometry used in the theoretical derivation. The heterogeneous layer is bounded between depth 0 and L. Seismic receivers (triangles) are placed on depth zero. Ray 1 connects to station $\vec{x}_1$ and it has the same slowness as plane wave 1. $\vec{r}_1(\xi)$ is the transverse distance between ray 1 at depth ξ to station $\vec{x}_1$. Similar meaning applies to ray 2 and $\vec{r}_2(\xi)$.

$$I_2(q, z') = \int_0^{z'} \left[c^{-2}(z) - q^2\right]^{-3/2} dz. \tag{29}$$

Obviously, we have $I_1, I_2 > 0$. The following replacement

$$\left(\vec{\kappa} \cdot \vec{q}\right)^2 \to \kappa^2 q^2 \tag{30}$$

in (27) will result in a more rapidly oscillatory $e^{i\omega\theta}$ with respect to κ. We can use the approximated phase [Eq. (26)] with replacement of Eq. (30) in (24).

Before we proceed further, let us introduce a τ function

$$\tau(p, z') = \int_0^{z'} \varsigma(z, p) dz, \tag{31}$$

where the vertical slowness at depth z is

$$\varsigma(z, p) = \left[c^{-2}(z) - p^2\right]^{1/2}. \tag{32}$$

This function has all the kinematic information we want (Buland and Chapman, 1983). For example, the horizontal distance traveled by a ray with ray parameter q from depth z' to depth zero is $\tau'_p|_{p=q} = -\partial\tau/\partial p|_{p=q}$. The second-order derivative of τ with respect to p contains information on the geometrical spreading of the wave front. The τ function is the Legendre transform of the travel time function. So the travel time can be easily computed using this function. Using (30) in (27), we obtain

$$\theta(\vec{\kappa}, \vec{q}, \omega, z') = \frac{1}{2} \tau'_{pp}\bigg|_{p=q} \frac{\kappa^2}{\omega^2}. \tag{33}$$

The subscript "$'pp$" denotes second-order derivative with respect to the argument p, that is, $\tau'_{pp} = \partial^2\tau/\partial p^2$. The derivative of θ with respect to depth is

$$\frac{\partial \theta(\vec{\kappa}, \vec{q}, \omega, z')}{\partial z'} = \frac{1}{2} \varsigma'_{pp} \frac{\kappa^2}{\omega^2}. \tag{34}$$

The complex phase (24) can be expressed as

$$\psi\left(\vec{q}, \vec{x}\right) \approx -(2\pi)^{-2} \int_0^L ia(q, z') dz' \iint_\kappa e^{i\omega\theta - i\vec{\kappa} \cdot \vec{r} + i\vec{\kappa} \cdot \vec{x}} dv\left(\vec{\kappa}, z'\right), \tag{35}$$

where

$$a(q, z') = \frac{k^2(z')}{k_z(q, z')}. \tag{36}$$

5. Coherence Function Between Two Plane Waves

Considering two plane waves with slowness vectors, $\vec{q}_1$ and $\vec{q}_2$, we can express the corresponding complex phases ψ_1 and ψ_2 at depth zero as:

$$\psi_1 = \psi(\vec{q}_1, \vec{x}_1) \approx -(2\pi)^{-2} \int_0^L ia_1(z')dz' \iint_\kappa e^{i\omega\theta_1(z') - i\vec{\kappa}_1 \cdot \vec{r}_1(z') + i\vec{\kappa}_1 \cdot \vec{x}_1} dv(\vec{\kappa}_1, z'), \tag{37}$$

and

$$\psi_2 = \psi(\vec{q}_2, \vec{x}_2) \approx -(2\pi)^{-2} \int_0^L ia_2(z'')dz'' \iint_\kappa e^{i\omega\theta_2(z'') - i\vec{\kappa}_2 \cdot \vec{r}_2(z'') + i\vec{\kappa}_2 \cdot \vec{x}_2} dv(\vec{\kappa}_2, z''). \tag{38}$$

The symbols are

$$a_1(z') = a(q_1, z'), \quad a_2(z'') = a(q_2, z''), \tag{39}$$

$$\theta_1(z') = \theta(\vec{\kappa}_1, \vec{q}_1, \omega, z') = \frac{1}{2}\tau_{pp}|_{p=q_1}\frac{\kappa_1^2}{\omega^2}, \tag{40}$$

$$\theta_2(z'') = \theta(\vec{\kappa}_2, \vec{q}_2, \omega, z'') = \frac{1}{2}\tau_{pp}|_{p=q_2}\frac{\kappa_2^2}{\omega^2}, \tag{41}$$

and

$$\vec{r}_1(z') = \vec{r}(\vec{q}_1, z'), \quad \vec{r}_2(z'') = \vec{r}(\vec{q}_2, z''). \tag{42}$$

The coherence function between ψ_1 and ψ_2 is

$$\langle \psi_1 \psi_2^* \rangle = (2\pi)^{-4} \int_0^L \int_0^L a_1(z')a_2(z'')dz'dz''$$
$$\times \iint_{\kappa_2} \iint_{\kappa_1} e^{i\omega[\theta_1(z') - \theta_2(z'')] + i\vec{\kappa}_2 \cdot [\vec{r}_2(z'') - \vec{x}_2] - i\vec{\kappa}_1 \cdot [\vec{r}_1(z') - \vec{x}_1]}$$
$$\times \langle dv(\vec{\kappa}_1, z')dv^*(\vec{\kappa}_2, z'') \rangle, \tag{43}$$

where $\langle \cdot \rangle$ is the ensemble average from multiple realizations of the random medium. For a brief introduction on random variables and random function, and several useful correlation functions and their spectral representations, see Appendix. It can be shown that in the 3-D case (Tatarskii, 1971) that

$$\langle dv(\vec{\kappa}_1, z')dv^*(\vec{\kappa}_2, z'') \rangle = (2\pi)^2 W(\vec{\kappa}_1, z', z'')\delta(\vec{\kappa}_1 - \vec{\kappa}_2)d^2\vec{\kappa}_1 \, d^2\vec{\kappa}_2. \tag{44}$$

In the 2-D case, we need replace $(2\pi)^2$ by 2π at the right-hand side of Eq. (44). In view of the simple substitution of dv by $vd^2\vec{\kappa}$ in Section 4, we see that $W(\vec{\kappa}_1, z', z'')$ is the correlation function of the two spectral fields $v(\vec{\kappa}_1, z')$ and $v(\vec{\kappa}_1, z'')$ at depths z' and z'', respectively. Applying the coordinate transformation

$$\eta = z' - z'', \quad \xi = \frac{z' + z''}{2} \tag{45}$$

in (43) and if the correlation function of the heterogeneity is slowly varying with depth, we can approximate W as

$$W(\vec{\kappa}, z', z'') \approx W(\xi, \vec{\kappa}, \eta). \tag{46}$$

Equation (43) can be simplified as

$$\langle \psi_1 \psi_2^* \rangle = (2\pi)^{-2} \int_0^L \int_0^L a_1\left(\xi + \frac{\eta}{2}\right) a_2\left(\xi - \frac{\eta}{2}\right) d\xi \, d\eta$$
$$\times \iint_{\vec{\kappa}} e^{i\omega[\theta_1(\xi + \frac{\eta}{2}) - \theta_2(\xi - \frac{\eta}{2})] + i\vec{\kappa} \cdot [\vec{r}_2(\xi - \frac{\eta}{2}) - \vec{x}_2] - i\vec{\kappa} \cdot [\vec{r}_1(\xi + \frac{\eta}{2}) - \vec{x}_1]} W(\xi, \vec{\kappa}, \eta) d^2\vec{\kappa}$$
$$\tag{47}$$

Because the correlation function W decreases rapidly with the vertical separation distance $|\eta|$, we can extend the integration limit of η from $-\infty$ to $+\infty$ without introducing much error. We also make following approximations:

$$a_1\left(\xi + \frac{\eta}{2}\right) \approx a_1(\xi), \quad a_2\left(\xi - \frac{\eta}{2}\right) \approx a_2(\xi), \tag{48}$$

$$\vec{r}_1\left(\xi + \frac{\eta}{2}\right) \approx \vec{r}_1(\xi), \quad \vec{r}_2\left(\xi - \frac{\eta}{2}\right) \approx \vec{r}_2(\xi). \tag{49}$$

Using Eqs. (48) and (49) in (47), we obtain

$$\langle \psi_1 \psi_2^* \rangle = (2\pi)^{-2} \int_0^L d\xi \int_{-\infty}^{+\infty} a_1(\xi) a_2(\xi) d\eta$$
$$\times \iint_{\vec{\kappa}} e^{i\omega[\theta_1(\xi + (\eta/2)) - \theta_2(\xi - (\eta/2))] + i\vec{\kappa} \cdot [\vec{r}_2(\xi) - \vec{x}_2] - i\vec{\kappa} \cdot [\vec{r}_1(\xi) - \vec{x}_1]} W(\xi, \vec{\kappa}, \eta) d^2\vec{\kappa}.$$
$$\tag{50}$$

The integral value (50) is nonnegligible only when $|\eta|$ is small, so we can have the following expansion

$$\theta_1\left(\xi + \frac{\eta}{2}\right) - \theta_2\left(\xi - \frac{\eta}{2}\right) \approx \theta_1(\xi) - \theta_2(\xi) + \frac{\dot\theta_1 + \dot\theta_2}{2}\eta, \tag{51}$$

where

$$\dot\theta_1 = \left.\frac{\partial \theta_1(z)}{\partial z}\right|_{z=\xi}, \quad \dot\theta_2 = \left.\frac{\partial \theta_2(z)}{\partial z}\right|_{z=\xi}. \tag{52}$$

Taking into account Eq. (51), Eq. (50) can be written as

$$\langle \psi_1 \psi_2^* \rangle = (2\pi)^{-2} \int_0^L a_1(\xi) a_2(\xi) \mathrm{d}\xi \iint_{\vec{\kappa}} \mathrm{d}^2 \vec{\kappa} \, e^{i\omega[\theta_1(\xi) - \theta_2(\xi)] + i\vec{\kappa} \cdot [\vec{r}_2(\xi) - \vec{x}_2] - i\vec{\kappa} \cdot [\vec{r}_1(\xi) - \vec{x}_1]}$$
$$\times \int_{-\infty}^{+\infty} \mathrm{d}\eta \, e^{i(\omega/2)(\dot{\theta}_1 + \dot{\theta}_2)\eta} W\left(\xi, \vec{\kappa}, \eta\right). \tag{53}$$

The depth-dependent power spectrum P and correlation function W are Fourier transform pairs, which can be shown as

$$W\left(\xi, \vec{\kappa}, \eta\right) = \frac{1}{2\pi} \int_{-\infty}^{+\infty} P\left(\xi, \vec{\kappa}, \kappa_z\right) e^{i\kappa_z \eta} \mathrm{d}\kappa_z. \tag{54}$$

Combination of Eqs. (53) and (54) yields

$$\langle \psi_1 \psi_2^* \rangle \approx (2\pi)^{-2} \int_0^L \mathrm{d}\xi \, a_1(\xi) a_2(\xi) \iint_{\vec{\kappa}} \mathrm{d}^2 \vec{\kappa} \, e^{i\vec{\kappa} \cdot [\vec{r}_2(\xi) - \vec{x}_2] - i\vec{\kappa} \cdot [\vec{r}_1(\xi) - \vec{x}_1]}$$
$$\times e^{i\omega[\theta_1(\xi) - \theta_2(\xi)]} P\left[\xi, \vec{\kappa}, \frac{\omega(\dot{\theta}_1 + \dot{\theta}_2)}{2}\right]. \tag{55}$$

We can also derive

$$\langle \psi_1 \psi_2 \rangle \approx -(2\pi)^{-2} \int_0^L \mathrm{d}\xi \, a_1(\xi) a_2(\xi) \iint_{\vec{\kappa}} \mathrm{d}^2 \vec{\kappa} \, e^{i\vec{\kappa} \cdot [\vec{r}_2(\xi) - \vec{x}_2] - i\vec{\kappa} \cdot [\vec{r}_1(\xi) - \vec{x}_1]}$$
$$\times e^{i\omega[\theta_1(\xi) + \theta_2(\xi)]} P\left[\xi, \vec{\kappa}, \frac{\omega(\dot{\theta}_1 - \dot{\theta}_2)}{2}\right]. \tag{56}$$

The Log amplitude u and phase ϕ can be expressed by the complex phase

$$u = \frac{\psi + \psi^*}{2} \quad \text{and} \quad \phi = \frac{\psi - \psi^*}{2i}. \tag{57}$$

Using this relation to the two plane waves, 1 and 2, we have log A coherence function

$$\langle u_1 u_2 \rangle = \frac{1}{2} \mathrm{Re} \langle \psi_1 \psi_2^* \rangle + \frac{1}{2} \mathrm{Re} \langle \psi_1 \psi_2 \rangle, \tag{58}$$

and the phase coherence function

$$\langle \phi_1 \phi_2 \rangle = \frac{1}{2} \mathrm{Re} \langle \psi_1 \psi_2^* \rangle - \frac{1}{2} \mathrm{Re} \langle \psi_1 \psi_2 \rangle. \tag{59}$$

We can also form the log$A$-phase coherence function

$$\langle u_1 \phi_2 \rangle = \frac{1}{2} \text{Im} \langle \psi_1 \psi_2 \rangle - \frac{1}{2} \text{Im} \langle \psi_1 \psi_2^* \rangle. \tag{60}$$

If $\vec{q}_1 = \vec{q}_2$, the two plane waves coming from same direction, we obtain TCFs and they depend only on the spatial lag $\vec{x}_1 - \vec{x}_2$. Note that these waves are not necessarily vertical incidences as in the Chernov theory. If $\vec{x}_1 = \vec{x}_2$, the ACFs are obtained for plane waves, $\vec{q}_1$ and $\vec{q}_2$. Because the background velocity profile is depth dependent, these ACFs depend on $\vec{q}_1$ and $\vec{q}_2$ independently, not necessarily the difference $\vec{q}_1 - \vec{q}_2$. The most general case is JTACF, $\vec{x}_1 \neq \vec{x}_2, \vec{q}_1 \neq \vec{q}_2$.

For 3-D isotropic heterogeneous media, the coherence functions can be simplified using the following identity (Abramowitz and Stegun, 1965)

$$2\pi J_0(\kappa R) = \int_0^{2\pi} e^{-i\kappa R \cos \alpha} d\alpha, \tag{61}$$

where J_0 is the 0th order Bessel function. Therefore, we arrive at

$$\langle \psi_1 \psi_2^* \rangle \approx (2\pi)^{-1} \int_0^L d\xi a_1(\xi) a_2(\xi) \times \int_0^\infty \kappa d\kappa J_0[\kappa R(\xi)] e^{i\omega[\theta_1(\xi) - \theta_2(\xi)]}$$

$$\times P\left[\xi, \kappa, \frac{\omega(\dot{\theta}_1 + \dot{\theta}_2)}{2}\right], \tag{62}$$

where $R(\xi) = |\vec{r}_2(\xi) - \vec{x}_2 - \vec{r}_1(\xi) + \vec{x}_1|$ has an obvious meaning, the transverse distance between two rays at depth ξ, with slownesses $\vec{q}_1$ and $\vec{q}_2$, respectively (see Fig. 1).

Likewise,

$$\langle \psi_1 \psi_2 \rangle \approx -(2\pi)^{-1} \int_0^L d\xi a_1(\xi) a_2(\xi) \times \int_0^\infty \kappa d\kappa J_0[\kappa R(\xi)] e^{i\omega[\theta_1(\xi) + \theta_2(\xi)]}$$

$$\times P\left[\xi, \kappa, \frac{\omega(\dot{\theta}_1 - \dot{\theta}_2)}{2}\right]. \tag{63}$$

For 2-D case, Eqs. (55) and (56) are

$$\langle \psi_1 \psi_2^* \rangle \approx (2\pi)^{-1} \int_0^L d\xi a_1(\xi) a_2(\xi) \int_\kappa d\kappa e^{i\kappa[r_2(\xi) - x_2] - i\kappa[r_1(\xi) - x_1]} \times e^{i\omega[\theta_1(\xi) - \theta_2(\xi)]}$$

$$\times P\left[\xi, \kappa, \frac{\omega(\dot{\theta}_1 + \dot{\theta}_2)}{2}\right], \tag{64}$$

$$\langle \psi_1 \psi_2 \rangle \approx -(2\pi)^{-1} \int_0^L d\xi a_1(\xi) a_2(\xi) \int_\kappa d\kappa e^{i\kappa[r_2(\xi) - x_2] - i\kappa[r_1(\xi) - x_1]}$$

$$\times e^{i\omega[\theta_1(\xi) + \theta_2(\xi)]} P\left[\xi, \kappa, \frac{\omega(\dot{\theta}_1 - \dot{\theta}_2)}{2}\right]. \tag{65}$$

6. Coherence Functions Using Delta-Correlated Assumption

The delta-correlated assumption between two depths is often invoked to simplify computation for wave propagation in stochastic media (Tatarskii, 1971; Ishimaru, 1978; Wu and Flatté, 1990)

$$\langle dv(\vec{\kappa}_1', z') dv^*(\vec{\kappa}_2', z'') \rangle = (2\pi)^m W(\vec{\kappa}_1', z', z'') \delta(\vec{\kappa}_1' - \vec{\kappa}_2') \delta(z' - z'') d\vec{\kappa}_1' d\vec{\kappa}_2', \quad (66)$$

in which $m = 1$ for 2-D case and $m = 2$ for 3-D case. Under this assumption, the vertical wave number of the power spectrum P is zero, thus Eqs. (55) and (56) reduce to

$$\langle \psi_1 \psi_2^* \rangle \approx (2\pi)^{-2} \int_0^L d\xi a_1(\xi) a_2(\xi) \iint_{\vec{\kappa}'} d^2 \vec{\kappa} \, e^{i\kappa[r_2(\xi) - x_2] - i\kappa[r_1(\xi) - x_1]} \\ \times e^{i\omega[\theta_1(\xi) - \theta_2(\xi)]} P(\xi, \vec{\kappa}, 0), \quad (67)$$

and

$$\langle \psi_1 \psi_2 \rangle \approx -(2\pi)^{-2} \int_0^L d\xi a_1(\xi) a_2(\xi) \iint_{\vec{\kappa}'} d^2 \vec{\kappa} \, e^{i\kappa[r_2(\xi) - x_2] - i\kappa[r_1(\xi) - x_1]} \\ \times e^{i\omega[\theta_1(\xi) + \theta_2(\xi)]} P(\xi, \vec{\kappa}, 0). \quad (68)$$

Various coherence functions according to Eqs. (58)–(60) can be formed as the following:

$$\langle u_1 u_2 \rangle \approx (2\pi)^{-2} \int_0^L d\xi a_1(\xi) a_2(\xi) \\ \times \iint_{\vec{\kappa}'} e^{i\vec{\kappa} \cdot [\vec{r}_2(\xi) - \vec{x}_2] - i\vec{\kappa} \cdot [\vec{r}_1(\xi) - \vec{x}_1]} \sin[\omega\theta_1(\xi)] \sin[\omega\theta_2(\xi)] P(\xi, \vec{\kappa}, 0) d^2 \vec{\kappa}, \quad (69)$$

$$\langle \phi_1 \phi_2 \rangle \approx (2\pi)^{-2} \int_0^L d\xi a_1(\xi) a_2(\xi) \\ \times \iint_{\vec{\kappa}'} e^{i\vec{\kappa} \cdot [\vec{r}_2(\xi) - \vec{x}_2] - i\vec{\kappa} \cdot [\vec{r}_1(\xi) - \vec{x}_1]} \cos[\omega\theta_1(\xi)] \cos[\omega\theta_2(\xi)] P(\xi, \vec{\kappa}, 0) d^2 \vec{\kappa}, \quad (70)$$

$$\langle u_1 \phi_2 \rangle \approx -(2\pi)^{-2} \int_0^L d\xi a_1(\xi) a_2(\xi) \\ \times \iint_{\vec{\kappa}'} e^{i\vec{\kappa} \cdot [\vec{r}_2(\xi) - \vec{x}_2] - i\vec{\kappa} \cdot [\vec{r}_1(\xi) - \vec{x}_1]} \sin[\omega\theta_1(\xi)] \cos[\omega\theta_2(\xi)] P(\xi, \vec{\kappa}, 0) d^2 \vec{\kappa}. \quad (71)$$

Using identity (61), we can explicitly obtain three types of coherence functions in 3-D:

$$\langle u_1 u_2 \rangle \approx (2\pi)^{-1} \int_0^L d\xi a_1(\xi) a_2(\xi) \int_0^\infty J_0[\kappa R(\xi)] \sin[\omega \theta_1(\xi)] \sin[\omega \theta_2(\xi)] P(\xi, \vec{\kappa}, 0) \kappa d\kappa \tag{72}$$

$$\langle \phi_1 \phi_2 \rangle \approx (2\pi)^{-1} \int_0^L d\xi a_1(\xi) a_2(\xi) \int_0^\infty J_0[\kappa R(\xi)] \cos[\omega \theta_1(\xi)] \cos[\omega \theta_2(\xi)] P(\xi, \vec{\kappa}, 0) \kappa d\kappa \tag{73}$$

$$\langle u_1 \phi_2 \rangle \approx -(2\pi)^{-1} \int_0^L d\xi a_1(\xi) a_2(\xi) \int_0^\infty J_0[\kappa R(\xi)] \sin[\omega \theta_1(\xi)] \cos[\omega \theta_2(\xi)] P(\xi, \vec{\kappa}, 0) \kappa d\kappa. \tag{74}$$

7. Coherence Functions in a Constant Background Medium

To compare our results with those in Wu and Flatté (1990), we assume that the background velocity model is homogeneous and the incident angle is small. Under these conditions, the τ function can be approximated as

$$\tau(\xi, p) = \int_0^\xi [c^{-2} - p^2]^{1/2} dz \approx c^{-1}\left(1 - \frac{1}{2} p^2 c^2\right)\xi. \tag{75}$$

Phase functions θ_1 and θ_2 can be explicitly solved

$$\omega \theta_1 = -\frac{1}{2} c \xi \frac{\kappa^2}{\omega} = -\frac{\xi}{2k} \kappa^2. \tag{76}$$

The amplitudes can be approximated as

$$a_1(\xi) a_2(\xi) \approx k^2. \tag{77}$$

Substituting Eqs. (75)–(77) into Eqs. (72)–(74), the generalized results in this chapter reduce to those contained in Wu and Flatté (1990) for a homogeneous background medium.

8. Numerical Examples

To demonstrate the validity of the theory proposed in this chapter, we compared the coherence functions from numerical simulations and theoretical predictions. We constructed a 2-D random model (Fig. 2a) with a background velocity profile same as the iasp91 model (Fig. 2b; Kennett and Engdahl, 1991). Two random layers are superimposed on the background model. The top layer is from 0 to 120 km in depth and it has a Gussian correlation function with correlation length 10 km in both horizontal and vertical directions, and rms 1% of the background velocity for those random velocity perturbations. The bottom layer extending from 120 to 310 km depth also has a Gussian

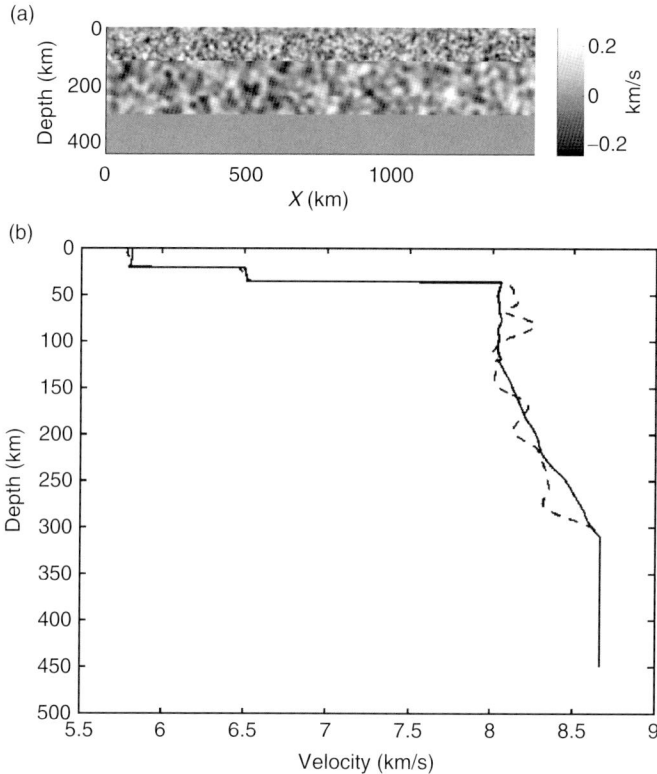

FIG. 2. (a) Random velocity model used in the 2-D numerical modeling; (b) 1-D iasp91 background velocity model (solid line) and a vertical velocity profile (dash line) at location $X = 600$ km, to indicate the randomness of the velocity fluctuation.

correlation function with correlation length 20 km in both directions and rms 1% of the background velocity. For both layers, we constrain the velocity perturbations not exceeding ±3% of the background velocity. Below 310 km depth, the medium is homogeneous and has no random perturbations and the velocity is same as the one at the 310 km depth. We used a full-wave finite difference code (Xie, 1988) to simulate the acoustic plane wave propagation in the random medium. The spectral amplitude and phase fluctuations at a given frequency are extracted from the waveforms recorded at the surface (Zheng and Wu, 2005). Then coherence functions are formed using those fluctuations and averaged over an ensemble of 100 different stochastic model realizations. The log$A$ (Fig. 3) and phase (Fig. 4) coherence functions at 0.5 Hz from the numerical simulations are well predicted by our formula for four different incidence geometries between two plane waves. The incidence angle is measured at the surface for consistency. The log$A$ coherence functions in Fig. 3 are sensitive to the angular separation between two incident plane waves. However, phase coherence functions are relatively simple (Fig. 4) in shape. Another salient feature is that the maximum amplitude of the coherence function (for both log$A$ and phase) is decreasing with increasing angular separation between the two plane waves, which is expected. It is also interesting to notice

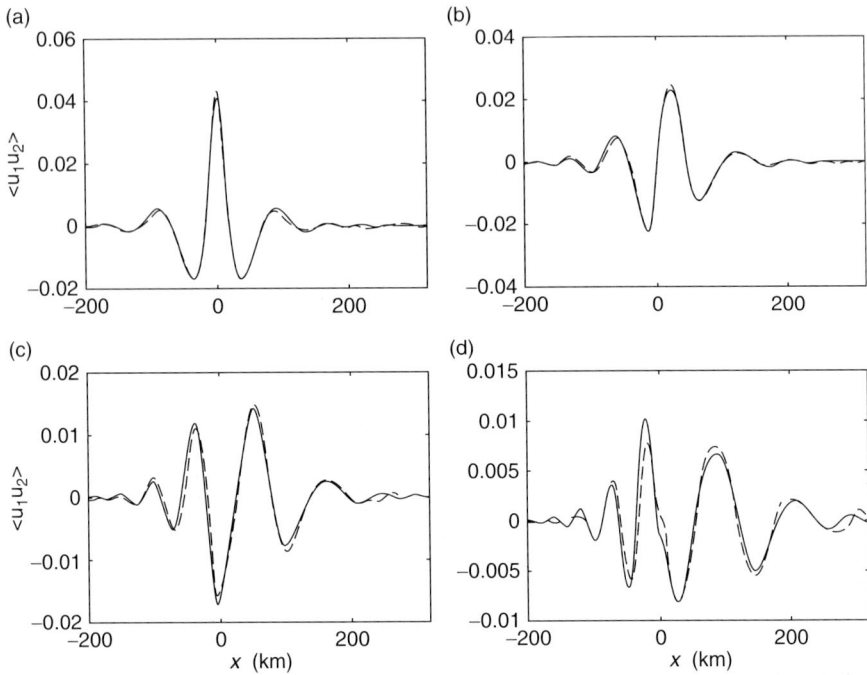

FIG. 3. log A coherence functions $\langle u_1 u_2 \rangle$ for two plane waves with incidence angles, $0°$ and $0°$ (a), $0°$ and $5°$ (b), $0°$ and $10°$ (c), and $0°$ and $15°$ (d). Numerical results are indicated as dash lines and the theoretical predictions are the solid lines.

that there are several velocity discontinuities in the iasp91 background model and this seems to pose difficulty to using the WKBJ Green's function as discontinuities can produce reflected waves. However, because we are using plane wave incidence, those discontinuities affect all receivers in a similar fashion and such effect is removed during coherence function formation.

9. Validity of the Delta-Correlated Assumption

In Section 6, we see that the delta-correlated assumption significantly simplifies the mathematical derivation of the coherence functions. This assumption is equivalent to certain conditions under which the result of the coherence function is same to that obtained as if the medium is delta-correlated along the depth direction. The condition is easily found to be $\kappa^2 \eta / 2k \ll 1$ in view of Eqs. (53) and (76) for a homogeneous background medium. We assume that Λ is the smallest correlation length at which $W(\xi, \vec{\kappa}, \eta)$ is only slightly different from zero. We also assume that $P(\xi, \kappa) \sim 0$ if $\kappa > \kappa_m = 2\pi/\ell_0$, with ℓ_0 being the inner scale of the random medium. Therefore, if condition $\lambda \Lambda \ll \ell_0$ is satisfied, the delta-correlated assumption will hold. Here λ is the wavelength. Of course, we generally further require $\ell_0 < L$. To assess the error

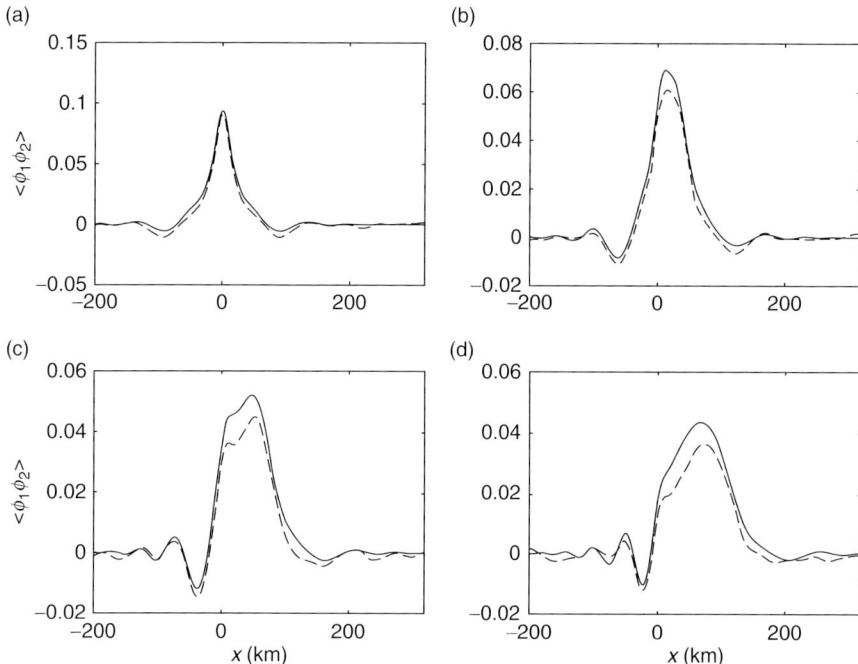

FIG. 4. Phase coherence functions $\langle \phi_1 \phi_2 \rangle$ for two plane waves with incidence angles, $0°$ and $0°$ (a), $0°$ and $5°$ (b), $0°$ and $10°$ (c), and $0°$ and $15°$ (d). Numerical results are indicated as dash lines and the theoretical predictions are the solid lines.

introduced by this assumption, we need numerical verification for a given random model. For a homogeneous background medium, Zheng et al. (2007) numerically found that the delta-correlated assumption will cause little error if the layer thickness of the random medium is large for short wave propagation. Using the model presented in Fig. 2, we computed coherence functions with (Section 6) and without (Section 5) the delta-correlated assumption and we found the results are basically the same. The delta-correlated assumption will not significantly simplify the coherence function calculation. However, it does provide convenience for us to invert for the depth-dependent random spectrum. With the delta-correlated assumption, we can specify the unknowns easily without concerning the vertical wave number in the spectrum P.

10. Discussions and Conclusions

Theory of the coherence function for both log A and phase in a depth-dependent background velocity model is important to draw correct inference on the random medium property. In the past, coherence functions have been theorized in the context of using a constant background medium, thus a simple Green's function. In this chapter, the natural spectral representation of the Green's function in the wave number domain allows us to extend the previous theory to a scenario where the background medium is depth-dependent.

The agreement between the numerical simulation and the theoretical prediction shows the correctness of the theory and the potential to apply the theory to real seismic data. We formulated the coherence function both in 3-D case and in 2-D case, with and without the delta-correlated assumption. The delta-correlation does not lead to discernable differences for the model we used to do the simulation. However, with the delta-correlated assumption, the parameterization for the inverse problem can be easier.

The previous theory on the coherence function C for the log A or for the phase depends on the spatial lag between stations, and the angle separation between the two plane waves, $C = C(\vec{x}_2 - \vec{x}_1, \vec{q}_2 - \vec{q}_1)$. However, this is valid only when the slowness vectors of the two plane waves are close, that is $\vec{q}_1 \approx \vec{q}_2$. This condition cannot always be true and it also limits the depth resolution of the coherence function, resulting the spectral smearing along the depth direction. Our current theory is capable of dealing with much larger angular separation between the two plane waves and the mathematical formulation for the coherence function depends on $\vec{q}_1$ and $\vec{q}_2$ individually, and it takes a function form of $C = C(\vec{x}_2 - \vec{x}_1, \vec{q}_2, \vec{q}_1)$.

ACKNOWLEDGMENTS

Comments from Professor Haruo Sato, Dr. Tatsuhiko Saito, and an anonymous reviewer have greatly improved the clarity of the chapter. We also thank Dr. Xiao-bi Xie for reading our manuscript and discussions. Y. Zheng is also grateful to Professor Thorne Lay for his encouragement. We thank the W.M. Keck Foundation for facility support. This work is funded in part by DOE/Basic Sciences, NSF EAR0635570, and the Wavelet Transform on Propagation and Imaging Consortium/UCSC. This is Contribution Number 495 of CSIDE, IGPP, University of California, Santa Cruz.

APPENDIX: RANDOM VARIABLES, RANDOM FUNCTIONS

The concept of random variables (RVs) occupied a central role in stochastic analysis. Excellent books on this topic are by *Papoulis* (1965) and *Yaglom* (1962). Here we give a brief overview of this concept. In probability theory, the outcomes of experiments conducted under identical condition form a set, denoted by $\Omega = \{\zeta_1, \zeta_2, \cdots\}$. Ω' can be denumerable or not. If we associate each event ζ_i a numerical value $v(\zeta_i)$, the statistic like the mean $\langle v(\zeta_i) \rangle$ and the variance can be computed. The values of v can be either discrete or continuous. Quite often, people have suppressed the explicit dependence of v on the experiment and its possible outcomes ζ_i's. This simplified notation frequently caused confusion. A random function is just a collection of RVs typically varying with time or location, $f(\vec{x})$. If we choose one element from the event set Ω at each location $\vec{x}$, we obtain a realization. We can produce many realizations and calculate the statistic (most interestingly the coherence function to us) using the ensemble average.

The spectral theory of the random field involves stochastic Fourier–Stieltjes integral theory to overcome some theoretical problems, like the absolute integrability of the random functions. In practice, this is not a serious problem if we assume that the random medium has spatial periodicity. Usually the correlation function is used to characterize a random medium. We assume that the mean of the ensemble is zero. If the mean is not vanishing, we can first subtract the mean. The correlation function $B(\vec{r}_1, \vec{r}_2)$ is defined as

$B(\vec{r}_1 - \vec{r}_2) = \langle f(\vec{r}_1)f(\vec{r}_2) \rangle$. Symbol $\langle \cdot \rangle$ is used to denote the ensemble average. By stationary random medium we mean that the statistic does not change by a translation $\vec{d}$, for example, $B(\vec{r}_1, \vec{r}_2) = B(\vec{r}_1 + \vec{d}, \vec{r}_2 + \vec{d})$. If the correlation function only depends on the Euclid distance between $\vec{r}_1$ and $\vec{r}_2$, the medium is called isotropic, that is, $B(\vec{r}_1, \vec{r}_2) = B(|\vec{x}_1 - \vec{x}_2|)$. To obtain the isotropic spectrum $P(\kappa)$, we use a Fourier transform of the correlation function, that is, $P(\kappa) = \int B(r) e^{-i\kappa r} dr$. In numerical studies, it is practical to produce realizations of random media of certain correlation function and this can be done in the spectral domain (Shapiro and Kneib, 1993). We first assign the spectral amplitude at each wave number then generate random phases in range $[-\pi, \pi]$. Of course, the complex conjugate symmetry has to be used if a real random medium is desired. We list two most common correlation functions and their spectra. The Gaussian correlation function reads $B(r) = \varepsilon^2 \exp(-r^2/r_0^2)$. ε^2 is the perturbation strength and r_0 is the correlation length. Its Fourier transform in N dimension is $P(\kappa) = \varepsilon^2 r_0^N \pi^{N/2} \exp(-\kappa^2 r_0^2/4)$. The exponential correlation function can be expressed as $B(r) = \varepsilon^2 e^{-|r/r_0|}$. Its Fourier transforms are $P(\kappa) = 2\varepsilon^2 r_0 / (1 + \kappa^2 r_0^2)$ in the 1-D case and $P(\kappa) = 8\pi^3 \varepsilon^2 r_0^3 / (1 + \kappa^2 r_0^2)^2$ in 3-D case. These results can be easily generalized to anisotropic cases, in which the correlation lengths in different directions (with correlation lengths, r_{0x}, r_{0y}, and r_{0z}, in x, y, and z directions, respectively) can be different, $r_{0x} \neq r_{0y} \neq r_{0z}$. The simplest way is to do the following replacement, $\kappa^2 \sim \kappa_x^2 + \kappa_y^2 + \kappa_z^2$; $r_0^2 \sim r_{0x} r_{0y}$ in 2-D and $r_0^3 \sim r_{0x} r_{0y} r_{0z}$ in 3-D case.

REFERENCES

Abramowitz, M., Stegun, I.A. (1965). Handbook of Mathematical Functions with Formulas, Graphs, and Mathematical Tables. Dover, New York.
Aki, K. (1973). Scattering of P waves under the Montana LASA. *J. Geophys. Res.* **78**(8), 1334–1346.
Aki, K., Lee, W.H.K. (1976). Determination of three-dimensional velocity anomalies under a seismic array using first P arrival times from local earthquakes; 1, A homogeneous initial model. *J. Geophys. Res.* **81**(23), 4381–4399.
Aki, K., Christoffersson, A., Husebye, E.S. (1976). Three-dimensional seismic structure of the lithosphere under Montana LASA. *Bull. Seismol. Soc. Am.* **66**(2), 501–524.
Aki, K., Christoffersson, A., Husebye, E.S. (1977). Determination of the three-dimensional seismic structure of the lithosphere. *J. Geophys. Res.* **82**(2), 277–296.
Aki, K., Richards, P.G. (2002). Quantitative Seismology, University Sciences Books, Sausalito, California.
Bataille, K.D., Flatté, S.M. (1988). Inhomogeneities near the core-mantle boundary inferred from short-period scattered PKP waves recorded at the global digital seismograph network. *J. Geophys. Res.* **93**(B12), 15057–15064.
Bataille, K.D., Wu, R.S., Flatté, S.M. (1990). Inhomogeneities near the core-mantle boundary evidenced from scattered waves; a review. *Pure Appl. Geophys.* **132**(1–2), 151–173.
Berteussen, K.A. (1975). Crustal structure and P-wave travel time anomalies at NORSAR. *J. Geophysics Zeitschrift fuer Geophysik* **41**(1), 71–84.
Berteussen, K.A., Christoffersson, A., Husebye, E.S., Dahle, A. (1975). Wave scattering theory in analysis of P-wave anomalies at NORSAR and LASA. Tenth International Symposium on Mathematical Geophysics. *Geophys. J. R. Astron. Soc.* **42**(2), 403–417.
Berteussen, K.A., Husebye, E.S., Mereu, R.F., Ram, A. (1977). Quantitative assessment of the crust-upper mantle heterogeneities beneath the Gauribidanur seismic array in southern India. *Earth Planet. Sci. Lett.* **37**(2), 326–332.

Buland, R., Chapman, C.H. (1983). The computation of seismic travel times. *Bull. Seismol. Soc. Am.* **73**(5), 1271–1302.
Capon, J. (1974). Characterization of crust and upper mantle structure under LASA as a random medium. *Bull. Seismol. Soc. Am.* **64**(1), 235–266.
Capon, J., Berteussen, K.A. (1974). A random medium analysis of crust and upper mantle structure under NORSAR. *Geophys. Res. Lett.* **1**(7), 327–328.
Chen, X., Aki, K. (1991). General coherence functions for amplitude and phase fluctuations in a randomly heterogeneous medium. *Geophys. J. Int.* **105**(1), 155–162.
Chernov, L.A. (1960). Wave Propagation in a Random Medium. McGraw-Hill, New York.
Cleary, J.R., Haddon, R.A.W. (1972). Seismic wave scattering near the core-mantle boundary: A new interpretation of precursors to PKP. *Nature* **240**(5383), 549–551.
Flatté, S.M., Wu, R.-S. (1988). Small-scale structure in the lithosphere and asthenosphere deduced from arrival time and amplitude fluctuations at NORSAR. *J. Geophys. Res.* **93**(B6), 6601–6614.
Flatté, S.M., Xie, X.-B. (1992). The transverse coherence function at NORSAR over a wide range of separations. *Geophys. Res. Lett.* **19**(6), 557–560.
Flatté, S.M., Wu, R.-S., Shen, Z. (1991a). Nonlinear inversion of phase and amplitude coherence functions at NORSAR for a model of nonuniform heterogeneities. *Geophys. Res. Lett.* **18**(7), 1269–1272.
Flatté, S.M., Xie, X.B., Wong, I.G., Sullivan, R. (1991b). Calculation of coherence functions; combining data from arrays of different apertures. *In Seismol. Res. Lett.* edited, p. 20.
Goff, J.A., Holliger, K. (1999). Nature and origin of upper crustal seismic velocity fluctuations and associated scaling properties; combined stochastic analyses of KTB velocity and lithology logs. *J. Geophys. Res.* **104**(B6), 13169–13182.
Hedlin, M.A.H., Shearer, P.M. (2002). Probing mid-mantle heterogeneity using PKP coda waves. *Phys. Earth Planet. Inter.* **130**(3–4), 195–208.
Hedlin, M.A.H., Shearer, P.M., Earle, P.S. (1997). Seismic evidence for small-scale heterogeneity throughout the Earth's mantle. *Nature* **387**(6629), 145–150.
Hong, T.-K., Wu, R.-S., Kennett, B.L.N. (2005). Stochastic features of scattering. *Phys. Earth Planet. Inter.* **148**(2–4), 131–148.
Huang, L.-J., Fehler, M.C., Roberts, P.M., Burch, C.C. (1999). Extended local Rytov Fourier migration method. *Geophysics* **64**(5), 1535–1545.
Ishimaru, A. (1978). Wave Propagation and Scattering in Random Media. Academic Press, New York.
Jones, A.G., Holliger, K., Haak, V., Jones, A.G. (1997). Spectral analyses of the KTB sonic and density logs using robust nonparametric methods. *J. Geophys. Res.* **102**(B8), 18391–18403.
Kennett, B.L.N., Engdahl, E.R. (1991). Travel times for global earthquake location and phase identification. *Geophys. J. Int.* **105**(2), 429–465.
Line, C.E.R., Hobbs, R.W., Hudson, J.A., Snyder, D.B. (1998a). Statistical inversion of controlled-source seismic data using parabolic wave scattering theory. *Geophys. J. Int.* **132**(1), 61–78.
Line, C.E.R., Hobbs, R.W., Snyder, D.B. (1998b). Estimates of upper-crustal heterogeneity in the Baltic Shield from seismic scattering and borehole logs. *Tectonophysics*, **286**, 171–184.
Liu, X., Wu, R.-S., Xie, X.-B. (1994). Joint coherence function analysis of seismic travel times and amplitudes fluctuations observed on Southern California seismographic network and its geophysical significance. *Eos Trans. AGU* **75**, 482.
Papoulis, A. (1965). Probability, Random Variables, and Stochastic Processes. McGraw-Hill, New York.
Sato, H., Fehler, M. (1998). Seismic Wave Propagation and Scattering in the Heterogeneous Earth. Springer-Verlag, New York.
Sato, H., Fehler, M., Wu, R.-S. (2002). Scattering and attenuation of seismic waves in the lithosphere, Chapter 13 of International Handbook of Earthquake and Engineering Seismology. In: Lee, W.H.K., Kanamori, H., Jennings, P.C., Kisslinger, C. (Eds.), Academic Press, New York, 195–208.

Shapiro, S.A., Kneib, G. (1993). Seismic attenuation by scattering; theory and numerical results. *Geophys. J. Int.* **114**(2), 373–391.

Tatarskii, V.I. (1971). The Effects of the Turbulent Atmosphere on Wave Propagation. US Department of Commerce, Springfield, Virginia.

Wu, R.S. (2002). Spatial coherences of seismic data and the application to characterization of small-scale heterogeneities. In: Goff, J.A., Holliger, K., (Eds.), *Characterization in the Crust and Upper Mantle: Nature, Scaling and Seismic Properties*. Kluwer Academic, New York, pp. 321–344.

Wu, R.S., Cao, J. (2005). WKBJ solution and transparent propagators. EAGE 67th Annual International Meeting, EAGE, Expanded abstracts. pp. 167–170.

Wu, R.-S., Flatté, S.M. (1990). Transmission fluctuations across an array and heterogeneities in the crust and upper mantle. *Pure Appl. Geophys.* **132**(1–2), 175–196.

Wu, R.-S., Xie, X.-B. (1991). Numerical tests of stochastic tomography. *Phys. Earth Planet. Inter.* **67**, 180–193.

Wu, R.-S., Xie, X.-B., Liu, X.-P. (1994a). Numerical simulations of joint coherence functions observed on seismic arrays and comparison with NORSAR data. *Eos, Trans. AGU* **75**, 478.

Wu, R.-S., Xu, Z., Li, X.-P. (1994b). Heterogeneity spectrum and scale-anisotropy in the upper crust revealed by the German Continental Deep-Drilling (KTB) holes. *Geophys. Res. Lett.* **21** (10), 911–914.

Wu, R.-S., Liu, X., Zhang, L. (1995). Random layers found by joint coherence analyses of array data observed at NORSAR and SCSN. *Eos Trans. AGU* **76**, 384.

Xie, X.B. Yao, Z.X. (1988). P-SV wave responses for a point source in twoHypdimensional heterogeneous media: finite-difference method, *Chinese J. Geophys.*, **31**, 473–493.

Yaglom, A.M. (1962). Stationary Random Medium. Prentice-Hall, Englewood Cliffs, New Jersey.

Zheng, Y., Wu, R.-S. (2005). Measurement of phase fluctuations for transmitted waves in random media. *Geophys. Res. Lett.* **32**, L14314, doi: 10.1029/2005GL023179.

Zheng, Y., Wu, R.-S., Lay, T. (2007). Inverting the power spectrum for a heterogeneous medium. *Geophys. J. Int.* **168**(3), 1005–1010.

SYNTHESIS OF VECTOR-WAVE ENVELOPES IN RANDOM ELASTIC MEDIA ON THE BASIS OF THE MARKOV APPROXIMATION

Haruo Sato[1] and Michael Korn

Abstract

High-frequency seismograms of earthquakes are complex mainly caused by scattering due to the lithospheric inhomogeneity. Disregarding phase information, seismologists have often focused on the characteristics of seismogram envelopes. The delay time of the maximum amplitude arrival from the onset and the apparent duration time are good measures of scattering caused by random velocity inhomogeneities. There is a stochastic method to directly simulate wave envelopes in random media. The Markov approximation for the parabolic equation is known to be powerful for the direct synthesis of scalar wave envelopes when the wavelength is shorter than the correlation length of random media. It leads to the master equation for the two-frequency mutual coherence function (TFMCF) of waves, of which the Fourier transform gives the time trace of the wave intensity. It well predicts the peak delay and the broadening of wave envelopes with increasing travel distance for an impulsive source. In this chapter, we extend this approximation to vector waves in random elastic media. When the medium inhomogeneity is weak and the wavelength is shorter than the correlation distance, P- and S-waves can be separately treated by using potentials since conversion scattering between them is weak. Applying the Markov approximation to the TFMCF of potential field, we are able to synthesize vector-wave envelopes. Vector-wave envelopes are analytically derived for plane wavelet incidence onto random media and for wavelet radiation from a point source in random media characterized by a Gaussian autocorrelation function. For P-waves, this approximation predicts not only the peak delay and envelope broadening in the longitudinal component but also the excitation of wave amplitude in the transverse component due to ray bending. The ratio of the mean square (MS) fractional velocity fluctuation to the correlation distance ε^2/a is the key parameter characterizing these vector-wave envelopes. The relation between the time integral of the transverse-component MS amplitude against travel distance gives this ratio. S-wave envelopes can be synthesized with an analogous mathematical approach. For the same randomness, the envelope broadening of S-wavelet is larger than that of P-wavelet by a factor of the ratio of their wave velocities. The validity of the direct envelope synthesis with the Markov approximation is confirmed by a comparison with vector-wave envelopes calculated from finite difference simulations in two dimensions. The direct syntheses of vector-wave envelopes developed here could serve for the mathematical interpretation of observed seismograms in terms of lithospheric inhomogeneity.

Key Words: Seismology, body waves, heterogeneity, scattering, random media, stochastic method, envelope. © 2008 Elsevier Inc.

---

[1] Author thanks e-mail: sato@zisin.geophys.tohoku.ac.jp; mikorn@uni-leipzig.de

1. INTRODUCTION

1.1. Markov Approximation for the Wave Envelope Synthesis

Recorded seismograms of earthquakes are complex reflecting the lithospheric inhomogeneity. Various types of inversion methods have been developed for the quantitative description of the lithospheric structure such as ray-based velocity tomography using first arrival-time readings, reflection survey using array records from artificial explosion sources, and receiver function analysis using PS conversion phases of teleseismic waves at velocity boundaries. The target of these methods is to deterministically estimate the inhomogeneous structure. There are different approaches to describe the lithospheric inhomogeneity statistically. Envelope analysis is known as one of the best methods for that purpose. It is useful for the analysis of complex seismograms especially at frequencies above 1 Hz. Disregarding complex phases and focusing on stable envelopes of bandpass filtered traces, we are able to estimate statistically the spectral structure of random inhomogeneity of the lithosphere. There have been rapid developments in the envelope synthesis in random media since the pioneering work on the coda envelope modeling by Aki (1969) [see a review by Sato and Fehler (1998)].

Analyzing S seismograms of microearthquakes in Japan at frequencies higher than 1 Hz, Sato (1989) found that the envelope width and the peak arrival delay from the onset increase as the travel distance increases. He interpreted this phenomena caused by scattering due to distributed velocity inhomogeneities. If the wavelength is much shorter than the characteristic scale of medium inhomogeneity, we may use the parabolic approximation to solve the wave equation, which means the dominance of scattering in a small cone around the forward direction. There is a simple deterministic way known as the phase screen method to solve the parabolic wave equation (e.g., Jensen *et al.*, 1994). The Markov approximation is a stochastic extension of the phase screen method for the direct synthesis of wave envelopes in random media. It derives the stochastic master equation for the two-frequency mutual coherence function (TFMCF), of which the Fourier transform gives the mean square (MS) of band-pass filtered trace. There were simulations of MS envelopes of scalar waves for specific cases (e.g., Shishov, 1974; Sreenivasiah *et al.*, 1976). This approximation successfully derives broadened envelopes with travel distance increasing: the characteristic time of envelope is proportional to the product of the MS fractional fluctuation of velocity and the square of travel distance over the correlation distance and the average velocity for the case of Gaussian autocorrelation function (ACF).

Sato (1989) used the solution of the Markov approximation for explaining the peak delay and the envelope broadening of observed S-seismograms of regional earthquakes. Since then, there have been attempts to derive envelopes in random media characterized by a von Kármán-type ACF for representing more realistic inhomogeneity (e.g., Lambert and Rickett, 1999; Saito *et al.*, 2002). The validity of the Markov approximation was confirmed by a comparison with numerically simulated scalar wave envelopes in 2-D random media (Fehler *et al.*, 2000). Using the Markov approximation as a propagator in the radiative transfer theory (RTT), Saito *et al.* (2003) and Sato *et al.* (2004) proposed models explaining not only the early envelope but also the whole envelope from the onset to coda for the case of von Kármán-type random media in two dimensions (Sato and Fehler, 2007).

1.2. Analyses of Seismogram Envelopes

1.2.1. Regional and Local Earthquake Seismograms

Applying the Markov approximation solution for the Gaussian ACF to the travel distance dependence of the peak delay and envelope width of band-pass filtered S-seismograms of local small earthquakes in Kanto, Japan in the frequency range from 2 to 16 Hz, Scherbaum and Sato (1991) simultaneously estimated attenuation factor and the ratio $\varepsilon^2/a \approx 0.00054$ km^{-1}. Using the first-order RTT with the Born scattering amplitudes to coda excitation and scattering attenuation of S-waves, Sato (1984) estimated the lithospheric inhomogeneity as $\varepsilon^2 = 0.01$ and $a = 2$ km for an exponential ACF. The inhomogeneity estimated from coda is larger than that estimated from envelope broadening.

Analyzing the frequency dependence of S-seismogram envelopes of microearthquakes observed in northern Honshu, Japan, Saito et al. (2002) found that a von Kármán-type ACF having rich short-wavelength spectrum is preferable to a Gaussian ACF. Gusev and Abubakirov (1996) studied how nonisotropic scattering affects the envelope based on the RTT. Petukhin and Gusev (2003) averaged seismogram envelopes of small earthquakes recorded in Kamchatka and compared the shapes with those calculated for various types of random media. They concluded that random media whose short-wavelength inhomogeneity power spectrum decreases as wave number to the power of –3.5 to –4 are appropriate.

Obara and Sato (1995) found regional differences of S-wave envelope characteristics of microearthquakes observed in Kanto and Tokai, Japan, where the Pacific plate subducts from east to west beneath the Japan arc and the volcanic front runs from north to south: peak delay from the onset and envelope broadening are weak and frequency independent in the fore-arc side of the volcanic front; however, they are large and frequency dependent in the back-arc side. Precisely examining S-wave seismograms in northern Japan, Takahashi et al. (2007) recently found that the peak delay depends on the ray path: peak delays observed in the back-arc side of the volcanic front are larger for rays that propagate beneath Quaternary volcanoes; however, peak delays for rays that propagate between them are as short as those observed in the fore-arc side (see Fig. 1). These observations suggest that the structure beneath Quaternary volcanoes is characterized not only by low velocity and large intrinsic absorption revealed from tomography analysis but also by strong inhomogeneity.

1.2.2. Teleseismic Waves

Aki (1973) first focused on the correlation between the log-amplitude and phase of teleseismic P-waves for measuring the lithospheric inhomogeneity. Applying the theoretical correlation predicted from the parabolic equation solution to array records of teleseismic P-waves of 0.6 Hz at LASA, Montana, arriving from near vertical incidence, he estimated $a = 10$ km and $\varepsilon^2 = 0.0016$ with thickness 60 km. Analyzing travel-time fluctuations of teleseismic P-waves of dominant frequency near 1 Hz observed in southern California, Powell and Meltzer (1984) inferred that $a = 25$ km with $\varepsilon^2 = 0.001$ to depths of at least 119 km. Analyzing array data obtained at NORSAR by using angular correlation functions, Flatté and Wu (1988) suggested that the von Kármán type ACF is more appropriate than the Gaussian ACF. They proposed a model for lithospheric and asthenospheric inhomogeneities that consists of two overlapping layers, where small-scale inhomogeneities dominate near the surface compared with the deeper portions.

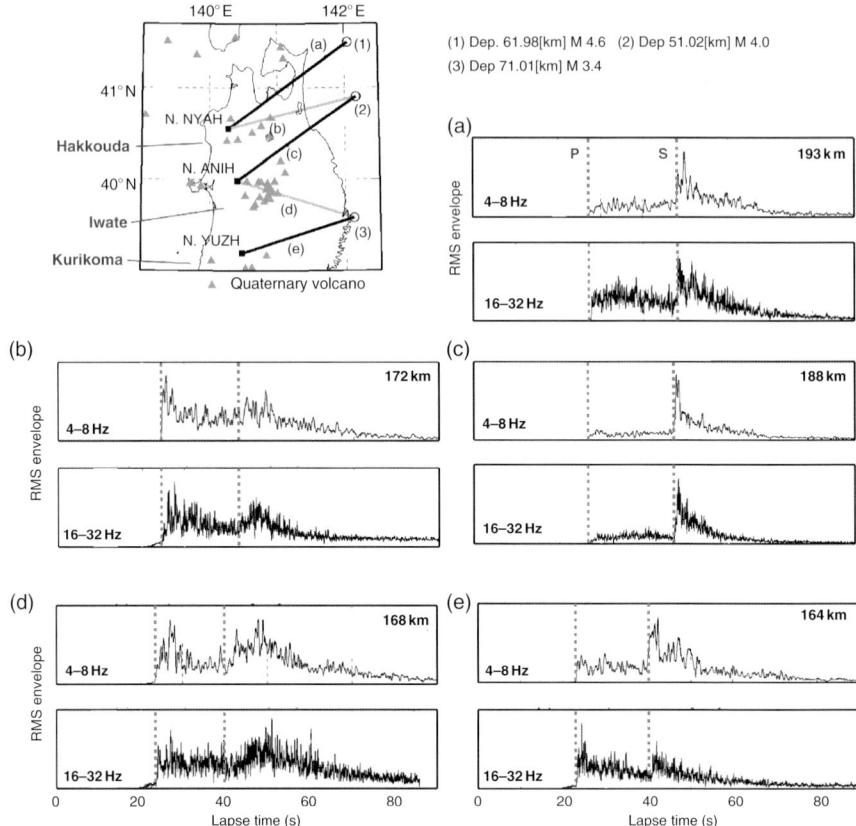

FIG. 1. Path dependence of S-wave envelopes registered in the back-arc side of the volcanic front in northern Honshu, Japan. RMS envelopes for 4–8 Hz and 16–32 Hz for three earthquakes (1)–(3) and five ray paths (a–e) are shown. Gray lines in the inserted map represent ray paths with long peak delay times. Gray triangles are Quaternary volcanoes (Takahashi et al., 2007).

Examining the long propagation distance P-wave signals from an explosion recorded at the NORSAR array, McLaughlin and Anderson (1987) found that 5 Hz band signals arrive later than those in the 1 Hz band. Analyzing differences in frequency-dependent intensities of the mean wave and the fluctuation part of teleseismic P-waves observed in Massif Central, France, Ritter et al. (1998) explained observed wave field fluctuations in the frequency range 0.3-3 Hz by scattering of the teleseismic P-wave front at elastic inhomogeneities in the lithosphere: 70 km in thickness with $\varepsilon^2 \approx 0.0009 - 0.005$ and $a \approx 1 - 16$ km.

Analyzing teleseismic P records in the world by using the extended energy flux model (Korn, 1990), Korn (1993) found relatively weak scattering on stable continental areas and strong scattering at plate boundaries. Applying the energy-flux model and the teleseismic fluctuation wave field method to interpret the teleseismic P coda observed in northern and central Europe, Hock et al. (2004) estimated lithospheric heterogeneity beneath the receivers. Figure 2 shows the geographical locations of the studied

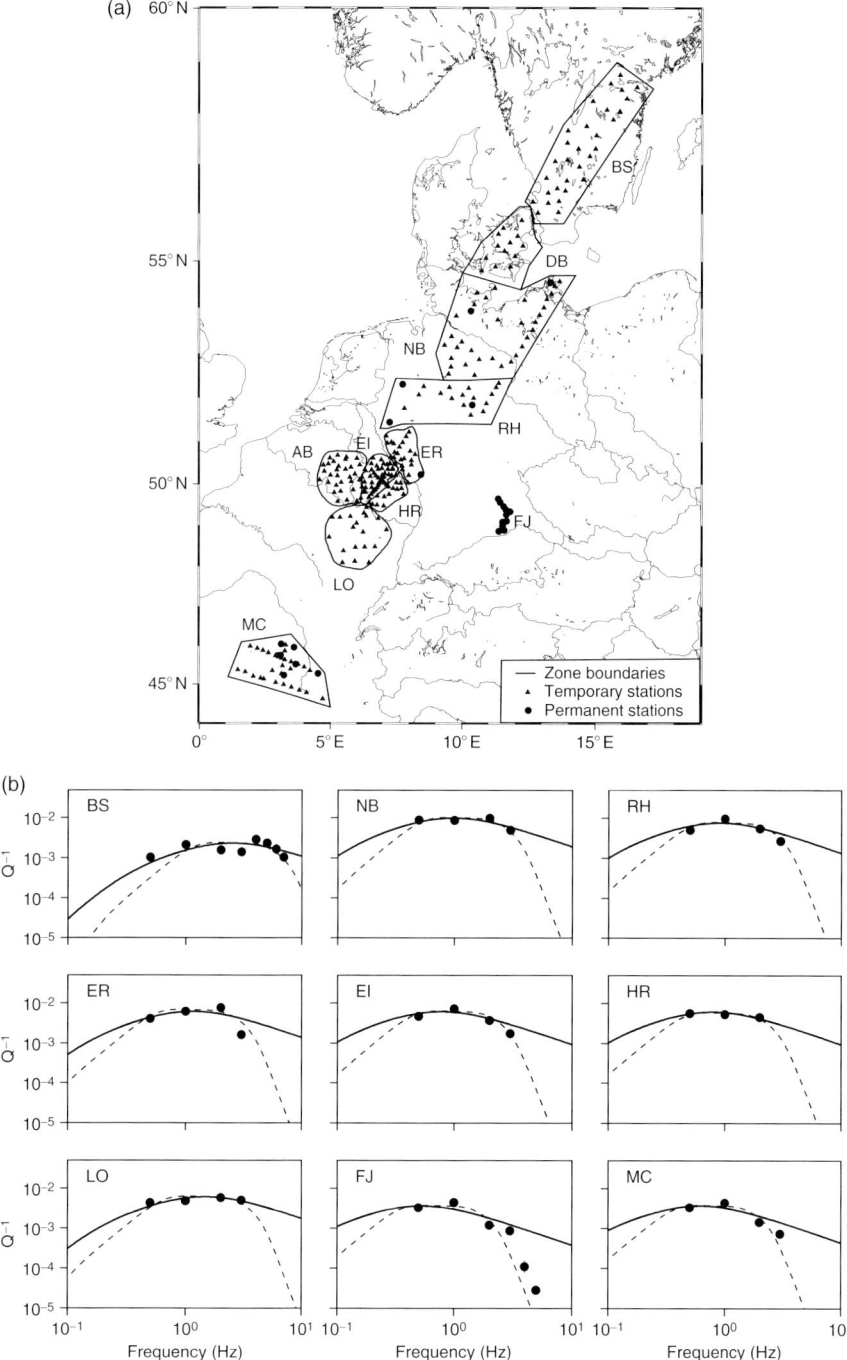

FIG. 2. (a) Map of central Europe with investigated subregions for the analysis of the energy-flux model and the teleseismic fluctuation wave field method. Triangles and circles mark seismic stations used in the analysis. (b) Scattering attenuation Q^{-1} between 0.5 and 5 Hz (black dots) derived from the P coda for nine subregions. Solid (dashed) lines show the best least-square fit for exponential (Gaussian) ACF of random media (Hock et al., 2004).

geological subregions and obtained values of scattering attenuation. It appears that there are clear differences in scattering attenuation between regions both in size and in frequency dependence, reflecting a spatial variation of the scattering properties of the lithosphere on regional scales. The largest scattering Q^{-1} was found in the northern German basin (NB) (0.01 at the peak frequency about 1 Hz) and the smallest scattering Q^{-1} in the Baltic shield (BS) (0.0022 at the peak frequency about 4 Hz). In most cases, an exponential ACF as well as a Gaussian ACF fit the data equally well except for the Frankonian Jura (FJ) where only a Gaussian ACF can fit the Q^{-1} values. For the frequency range from 0.5 to 5 Hz, a of 1–7 km and $\varepsilon^2 \approx 0.0009 - 0.005$ are obtained.

Nishimura et al. (2002) analyzed the transverse amplitude of teleseismic P-wave to evaluate lateral heterogeneity in the lithosphere in western Pacific region and showed that strong inhomogeneity is present in and around the tectonically active regions. Figure 3a shows stacked envelopes of teleseismic P-waves registered at station PMG of IRIS as an example. Kubanza et al. (2006) systematically characterize the medium inhomogeneity of the lithosphere by analyzing the relative partition of energy to the transverse component of teleseismic P-waves in short periods from 0.5 Hz to 4 Hz. They found significant regional differences in lateral heterogeneity of the lithosphere as shown in Fig. 3b, where small transverse amplitudes are observed at stations on stable continents, whereas seismically active regions such as island arcs or collision zones are indicated by large transverse amplitudes. These spatial changes are consistent with the tectonic settings of each station; however, large transverse amplitudes are also observed in regions of very low seismicity, as well as at regions where no seismic activity is recognized, which may indicate the existence of medium heterogeneity in the lithosphere that has been formed in ancient times.

Envelopes simulated by using the RTT with the Born scattering amplitudes were used in the analyses of precursors of PKP by Margerin and Nolet (2003) They suggested that the whole mantle scattering may be significant even though ε^2 is as small as $10^{-6} \sim 6 \times 10^{-6}$, contrary to previous suggestions that mantle scattering occurs primarily in the vicinity of the D″ layer. Analyzing stacked envelopes of teleseismic P-waves and subsequent modeling using a Monte Carlo simulation, Shearer and Earle (2004) pointed out the importance of scattering due to lower mantle inhomogeneity ($\varepsilon^2 \approx 3 \times 10^{-5}$ and $a = 8$ km) even though it is smaller than that in the upper mantle ($\varepsilon^2 \approx 0.001$ and $a = 4$ km).

Table 1 enumerates statistical parameters of the lithospheric inhomogeneity reported in the world; however, there are large differences between different measurements.

1.3. Objectives

Seismogram envelopes of both regional and distant earthquakes have been extensively analyzed for the study of lithospheric inhomogeneity. Recent observations show the importance of envelope analysis of vector component seismograms; however, there have been few theoretical approaches for the simulation of vector-wave envelopes. The Markov approximation has been known as a powerful stochastic method to directly simulate wave envelopes in random media. Although it is only applicable to envelopes around the direct phase for the short-wavelength case, it has an advantage over RTT: numerical calculations are easy since the master equation is parabolic, and there are analytic solutions for some specific cases. But most of synthetic methods were not for vector waves but for scalar waves.

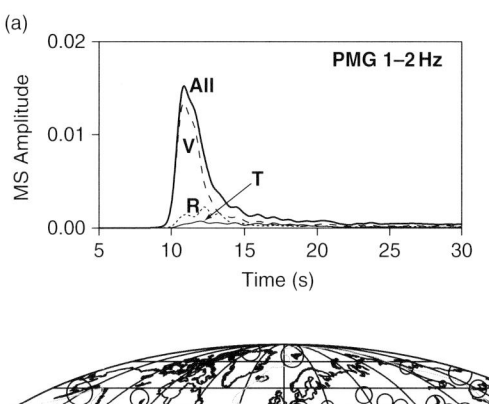

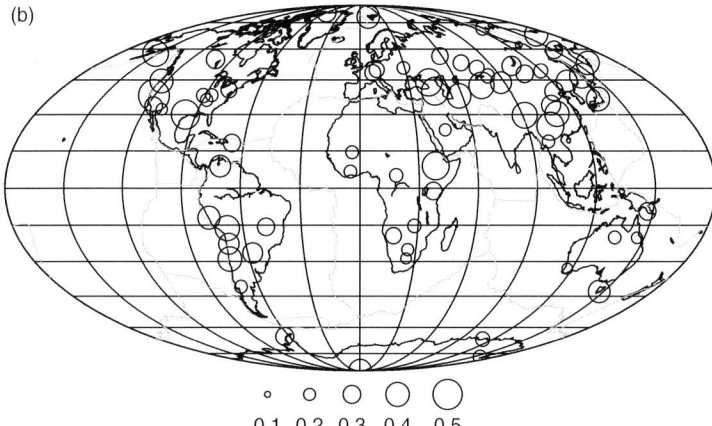

FIG. 3. (a) Stacked MS envelopes of teleseismic P-waves recorded at PMG, New Guinea at the 1–2 Hz band: V, R, T, and All for vertical-, radial-, transverse components, and for the total, respectively (Kubanza et al., 2007). (b) Spatial variation of the relative partition of transverse-component energy to the total energy in teleseismic P-waves at 1–2 Hz band. Circles with larger radius show larger partition of energy into transverse component caused by scattering due to lithospheric inhomogeneity (Kubanza et al., 2006).

In this chapter, summarizing our recent works (Korn and Sato, 2005; Sato, 2006, 2007; Sato and Korn, 2007), we introduce mathematical formulation of vector-wave envelope synthesis based on the Markov approximation for the case that random media are characterized by a Gaussian ACF. Wave propagation characteristics are case sensitive; therefore, we first introduce the formulation for the plane wavelet incidence to random media in Section 2, and then that for the wavelet radiation from a point source in random media in Section 3. The validity of this approximation is confirmed by through comparison with finite difference (FD) numerical simulations in 2-D case. In Section 4, we discuss a comparison with the RTT employing the Born approximation scattering coefficients, a possible extension to random media having more realistic ACFs, and possible developments in the future. In Section 5, we summarize the findings.

TABLE 1. Statistical parameters of the lithospheric random inhomogeneity measured in the world

Area	Thickness [km]	ε^2	a (km^{-1})	ε^2/a (km^{-1})	f (Hz)	Model (Reference)
Average	–	0.01	2	0.005	1–20	RTT+Born approximation for local S coda (Sato, 1984)
Kanto, Japan	–	–	–	0.00054	2–16	Markov approximation for local S envelope (Scherbaum and Sato, 1991)
Tohoku, Japan	–	–	–	$\varepsilon^{2.2}/a = 0.00027$	2–32	Markov approximation (von Kármán type κ=0.6) for local S envelope (Saito et al., 2002)
Montana, USA	60	0.0016	10	0.00016	0.6	Amplitude phase correlation for teleseismic P (Aki, 1973)
California, USA	119	0.001	25	0.00004	1	Amplitude phase correlation of teleseismic P (Powell and Meltzer, 1984)
Massif C., France	70	0.0009–0.005	1–16	–	0.3–3	Amplitude fluctuation of teleseismic P (Ritter et al., 1998)
Europe	–	0.0009–0.005	1–7	–	0.5–5	Energy flux model for teleseismic P (Hock et al., 2004)
Average	100	–	–	0.0002–0.0008	0.5–4	Markov approximation for the ratio of transverse amplitude of teleseismic P (Kubanza et al., 2007)

2. Vector-Wave Envelopes for the Plane Wavelet Incidence

2.1. Three-Dimensional Random Elastic Media

We first introduce the vector-wave envelope synthesis for the incidence of a plane wavelet to random media in three dimensions on the basis of the Markov approximation (Sato, 2006).

2.1.1. Wave Equations for Potentials in Inhomogeneous Media

In a locally isotropic inhomogeneous elastic medium displacement vector **u** is governed by

$$\rho \ddot{u}_i = \left(\lambda u_{l,l}\right)_{,i} + \left[\mu\left(u_{i,j} + u_{j,i}\right)\right]_{,j}, \tag{1}$$

where $\rho(\mathbf{x})$, $\lambda(\mathbf{x})$, and $\mu(\mathbf{x})$ are mass density and Lamé elastic coefficients. When these parameters have small fluctuations around their average values and the wavelength is much smaller than the characteristic scale a of medium inhomogeneity, we may neglect all the spatial derivatives of the medium parameters. The displacement vector is written as $\mathbf{u} = \nabla \phi + \nabla \times \mathbf{B}$ by using scalar potential ϕ and vector potential $\mathbf{B}$. Since there is little conversion scattering between P- and S-wave modes in this case, each potential is independently governed by a wave equation:

$$\Delta \phi - \frac{1}{\alpha(\mathbf{x})^2} \ddot{\phi} = 0 \quad \text{and} \quad \Delta \mathbf{B} - \frac{1}{\beta(\mathbf{x})^2} \ddot{\mathbf{B}} = 0, \tag{2}$$

where $\alpha(\mathbf{x})$ and $\beta(\mathbf{x})$ are P- and S-wave velocities, respectively.

2.1.2. Plane P-Wavelet

We imagine a 3-D elastic medium that is divided into two regions: the P-wave velocity is constant $\alpha = V_0$ for $z < 0$ and $\alpha(\mathbf{x}) = V_0(1 + \xi(\mathbf{x}))$ for $z > 0$, where $\xi(\mathbf{x})$ is a fractional fluctuation around the average velocity V_0. We study the vertical incidence of a plane P-wavelet from the homogeneous medium to the inhomogeneous medium, where the amplitude of the incident wavelet is uniform in the transverse plane ($x - y$ plane), orthogonal to the ray direction in the z-axis as schematically illustrated in Fig. 4.

When the fractional fluctuation is small ($|\xi| \ll 1$), the wave equation for the scalar potential is written as

$$\Delta \phi - \frac{1}{V_0^2} \ddot{\phi} + \frac{2}{V_0^2} \xi(\mathbf{x}) \ddot{\phi} = 0. \tag{3}$$

Scalar potential is written as

$$\phi(\mathbf{x}_\perp, z, t) = \frac{1}{2\pi} \int_{-\infty}^{\infty} d\omega e^{ik_0 z - i\omega t} \frac{U(\mathbf{x}_\perp, z, \omega)}{ik_0}, \tag{4}$$

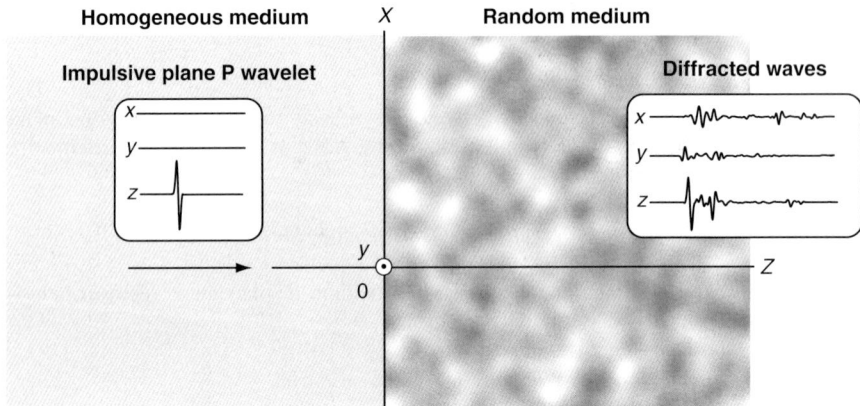

FIG. 4. Schematic illustration of vector-wave propagation through a random medium spreading over a half space $z > 0$ in three dimensions for the incidence of an impulsive plane P-wavelet, where the medium is homogenous for $z < 0$ (Sato, 2006).

where ω is angular frequency, $k_0 = \omega/V_0$ is wave number, and $\mathbf{x}_\perp = (x, y)$ is transverse coordinate. Since the variation of U for an increment in the z-direction is small because $ak_0 \gg 1$, field U is governed by the parabolic wave equation:

$$2ik_0 \partial_z U + \Delta_\perp U - 2k_0^2 \xi(\mathbf{x}) U = 0, \tag{5}$$

where $\Delta_\perp \equiv \partial_x^2 + \partial_y^2$ is a transverse Laplacian. Waves just after the direct arrival are mostly composed of waves scattered in a small angle around the forward direction, which are well described by the parabolic wave equation.

2.1.3. Ensemble of Random Media

We first imagine an ensemble of random media $\{\xi(\mathbf{x})\}$, where small fractional fluctuation $\xi(\mathbf{x})$ is assumed to be a statistically random function of space coordinate $\mathbf{x}$, and $\langle \xi(\mathbf{x}) \rangle = 0$, where angular brackets mean the average over the ensemble. The statistical measure of randomness is quantitatively described by ACF, $R(\mathbf{x}) \equiv \langle \xi(\mathbf{x}') \xi(\mathbf{x}' + \mathbf{x}) \rangle$, which is characterized by a correlation distance a and an MS fractional fluctuation $\varepsilon^2 \equiv \langle R(\mathbf{x} = 0) \rangle$. The randomness is supposed to be statistically homogeneous and isotropic, which means that ACF is a function of lag distance $|\mathbf{x}|$ only.

2.1.4. Stochastic Master Equation for the TFMCF

We define the TFMCF as correlation of U between two different locations on the transverse plane at distance z and two angular frequencies (e.g., Ishimaru, 1978),

$$\Gamma_2(\mathbf{x}_{\perp c}, \mathbf{x}_{\perp d}, z, \omega_c, \omega_d) \equiv \langle U(\mathbf{x}'_\perp, z, \omega') U(\mathbf{x}''_\perp, z, \omega'')^* \rangle, \tag{6}$$

where $\mathbf{x}_{\perp c} = (\mathbf{x}'_\perp + \mathbf{x}''_\perp)/2$ and $\mathbf{x}_{\perp d} = \mathbf{x}'_\perp - \mathbf{x}''_\perp$ are center of mass and difference transverse coordinates, and $\omega_c = (\omega' + \omega'')/2$ and $\omega_d = \omega' - \omega''$ are center and difference angular frequencies, respectively. For the incidence of a plane wavelet to the z-direction, we may expect that waves have statistically uniform amplitude in the transverse plane since the randomness is homogeneous. That is, TFMCF is independent of $\mathbf{x}_{\perp c}$ and becomes a function of $r_{\perp d} \equiv |\mathbf{x}_{\perp d}|$ only because of isotropy. In the following text, $\mathbf{x}_{\perp c}$ is dropped from the arguments of TFMCF.

If the phase fluctuation for an increment a is small, there exists an intermediate scale of increment Δz, which is larger than a but smaller than the scale of variation of U. Using causality and neglecting backscattering, we can derive the master equation for the TFMCF. For the case of quasi-monochromatic waves $|\omega_d| \ll |\omega_c|$, the stochastic master equation for TFMCF is given by

$$\partial_z \Gamma_2 + i\frac{k_d}{2k_c^2}\Delta_{\perp d}\Gamma_2 + k_c^2[A(0) - A(r_{\perp d})]\Gamma_2 + \frac{k_d^2}{2}A(0)\Gamma_2 = 0, \tag{7}$$

where $k_c = \omega_c/V_0$ and $k_d = \omega_d/V_0$. The third and fourth terms represent the interaction with the medium inhomogeneity. Function A is the longitudinal integral of ACF along the z-axis:

$$A(r_\perp) \equiv \int_{-\infty}^{\infty} dz R(\mathbf{x}_\perp, z), \tag{8}$$

where $r_\perp \equiv |\mathbf{x}_\perp|$. At a long travel distance, the dominant contribution of incoherent diffracted waves is strongly controlled by A at a short offset in the transverse plane, $r_{\perp d} \ll a$ (see Lee and Jokipii, 1975a; Ishimaru, 1978).

This approximation is called the Markov approximation. Precise derivation of Eq. (7) is shown in Lee and Jokipii (1975a) and an alternative derivation by using functional differentiation is given by Rytov et al. (1987) and Ishimaru (1978). In addition to the applicability condition of the parabolic approximation, $ak_0 \gg 1$, an additional condition for the Markov approximation is that the MS phase fluctuation over a correlation distance is small, $A(0)k_0^2 a \ll 1$, which is $\varepsilon^2 a^2 k_0^2 \ll 1$ for the case of Gaussian ACF (e.g., Shishov, 1974; Rytov et al. 1987, p. 110).

TFMCF Γ_2 can be factorized into two terms as

$$\Gamma_2 = e^{-A(0)k_d^2 z/2}{}_0\Gamma_2 = \tilde{w}(z, \omega_d)_0\Gamma_2, \tag{9}$$

where $\tilde{w}(z, \omega_d) = \exp\left[-A(0)\omega_d^2 z/(2V_0^2)\right]$. Then Eq. (7) is written as the master equation for ${}_0\Gamma_2$ as

$$\partial_z {}_0\Gamma_2 + i\frac{k_d}{2k_c^2}\Delta_{\perp d\, 0}\Gamma_2 + k_c^2[A(0) - A(r_{\perp d})]_0\Gamma_2 = 0. \tag{10}$$

We solve this differential equation under the following initial condition:

$$\Gamma_2(\mathbf{x}_{\perp d}, z=0, \omega_c, \omega_d) = {}_0\Gamma_2(\mathbf{x}_{\perp d}, z=0, \omega_c, \omega_d) = 1. \tag{11}$$

2.1.5. Intensity Spectral Densities

The ensemble average of the square of displacement vector component u_i gives the intensity of ith vector component I_i. For the case of P-waves, the x-component intensity is given by

$$I_x^P(\mathbf{x}_\perp, z, t) \equiv \langle |u_x(\mathbf{x}_\perp, z, t)|^2 \rangle = \langle \partial_{x'}\phi(\mathbf{x}'_\perp, z, t)\partial_{x''}\phi(\mathbf{x}''_\perp, z, t)^* \rangle_{\mathbf{x}'=\mathbf{x}''}$$

$$= \frac{1}{2\pi}\int_{-\infty}^{\infty} d\omega' e^{ik'_0 z - i\omega' t} \left[\frac{1}{2\pi}\int_{-\infty}^{\infty} d\omega'' e^{-ik''_0 z + i\omega'' t} \frac{1}{k'_0 k''_0} \langle \partial_{x'}U'\partial_{x'}U''^* \rangle_{\mathbf{x}'=\mathbf{x}''}\right]$$

$$= \frac{1}{2\pi}\int_{-\infty}^{\infty} d\omega_c I_x^P(z, t, \omega_c), \tag{12}$$

where representations U' and U'' mean that their arguments are $(\mathbf{x}'_\perp, \omega')$ and $(\mathbf{x}''_\perp, \omega'')$, respectively. In the second line, the integrand in the angular brackets gives the definition of the intensity spectral density (ISD) $I_x^P(z, t, \omega_c)$, where the transverse coordinate is dropped from the argument because of homogeneity. MS envelope of band-pass filtered trace at central angular frequency ω_c with frequency bandwidth Δf is given by $I_x^P(z, t, \omega_c)\Delta f$. Putting $\partial_{x'} = \partial_{x_d}$ and $\partial_{x''} = -\partial_{x_d}$ since Γ_2 is practically independent of the center of mass coordinate and using $1/(k'_0 k''_0) \approx 1/k_c^2$, we may write the x-component ISD as

$$I_x^P(z, t, \omega_c) = \frac{1}{2\pi}\int_{-\infty}^{\infty} d\omega_d e^{-i\omega_d(t - z/V_0)} \left[-\frac{1}{k_c^2}\partial_{x_d}^2 \Gamma_2(\mathbf{x}_{\perp d}, z, \omega_c, \omega_d)\right]_{\mathbf{x}_{\perp d}=0}. \tag{13}$$

Taking the same procedure, we have the y-component ISD with replacing x by y. Because of the isotropy of TFMCF in the transverse plane, we have $I_x^P(z, t, \omega_c) = I_y^P(z, t, \omega_c)$. The z-component ISD is given by

$$I_z^P(z, t, \omega_c) = \frac{1}{2\pi}\int_{-\infty}^{\infty} d\omega_d e^{-i\omega_d(t - z/V_0)} \left\langle \left(U' + \frac{\partial_z U'}{ik'_0}\right)\left(U''^* - \frac{\partial_z U''^*}{ik''_0}\right)\right\rangle$$

$$= \frac{1}{2\pi}\int_{-\infty}^{\infty} d\omega_d e^{-i\omega_d(t - z/V_0)}\left[\left(1 + \frac{\Delta_{\perp d}}{k_c^2}\right)\Gamma_2(\mathbf{x}_{\perp d}, z, \omega_c, \omega_d)\right]_{\mathbf{x}_{\perp d}=0}, \tag{14}$$

where the leading term $\partial_z U \approx (i/2k_0)\Delta_\perp U$ of the parabolic Eq. (5) is used, and $\Delta'_\perp = \Delta''_\perp = \Delta^2_{\perp d}$ and $1/k_0'^2 + 1/k_0''^2 \approx 2/k_c^2$ are used. We also assume to neglect a product $\Delta'_\perp U' \Delta'_\perp U''^*$.

Here we define the reference ISD as

$$I^R(z, t, \omega_c) = \frac{1}{2\pi}\int_{-\infty}^{\infty} d\omega_d e^{-i\omega_d(t - z/V_0)}\Gamma_2(\mathbf{x}_{\perp d} = 0, z, \omega_c, \omega_d), \tag{15}$$

which is literally the ISD of scalar potential (see Sato and Fehler, 1998). By using Eqs. (13) and (15), we may write Eq. (14) as

$$I_z^P(z,t,\omega_c) = I^R(z,t,\omega_c) - I_x^P(z,t,\omega_c) - I_y^P(z,t,\omega_c)$$
$$= I^R(z,t,\omega_c) - 2I_x^P(z,t,\omega_c). \quad (16)$$

That is, the longitudinal-component ISD can be represented by using the reference ISD and the transverse-component ISD.

The initial condition [Eq. (11)] for TFMCF leads to

$$I_x^P(z,t,\omega_c) = I_y^P(z,t,\omega_c) = 0 \quad \text{and} \quad I_z^P(z,t,\omega_c) = \delta\left(t - \frac{z}{V_0}\right) \quad \text{at} \quad z = 0, \quad (17)$$

which mean that the z-component incident plane P-wavelet is a δ function pulse.

2.1.6. Wandering Effect

The Fourier transform of the factor $\tilde{w}$ in Eq. (9) with respect to ω_d gives a Gaussian distribution in the time domain:

$$w\left(z, t - \frac{z}{V_0}\right) \equiv \frac{1}{2\pi} \int_{-\infty}^{\infty} d\omega_d\, \tilde{w}(z, \omega_d) e^{-i\omega_d(t-(z/V_0))}$$
$$= \frac{V_0}{\sqrt{2\pi A(0)z}} e^{-\frac{V_0^2}{2A(0)z}(t-(z/V_0))^2}, \quad (18)$$

where $\int_{-\infty}^{\infty} dt\, w(z, t - z/V_0) = 1$ since $\tilde{w}(\omega_d = 0) = 1$ and $w(z, t - z/V_0) \to \delta(t)$ as $z \to 0$. This function does not mean the broadening of individual wave packets but it shows the wandering effect of the travel time fluctuations of different rays at a travel distance z (Lee and Jokipii, 1975b).

Referring to Eq. (15), we define the Fourier transform of $_0\Gamma_2$ as

$$I_0^R(z,t,\omega_c) = \frac{1}{2\pi} \int_{-\infty}^{\infty} d\omega_d e^{-i\omega_d(t-(z/V_0))}\, _0\Gamma_2(\mathbf{x}_{\perp d} = 0, z, \omega_c, \omega_d), \quad (19)$$

which is the reference ISD without wandering effect. The convolution of I_0^R and w gives the reference ISD for the incidence of a unit impulsive plane P-wavelet: $I^R = I_0^R * w$. In the following, ISDs with subscript "zero" mean ISDs without wandering effect as I_{x0}^P, I_{y0}^P, and I_{z0}^P. Convolution with the wandering term gives ISD as $I_x^P = I_{x0}^P * w$, $I_y^P = I_{y0}^P * w$, and $I_z^P = I_{z0}^P * w$. For practical comparison with the ensemble averaged intensity calculated from numerical simulations, it is necessary to convolve ISD with the source power time function i for a given angular frequency as $I_x^P * i$, $I_y^P * i$, and $I_z^P * i$.

2.1.7. Angular Spectrum

Taking the Fourier transform of TFMCF in the transverse plane, we get the angular spectrum

$$\tilde{\Gamma}_2(\mathbf{k}_\perp, z, \omega_c, \omega_d) = \int_{-\infty}^{\infty} d\mathbf{x}_{\perp d} e^{-i\mathbf{k}_\perp \mathbf{x}_{\perp d}} \Gamma_2(\mathbf{x}_{\perp d}, z, \omega_c, \omega_d), \qquad (20)$$

which means the distribution of ray directions. Taking the Fourier transform of the angular spectrum with respect to ω_d, we get the angular spectrum in time domain,

$$I^A(\mathbf{k}_\perp, z, \omega_c, t) = \frac{1}{2\pi} \int_{-\infty}^{\infty} d\omega_d e^{-i\omega_d(t-z/V_0)} \tilde{\Gamma}_2(\mathbf{k}_\perp, z, \omega_c, \omega_d). \qquad (21)$$

The initial condition [Eq. (11)] leads to $\tilde{\Gamma}_2 = (2\pi)^2 \delta(\mathbf{k}_\perp)$ and $I^A = (2\pi)^2 \delta(\mathbf{k}_\perp) \delta(t - z/V_0)$, which mean that all the rays are parallel to the z-direction, $\mathbf{k}_\perp = 0$ at the incidence.

2.1.8. Gaussian ACF

A Gaussian ACF is selected to characterize random media for our study:

$$R(\mathbf{x}) = \varepsilon^2 e^{-r^2/a^2}, \qquad (22)$$

where $r \equiv |\mathbf{x}|$. The longitudinal integral [Eq. (8)] for small transverse distances is

$$A(r_{\perp d}) \approx \sqrt{\pi} \varepsilon^2 a \left(1 - \frac{r_{\perp d}^2}{a^2} \right) \quad \text{for } r_{\perp d} \ll a. \qquad (23)$$

Master equation for TFMCF: The master equation (10) is explicitly given by

$$\partial_z {}_0\Gamma_2 + i \frac{k_d}{2k_c^2} \Delta_{\perp d} {}_0\Gamma_2 + \frac{k_c^2 \sqrt{\pi} \varepsilon^2 r_{\perp d}^2}{a} {}_0\Gamma_2 = 0, \qquad (24)$$

where we define the characteristic time

$$t_M = \frac{\sqrt{\pi} \varepsilon^2 z^2}{2 V_0 a} \qquad (25)$$

and a parameter

$$s_0 = 2 e^{i(\pi/4)} \sqrt{t_M \omega_d} . \qquad (26)$$

The characteristic time is independent of central frequency. Sreenivasiah *et al.* (1976) solved the differential equation [Eq. (24)] for the initial condition [Eq. (11)]. The solution of Eq. (24) [see Eq. (A.5) in Appendix] is written as

$$_0\Gamma_2(\mathbf{x}_{\perp d}, z, \omega_c, \omega_d) = \frac{1}{\cos s_0} e^{-(\tan s_0/s_0)\left((2V_0 k_c^2 t_M)/z\right)(x_d^2+y_d^2)}$$
$$= \frac{1}{\cos s_0} e^{-(\tan s_0/s_0)\left((x_d^2+y_d^2)/a_\perp^2\right)}, \quad (27)$$

where

$$a_\perp \equiv \sqrt{\frac{z}{2V_0 t_M k_c^2}} = \sqrt{\frac{a}{\sqrt{\pi}\varepsilon^2 z k_c^2}} \quad (28)$$

is the coherence radius.

ISDs of P-wavelet: The reference ISD without wandering effect [Eq. (19)] is analytically solved as follows:

$$I_0^R(z, t, \omega_c) = \frac{1}{2\pi}\int_{-\infty}^{\infty} d\omega_d e^{-i\omega_d(t-(z/V_0))} \frac{1}{\cos s_0}$$
$$= \frac{1}{2\pi}\int_{-\infty}^{\infty} d\omega_d e^{-i\omega_d(t-(z/V_0))} \left[\pi \sum_{k=0}^{\infty} \frac{(-1)^k(2k+1)}{\pi^2(k+(1/2))^2 - 4it_M\omega_d}\right]$$
$$= \frac{\pi}{4t_M}\sum_{k=0}^{\infty}(-1)^k(2k+1)e^{-(\pi^2/(4t_M))(k+(1/2))^2(t-(z/V_0))}H\left(t-\frac{z}{V_0}\right)$$
$$= \frac{1}{t_M}\frac{\pi}{8}\vartheta_1'\left(0, e^{-(\pi^2/4)((t-z/V_0)/t_M)}\right)H\left(t-\frac{z}{V_0}\right), \quad (29)$$

where function $\vartheta_1'(v, q) \equiv \partial_v \vartheta_1(v, q)$ is the derivative of the elliptic theta function of the first kind ϑ_1 with respect to v, where $\vartheta_1(v, q) \equiv 2\sum_{n=0}^{\infty}(-1)^n q^{(n+1/2)^2}\sin[(2n+1)v]$ (e.g., Weisstein, 2005). Williamson (1972) derived the same representation as Eq. (29) based on the stochastic ray path method. A solid curve in Fig. 5 shows the plot of I_0^R against reduced time normalized by the characteristic time. Function I_0^R has a broadened envelope with a long tail, where the maximum peak is numerically about $0.46/t_M$ at $(t - z/V_0)/t_M \approx 0.67$. The peak values of I_0^R decay according to the inverse square of travel distance.

Substituting Eq. (27) into Eq. (20), we have the angular spectrum as

$$_0\bar{\Gamma}_2(\mathbf{k}_\perp, z, \omega_c, \omega_d) = \frac{\pi a_\perp^2 s_0}{\sin s_0} e^{-s_0/\tan s_0\left((a_\perp^2 k_\perp^2)/4\right)}. \quad (30)$$

Angular spectrum in time domain $I_0^A(\mathbf{k}_\perp, z, \omega_c, t)$ is numerically calculated by using FFT. Figure 6 shows angular spectra I_0^A at different reduced times. The angular spectrum has a peak around $k_\perp = 0$ just after the direct arrival; however, the spectrum is flattened out with reduced time increasing. That is, the distribution of ray directions spreads over a wide angle with reduced time increasing.

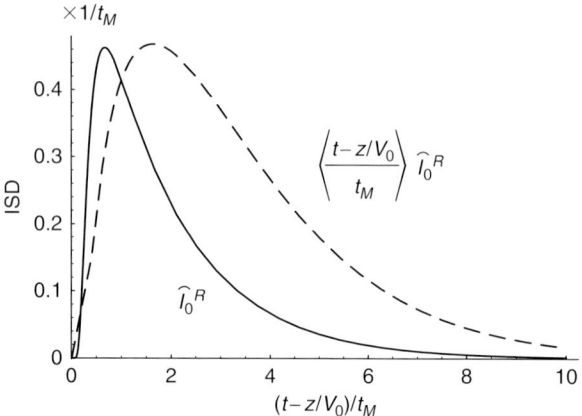

FIG. 5. Plots of reference ISD without wandering effect $\hat{I}_0^R$ (solid curve) and $((t-z/V_0)/t_M)\hat{I}_0^R$ (broken curve) against normalized reduced time for the incidence of an impulsive plane wavelet to 3-D random media.

Replacing Γ_2 with $_0\Gamma_2$ in Eq. (13) and substituting the solution [Eq. (27)] into them, we have transverse-component ISDs without wandering effect as

$$I_{x0}^P(z,t,\omega_c) = I_{y0}^P(z,t,\omega_c) = \frac{4V_0 t_M}{z}\frac{1}{2\pi}\int_{-\infty}^{\infty} d\omega_d e^{-i\omega_d(t-(z/V_0))}\frac{\tan s_0}{s_0 \cos s_0}$$

$$= \frac{4V_0 t_M}{z}\frac{1}{2\pi}\int_{-\infty}^{\infty} d\omega_d e^{-i\omega_d(t-(z/V_0))}\frac{1}{i2t_M}\frac{\partial}{\partial \omega_d}\left(\frac{1}{\cos s_0}\right)$$

$$= \frac{V_0 t_M}{z} 2\frac{(t-(z/V_0))}{t_M}I_0^R(z,t,\omega_c), \tag{31}$$

since $uF(u) = (2\pi)^{-1}\int_{-\infty}^{\infty} d\omega_d e^{-i\omega_d u}[-i\partial_{\omega_d}\tilde{F}(\omega_d)]$ for $F(u) \equiv (2\pi)^{-1}\int_{-\infty}^{\infty} d\omega_d e^{-i\omega_d u}\tilde{F}(\omega_d)$. A broken curve in Fig. 5 shows the plot of $((t-z/V_0)/t_M)\hat{I}_0^R$ against normalized reduced time. The broken curve has numerically the maximum peak of about $0.47/t_M$ at $(t-z/V_0)/t_M \approx 1.63$, which is much later than the peak delay of $\hat{I}_0^R$.

Putting Eq. (31) into Eq. (16), we have an explicit representation of the z-component ISD without wandering effect as

$$I_{z0}^P(z,t,\omega_c) = \left[1 - 4\frac{V_0}{z}\left(t - \frac{z}{V_0}\right)\right]I_0^R(z,t,\omega_c). \tag{32}$$

The calculated longitudinal (z)-component ISD becomes negative for lapse times as larger than $t > 1.25(z/V_0)$, which means the breakdown of this approximation. The applicable range for the reduced time $0 < t - z/V_0 < z/(4V_0)$ is the additional condition of this approximation especially for vector-wave envelopes. The reduction of the applicable reduced time range may come from the neglect of the product of second derivative terms in Eq. (14).

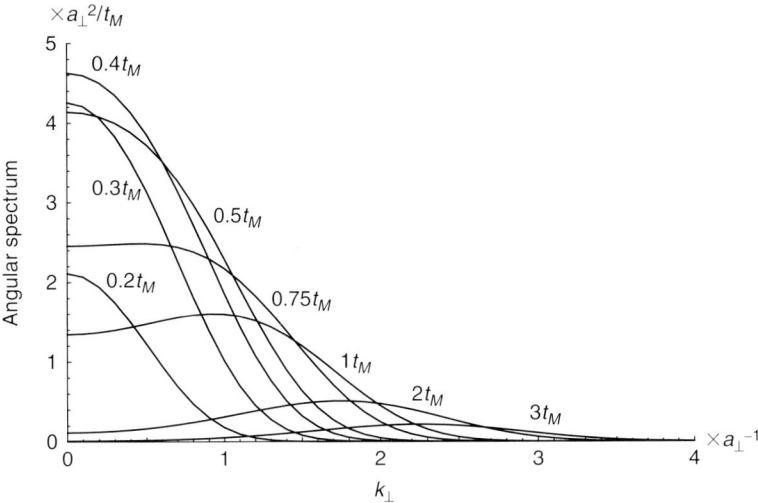

FIG. 6. Plots of angular spectra at different reduced times for the plane wavelet incident to 3-D random media.

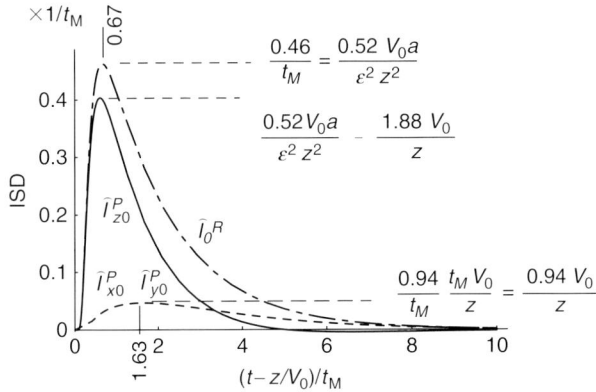

FIG. 7. Plots of ISDs without wandering effect, I^P_{x0}, I^P_{y0}, and I^P_{z0} with I^R_0 against normalized reduced time at $V_0 t_M/z = 0.05$ for the incidence of an impulsive plane P-wavelet to 3-D random media (Sato, 2006).

Figure 7 shows plot of I^P_{x0} and I^P_{y0} (broken line), I^P_{z0} (solid line), and I^R_0 (chained line) against normalized reduced time at a distance of $V_0 t_M/z = 0.05$. When $\varepsilon^2 z/a \ll 1$, the peak height of I^P_{z0} is approximately equal to I^R_0, which decays according to the inverse square of travel distance, and the peak ratio of transverse component to longitudinal component is proportional to $\varepsilon^2 z/a$. As reduced time increases, transverse-component ISD exceeds the longitudinal-component ISD, $I^P_{0x} > I^P_{0z}$ for $t - z/V_0 > z/(6V_0)$, where large incident angle rays dominate over small incident angle rays as shown in Fig. 6.

We note that the wandering term (18) with (23) is Gaussian as

$$w\left(z, t - \frac{z}{V_0}\right) = \frac{V_0}{\sqrt{2\pi\sqrt{\pi\varepsilon^2 a z}}} e^{-\left(V_0^2/\left(2\sqrt{\pi}\varepsilon^2 a z\right)\right)(t-(z/V_0))^2}, \qquad (33)$$

where the time width of the wandering effect is proportional to the square root of travel distance.

Characteristics of P-wave envelopes: We take the following values typically representing thick inhomogeneous lithosphere: the average velocities $\alpha_0 = 7.8$ km/s and $\beta_0 = 4.5$ km/s ($\alpha_0/\beta_0 = 1.73$) and the random inhomogeneity characterized by $\varepsilon = 0.05$ and $a = 8$ km, that is, the ratio $\varepsilon^2/a \approx 0.00031$ km^{-1}. Figure 8 shows ISDs without wandering effect (black curves) and those with wandering effect (gray curves) at 100 km distance for the incidence of an impulsive plane P-wavelet. Scattering produces envelope broadening of a large peak in the longitudinal component and a small peak in the transverse component; however, the wandering effect causes a collapse of the sharp peak of $\hat{I}^P_{z0}$ but causes little change in the small peak with slow variation of $\hat{I}^P_{x0}$. The maximum peak of the transverse component appears later than the maximum peak arrival of the longitudinal component.

In Fig. 9a, the upper panel shows ISD time traces at four travel distances for the incidence of an impulsive plane P-wavelet. The upper panel of Fig. 9b enlarges ISD time traces of the transverse component. Envelope broadening becomes apparent as travel distance increases. The wandering effect is stronger at longer travel distances; however, apparent contribution to the envelope broadening is stronger at shorter distances since ISDs without wandering effect have sharper peaks at shorter distances. The wandering effect makes little change in envelopes of the transverse component.

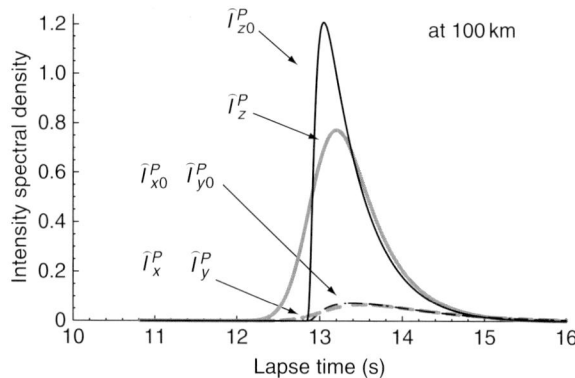

FIG. 8. Time traces of ISDs without wandering effect (black curves) and those with wandering effect (gray curves) at a travel distance of 100 km in 3-D random media for the incidence of an impulsive plane P-wavelet, where the average P-wave velocity is 7.8 km/s, and Gaussian ACF with $\varepsilon = 0.05$ and $a = 8$ km (Sato, 2006).

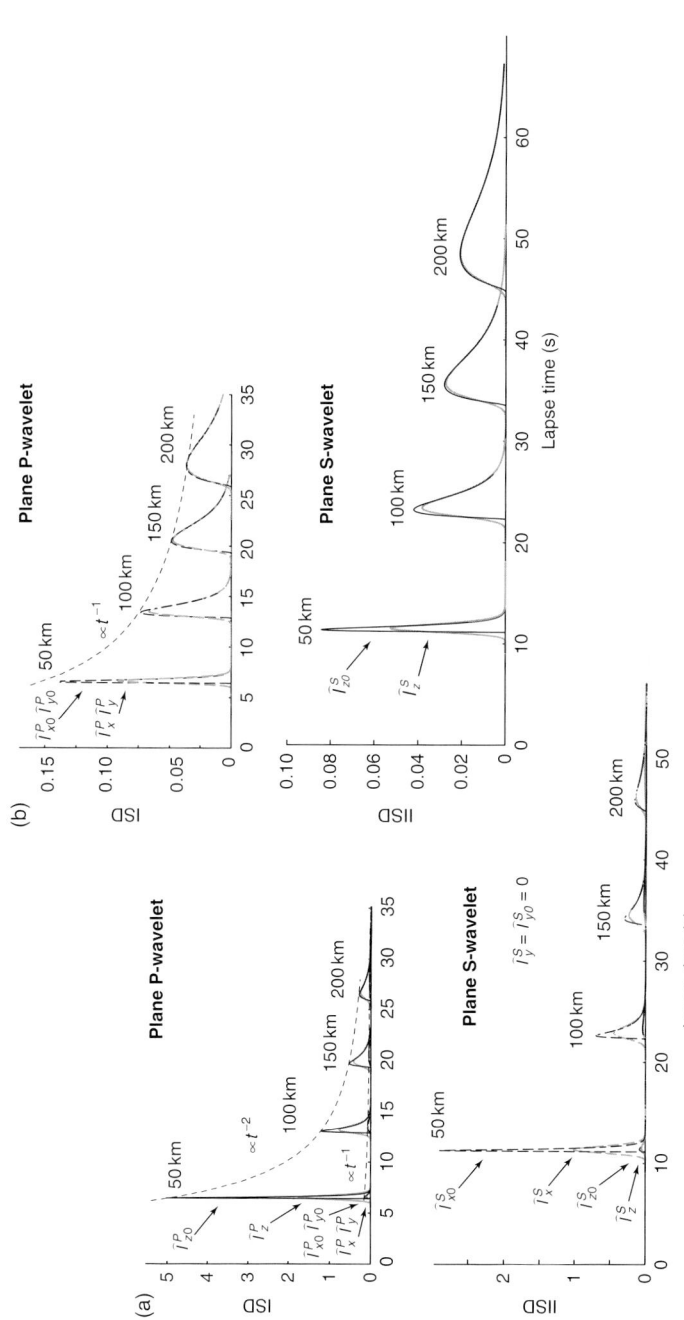

FIG. 9. (a) Upper panel shows ISD time traces in 3-D random media for the incidence of an impulsive plane P-wavelet, where the average P- and S- wave velocities are 7.8 and 4.5 km/s, respectively, and Gaussian ACF with $\varepsilon = 0.05$ and $a = 8$ km. Lower panel shows those for the incidence of an impulsive plane S-wavelet polarized to the x-direction. (b) Zoom up of ISD time traces (Sato, 2006).

The time integral of the transverse-component ISD without wandering effect is

$$\int_{z/V_0}^{\infty} I_{x0}^P(z,t,\omega_c)dt = \int_{z/V_0}^{\infty} I_{y0}^P(z,t,\omega_c)dt = \frac{4V_0 t_M}{z} = \frac{2\sqrt{\pi}\varepsilon^2}{a}z \qquad (34)$$

since $\lim_{\omega_d \to 0} \tan s_0/(s_0 \cos s_0) = 1$. The time integral of ISD with wandering effect has the same value. That is, the time integral of MS envelope of the transverse-component displacement linearly increases with travel distance increasing. The peak values of I_{x0}^P and I_{y0}^P decay according to the inverse of lapse time as shown by a thin broken curve in the upper panel of Fig. 9b since the envelope width increases with the square of travel distance. The time integral of the longitudinal-component displacement is

$$\int_{z/v_0}^{\infty} I_{z0}^P(z,t,\omega_c)dt = 1 - \frac{8V_0 t_M}{z} = 1 - \frac{4\sqrt{\pi}\varepsilon^2}{a}z \qquad (35)$$

since $\int_{z/V}^{\infty} I_0^R(z,t,\omega_c)dt = 1$ because $\lim_{\omega_d \to 0} 1/\cos s_0 = 1$. When $\varepsilon^2 z/a \ll 1$, the second term is negligible, therefore, we may roughly say that the peak intensity of the z-component decreases with the inverse square of travel distance as plotted by thin broken curve in the upper panel of Fig. 9a.

Analysis of observed teleseismic P-wave envelopes: Analyzing teleseismic P-waves registered by IRIS stations in the world, Kubanza et al. (2007) measured the ratio of the peak intensity of transverse component to the peak value of the total intensity at three frequency bands: 0.040 at 0.5–1 Hz, 0.087 at 1–2 Hz, and 0.141 at 2–4 Hz. Assuming the thickness of the inhomogeneous lithosphere to be 100 km and applying the above theoretical prediction for a Gaussian ACF, they estimate ε^2/a to be 2.23×10^{-4} km^{-1} at 0.5–1 Hz, 4.86×10^{-4} km^{-1} at 1–2 Hz, and 7.81×10^{-4} km^{-1} at 2–4 Hz. Assuming $a = 5$ km for all frequency bands, they estimate ε as small as 2–4% for the lithosphere beneath stable continents; however, stations on Japan and collision zones of Indian and Eurasian continents show large ε of 5–10%. We should note that the increase of envelope broadening with increasing frequency has some contradiction to the frequency independence predicted from Gaussian ACF.

2.1.9. Plane S-Wavelet

Parabolic wave equation: Three components of vector potential of S-wave are independent of each other as written by Eq. (2). If vector potential has y-component only, displacement vector $\mathbf{u} = (-\partial_z B_y, 0, \partial_x B_y)$. That is, the displacement vector is always in the $x - z$ plane; however, the ray vector is not confined in the $x - z$ plane. If we write the y-component of vector potential $B_y = \phi$ and the S-wave velocity as $\beta(\mathbf{x}) = V_0(1 + \xi(\mathbf{x}))$, where V_0 is the average S-wave velocity and ξ is fractional fluctuation, vector potential y-component ϕ satisfies the inhomogeneous wave Eq. (3). We study S-wave propagation through a random medium spreading over a half space ($z > 0$) for the vertical incidence of a plane S-wavelet having a polarization to the x-axis from a homogeneous medium ($z < 0$). When the wavelength is supposed to be smaller than a, field U satisfies the parabolic wave Eq. (5) if we write ϕ as a superposition of plane waves as Eq. (4).

Intensity spectral densities: Taking the same procedure for P-waves, we can derive intensities and ISDs for S-waves. In this case, the y-component ISD is always zero,

$$I_y^S(z,t,\omega_c) = 0. \tag{36}$$

The z-component ISD is

$$I_z^S(z,t,\omega_c) = \frac{1}{2\pi}\int_{-\infty}^{\infty} d\omega_d e^{-i\omega_d(t-(z/V_0))} \left[-\frac{\partial_{x_d}^2}{k_c^2}\Gamma_2(\mathbf{x}_{\perp d},z,\omega_c,\omega_d)\right]_{\mathbf{x}_{\perp d}=0} \tag{37}$$

and the x-component ISD is

$$\begin{aligned}I_x^S(z,t,\omega_c) &= \frac{1}{2\pi}\int_{-\infty}^{\infty} d\omega_d e^{-i\omega_d(t-(z/V_0))}\left[\left(1+\frac{\Delta_{\perp d}}{k_c^2}\right)\Gamma_2(\mathbf{x}_{\perp d},z,\omega_c,\omega_d)\right]_{\mathbf{x}_{\perp d}=0}\\ &= I^R(z,t,\omega_c) - 2I_z^S(z,t,\omega_c),\end{aligned} \tag{38}$$

where the relation $\partial_{x_d}^2\Gamma_2 = \partial_{y_d}^2\Gamma_2$ is used. Replacing V_0 with the average S-wave velocity in Eqs. (15) and (19), we have the reference ISDs for S-wave. In the case of quasi-monochromatic waves, TFMCF Γ_2 is factorized into $_0\Gamma_2$ and $\tilde{w}$. Taking the same procedure as for P-waves, we solve the master equation (10) with the initial condition (11), which means the incidence of an impulsive plane S-wavelet polarized to the x-direction propagating to the z-direction:

$$I_x^S(z,t,\omega_c) = \delta\left(t-\frac{z}{V_0}\right) \text{ and } I_y^S(z,t,\omega_c) = I_z^S(z,t,\omega_c) = 0 \text{ at } z=0. \tag{39}$$

Gaussian ACF: For the case of random media characterized by a Gaussian ACF [Eq. (22)], substituting the solution [Eq. (27)] into Eqs. (37) and (38), we obtain the z-component ISD without wandering effect as

$$I_{z0}^S(z,t,\omega_c) = \frac{V_0 t_M}{z} 2\frac{t-(z/V_0)}{t_M} I_0^R(z,t,\omega_c) \tag{40}$$

and the x-component ISD without wandering effect as

$$I_{x0}^S(z,t,\omega_c) = \left[1 - \frac{V_0 t_M}{z}\frac{4(t-z/V_0)}{t_M}\right] I_0^R(z,t,\omega_c). \tag{41}$$

At large lapse times as $t > (5/4)(z/V_0)$, I_{x0}^S becomes negative, which means the breakdown of the approximation.

The lower panel of Fig. 9a shows ISDs for the incidence of an impulsive plane S-wavelet polarized to the x-direction. The lower panel of Fig. 9b enlarges ISD traces of the z-component. Peak delay and envelope broadening are seen in both components; however, the peak delay of the z-component is larger than that of the x-component and the peak value of z-component is smaller than that of x-component at each travel distance. For the same fractional fluctuation for both P- and S-wave velocities, the envelope width of S-wave is 1.73 times larger than that of P-waves since the characteristic time is proportional to the reciprocal of the average velocity of the wave type.

The time integral of the longitudinal-component ISD linearly increases with travel distance and the linear coefficient is the ratio of the MS fractional fluctuation to the correlation distance since

$$\int_{z/V_0}^{\infty} I_{z0}^S(z,t,\omega_c)dt = \frac{4V_0 t_M}{z} = \frac{2\sqrt{\pi}\varepsilon^2}{a}z. \quad (42)$$

The time integral of transverse-component ISD without wandering effect is

$$\int_{z/V_0}^{\infty} I_{x0}^S(z,t,\omega_c)dt = 1 - \frac{8V_0 t_M}{z} = 1 - \frac{4\sqrt{\pi}\varepsilon^2}{a}z. \quad (43)$$

2.2. Two-Dimensional Random Elastic Media

In order to examine the validity of the Markov approximation, we compare the envelopes directly simulated with envelopes numerically simulated based on the FD simulations in two dimensions according to Korn and Sato (2005).

2.2.1. Wave Envelopes for a Gaussian ACF

Elastic media in 2-D space ($x - z$ space) is divided into two: a half space of $z < 0$ is homogenous and another half space of $z > 0$ is randomly inhomogeneous. For the vertical incidence of a plane P-wavelet from the homogeneous medium, scalar potential in the inhomogeneous space is written by using a superposition of plane waves [Eq. (4)]. We define the TFMCF of field U at distance z as $\Gamma_2(x_d, z, \omega_c, \omega_d) \equiv \langle U(x', z, \omega')$ $U(x'', z, \omega'')^* \rangle$. For the case of Gaussian ACF in 2-D $R(\mathbf{x}) = \varepsilon^2 \exp[-(x^2 + z^2)/a^2]$, replacing $\Delta_{\perp d}$ with $\partial_{x_d}^2$ and $r_{\perp d}$ with x_d in Eq. (24), we have the master equation for TFMCF as

$$\partial_z {}_0\Gamma_2 + i\frac{k_d}{2k_c^2} \partial_{x_d}^2 {}_0\Gamma_2 + k_c^2 \sqrt{\pi}\varepsilon^2 a\left(\frac{x_d}{a}\right)^2 {}_0\Gamma_2 = 0. \quad (44)$$

The solution (A.10) for the initial condition (11) is given by

$${}_0\Gamma_2(x_d, z, \omega_c, \omega_d) = \frac{e^{-(\tan s_0/s_0)\left((2V_0 k_c^2 t_M)/z\right) x_d^2}}{\sqrt{\cos s_0}} w \quad (45)$$

according to Korn and Sato (2005). Substituting it into Eq. (19), we have the reference ISD without wandering effect as

$$I_0^R(z,t;\omega_c) = \frac{1}{2\pi}\int_{-\infty}^{\infty} d\omega_d \frac{1}{\sqrt{\cos s_0}} e^{-i\omega_d(t-(z/V_0))}. \quad (46)$$

Different from the 3-D case, this function has to be evaluated numerically, where we take the branch so that the integrand is a continuous function of ω_d. The x-component ISD without wandering effect is given by Eq. (13) as

$$I^P_{x0}(z,t;\omega_c) = \frac{4V_0 t_M}{z}\frac{1}{2\pi}\int_{-\infty}^{\infty} d\omega_d e^{-i\omega_d(t-(z/V_0))}\frac{\tan(s_0)}{s_0\sqrt{\cos(s_0)}}$$

$$= \frac{4V_0}{z}\frac{1}{2\pi}\int_{-\infty}^{\infty} d\omega_d e^{-i\omega_d(t-(z/V_0))}\frac{\partial}{i\partial\omega_d}\frac{1}{\sqrt{\cos s_0}}$$

$$= \frac{4V_0}{z}\left(t-\frac{z}{V_0}\right)I^R_0(z,t;\omega_c) \qquad (47)$$

and then the z-component ISD without wandering effect is given by the residual [Eq. (16)] as

$$I^P_{z0}(z,t;\omega_c) \equiv I^R_0(z,t;\omega_c) - I^P_{x0}(z,t;\omega_c)$$

$$= \left(1 - \frac{4V_0 t_M}{z}\frac{(t-(z/V_0))}{t_M}\right)I^R_0(z,t;\omega_c). \qquad (48)$$

Figure 10 shows plots of I^P_{x0} (light gray curve), I^P_{z0} (black curve), and I^R_0 (dark gray curve) against reduced time at a distance $4V_0 t_M/z = 0.177$. The peak height of I^P_{x0} is small compared with that of I^P_{z0}. The partition of energy into two components depends on the lapse time: $I^P_{z0} \approx I^R_0 \gg I^P_{x0}$ around the peak arrival when $\varepsilon^2 z/a \ll 1$; however, $I^P_{x0} > I^P_{z0}$ at reduced times larger than about $3t_M$ in this case. The time integral of ISD depends on travel distance:

$$\int_{z/V_0}^{\infty} I^P_{x0}(z,t;\omega_c)dt = \frac{4V_0 t_M}{z} = \frac{2\sqrt{\pi}\varepsilon^2}{a}z \qquad (49)$$

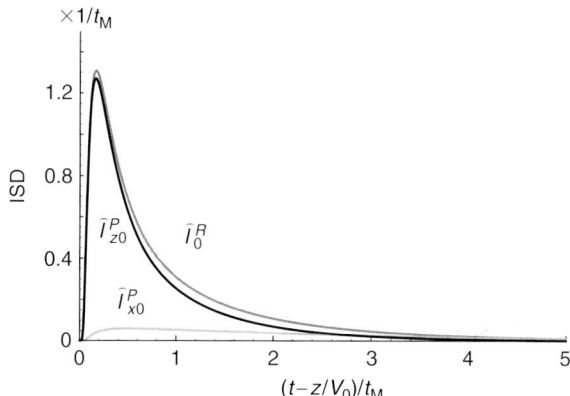

FIG. 10. Plots of I^P_{x0} and I^P_{z0} at a distance $4V_0 t_M/z = 0.177$ for the incidence of a plane P-wavelet to 2-D random media (Korn and Sato, 2005).

and

$$\int_{z/V_0}^{\infty} I_{z0}^P(z,t;\omega_c)dt = 1 - \frac{4V_0 t_M}{z} = 1 - \frac{2\sqrt{\pi}\varepsilon^2}{a}z. \quad (50)$$

For the incidence of a plane S-wavelet, exchanging x with z in Eqs. (46)–(48), we have S-wave ISDs, I_0^R, I_{z0}^S, and I_{x0}^S, respectively.

2.2.2. Comparison with Numerically Simulated Envelopes

FD simulations: For the computation of theoretical waveforms of vector waves in various realizations of 2-D random media, a standard FD technique in space-time domain is employed. We use a scheme where the equations for particle velocities and stresses in an isotropic inhomogeneous elastic medium are solved on a staggered grid (Levander, 1988). The accuracy is second-order in time and fourth-order in space.

The size of the rectangular model is 300 km by 250 km, where average P- and S-velocities are chosen as $\alpha_0 = 6$ km/s and $\beta_0 = 3.46$ km/s. The average mass density is 2800 kg/m^3 and the fractional fluctuation of mass density is chosen as $0.8\xi(\mathbf{x})$ according to Birch's law (e.g., Sato and Fehler, 1998). The medium is homogeneous for -50 km $< z < 0$. Between $z = 0$ and 200 km, random fractional velocity fluctuation $\xi(\mathbf{x})$ is characterized by a Gaussian ACF with $a = 5$ km and $\varepsilon = 0.05$. Periodic boundary conditions are used at $x = 0$ and 300 km and absorbing boundary conditions at $z = -50$ and 200 km. Absorbing boundary conditions are based on the paraxial approximation of the wave equation representing only the outgoing part of the wave field (Reynolds, 1978).

A plane P- or S-wavelet propagating parallel to the grid in the z-direction is initialized at $z = 0$. The pulse shape of the plane wavelet is given by

$$h(t) = \sin\frac{N\pi t}{T} - \left(\frac{N}{N+2}\right)\sin\frac{(N+2)\pi t}{T} \quad 0 \le t \le T, \quad (51)$$

where T is the duration of the wavelet and N is a parameter indicating the number of maxima and minima of the wavelet. Here, we choose $N = 2$ and $T = 0.5$ s, which gives a wavelet of nearly sinusoidal shape with dominant frequency of 2 Hz and band-limited spectrum of half-width between 0.8 and 4.1 Hz. The dominant wavelength is smaller than the correlation distance, which ensures the validity of the parabolic approximation. Elastic waves are recorded at three line arrays parallel to the initial wave front at $z = 50$, 100, and 150 km. Each line array consists of 200 receivers at 1 km intervals between $x = 50$ and 250 km. The spatial discretization in the FD scheme is 0.1 km and the temporal discretization is 8 ms, slightly below the stability limit of the numerical scheme. This choice ensures that the numerical errors remain small. In a homogeneous medium with the mean velocities, the phase velocity error introduced by grid dispersion would be 0.04% at the dominant frequency and 0.15% at the upper half width frequency.

Figure 11 shows examples of waveforms after traveling 100 km through a random medium for P-wavelet incidence and S-wavelet incidence. Distortions of pulse shapes and fluctuations of travel times and amplitudes along the direct wave front are clearly visible, as well as excitation of secondary arrivals after the primary wave resulting in an effective broadening of the wave front. For P (S)-wavelet incidence, there is significant energy arriving on the transverse (longitudinal) component. This indicates that the wave

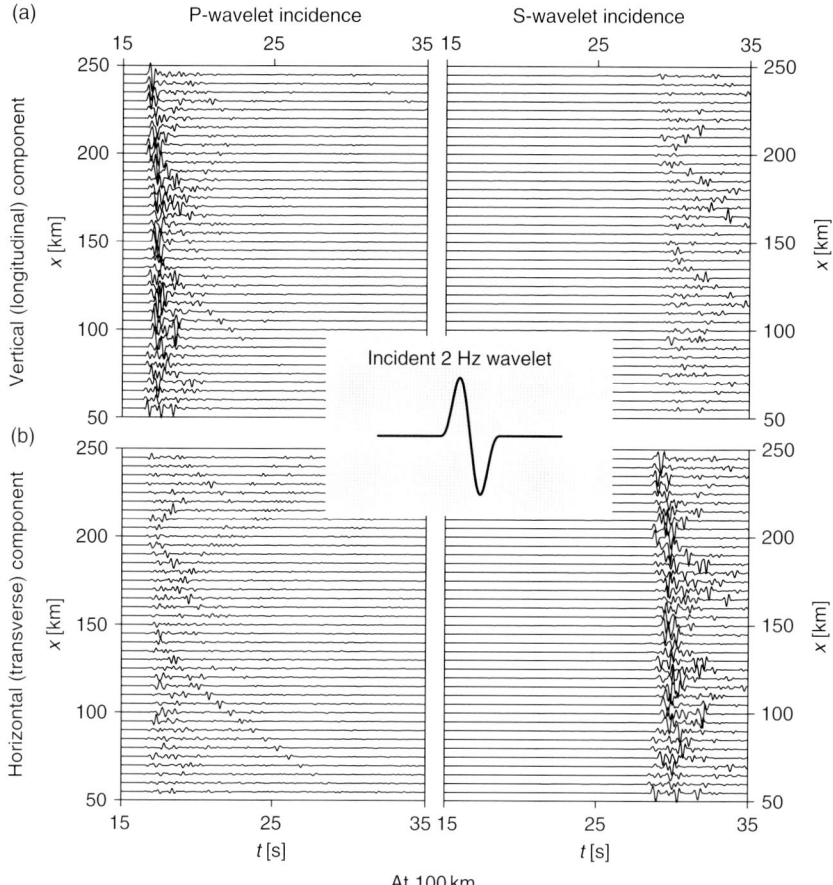

FIG. 11. Examples of FD simulation traces at a distance of 100 km in a 2-D random medium for the incidence of plane P- and S-wavelet, where the average P- and S-velocities and the mass density are 6 km/s, 3.46 km/s, and 2800 kg/m^3, and a Gaussian ACF with $a = 5$ km and $\varepsilon = 0.05$: (a) longitudinal (z)-component traces and (b) transverse (x)-component traces for P- and S-wave incidence. Only every fifth trace is plotted (Korn and Sato, 2006).

front gets strongly distorted by the inhomogeneity of the medium and the propagation direction locally deviates from the straight ray paths. On the other hand, only very little energy arrives before the S-wave front in the case of plane S-wave incidence. This indicates that conversion scattering from S to P is extremely weak. In the same way, the lack of energy at longer lapse times for P-wave incidence suggests that both conversion scattering and large-angle scattering are small. In general, the most effective scattering mechanism for the case studies here is scattering of the same wave type into the near forward direction.

The single-component traces of 200 receivers are first squared and then averaged to obtain single-component MS envelope. Additionally, the MS envelopes obtained from

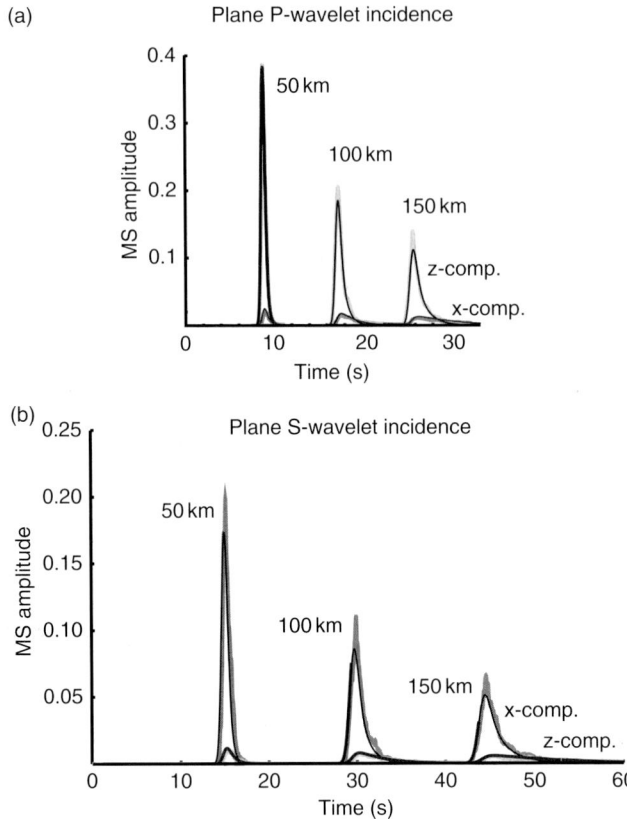

FIG. 12. Comparison of FD envelopes (gray curves) with Markov envelopes (black curves) in 2-D random media (a) for incidence of 2 Hz plane P-wavelet and (b) for incidence of 2 Hz plane S-wavelet (Korn and Sato, 2005).

10 different realizations of the random medium were again averaged to get the ensemble averaged MS envelope. In Fig. 12, gray curves show MS envelopes numerically simulated by the FD method (FD envelopes). Envelope broadening and peak amplitude decay with increasing propagation distance are obvious. For P-wavelet incidence, the ratio of transverse to longitudinal-component envelopes increases with time after the first arrival. A few seconds after the peak, transverse-component energy becomes larger than longitudinal-component energy irrespective of travel distance (see traces at 150 km distance in Fig. 12a). The opposite feature is observed for S-wavelet incidence. Even after stacking 2000 single traces, the standard deviation is still large around the peak amplitude, but is small at later times. This indicates that peak amplitudes only are not a stable measure to be used in envelope interpretation. It is better to use the whole envelope shape for interpretation.

Comparison of Markov and FD envelopes: In Fig. 12, fine black curves are MS envelopes theoretically predicted by the Markov approximation (Markov envelopes),

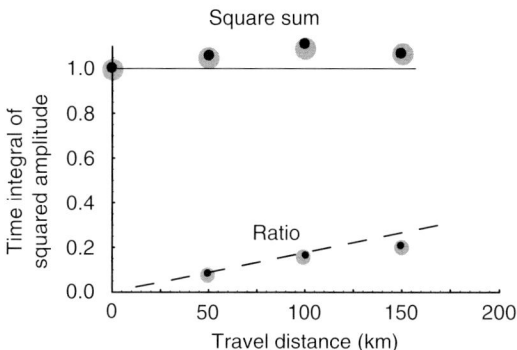

Fig. 13. Plot of the time integral of squared amplitude against travel distance in 2-D random media for the incidence of a 2-Hz plane wavelet. A large black (gray) circle is the time integral of the sum of squared traces for P (S)-wave incidence. A small black (gray) circle is the ratio of the time integral of squared amplitude of the x- (z-) component to the time integral of the sum of squared amplitude for the incidence of P (S-) waves. A broken line is the theoretical prediction [Eq. (49)] by the Markov approximation (Korn and Sato, 2005).

where we used the convolution of the ISD with wandering effect and the MS trace of the incident wavelet i. The peak value of FD envelope is a little larger than the smooth peak of Markov envelopes at each distance; however, the whole envelope shapes well coincide with each other. The Markov envelope well explains the delay of peak arrival from the onset and the envelope broadening of each component at each travel distance for both of P- and S-wavelet incidence.

In Fig. 13, we plot the time integral of squared amplitude against travel distance for the incidence of a 2-Hz plane wavelet. Black and gray circles are calculated from FD simulations and a broken line is the theoretical prediction [Eq. (49)] by the Markov approximation. Good coincidence between FD simulations and the Markov approximation suggests that the partition of energy to the transverse component for P-wavelet incidence and that to the longitudinal component for S-wavelet incidence are good measure of the medium inhomogeneity.

3. Vector-Wave Envelopes for the Radiation from a Point Source

3.1. Three-Dimensional Random Elastic Media

Here we introduce the Markov approximation for waves radiated from a point source in 3-D random media since there is a difference in geometrical spreading between spherical waves and plane waves (Sato, 2007). We study the propagation of elastic vector waves radiated impulsively from a point source through a weakly inhomogeneous elastic medium in 3-D infinite space (see Fig. 14). When the wavelength is much smaller than the correlation distance a of medium inhomogeneity $ak_0 \gg 1$, there is little conversion scattering between P- and S-waves, which can be treated separately in the same way as for the case of plane waves.

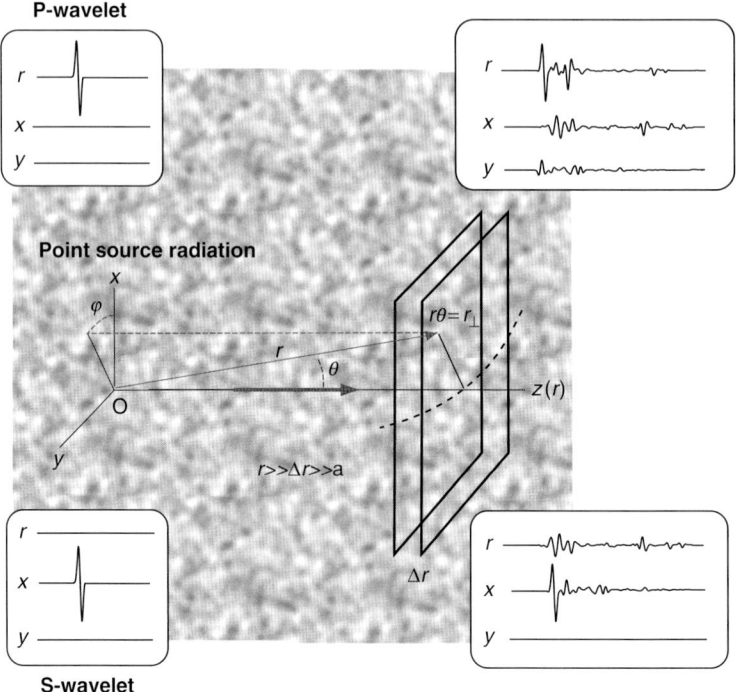

FIG. 14. Geometry of a transverse plane perpendicular to the global ray direction along the z-axis at a long travel distance for a point source radiation in 3-D random media.

3.1.1. Spherically Outgoing P-Wavelet

For isotropic radiation of P-wavelet from a point source at the origin, scalar potential is written as a superposition of spherically outgoing harmonic waves as

$$\phi = \frac{1}{2\pi}\int_{-\infty}^{\infty} \tilde{\phi} e^{-i\omega t} d\omega = \frac{1}{2\pi}\int_{-\infty}^{\infty} \frac{U(\theta,\varphi,r,\omega)}{ik_0 r} e^{ik_0 r - i\omega t} d\omega, \qquad (52)$$

where r is a distance from the source, angles θ and φ are measured from the z-axis and x-axis, respectively. Neglecting the second derivative with respect to radius, we have the parabolic equation for field U as

$$2ik_0 \partial_r U + \Delta_\perp U - 2k_0^2 \xi(\mathbf{x}) U = 0. \qquad (53)$$

In the vicinity of the z axis ($|\theta| \ll 1$), transverse coordinates are $x \approx r\theta \cos\varphi$ and $y \approx r\theta \sin\varphi$, and transverse Laplacian

$$\Delta_\perp \equiv \partial_x^2 + \partial_y^2 \approx r^{-2}\partial_\theta^2 + r^{-2}\theta^{-1}\partial_\theta + r^{-2}\theta^{-2}\partial_\varphi^2.$$

3.1.2. Stochastic Master Equation for TFMCF

Taking the same procedure for the plane wave case, we define the TFMCF of U between two different locations and different angular frequencies at ω' and ω'' on the transverse plane at a distance r in the vicinity of the z-axis as

$$\Gamma_2(\mathbf{x}_{\perp c}, \mathbf{x}_{\perp d}, r, \omega_c, \omega_d) \equiv \left\langle U\left(\mathbf{x}'_\perp, r, \omega'\right) U\left(\mathbf{x}''_\perp, r, \omega''\right)^* \right\rangle. \tag{54}$$

TFMCF is independent of $\mathbf{x}_{\perp c}$ and practically becomes a function of $r_{\perp d} \equiv |\mathbf{x}_{\perp d}| = r\theta_d$ because of the isotropic source radiation and homogeneity and isotropy of randomness. In the case of quasi-monochromatic waves $|\omega_d| \ll |\omega_c|$, using causality and neglecting back scattering, we derive the same master equation (7) for TFMCF. Replacing z with r in the decomposition [Eq. (9)], where $\tilde{w}(r, \omega_d) = \exp\left[-A(0)\omega_d^2 r / (2V_0^2)\right]$, we have the master equation for $_0\Gamma_2$ as

$$\partial_r \,_0\Gamma_2 + i\frac{k_d}{2k_c^2}\Delta_{\perp d} \,_0\Gamma_2 + k_c^2[A(0) - A(r_{\perp d})]_0\Gamma_2 = 0. \tag{55}$$

Scalar potential in the angular frequency domain $\tilde{\phi} \approx e^{ik_0 r}/(ik_0\sqrt{4\pi}r)$, where $U = 1/\sqrt{4\pi}$, leads to isotropic displacement vector components as $\tilde{u}_r \approx e^{ik_0 r}/\sqrt{4\pi}r$, $\tilde{u}_\theta = 0$, and $\tilde{u}_\varphi = 0$. Therefore, we put the initial condition for isotropic P-wave radiation as

$$_0\Gamma_2(\mathbf{x}_{\perp d}, r = 0, \omega_c, \omega_d) = \frac{1}{4\pi}. \tag{56}$$

3.1.3. ISDs

Taking the same procedure as for plane wave case, we derive three component ISDs, where Cartesian coordinates are used for vector calculation for small transverse distances around the z-axis. The x-component ISD is

$$I_x^P(r, t, \omega_c) = \frac{1}{2\pi r^2}\int_{-\infty}^{\infty} d\omega_d e^{-i\omega_d(t - r/V_0)} \tilde{w}(r, \omega_d) \left[-\frac{1}{k_c^2}\partial_{x_d,0}^2 \Gamma_2(\mathbf{x}_{\perp d}, r, \omega_c, \omega_d)\right]_{\mathbf{x}_{\perp d}=0}, \tag{57}$$

where a factor r^{-2} represents geometrical spreading and the spatial derivative of this factor is neglected. The x- and y-component ISDs are identical because of the isotropy of TFMCF in the transverse plane, $I_y^P(r, t, \omega_c) = I_x^P(r, t, \omega_c)$.

Here we define the reference ISD for spherical waves as

$$I^R(r, t, \omega_c) \equiv \frac{1}{2\pi r^2}\int_{-\infty}^{\infty} d\omega_d e^{-i\omega_d(t-(r/V_0))} \tilde{w}(r, \omega_d) \,_0\Gamma_2(\mathbf{x}_{\perp d} = 0, r, \omega_c, \omega_d). \tag{58}$$

Reference ISD without wandering effect is obtained by taking $\tilde{w} = 1$ in Eq. (58). For the derivation of the longitudinal (z)-component ISD, taking the same procedure as for plane waves, we have

$$I_z^P(r,t,\omega_c) = \frac{1}{2\pi r^2} \int_{-\infty}^{\infty} d\omega_d e^{-i\omega_d(t-r/V_0)} \tilde{w}(r,\omega_d)$$
$$\left[\left(1 + \frac{\Delta_{\perp d}}{k_c^2}\right){}_0\Gamma_2(\mathbf{x}_{\perp d}, r, \omega_c, \omega_d)\right]_{\mathbf{x}_{\perp d}=0}$$
$$= I^R(z,t,\omega_c) - 2I_x^P(z,t,\omega_c). \tag{59}$$

The initial condition [Eq. (56)] practically gives ISDs as

$$I_x^P = I_y^P = 0 \text{ and } I_z^P = I^R = \frac{1}{4\pi r^2}\delta\left(t - \frac{r}{V_0}\right) \text{ as } r \to 0 \text{ along the } z-\text{axis} \tag{60}$$

that mean isotropic radiation of an impulsive P-wavelet with unit intensity from a point source at the origin.

3.1.4. Gaussian ACF

ISDs of P-wavelet: When 3-D random media are statistically characterized by a Gaussian ACF [Eq. (22)], we have the following differential equation from Eq. (55):

$$\partial_r {}_0\Gamma_2 + i\frac{k_d}{2k_c^2 r^2}\left(\frac{\partial^2}{\partial \theta_d^2} + \frac{1}{\theta_d}\frac{\partial}{\partial \theta_d}\right){}_0\Gamma_2 + \frac{\sqrt{\pi}\varepsilon^2 k_c^2}{a}r^2\theta_d^2 {}_0\Gamma_2 = 0 \tag{61}$$

since ${}_0\Gamma_2$ is axial symmetric with respect to $\theta_d = 0$. According to Shishov (1974), solving Eq. (61) with the initial condition [Eq. (56)], we have the analytical solution [Eq. (A.15)] as

$${}_0\Gamma_2(\mathbf{x}_{\perp d}, r, \omega_c, \omega_d) = \frac{1}{4\pi}\frac{s_0}{\sin s_0}e^{-[1/s_0^2 - ((\cot s_0)/s_0)](2V_0 k_c^2 t_M/r)r_{\perp d}^2}, \tag{62}$$

where $r_{\perp d} = r\theta_d$, the characteristic time $t_M = \sqrt{\pi}\varepsilon^2 r^2/(2V_0 a)$, and a parameter $s_0 = 2e^{i\pi/4}\sqrt{t_M \omega_d}$ are given by replacing z with r in Eqs. (25) and (26). Substituting Eq. (62) into Eq. (58) with $\tilde{w}=1$, we get ISD without wandering effect as

$$I_0^R(r,t,\omega_c) = \frac{1}{2\pi r^2}\int_{-\infty}^{\infty}d\omega_d e^{-i\omega_d(t-(r/V_0))}\frac{1}{4\pi}\frac{s_0}{\sin s_0}$$
$$= \frac{1}{4\pi r^2}\frac{1}{2\pi}\int_{-\infty}^{\infty}d\omega_d e^{-i\omega_d(t-(r/V_0))}\left(1 + \sum_{k=1}^{\infty}(-1)^k\frac{2s_0^2}{s_0^2 - \pi^2 k^2}\right)$$
$$= \frac{\pi}{2r^2}\sum_{k=1}^{\infty}(-1)^{k+1}k^2\left[\frac{1}{2\pi}\int_{-\infty}^{\infty}d\omega_d\frac{e^{-i\omega_d(t-(r/V_0))}}{\pi^2 k^2 - 4it_M\omega_d}\right], \tag{63}$$
$$= \frac{1}{4\pi r^2}H\left(t - \frac{r}{V_0}\right)\frac{\pi^2}{2t_M}\sum_{k=1}^{\infty}(-1)^{k+1}k^2 e^{-\left(\pi^2 k^2/(4t_M)\right)(t-(r/V_0))}$$
$$= \frac{1}{4\pi r^2}H\left(t - \frac{r}{V_0}\right)\frac{\pi^2}{16t_M}\vartheta_4'' e^{-(\pi^2/4)((t-r/V_0)/t_M)}$$

where we take $1 + 2\sum_{k=1}^{\infty}(-1)^k = 1 - 2 + 2 - 2 \ldots = 0$. In the last line, $\vartheta_4''(q) \equiv \partial_u^2 \vartheta_4(u,q)\big|_{u=0} = 8\sum_{n=1}^{\infty}(-1)^{n+1} n^2 q^{n^2}$, where $\vartheta_4(u,q) = 1 + 2\sum_{n=1}^{\infty}(-1)^n q^{n^2} \cos 2nu$ is the elliptic theta function of the fourth kind (e.g., Weisstein, 2005). Taking the inverse Fourier transform of Eq. (63) at $\omega_d = 0$, we have $\int_{-\infty}^{\infty} 4\pi r^2 I_0^R dt = 1$, which is satisfied by the expression using ϑ_4''. In Fig. 15 a black curve shows I_0^R for spherical wavelet given by Eq. (63) against reduced time $t - r/V_0$. The maximum peak value is numerically about $1.48/(4\pi r^2 t_M)$ at reduced time about $0.367 t_M$. It shows a broadened envelope having a delayed peak and a smoothly decaying tail. For comparison, reference ISD for plane wave [Eq. (29)] is shown by a gray curve. Two curves are normalized to satisfy that the time integral is equal to one. Broadening of a spherically outgoing wavelet envelope is smaller than that of a plane wavelet because of the convex curvature of the wave front.

Then, we calculate the transverse-component (x-component) ISD without wandering effect. Substituting Eq. (62) into Eq. (57) with $\tilde{w} = 1$, we have

$$\begin{aligned}
I_{x0}^P(r,t,\omega_c) &= I_{y0}^P(r,t,\omega_c) \\
&= \frac{1}{r^2} \frac{V_0}{\pi r} \left[\frac{1}{2\pi} \int_{-\infty}^{\infty} d\omega_d e^{-i\omega_d(t-(r/V_0))} \frac{t_M}{\sin s_0} \left(\frac{1}{s_0} - \cot s_0\right)\right] \\
&= \frac{2V_0}{r} \left[\frac{1}{2\pi r^2} \int_{-\infty}^{\infty} d\omega_d e^{-i\omega_d(t-(r/V_0))} \frac{1}{i} \frac{\partial}{\partial \omega_d} \left(\frac{s_0}{4\pi \sin s_0}\right)\right] \\
&= \frac{V_0 t_M}{r} 2 \frac{(t - (r/V_0))}{t_M} I_0^R(r,t,\omega_c).
\end{aligned} \qquad (64)$$

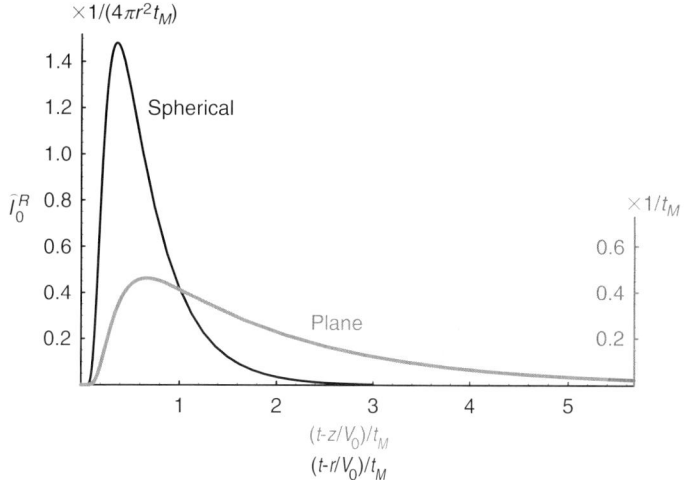

FIG. 15. Plots of reference ISDs without wandering effect for a spherically outgoing wavelet (black curve) and that for a plane wavelet (gray curve) in 3-D random media.

Putting Eqs. (63) and (64) into Eq. (59), we get an explicit representation of the longitudinal-component ISD without wandering effect as

$$I^P_{z0}(r,t,\omega_c) = \left[1 - 4\frac{V_0}{r}\left(t - \frac{r}{V_0}\right)\right] I^R_0(r,t,\omega_c). \quad (65)$$

The calculated longitudinal-component ISD becomes negative for large lapse time as $t > 1.25(r/V_0)$, which means the breakdown of this approximation. The applicable range for the reduced time $0 < t - r/V_0 < r/(4V_0)$ is the additional condition of this approximation especially for vector-wave envelopes, which is the same as the case of plane wavelet.

Characteristics of spherically outgoing P-wave envelopes: As an example, Fig. 16 shows simulated P-wave ISDs against lapse time at a travel distance of 200 km, where random media are characterized by $V_0 = 6$ km/s, $\varepsilon = 0.05$, and $a = 5$ km: $\hat{I}^R_0$ (fine black solid curve), $\hat{I}^P_{z0}$ (gray solid curve), $\hat{I}^P_z$ (black solid curve), $\hat{I}^P_{x0}$ (gray broken curve), and $\hat{I}^P_x$ (black broken curve). The P-wave onset time for the average velocity medium is 33.3 s as shown by a fine vertical line. We note that the y-component trace is the same as the x-component trace. The characteristic time is about 3.0 s in this case, the peak height of $\hat{I}^R_0$ is numerically about $1.48/(4\pi r^2 t_M) \approx 0.133 V_0 a/(\varepsilon^2 r^4)$ and the peak delay is about $0.367 t_M = 0.325 \varepsilon^2 r^2/(V_0 a)$. The peak height of $\hat{I}^P_{x0}$ is about $(1.30 V_0 t_M/r)/(4\pi r^2 t_M) \approx 0.104 V_0/r^3$ independent of randomness; however, the peak delay is about $0.536 t_M \approx 0.475 \varepsilon^2 r^2/(V_0 a)$, which depends on randomness. We may roughly approximate the peak height of $\hat{I}^P_{z0}$ as the peak difference $(1.48 - 2.60 \ V_0 t_M/r)/(4\pi r^2 t_M) \approx$

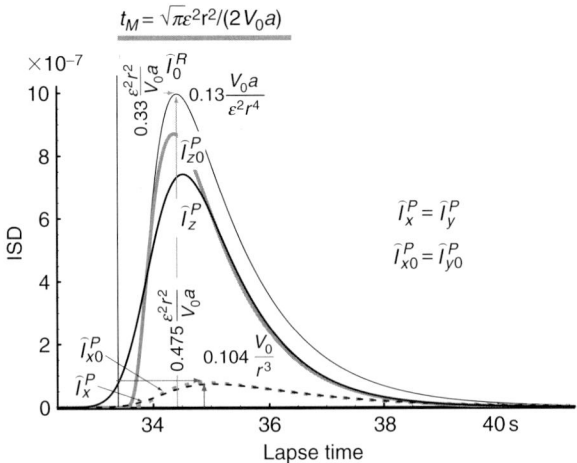

FIG. 16. ISD time traces at a distance of 200 km in 3-D random media for the isotropic P-wavelet radiation from a point source, where the gray horizontal bar shows the characteristic time. Random media are characterized by the average P-wave velocities of 6 km/s and Gaussian ACF with $\varepsilon = 0.05$ and $a = 5$ km (Sato, 2007).

$0.133V_0 a/(\varepsilon^2 r^4) - 0.208V_0/r^3$ neglecting their peak arrival times. When $\varepsilon^2 r/a \ll 1$, the peak height of I_{z0}^P approximately decays according to the minus fourth power of travel distance and the peak ratio of transverse component to longitudinal component is proportional to $\varepsilon^2 r/a$. All the components have broadened traces; however, the peak height of the transverse component is smaller than that of the longitudinal component and the peak delay of the transverse component is larger than that of the longitudinal component. As lapse time increases, the transverse-component ISD exceeds the longitudinal-component ISD: $I_{x0}^P > I_{z0}^P$ for $t - r_0/V_0 > r_0/(6V_0)$. At 200 km, it happens at lapse times larger than 38.9 s.

In Fig. 17 a, the first row shows ISDs at four travel distances: I_{z0}^P (gray solid curve), I_z^P (black solid curve), I_{x0}^P and I_{y0}^P (gray broken curve), and I_x^P and I_y^P (black broken curve). The second row shows those traces normalized by the maximum peak value of I_z^P. These traces clearly show increased envelope broadening with increasing travel distance. At large travel distances, the envelope broadening due to scattering dominates over the wandering effect. The ratio of the transverse- to the longitudinal-component amplitude relatively grows up with increasing travel distance even though randomness is small.

Using the inverse Fourier transform of Eq. (64) at $\omega_d = 0$, we have

$$\int_{r/V_0}^{\infty} I_{x0}^P(r, t, \omega_c) dt = \int_{r/V_0}^{\infty} I_{y0}^P(r, t, \omega_c) dt = \frac{1}{4\pi r^2} \frac{4V_0 t_M}{3r} = \frac{\varepsilon^2}{6\sqrt{\pi}\, ar} \quad (66)$$

since $\lim_{\omega_d \to 0} (1/s_0 - \cot s_0)/\sin s_0 = 1/3$. The time integral of MS envelope of the transverse component is proportional to the inverse of travel distance. Using $\int_{-\infty}^{\infty} 4\pi r^2 I_0^R dt = 1$, we have

$$\int_{r/V_0}^{\infty} I_{z0}^P(r, t, \omega_c) dt = \frac{1}{4\pi r^2}\left(1 - \frac{8}{3}\frac{V_0 t_M}{r}\right) = \frac{1}{4\pi r^2}\left(1 - \frac{4\sqrt{\pi}\varepsilon^2 r}{3a}\right). \quad (67)$$

The time integral of I_{z0}^P is almost proportional to the inverse square of travel distance whereas the second term is small. Chained curves in Fig. 17 approximately show peak decay curves of I_{z0}^P and I_{x0}^P: the former decreases with the minus fourth power of travel distance (lapse time) since the second power due to the envelope broadening and the second power due to geometrical decay; the latter decreases with the minus third power of travel distance (lapse time) since the first power due to the envelope broadening and the minus second power due to geometrical decay.

3.1.5. Spherically Outgoing S-Wavelet

Intensity spectral densities: Taking the same procedure as for plane S-waves, we formulate the envelope synthesis of S-wavelet radiated from a point source in random media in three dimensions in the case that the S-wave velocity $\beta(\mathbf{x}) = V_0(1 + \xi(\mathbf{x}))$ has a small fractional fluctuation $\xi(\mathbf{x})$ around the average S-wave velocity V_0 and the wavelength is much smaller than the correlation distance a of medium inhomogeneity. When we choose vector potential having only y-component, the displacement vector has polarization in the x-z plane. For isotropic radiation of the y-component S-wave potential ϕ from a point source located at the origin, ϕ can be written as a superposition of

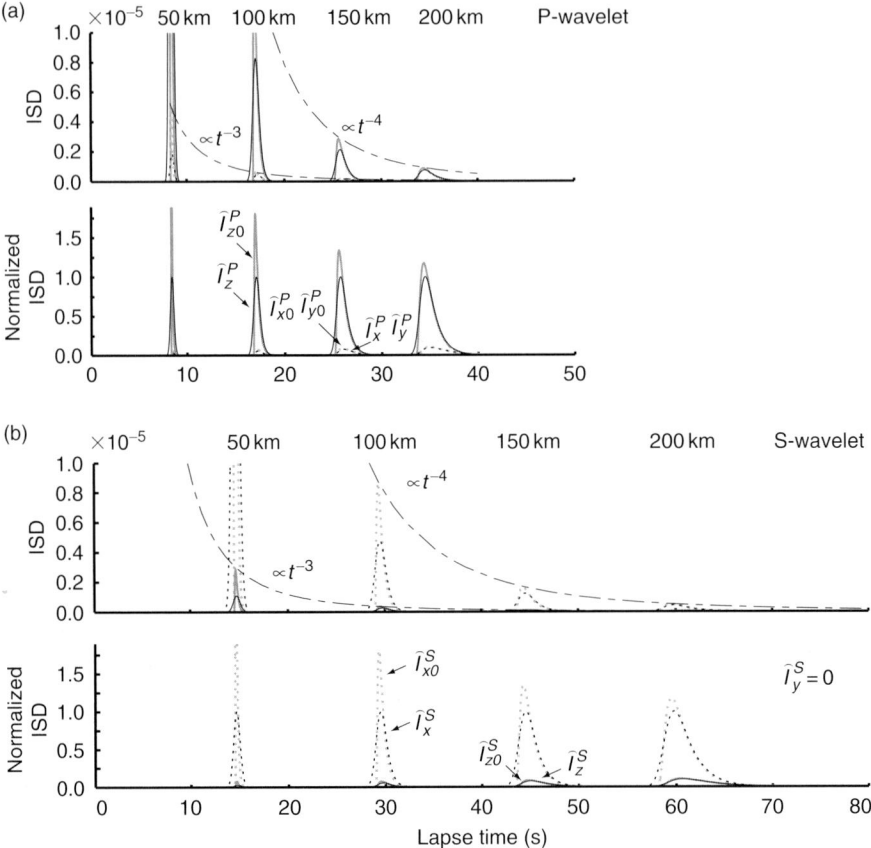

FIG. 17. MS envelopes at four travel distances for wavelet radiation from a point source in 3-D random elastic media, which are characterized by the average P- and S-wave velocities of 6 and 3.46 km/s, respectively, and Gaussian ACF with $\varepsilon = 0.05$ and $a = 5$ km. Both ISDs with and without wandering effect are shown. Chained curves show asymptotic peak decay curves. (a) ISD traces and normalized ISD traces for P-waves. Normalization uses the maximum peak value of the longitudinal component. (b) ISD traces and normalized ISD traces for S-waves with polarization in the x-z plane. Normalization uses the maximum peak value of the x-component (Sato, 2007).

spherical outgoing harmonic waves as Eq. (52). Field U is governed by the parabolic equation (53) around small transverse distance from the z-axis since the wavelength is shorter than the correlation distance $ak_0 \gg 1$.

By using Cartesian coordinates in a small region around the z-axis at a long distance from the source, we calculate ISDs by using the TFMCF of field U. The y-component ISD is always zero,

$$\hat{I}_y^S(r, t, \omega_c) = 0 \tag{68}$$

and the longitudinal (z)-component ISD is

$$I_z^S(r,t,\omega_c) = \frac{1}{2\pi r^2}\int_{-\infty}^{\infty} d\omega_d e^{-i\omega_d(t-r/V_0)}\tilde{w}(r,\omega_d)\left[-\frac{\partial_{x_{\perp d}}^2}{k_c^2}{}_0\Gamma_2(\mathbf{x}_{\perp d},r,\omega_c,\omega_d)\right]_{\mathbf{x}_{\perp d}=0}.$$
(69)

By using the reference ISD I^R defined by Eq. (58), we have the x-component ISD as

$$\begin{aligned}I_x^S(r,t,\omega_c) &= \frac{1}{2\pi r^2}\int_{-\infty}^{\infty} d\omega_d e^{-i\omega_d(t-(r/V_0))}\tilde{w}(r,\omega_d)\\ &\times\left[\left(1+\frac{\Delta_{\perp d}}{k_c^2}\right){}_0\Gamma_2(\mathbf{x}_{\perp d},r,\omega_c,\omega_d)\right]_{\mathbf{x}_{\perp d}=0}\\ &= I^R(r,t,\omega_c) - 2I_z^S(r,t,\omega_c),\end{aligned}$$
(70)

where $\partial_{x_d}^2{}_0\Gamma_2 = \partial_{y_d}^2{}_0\Gamma_2$ is used because of the isotropy of TFMCF in the transverse plane. When $U = -1/\sqrt{4\pi}$ in Eq. (52), then the y-component vector potential in the angular frequency domain $\tilde{\phi} \approx -e^{ik_0 r}/(ik_0\sqrt{4\pi}r)$ is isotropic; however, the corresponding displacement vector components are $\tilde{u}_x \approx (z/r)e^{ik_0 r}/\sqrt{4\pi}r$, $\tilde{u}_y = 0$, and $\tilde{u}_z \approx -(x/r)e^{ik_0 r}/\sqrt{4\pi}r$, which is axially symmetric around the y-axis. That is, the initial condition [Eq. (56)] means the axially symmetric radiation of an S-wavelet with the polarization perpendicular to the y-axis. We may take the polarization of S-wavelet in the x-direction at the origin since we study wave propagation along the z-axis. The corresponding initial ISDs are written by

$$I_y^S = I_z^S = 0 \quad\text{and}\quad I_x^S = I^R = \frac{1}{4\pi r^2}\delta\left(t-\frac{r}{V_0}\right)\quad\text{as } r\to 0 \text{ along the } z\text{-direction}.$$
(71)

Characteristics of S-wave envelopes for Gaussian ACF: For the case of Gaussian ACF, using the explicit representation of the reference ISD without wandering effect [Eq. (63)], we have S-wave ISD of each component without wandering effect as

$$I_{z0}^S(r,t,\omega_c) = \frac{2V_0}{r}\left(t-\frac{r}{V_0}\right)I_0^R(r,t,\omega_c)$$
(72)

and

$$I_{x0}^S(r,t,\omega_c) = \left[1-4\frac{V_0}{r}\left(t-\frac{r}{V_0}\right)\right]I_0^R(r,t,\omega_c).$$
(73)

That is, by replacing the P-wave average velocity with S-wave average velocity in the definition of characteristic time, the longitudinal (z) and x-component S-wave ISDs are given by the transverse and longitudinal-component P-wave ISDs, respectively.

In Fig. 17b, the fist row shows ISD traces at four travel distances for radiation of S-wavelet from a point source: I_{z0}^S (gray solid curve), I_z^S (black solid curve), I_{x0}^S (gray broken curve), and I_x^S (black broken curve), where the random media are characterized by the average S-wave velocity of 3.46 km/s and Gaussian ACF with $\varepsilon = 0.05$ and $a = 5$ km. The second row shows those normalized by the maximum peak value of I_x^S. Envelope broadening becomes apparent with travel distance increasing, where the envelope width is 1.73 times larger than that of P-wavelet at each travel distance. Chained curves approximately show peak decay curves of I_{x0}^S and I_{z0}^S. The second row shows that the ratio of the longitudinal- to the transverse-component amplitude grows with travel distance.

So far it is difficult to synthesize envelopes for nonisotropic radiation from a point shear dislocation source; however, it might be helpful for understanding seismograms of microearthquakes to simulate envelopes in random media for simultaneous isotropic radiation of P- and S-wavelets from a point source. In Fig. 18, we plot ± square root of ISDs at four travel distances, where the S- to P-wave radiation energy ratio W_S/W_P is chosen as that of a point shear dislocation source, 23.4 and the random media are the same as used in Fig. 17. RMS envelopes are normalized by the maximum peak value of S-wave envelope in the x-component at each travel distance. For S-waves, the appearance of scattered waves having long duration is prominent not only in the transverse component but also in the longitudinal component at large travel distances. P coda has long duration in each of three components at large travel distances. These features of theoretical RMS envelopes in three components qualitatively explain the characteristics of observed seismograms at high frequencies.

3.2. Two-Dimensional Random Elastic Media

We examine the validity of the Markov approximation for P-wavelet radiation from a point source by a comparison with the FD simulation in two dimensions (Sato and Korn, 2007).

3.2.1. Wave Envelopes for a Gaussian ACF

For P-waves radiated from a point source at the origin, a cylindrical solution of the homogeneous equation is $H_0(k_0 r)$, which has an asymptotic solution $r^{-1/2} \exp(i k_0 r - \omega t)$ in the far field ($r \gg 1/k_0$). At a long distance from the point source, we may write scalar potential for outgoing P-wavelet in inhomogeneous media as a sum of harmonic cylindrical waves of angular frequency ω in polar coordinate (r, θ) as

$$\phi = \frac{1}{2\pi} \int_{-\infty}^{\infty} d\omega \frac{U(\theta, r, \omega)}{i k_0 \sqrt{r}} e^{i(k_0 r - \omega t)}. \tag{74}$$

Substituting Eq. (74) into Eq. (3), we obtain the parabolic wave equation for U as

$$2 i k_0 \partial_r U + \frac{\partial_\theta^2 U}{r^2} - 2 k_0^2 \xi U = 0. \tag{75}$$

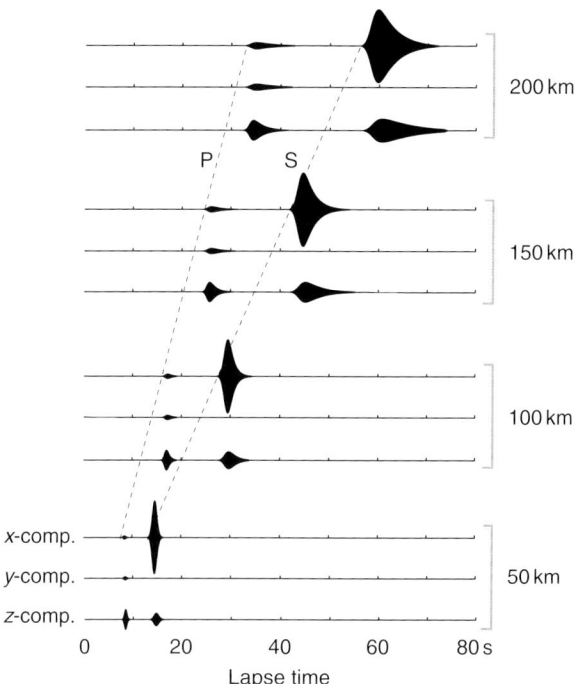

FIG. 18. Schematic illustration of three-component RMS envelopes at four distances along the z-axis for isotropic radiation of wavelet from a point source, where the ratio of S- to P-wave source energy is 23.4, S-waves polarization is in the x-z plane, and the randomness is the same as for Fig. 4. Traces are normalized by the maximum peak of the x-component amplitude at each distance (Sato, 2007).

The TFMCF is defined as the correlation of field U between two locations θ' and θ'' within a small distance on the transverse line at a distance r at different angular frequencies ω' and ω'', $\Gamma_2(\theta_d, r, \omega_c, \omega_d) \equiv \left\langle U\left(\theta', r, \omega'\right) U\left(\theta'', r, \omega''\right)^* \right\rangle$, where difference angle $\theta_d = \theta' - \theta''$ [Eq. (22)]. When random media in two dimensions are statistically characterized by a Gaussian ACF $R(x) = \varepsilon^2 \exp[-(x^2 + z^2)/a^2]$, for quasi-monochromatic waves, we derive the master equation for $_0\Gamma_2$ by using the Markov approximation as

$$\partial_r \; _0\Gamma_2 + i \frac{k_d}{2k_c^2 r^2} \partial^2_{\theta_d} \; _0\Gamma_2 + \frac{\sqrt{\pi}\varepsilon^2 k_c^2 r^2 \theta_d^2}{a} \; _0\Gamma_2 = 0 \qquad (76)$$

for small difference angle $|\theta_d| \ll 1$. Solving Eq. (76) with the initial condition

$$_0\Gamma_2(\theta_c, \theta_d, r = 0, \omega_c, \omega_d) = \frac{1}{2\pi} \qquad (77)$$

according to Fehler et al. (2000), we have the analytical solution (A.20) as

$$_0\Gamma_2(\theta_d, r, \omega_c, \omega_d) = \frac{1}{2\pi}\sqrt{\frac{s_0}{\sin s_0}} e^{-\left[(1/s_0^2) - (\cot s_0/s_0)\right]\left((2V_0 k_c^2 t_M)/r\right) r_{\perp d}^2}, \quad (78)$$

where $r_{\perp d} = r\theta_d$.

Taking the same procedure as for 3-D case, we define the reference ISD without wandering effect in two dimensions as

$$I_0^R(r, t; \omega_c) = \frac{1}{2\pi r}\frac{1}{2\pi}\int_{-\infty}^{\infty} d\omega_d \sqrt{\frac{s_0}{\sin s_0}} e^{-i\omega_d(t - (r/V_0))}. \quad (79)$$

The angular-component ISD without wandering effect is given by

$$I_{\theta\theta}^P(r, t; \omega_c) \equiv \frac{1}{2\pi r}\int_{-\infty}^{\infty} d\omega_d \left[-\frac{1}{k_c^2 r^2}\partial_{\theta_d}^2 {}_0\Gamma_2(\theta_c, \theta_d, r, \omega_c, \omega_d)\right]_{\theta_d=0} e^{-i\omega_d(t-r/V_0)}$$

$$= \frac{1}{2\pi r}\frac{4V_0 t_M}{r}\frac{1}{2\pi}\int_{-\infty}^{\infty} d\omega_d e^{-i\omega_d(t-(r/V_0))}\frac{1-s_0\cot s_0}{s_0^2}\sqrt{\frac{s_0}{\sin s_0}}$$

$$= \frac{4V_0}{r}\frac{1}{2\pi r}\frac{1}{2\pi}\int_{-\infty}^{\infty} d\omega_d e^{-i\omega_d(t-(r/V_0))}\frac{\partial}{i\partial\omega_d}\sqrt{\frac{s_0}{\sin s_0}}$$

$$= \frac{4V_0 t_M}{r}\frac{(t-(r/V_0))}{t_M} I_0^R(r, t; \omega_c). \quad (80)$$

where t_M and s_0 are the same as those in 3-D case. Using the leading term $\partial_r U \approx i\partial_{\theta_0}^2 U/(2k_0 r^2)$ in Eq. (75), we have the radial component ISD without wandering effect as

$$I_{r0}^P(r, t; \omega_c) = \frac{1}{2\pi r}\int_{-\infty}^{\infty} d\omega_d \left[\left(1 + \frac{1}{r^2 k_c^2}\partial_{\theta_d}^2\right){}_0\Gamma_2(\theta_c, \theta_d, r, \omega_c, \omega_d)\right]_{\theta_d=0} e^{-i\omega_d(t-r/V_0)}$$

$$= I_0^R(r, t; \omega_c) - I_{\theta\theta}^P(r, t; \omega_c)$$

$$= \left[1 - \frac{4V_0}{r}\left(t - \frac{r}{V_0}\right)\right] I_0^R(r, t; \omega_c). \quad (81)$$

The initial condition [Eq. (77)] represents an isotropic radiation of P-wavelet given by a δ function for the source power time function:

$$I_0^R(r, t; \omega_c) = I_{r0}^P(r, t; \omega_c) = \frac{1}{2\pi r}\delta\left(t - \frac{r}{V_0}\right) \text{ and } I_{\theta\theta}^P(r, t; \omega_c) = 0 \text{ as } r \to 0. \quad (82)$$

We can numerically evaluate the reference ISD without wandering effect [Eq. (79)] by using an FFT. In Fig. 19 we plot I_0^R by a gray curve against reduced time $t - r/V_0$.

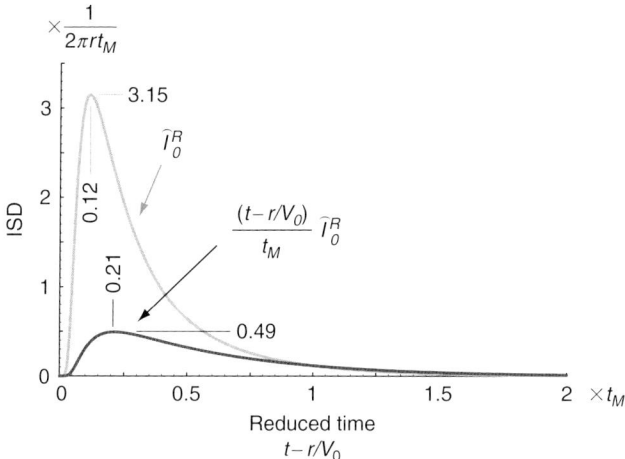

FIG. 19. Temporal change in $\hat{I}_0^R$ and $(t - r/V_0)\hat{I}_0^R/t_M$ against the reduced time $t - r/V_0$ at a distance r from a point source in 2-D random elastic media (Sato and Korn, 2007).

It takes the maximum value about $3.15/(2\pi r t_M)$ at reduced time about $0.12 t_M$. The peak height of $\hat{I}_0^R$ is proportional to the inverse cube of travel distance since the characteristic time is proportional to the square of distance. A black curve shows $(t - (r/V_0))\hat{I}_0^R/t_M$ that has the maximum value about $0.49/(2\pi r t_M)$ at reduced time about $0.21 t_M$, which is nearly twice the peak delay of $\hat{I}_0^R$. It means that I_{00}^P has the maximum value about $0.31 V_0/r^2$. The peak height of I_{0r}^P is nearly proportional to the inverse square of travel distance as I_0^R when the peak height of I_{00}^P is negligible. We note that the time integral of I_{00}^P is independent of travel distance as

$$\int_{r/V_0}^{\infty} dt I_{00}^P = \varepsilon^2 / (3\sqrt{\pi} a) \tag{83}$$

since $\int_{r/V_0}^{\infty} dt I_0^R = 1/(2\pi r)$.

For the case of isotropic radiation of an S-wavelet from a point source in 2-D random media, replacing P with S and substituting the average S-wave velocity into V_0, I_r^P and I_0^P represent the angular and radial component ISDs I_θ^S and I_r^S, respectively.

3.2.2. Comparison of Markov and FD Envelopes

The envelopes directly simulated by using the Markov approximation are compared with FD simulations of P-wave traces. The practical scheme is the same as mentioned in Section 2.2.2. The size of the model is 450 km by 450 km, where absorbing boundary conditions are implemented. In the following simulation, we put $\varepsilon = 0.05$ and $a = 5$ km. Average P- and S-wave velocities and mass density are 6 km/s, 3.46 km/s, and 2800 kg/m^3, respectively. The fractional fluctuation of density is chosen as $0.8\xi(\mathbf{x})$ according to Birch'fs law.

The far-field pulse shape of the outgoing P-wavelet in a homogenous medium radiated isotropically from a source located at the center is given by the convolution $u_r = g_2 h$, where $g_2(r, t)$ is the 2-D Green function, and the source time function $h(t)$ is given by Eq. (51) with $N = 2$ and $T = 0.5$ s for a 2 Hz wavelet. Around the source, a homogeneous space of 1 km width is introduced to ensure pure isotropic P-wavelet radiation, where the source time function is scaled to satisfy $\int_0^T 2\pi r |u_r|^2 dt = 1$ near the source. The wave field is recorded at four circular arrays at $r = 50$, 100, 150, and 200 km. Each circular array consists of 72 receivers at 5° intervals. The spatial discretization in the FD scheme is 0.1 km and the temporal discretization is 6 ms. Simulated wave traces are transferred into radial and angular component traces.

Figure 20 shows examples of FD waveforms after traveling 150 km through one realization of random medium. Strong distortions of pulse shape and travel time fluctuations are clearly seen. P-wave is followed by scattered waves in radial-component traces and scattered waves also appear on angular-component traces. At each travel distance, averaging the square of wave traces over 72 receivers along the circular array in 5 realizations of random media, smoothing with time constant 0.5 s, and taking the square root, we obtain the ensemble-averaged trace. Gray curves in Fig. 21 show RMS traces at four travel distances calculated from FD simulations (FD envelopes). These traces clearly show that the peak delay and the time width of envelope increase with travel distance increasing in both radial and angular components. Wave trains in the angular component are a clear evidence of scattering caused by random inhomogeneity. At each travel distance, the peak amplitude of the angular component is smaller than that of the radial component; however, the former amplitude decreases more slowly than that of the later amplitude. The peak delay of angular component is a little larger than that of radial component at each distance.

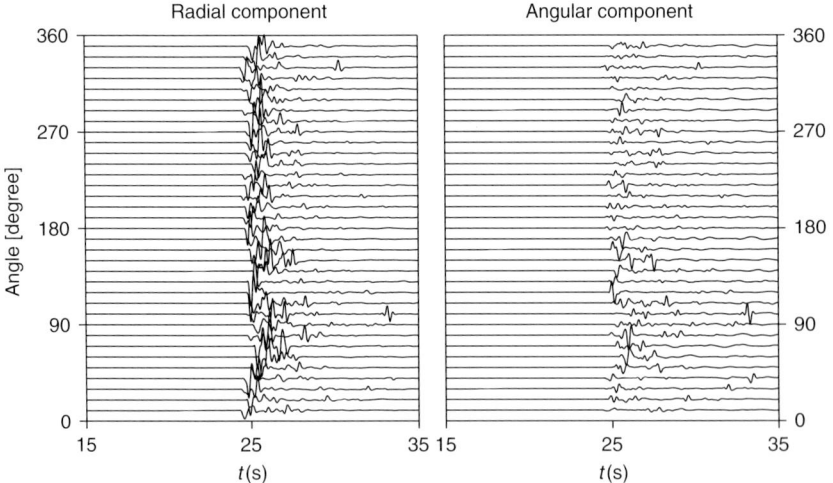

FIG. 20. Examples of FD simulation traces on a circular array at a distance of 150 km in a 2-D random medium for the isotropic radiation of a 2 Hz P-wavelet from the center. Only every second trace is plotted. Random elastic media are characterized by Gaussian ACF with $\varepsilon = 0.05$ and $a = 5$ km, where the average P- and S-wave velocities and mass density are 6 km/s, 3.46 km/s, and 2800 kg/m^3, respectively (Sato and Korn, 2007).

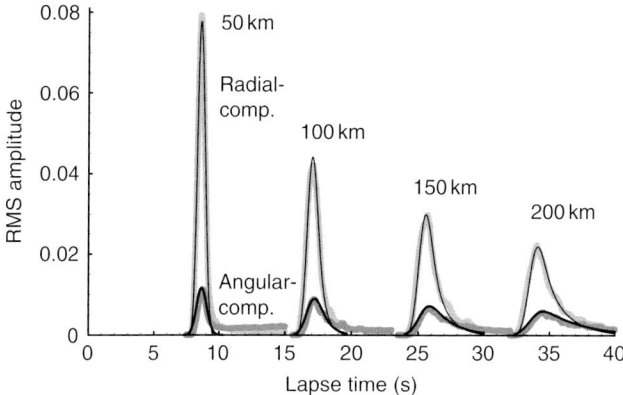

FIG. 21. Comparison of RMS envelopes of FD-simulated waves (light gray curves and dark gray curves for radial and angular components) with theoretical envelopes predicted by the Markov approximation (black curves) in 2-D random elastic media for the isotropic radiation of a 2 Hz P-wavelet from the point source (Sato and Korn, 2007).

Using ISDs with wandering effect calculated by the Markov approximation, we perform the convolution with the source time function's square of the 2 Hz P-wavelet $I^P_{\theta 0}*w*i$ and $I^P_{r0}*w*i$, and taking square root, we obtain RMS envelopes (Markov envelopes), where we practically put $i(t) = 2\pi r |u_r|^2$ of the FD simulation near the source. In Fig. 21, RMS Markov envelopes are plotted by black curves together with FD envelopes in gray curves. We find that the Markov envelope well explains the peak height, the delay of the peak arrival from the onset, and the envelope broadening of FD envelope in each component at each travel distance. We find a small discrepancy between them as the reduced time increases at each travel distance since FD envelopes contain large-angle scattering and conversion scattering that the Markov approximation neglected. Except for the coda portion, FD envelopes are quantitatively well explained by Markov envelopes. We confirmed the constancy of the time integral of the MS amplitude of the angular component independent of travel distances with a relative error less than 4%.

4. Discussions

4.1. RTT with the Born Approximation Scattering Coefficients

We have limited ourselves to the case that wavelength is shorter than correlation distance that leads to little conversion scattering between P- and S-waves and little large-angle scattering since the basic equation is parabolic. The Markov approximation cannot be used for modeling the later parts of coda behind the primary arrival of P- or S-waves, where multiple large-angle scattering dominate. Therefore, it is worthwhile to compare the Markov approximation with other approaches to the envelope simulation problem.

A convenient way to describe the single scattering interaction of a wave field with random medium fluctuations is the Born approximation. One problem with the Born approximation is that it does not conserve energy and that it breaks down for the case of strong forward scattering. This problem was partly overcome by the introduction of the travel time-corrected Born approximation (Sato, 1984) that excludes the near forward scattered energy by the introduction of a minimum scattering (cutoff) angle. Within this concept, energy scattered at small angles around the forward direction is not counted as energy loss of the primary wave front but only contributes to its travel time fluctuations. This concept, however, is somewhat heuristic and the size of the cutoff angle has been a matter of dispute for a long time (Frenje and Juhlin, 2000). There is a theoretical approach by Kawahara (2002), who determined the cutoff angle on the basis of the Kramers–Krönig relation.

A more rigorous approach is the RTT (e.g., Rytov et al.,1987; Ryzhik et al., 1996; Margerin, 2005). The RTT describes energy transport through a scattering medium neglecting phase information, and is, in principle, capable of modeling both short and long lapse time coda. Thus, it is more general than the Markov approximation. It works well if the typical scale length of inhomogeneities and the wavelength are of comparable size and fluctuations are moderate. For the special case of isotropic scattering, there exists an analytical solution to the radiative transfer equations (e.g., Zeng et al., 1991; Paaschens, 1997). For more complex cases, they are usually solved by Monte Carlo methods (e.g., Hoshiba, 1994; Gusev and Abubakirov, 1996). However, with the isotropic scattering assumption, only the coda formation but not the envelope broadening and peak delay can be satisfactorily modeled. Nonisotropic scattering can be included by using Born scattering coefficients at the expense of extra computing time. Comparing the Markov approximation and the RTT with the Born scattering coefficient to FD simulations of the full wave equation in two dimensions, Wegler et al. (2006) find a good correspondence of the RTT to the Markov approximation for the case of multiple forward scattering. Przybilla et al. (2006) presented a Monte Carlo scheme for the solution of the coupled elastic transport equations, in which individual scattering events in a continuously fluctuating random medium are described by the angular-dependent Born scattering coefficients. The simulated envelopes in two dimensions have been compared to envelopes from full wave field FD simulations, much in the same way as was done in this chapter. Their general result is that the Monte Carlo solution yields accurate envelope shapes not only for the late coda but also for the time range of the initial P- and S-wave arrivals, that is forward and small-angle scattering. In the strong forward scattering regime, $ak_0 \gg 1$, broadening by travel time fluctuations of individual ray paths through the random medium is not explained within RTT, but can be taken into account by convolving the RTT envelopes with the wandering effect in the same way as it was done here with the ISDs of the Markov approximation.

Therefore, the Monte Carlo solution of RTT with angular-dependent Born scattering coefficients offers a unified approach to modeling MS envelope shapes in random media. However, it needs considerable computing resources. This is especially true for strong forward scattering, where many scattering interactions have to be computed to advance the energy particles along essentially straight ray paths. For $ak_0 \gg 1$, the large computing times of RTT may even become prohibitive. Therefore, in all cases where we have strong forward scattering and/or we are not interested in the later parts of the coda, the Markov approximation has the advantage of a very fast and efficient envelope simulation method.

4.2. Realistic ACFs for Random Media

The theoretically predicted envelopes are independent of frequency since a Gaussian ACF is used for characterizing random media for mathematical simplicity. In real data, the broadening of S-seismogram envelope depends on frequency in some areas (Obara and Sato, 1995) and the peak ratio of the transverse to longitudinal envelopes of teleseismic P-waves increases with frequency (Kubanza et al., 2007). For representing realistic inhomogeneity of the lithosphere, a power-law spectrum is more appropriate than a Gaussian spectrum (e.g., Goff and Holliger, 2000; Shiomi et al., 1997). As shown in the case of scalar waves (Saito et al., 2002), we will be able to derive vector-wave envelopes having strong frequency dependence if we use a von Kármán-type ACF, which has a power-law spectrum as the asymptote at large wave numbers. It will be useful especially for the study of S-wave envelopes observed in the back-arc side of the volcanic front in Japan, where the envelope broadening is larger in higher frequencies (Obara and Sato, 1995; Takahashi et al., 2007). Even for the case of a von Kármán-type ACF, it is basically possible to simulate vector-wave envelopes by numerically integrating the master equation for TFMCF. When the PSDF is rich in short wavelengths, however, it is necessary to compare with FD simulations since the contribution of large-angle scattering and conversion scattering between P- and S-waves becomes larger.

The mathematical formulation is restricted to specific initial conditions; however, the plane wave formulation is applicable to teleseismic P-wave envelopes and the spherical wavelet formulation is useful for the study of seismograms of explosions. There remains an important subject to formulate envelope synthesis of vector waves for nonisotropic radiation from a point shear dislocation source.

We may put a focus on the anisotropic characteristics of random inhomogeneity. Examining the lithological map of a typical exposure of the lower continental crust, Holliger and Levander (1992) found anisotropy of correlation distance with the aspect ratio of 3–5. Wu et al. (1994) reported a difference in correlation distances in vertical and horizontal from the analysis of the well-log data of KTB holes. Deep seismic soundings of the crust as well as geological observations often reveal the existence of elongated or preferentially oriented scattering structures. Furumura and Kennett (2005) proposed a random-medium oceanic slab as an efficient waveguide, where the correlation distance in thickness is shorter than that along the length of the subducting plate. They numerically showed frequency-selective propagation characteristics with a faster low-frequency phase followed by large high-frequency signals with very long coda. Saito (2006) studied the envelope broadening of scalar waves in 2-D random media characterized by an anisotropic ACF by using the Markov approximation. It predicts that envelopes increase in duration more rapidly in the horizontal propagation than in the vertical propagation when the media are characterized by long horizontal and short vertical correlation distances. Margerin (2006) directly simulated envelopes in anisotropic random media by solving the Bethe-Salpeter equation in the diffusion regime. He showed that simulated coda envelope decays are strongly controlled by eigenvalues of the diffusion tensor calculated from the anisotropic ACF. These simulations suggest that observed seismogram envelopes could be strongly controlled not only by travel distance but also by ray directions relative to the principal axes of anisotropic inhomogeneity. It is very necessary to simulate vector envelopes from the onset to coda in anisotropic random elastic media.

The envelope simulation method here developed assumed the homogeneity of randomness; however, the randomness varies from place to place as shown in the Japan arc in relation with Quaternary volcanoes. It is important to develop an envelope simulation method in the case that the statistical parameters are not spatially uniform. Recently, Saito *et al.* (2008) studied scalar wave propagation through two-layer random media in two dimensions characterized by weak and strong inhomogeneities on the basis of the Markov approximation. They confirmed the coincidence of the synthesized envelopes with envelopes calculated from FD simulations. It will be also necessary for us to develop envelope synthesis in random media with smoothly varying background velocity.

5. Summary

Observed high-frequency seismograms of local earthquakes are complex; however, it is known that their amplitude envelopes vary regionally and systematically, reflecting seismotectonic settings. It is useful to characterize the band-pass filtered seismogram of each vector component by its peak amplitude, peak delay from the onset, and apparent envelope width in relation to the travel distance and the ray path. As mathematical basis of the vector-wave envelope analysis for structure studies, we present a stochastic method considering an ensemble of random elastic media. We have introduced the Markov approximation method for the direct simulation of vector-wave envelopes for the case that the fractional fluctuation velocity is small and the wavelength is shorter than the correlation distance of random media characterized by a Gaussian ACF. Then, P- and S-waves are treated separately by using potentials since there is little conversion scattering between them. The Markov approximation leads to the stochastic master equation for the TFMCF of potential field, from which by use of the Fourier transform we are able to synthesize vector-wave envelope traces in time domain. We have shown the mathematical formulations for a plane wavelet incidence case and a point source radiation case in two and three dimensions. In the case of Gaussian ACF, TFMCF is analytically solved, and vector-wave envelopes in time domain are analytically expressed by using elliptic theta functions especially in three dimensions. The characteristics of resultant vector-wave envelopes are frequency independent. In addition to the applicability condition for the parabolic approximation $ak_0 \gg 1$ and that for the Markov approximation $a^2 k_0^2 \varepsilon^2 \ll 1$, there is an additional positive condition of all vector component intensities that the reduced time is shorter than the quarter of the travel time. The validity of the direct envelope synthesis with the Markov approximation is confirmed by a comparison with vector-wave envelopes calculated from FD simulations in 2-D random media with $a = 5$ km and ε up to 0.05 for a 2 Hz wavelet.

For P-wave, this approximation predicts not only the peak delay and the envelope broadening with a smoothly decaying tail in the longitudinal component but also the excitation of wave amplitude with a longer tail in the transverse component. Main mechanism is successive ray bending processes caused by random velocity fluctuations. The maximum peak of the transverse component is smaller than that of the longitudinal component; however, the peak delay time from the onset of the transverse component is longer than that of the longitudinal component. The decay of the maximum peak of the transverse component with travel distance increasing is weaker than that of the longitudinal component, that is the relative partition of energy into the transverse component becomes larger as travel distance increasing. The ratio of the MS fractional velocity fluctuation to the correlation distance is the key parameter characterizing the envelope broadening and the partition of energy into the transverse component. For plane wave case, the time integral of the MS amplitude of the transverse component linearly increases with travel distance increasing, where the linear coefficient gives this ratio. For spherical wave case, the time integral of the MS amplitude of the transverse component is inversely proportional to the travel distance, where the coefficient gives this ratio.

Characteristics of polarized S-wave envelopes are also studied. The maximum peak of the longitudinal component is smaller than that of the transverse component in the original polarization; however, the peak delay time from the onset of the longitudinal component is longer than that of the transverse component, where there is no excitation of amplitude in the direction normal to the global ray direction and the original polarization even though ray bending is possible in any direction. For the same randomness, the envelope broadening of S-wavelet is larger than that of P-wavelet by a factor of the ratio of their average velocities.

Developing the Markov approximation for random media having more realistic spectrum and background velocities, the stochastic direct syntheses of vector envelopes will serve for the mathematical interpretation of high-frequency seismograms in terms of lithospheric inhomogeneity.

Acknowledgments

The authors are grateful to Mike Fehler and two anonymous reviewers for their helpful suggestions and comments.

Appendix: Analytic Solutions of the Stochastic Master Equations for TFMCF

When random media are characterized by the average velocity V_0 and the Gaussian ACF with RMS fractional fluctuation ε and correlation distance a, the stochastic master equation for TFMCF based on the Markov approximation can be analytically solved. In the following text, we briefly show how to solve differential equations in different cases.

Plane Wave in Three Dimensions

For wave propagation through random media spreading over a 3-D half space ($z > 0$), we solve the following differential equation under the initial condition $_0\Gamma_2(r_{\perp d}, z = 0) = 1$:

$$\partial_z {}_0\Gamma_2 + i\frac{k_d}{2k_c^2}\Delta_{\perp d}\, {}_0\Gamma_2 + \frac{k_c^2\sqrt{\pi}\varepsilon^2 r_{\perp d}^2}{a}\, {}_0\Gamma_2 = 0, \tag{A.1}$$

where $\Delta_{\perp d}$ is a transverse Laplacian and $r_{\perp d} \equiv \sqrt{x_d^2 + y_d^2}$ is a transverse distance. Normalizing the z-coordinate by the receiver distance Z as $\tau = z/Z$, and scaling the transverse distance by using the coherent radius $a_\perp$ as $\chi \equiv r_{\perp d}/a_\perp = \sqrt{\sqrt{\pi}\varepsilon^2 k_c^2 Z/a}\, r_{\perp d} = \sqrt{(2V_0 k_c^2 t_M/Z)}\, r_{\perp d}$, we have

$$\partial_\tau {}_0\Gamma_2 + it_M\omega_d\left(\partial_\chi^2 + \chi^{-1}\partial_\chi\right){}_0\Gamma_2 + \chi^2 {}_0\Gamma_2 = 0, \tag{A.2}$$

where the characteristic time $t_M = \sqrt{\pi}\varepsilon^2 Z^2/(2V_0 a)$ and the initial condition is $_0\Gamma_2(\chi, \tau = 0) = 1$. Assuming the solution having the form $_0\Gamma_2(\chi, \tau) = e^{v(\tau)\chi^2}/w(\tau)$, we write Eq. (A.1) as

$$\left(\frac{dv}{d\tau} + s_0^2 v^2 + 1\right)\chi^2 + \left(s_0^2 v - \frac{1}{w}\frac{dw}{d\tau}\right) = 0, \tag{A.3}$$

where parameter $s_0 = 2e^{\pi i/4}\sqrt{t_M \omega_d}$. Each term in parentheses must be zero regardless of χ. Using the initial condition $v(\tau = 0) = 0$, we have $v(\tau) = -\tan s_0\tau/s_0$ as the solution of $v' + s_0^2 v^2 + 1 = 0$. For the initial condition $w(\tau = 0) = 1$, we have $w(\tau) = \cos s_0\tau$ as the solution of $s_0^2 v - w'/w = 0$. Thus, we obtain

$$_0\Gamma_2(\chi, \tau = 1) = \frac{e^{-(\tan s_0/s_0)\chi^2}}{\cos s_0}. \tag{A.4}$$

By using the original coordinates, we have TFMCF at distance Z as

$$_0\Gamma_2(r_{\perp d}, Z) = \frac{1}{\cos s_0} e^{-(\tan s_0/s_0)\left((2V_0 k_c^2 t_M)/z\right)(x_d^2 + y_d^2)}. \tag{A.5}$$

It was originally solved by Sreenivasiah *et al.* (1976).

Plane Wave in Two Dimensions

For wave propagation through random media spreading over a 2-D half space ($z > 0$), we solve the following differential equation under the initial condition $_0\Gamma_2(x_{\perp d}, z = 0) = 1$:

$$\partial_z\,_0\Gamma_2 + i\frac{k_d}{2k_c^2}\partial_{x_d}^2\,_0\Gamma_2 + k_c^2\sqrt{\pi}\varepsilon^2 a\left(\frac{x_d}{a}\right)^2\,_0\Gamma_2 = 0. \tag{A.6}$$

Taking the same scaling as was done for the 3-D case, $\tau = z/Z$ and $\chi = \sqrt{2V_0 k_c^2 t_M/Z}x_d$, we have the master equation in nondimensional form as

$$\partial_\tau\,_0\Gamma_2 + it_M\omega_d\partial_\chi^2\,_0\Gamma_2 + \chi^2\,_0\Gamma_2 = 0. \tag{A.7}$$

Assuming the solution having the form $_0\Gamma_2(\chi, \tau) = e^{v(\tau)\chi^2}/\omega(\tau)$, we write Eq. (A.6) as

$$\left(\frac{dv}{d\tau} + s_0^2 v^2 + 1\right)\chi^2 + \left(\frac{s_0^2}{2}v - \frac{1}{w}\frac{dw}{d\tau}\right) = 0. \tag{A.8}$$

In order to satisfy Eq. (A.8) for any χ, each quantity in parentheses should be zero. The solution of $v' + s_0^2 v^2 + 1 = 0$ satisfying the initial condition $v(\tau = 0) = 0$ is $v(\tau) = -\tan(s_0\tau)/s_0$. The solution of $(s_0^2/2)v - w'/w = 0$ satisfying the initial condition $w(\tau = 0) = 1$ is $w(\tau) = \sqrt{\cos(s_0\tau)}$. Then, we have

$$_0\Gamma_2(\chi, \tau = 1) = \frac{e^{-(\tan s_0/s_0)\chi^2}}{\sqrt{\cos s_0}}. \tag{A.9}$$

By using the original coordinates, we have TFMCF at distance Z as

$$_0\Gamma_2(x_{\perp d}, Z) = \frac{1}{\sqrt{\cos s_0}} e^{-(\tan s_0/s_0)\left((2V_0 k_c^2 t_M)/z\right) x_d^2}. \tag{A.10}$$

It was solved by Korn and Sato (2005).

Spherical Wave in Three Dimensions

For spherical wave isotropically radiated from a point source at the origin in 3-D random media, we solve the following differential equation:

$$\partial_r \, _0\Gamma_2 + i\frac{k_d}{2k_c^2}\frac{1}{r^2}\left(\frac{\partial^2}{\partial \theta_d^2} + \frac{1}{\theta_d}\frac{\partial}{\partial \theta_d}\right) _0\Gamma_2 + \frac{\sqrt{\pi}\varepsilon^2 k_c^2 r^2}{a}\theta_d^2 \, _0\Gamma_2 = 0, \tag{A.11}$$

under the initial condition $_0\Gamma_2(\theta_d, r = 0) = 1/(4\pi)$. Normalizing the radial distance r by the receiver distance r_0 as $\tau = r/r_0$ and scale the transverse distance as $\chi = \sqrt{\sqrt{\pi}\varepsilon^2 k_c^2 r_0/a r_0 \theta_d}$, we have

$$\frac{\partial}{\partial \tau} \, _0\Gamma_2 + it_M \omega_d \frac{1}{\tau^2}\left(\frac{\partial^2}{\partial \chi^2} + \frac{1}{\chi}\frac{\partial}{\partial \chi}\right) _0\Gamma_2 + \tau^2 \chi^2 \, _0\Gamma_2 = 0, \tag{A.12}$$

where the characteristic time $t_M = \sqrt{\pi}\varepsilon^2 r_0^2/(2V_0 a)$ and the initial condition is $_0\Gamma_2(\chi, \tau = 0) = 1/4\pi$. Assuming the solution having the form $_0\Gamma_2(\chi, \tau) = e^{v(\tau)\chi^2}/w(\tau)$, we write the master equation as

$$\left(\frac{2v}{\tau} + v' + s_0^2 v^2 + 1\right)\tau^2\chi^2 + \left(s_0^2 v - \frac{w'}{w}\right) = 0. \tag{A.13}$$

Each term in brackets must be zero regardless of χ. Solving $2v/\tau + v' + s_0^2 v^2 + 1 = 0$ under the initial condition $v(0) = 0$, we have the solution $v = (\cot s_0 \tau)/s_0 - 1/(s_0^2 \tau)$. The solution of $s_0^2 v - w'/w = 0$ is $w(\tau) = 4\pi(\sin s_0 \tau)/s_0 \tau$ for the initial condition $w(0) = 1$. Thus, we obtain

$$_0\Gamma_2(\chi, \tau = 1) = \frac{1}{4\pi}\frac{s_0}{\sin s_0} e^{-\left[(1/s_0^2) - (\cot s_0/s_0)\right]\chi^2}. \tag{A.14}$$

By using the original coordinates, we have TFMCF at distance r_0 as

$$_0\Gamma_2(r_{\perp d}, r_0) = \frac{1}{4\pi}\frac{s_0}{\sin s_0} e^{-\left[(1/s_0^2) - (\cot s_0/s_0)\right]((2V_0 k_c^2 t_M)/r_0) r_{\perp d}^2}, \tag{A.15}$$

where $r_{\perp d} = r_0 \theta_d$. It was first solved by Shishov (1974).

Cylindrical Wave in Two Dimensions

For cylindrical wave isotropically radiated from a point source at the origin in 2-D random media, we solve the following differential equation:

$$\partial_r {}_0\Gamma_2 + i\frac{k_\mathrm{d}}{2k_\mathrm{c}^2 r^2}\partial_{\theta_\mathrm{d}}^2 {}_0\Gamma_2 + \frac{\sqrt{\pi}\varepsilon^2 k_\mathrm{c}^2 r^2 \theta_\mathrm{d}^2}{a} {}_0\Gamma_2 = 0, \tag{A.16}$$

under the initial condition ${}_0\Gamma_2(\theta_\mathrm{d}, r = 0) = 1/(2\pi)$. Normalizing the radial distance r by the receiver distance r_0 as $\tau = r/r_0$ and scale the transverse distance as $\chi = \sqrt{\sqrt{\pi}\varepsilon^2 k_\mathrm{c}^2 r_0/a r_0 \theta_\mathrm{d}}$, we have

$$\frac{\partial}{\partial \tau}{}_0\Gamma_2 + it_\mathrm{M}\omega_\mathrm{d}\frac{1}{\tau^2}\frac{\partial^2}{\partial \chi^2}{}_0\Gamma_2 + \tau^2\chi^2 {}_0\Gamma_2 = 0, \tag{A.17}$$

where the initial condition is ${}_0\Gamma_2(\chi, \tau = 0) = 1/(2\pi)$. Assuming the solution having the form ${}_0\Gamma_2(\chi, \tau) = e^{v(\tau)\chi^2}/w(\tau)$, we write the master equation as

$$\left(\frac{2v}{\tau} + v' + s_0^2 v^2 + 1\right)\tau^2\chi^2 + \left(\frac{s_0^2}{2}v - \frac{w'}{w}\right) = 0. \tag{A.18}$$

Each term in brackets must be zero regardless of χ. The solution of $2v/\tau + v' + s_0^2 v^2 + 1 = 0$ for the initial condition $v(0) = 0$, we have $v = (\cot s_0\tau)/s_0 - 1/(s_0^2 \tau)$. The solution of $(s_0^2/2)v - w'/w = 0$ is $w(\tau) = 2\pi\sqrt{\sin s_0\tau/s_0\tau}$. for the initial condition $w(0) = 1$. Thus, we obtain

$${}_0\Gamma_2(\chi, \tau) = \frac{1}{2\pi}\sqrt{\frac{s_0}{\sin s_0}}e^{-[(1/s_0^2)-(\cot s_0/s_0)]\chi^2} \tag{A.19}$$

By using the original coordinates, we have TFMCF at distance r_0 as

$${}_0\Gamma_2(r_{\perp\mathrm{d}}, r_0) = \frac{1}{2\pi}\sqrt{\frac{s_0}{\sin s_0}}e^{-[(1/s_0^2)-(\cot s_0/s_0)]((2V_0 k_\mathrm{c}^2 t_\mathrm{M})/r_0)r_{\perp\mathrm{d}}^2}, \tag{A.20}$$

where $r_{\perp\mathrm{d}} = r\theta_\mathrm{d}$. It was solved by Fehler *et al.* (2000).

References

Aki, K. (1969). Analysis of seismic coda of local earthquakes as scattered waves. *J. Geophys. Res.* **74**, 615–631.

Aki, K. (1973). Scattering of P waves under the Montana LASA. *J. Geophys. Res.* **78**, 1334–1346.

Fehler, M., Sato, H., Huang, L.-J. (2000). Envelope broadening of outgoing waves in 2-D random media: A comparison between the Markov approximation and numerical simulations. *Bull. Seismol. Soc. Am.* **90**, 914–928.

Flatté, S.M., Wu, R.S. (1988). Small-scale structure in the lithosphere and asthenosphere deduced from arrival time and amplitude fluctuations at NORSAR. *J. Geophys. Res.* **93**, 6601–6614.

Frenje, L., Juhlin, C. (2000). Scattering attenuation: 2-D and 3-D finite-difference simulations versus theory. *J. Appl. Geophys.* **44**, 33–46.

Furumura, T., Kennett, B.L.N. (2005). Subduction zone guided waves and the heterogeneity structure of the subducted plate: Intensity anomalies in northern Japan. *J. Geophys. Res.* **110**, B10302, doi: 10.1029/2004JB003486.

Goff, J.A., Holliger, K. (2000). Nature and origin of upper crustal seismic velocity fluctuations and associated scaling properties: Combined stochastic analyses of KTB velocity and lithology logs. *J. Geophys. Res.* **104**, 13169–13182.

Gusev, A.A., Abubakirov, I.R. (1996). Simulated envelopes of non-isotropically scattered body waves as compared to observed ones: Another manifestation of fractal heterogeneity. *Geophys. J. Int.* **127**, 49–60.

Hock, S., Korn, M., Ritter, J., Rothert, E. (2004). Mapping random lithospheric heterogeneities in Northern and Central Europe. *Geophys. J. Int.* **157**, 251–264.

Holliger, K., Levander, A. (1992). A stochastic view of lower crustal fabric based on evidence from the Ivrea zone. *Geophys. Res. Lett.* **19**, 1153–1156.

Hoshiba, M. (1994). Simulation of coda wave envelope in depth dependent scattering and absorption structure. *Geophys. Res. Lett.* **21**, 2853–2856.

Ishimaru, A. (1978). Wave Propagation and Scattering in Random Media. vols. 1 and 2, Academic Press, San Diego, CA.

Jensen, F., Kuperman, W., Porter, M., Schmidt, H. (1994). Computational Ocean Acoustics. American Institute of Physics Press, New York.

Kawahara, J. (2002). Cutoff scattering angles for random acoustic media. *J. Geophys. Res.* **107**(B1), 2012, doi: 10.1029/2001JB000429.

Korn, M. (1990). A modified energy flux model for lithospheric scattering of teleseismic body waves. *Geophys. J. Int.* **102**, 165–175.

Korn, M. (1993). Determination of site-dependent scattering Q from P-wave coda analysis with an energy-flux model. *Geophys. J. Int.* **113**, 54–72.

Korn, M., Sato, H. (2005). Synthesis of plane vector-wave envelopes in 2-D random elastic media based on the Markov approximation and comparison with finite difference simulations. *Geophys. J. Int.* **161**, 839–848.

Kubanza, M., Nishimura, T., Sato, H. (2006). Spatial variation of lithospheric heterogeneity on the globe as revealed from transverse amplitudes of short-period teleseismic P-waves. *Earth, Planets and Space* **58**, e45–e48.

Kubanza, M., Nishimura, T., Sato, H. (2007). Evaluation of strength of heterogeneity in the lithosphere from peak amplitude analyses of teleseismic short-period vector P-waves. *Geophys. J. Int.* **171**, 390–398.

Lambert, H.C., Rickett, B.J. (1999). On the theory of pulse propagation and two-frequency field statistics in irregular interstellar plasmas. *Astrophys. J.* **517**, 299–317.

Lee, L.C., Jokipii, J.R. (1975a). Strong scintillations in astrophysics. I. The Markov approximation, its validity and application to angular broadening. *Astrophys. J.* **196**, 695–707.

Lee, L.C., Jokipii, J.R. (1975b). Strong scintillations in astrophysics. II. A theory of temporal broadening of pulses. *Astrophys. J.* **201**, 532–543.

Levander, A.R. (1988). Fourth-order finite-difference P-SV seismograms. *Geophysics* **53**, 1425–1436.

Margerin, L. (2005). Introduction to radiative transfer of seismic waves. *In* Seismic Earth: Array Analysis of Broad-band Seismograms, (A. Levander and G. Nolet, eds.), Geophysical Monograph Series, Vol. 157, Ch 14, AGU, Washington, pp. 229–252.

Margerin, L. (2006). Attenuation, transport and diffusion of scalar waves in textured random media. *Tectonophysics* **416**, 229–244.

Margerin, L., Nolet, G. (2003). Multiple scattering of high-frequency seismic waves in the deep Earth: PKP precursor analysis and inversion for mantle granularity. *J. Geophys. Res.* **108**(B11), 2514, doi: 10.1029/2003JB002455.

McLaughlin, K.L., Anderson, L.M. (1987). Stochastic dispersion of short-period P-waves due to scattering and multipathing. *Geophys. J. R. Astron. Soc.* **89**, 933–963.
Nishimura, T., Yoshimoto, K., Ohtaki, T., Kanjo, K., Purwana, I. (2002). Spatial distribution of lateral heterogeneity in the upper mantle around the western Pacific region as inferred from analysis of transverse components of teleseismic P-coda. *Geophys. Res. Lett.* **29**(23), 2137, doi: 10.1029/2002GL015606.
Obara, K., Sato, H. (1995). Regional differences of random inhomogeneities around the volcanic front in the Kanto-Tokai area, Japan, revealed form the broadening of S wave seismogram envelopes. *J. Geophys. Res.* **100**, 2103–2121.
Paaschens, J.C.J. (1997). Solution of the time-dependent Boltzmann equation. *Phys. Rev. E.* **56**, 1135–1141.
Petukhin, A.G., Gusev, A.A. (2003). The duration-distance relationship and average envelope shapes of small Kamchatka earthquakes. *Pure Appl. Geophys.* **160**, 1717–1743.
Powell, C.A., Meltzer, A.S. (1984). Scattering of P-waves beneath SCARLET in southern California. *Geophys. Res. Lett.* **11**, 481–484.
Przybilla, J., Korn, M., Wegler, U. (2006). Radiative transfer of elastic waves versus finite difference simulations in two-dimensional random media. *J. Geophys. Res.* **111**, B04305, doi: 10.1029/2005JB003952.
Reynolds, A.C. (1978). Boundary conditions for the numerical solution of wave propagation problems. *Geophysics* **43**, 1099–1110.
Ritter, J.R.R., Shapiro, S.A., Schechinger, B. (1998). Scattering parameters of the lithosphere below the Massif Central, France, from teleseismic wavefield records. *Geophys. J. Int.* **134**, 187–198.
Rytov, S.M., Kravtsov, Y.A., Tatarskii, V.I. (1987). *Principles of Statistical Radio Physics (Vol. 4) Wave Propagation Through Random Media*. Springer-Verlag, Berlin.
Ryzhik, L.V., Papanicolaou, G.C., Keller, J.B. (1996). Transport equations for elastic and other waves in random media. *Wave Motion* **24**, 327–370.
Saito, T. (2006). Synthesis of scalar-wave envelopes in two-dimensional weakly anisotropic random media by using the Markov approximation. *Geophys. J. Int.* **165**, 501–515, doi: 10.1111/j.1365-246X.2006.02896.x.
Saito, T., Sato, H., Ohtake, M. (2002). Envelope broadening of spherically outgoing waves in three-dimensional random media having power-law spectra. *J. Geophys. Res.* **107**, doi: 10.1029/2001JB000264.
Saito, T., Sato, H., Fehler, M., Ohtake, M. (2003). Simulating the envelope of scalar waves in 2D random media having power-law spectra of velocity fluctuation. *Bull. Seismol. Soc. Am.* **93**, 240–252.
Saito, T., Sato, H., Takahashi, T. (2008). Direct simulation methods for scalar-wave envelopes in two-dimensional layered random media based on the small-angle scattering approximation. *Commun. Comput. Phys.* **3**, 63–84.
Sato, H. (1984). Attenuation and envelope formation of three-component seismograms of small local earthquakes in randomly inhomogeneous lithosphere. *J. Geophys. Res.* **89**, 1221–1241.
Sato, H. (1989). Broadening of seismogram envelopes in the randomly inhomogeneous lithosphere based on the parabolic approximation: Southeastern Honshu, Japan. *J. Geophys. Res.* **94**, 17735–17747.
Sato, H. (2006). Synthesis of vector-wave envelopes in 3-D random elastic media characterized by a Gaussian autocorrelation function based on the Markov approximation I: Plane wave case. *J. Geophys. Res.* **111**, B06306, doi: 10.1029/2005JB004036.
Sato, H. (2007). Synthesis of vector-wave envelopes in 3-D random elastic media characterized by a Gaussian autocorrelation function based on the Markov approximation: Spherical wave case. *J. Geophys. Res.* **112**, B01301, doi: 10.1029/2006JB004437.
Sato, H., Fehler, M. (1998). *In* Seismic Wave Propagation and Scattering in the Heterogeneous Earth pp. 1–308. AIP Press/Springer Verlag, New York.

Sato, H., Fehler, M.C. (2007). Synthesis of seismogram envelopes in heterogeneous media. *In* Advances in Geophysics, Vol 48: Wave Propagation in Heterogeneous Earth (Series ed. Renata Dmowska),, (R.-S. Wu and V. Maupin, eds.), pp. 561–597. Elsevier, New York.

Sato, H., Korn, M. (2007). Synthesis of cylindrical vector-wave envelopes in 2-D random elastic media based on the Markov approximation. *Earth, Planets Space* **59**, 209–219.

Sato, H., Fehler, M., Saito, T. (2004). Hybrid synthesis of scalar wave envelopes in 2-D random media having rich short wavelength spectra. *J. Geophys. Res.* **109**, B06303, doi: 10.1029/2003JB002673.

Scherbaum, F., Sato, H. (1991). Inversion of full seismogram envelopes based on the parabolic approximation: Estimation of randomness and attenuation in southeast Honshu, Japan. *J. Geophys. Res.* **96**, 2223–2232.

Shearer, P.M., Earle, P.S. (2004). The global short-period wavefield modeled with a Monte Carlo seismic phonon method. *Geophys. J. Int.* **158**, 1103–1117.

Shiomi, K., Sato, H., Ohtake, M. (1997). Broad-band power-law spectra of well-log data in Japan. *Geophys. J. Int.* **130**, 57–64.

Shishov, V.L. (1974). Effect of refraction on scintillation characteristics and average pulsars. *Sov. Astron.* **17**, 598–602.

Sreenivasiah, I., Ishimaru, A., Hong, S.T. (1976). Two-frequency mutual coherence function and pulse propagation in a random medium: An analytic solution to the plane wave case. *Radio Sci.* **11**, 775–778.

Takahashi, T., Sato, H., Nishimura, T., Obara, K. (2007). Strong inhomogeneity beneath Quaternary volcanoes revealed from the peak delay analysis of S-wave seismograms of microearthquakes in northeastern, Japan. *Geophys. J. Int.* **168**, 90–99, doi: 10.1111/j.1365-246X.2006.03197.x.

Wegler, U., Korn, M., Przybilla, J. (2006). Modeling full seismogram envelopes using radiative transfer theory with Born scattering coefficients. *Pure Appl. Geophys.* **163**, 503–531, doi: 10.1007/s00024-005-0027-5.

Weisstein, E.W. (2005). MathWorld-A Wolfram Web Resource. http://mathworld.wolfram.com/JacobiThetaFunctions.html.

Williamson, I.P. (1972). Pulse broadening due to multiple scattering in the interstellar medium. *Mon. Not. R. Astron. Soc.* **157**, 55–71.

Wu, R.S., Xu, Z., Li, X.P. (1994). Heterogeneity spectrum and scale-anisotropy in the upper crust revealed by the German continental deep-drilling (KTB) holes. *Geophys. Res. Lett.* **21**, 911–914.

Zeng, Y., Su, F., Aki, K. (1991). Scattering wave energy propagation in a random isotropic scattering medium I. Theory. *J. Geophys. Res.* **96**, 607–619.

GEOMETRICAL OPTICS OF ACOUSTIC MEDIA WITH ANISOMETRIC RANDOM HETEROGENEITIES: TRAVEL-TIME STATISTICS OF REFLECTED AND REFRACTED WAVES

Ayse Kaslilar,[1] Yury A. Kravtsov and Serge A. Shapiro

Abstract

When deterministic methods become insufficient to resolve the complexity of a medium, statistical investigation becomes necessary. This helps to characterize the medium by its statistical parameters such as mean value, standard deviation, correlation function, and correlation distance. In this chapter, we develop a formalism based on geometrical optics (GO), which allows us to estimate the statistical parameters (the standard deviation and the inhomogeneity scale lengths in vertical and horizontal directions) from travel-time fluctuations of reflected and refracted seismic waves. We consider a three-dimensional random elastic medium with quasi-homogeneous statistics and anisometric (statistically anisotropic) inhomogeneities. We derive the covariance and the variance functions which are fundamental to estimate the statistical parameters. For the reflection geometry, we reconfirm the double passage effect (DPE) of the travel-time variance quadruplicating at zero offsets. For the refraction geometry, we observe a closely related but a new phenomenon—the reduction of travel-time variance at large offsets, which has not yet been described before. We propose a procedure for estimating the statistical parameters of the medium from travel-time fluctuations of refracted waves. The procedure is illustrated by the numerical simulations of the random refractive medium.

Key Words: Random media, geometrical optics, travel-time fluctuations, double passage effect, nonisometric random media, reflected and refracted waves. © 2008 Elsevier Inc.

1. Introduction

The geometrical optics (GO) method is the most popular approach used in global and exploration seismology to determine the velocity field from both active and passive seismic data. For estimating the velocity field of the large-scale geological structures, the travel-time tomography and velocity analysis methods are sufficient. When information on small-scale structures is necessary, as in the case of reservoir characterization, the statistical parameters of the medium such as the mean value, the standard deviation, the covariance function, and the correlation lengths in different directions might be helpful to obtain more information on the velocity structure. This information can be provided by considering the wave propagation in random media.

---
[1] Author thanks e-mail: kaslilar@itu.edu.tr; kravtsov@mail.am.szczecin.pl; kravtsov@iki.rssi.ru; shapiro@geophysik.fu-berlin.de.

In 1960s, the main applications of wave propagation in random media were related with the atmospheric wave propagation (optics, radio, and radar) and underwater sound propagation. The related theory was developed by Chernov (1960), Tatarskii (1961, 1967, 1971), Ishimaru (1978, 1997), and Rytov *et al.* (1989a,b) and was widely used in these fields. Random media concept in seismology appeared first around 1970s by Aki (1969) and by Aki and Chouet (1975) with their work on coda waves from local earthquakes. Following these pioneering works, wave propagation in random media became an important topic in seismology and was examined by many scientists. The developments over a few last decades in the field of seismic wave propagation and scattering in randomly inhomogeneous earth structure were given in Sato and Fehler (1998). Scattering, transmission, and reflection of elastic waves in randomly layered media have been investigated by Shapiro and Hubral (1998). Also, detailed information on ray perturbation theory for inhomogeneous media can be found in publications by Snieder and Sambridge (1992) and by Kravtsov (2005).

In a randomly inhomogeneous media, the fluctuations of the medium parameters affect the travel-time and amplitudes of the seismic waves. In such a medium, the perturbation theory can be used to estimate the statistical parameters of the medium from the travel-time information. The relation between the travel-time fluctuations and medium fluctuations has been investigated by Müller *et al.* (1992), Witte *et al.* (1996), and statistical inverse problems for estimating the statistical parameters of the medium has been studied by Roth (1997), Touati *et al.* (1999), Iooss *et al.* (2000), Kravtsov *et al.* (2003), and Kravtsov *et al.* (2005). Non-GO (based on Rytov Approximation) consideration of primary (ballistic) wave travel-time fluctuation and attenuation has been examined by Shapiro and Kneib (1993), Shapiro *et al.* (1996), and Müller *et al.* (2002).

In this chapter, the GO method is used to estimate the statistical properties of the medium by considering both reflection and refraction geometries. The ray trajectories for both geometries are illustrated in Fig. 1a and b. In the case of reflecting geometry (Fig. 1a), the seismic signal emitted by the source reflects from an interface *I*, and then returns to the Earth's surface. For refraction geometry, the seismic wave does not experience reflection and arrive to the surface due to regular refraction in the rocks (Fig. 1b). In both cases, the problem is to estimate the statistical parameters of the rocks

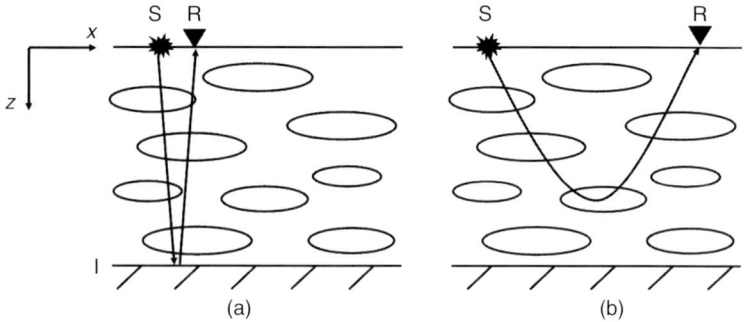

FIG. 1. The geometry of the reflected (a) and refracted (b) rays. S is the wave source, R is the receiver, and I is the reflecting interface.

from the observed time series. By refracted waves, we indicate the turning (or diving) type of refracted waves and exclude the head waves.

The chapter is organized as follows. General information on GO method as applied to seismic wave propagation in random media is given in Section 2. In Sections 3 and 4, the travel-time fluctuations are analyzed and the fundamental equations for estimating the medium statistical parameters are derived for reflected and refracted waves, respectively. In Section 5, a numerical simulation is performed in a random refractive medium and the estimation procedures of the medium statistical parameters are explained. Finally the conclusions are given in Section 6.

It should be noted that this chapter essentially uses materials from our recent publications Kravtsov *et al.* (2003) and Kravtsov *et al.* (2005).

2. Basic Elements of the GO Method

2.1. Basic Equations of the GO

The GO is an efficient method widely used in exploration seismology to solve the wave propagation problem both in homogeneous and inhomogeneous media. GO deals with the rays, which are perpendicular to the wave fronts. In this chapter, wave propagation in random media is considered and the main subject of interest is the statistical properties of the ray trajectories, their travel-times, and amplitudes. In what follows, we shall discuss mainly the statistics of travel-time fluctuations.

Let us consider the wave propagation in a time-independent, acoustic, isotropic, and nondispersive medium. The medium is characterized by the refractive index $n(\mathbf{r})$, which is the ratio of the reference velocity v_0 to the wave velocity $v(\mathbf{r})$ at a given location, $n(\mathbf{r}) = v_0/v(\mathbf{r})$. The zeroth-order approximation of the GO represents the wave-field $u(\mathbf{r}, t)$ in the form (Kravtsov and Orlov 1990; Born and Wolf 1999; Červený 2001):

$$u(\mathbf{r}, t) = A(\mathbf{r}) f\left[t - \frac{\psi(\mathbf{r})}{v_0}\right]. \tag{1}$$

Here f is an initial wave-form of the emitted seismic pulse, $A(\mathbf{r})$ is the amplitude of the leading term of the Debye expansion, and $\psi(\mathbf{r})$ is the "optical path" or "eikonal," which satisfies the eikonal equation

$$|\nabla \psi(\mathbf{r})|^2 = n^2(\mathbf{r}). \tag{2}$$

In the framework of GO, the travel-time t from the source to the observation point is given by

$$t = \frac{\psi(\mathbf{r})}{v_0}. \tag{3}$$

It follows from Eq. (2) that eikonal $\psi(\mathbf{r})$ can be obtained by integrating the refractive index over the ray

$$\psi(\mathbf{r}) = \int n(\mathbf{r}) ds. \tag{4}$$

Here ds is the differential elementary arc length along the ray and $\mathbf{r} = \mathbf{r}(s)$ is the ray trajectory, which obeys the following ray equations:

$$\frac{d\mathbf{r}}{ds} = \boldsymbol{\kappa}, \qquad \frac{d\boldsymbol{\kappa}}{ds} = \nabla_\perp n(\mathbf{r}) \equiv \nabla n(\mathbf{r}) - \boldsymbol{\kappa}(\boldsymbol{\kappa}\nabla n(\mathbf{r})). \tag{5}$$

In Eq. (5), $\boldsymbol{\kappa} = \dfrac{\nabla \psi}{n}$ is a unit vector, tangent to the ray and $\nabla_\perp n(\mathbf{r})$ is the transverse component of $\nabla n(\mathbf{r}) = \operatorname{grad} n(\mathbf{r})$. According to Eqs. (3) and (4), the travel-time can also be presented by an integral along the ray as

$$t = \frac{1}{v_0} \int n(\mathbf{r}) ds = \int \frac{ds}{v(\mathbf{r})} = \int \mu(\mathbf{r}) ds, \tag{6}$$

where $\mu(\mathbf{r}) = 1/v(\mathbf{r})$ is the wave slowness. Eq. (3) explains how to apply the known results obtained earlier for eikonal variations in the random media by Chernov (1960), Tatarskii (1961, 1967, 1971), Ishimaru (1978, 1997), and Rytov *et al.* (1989b) for the analysis of the travel-time fluctuations.

Following perturbation theory, let us represent the refractive index $n(\mathbf{r})$ in the random medium as a sum of the mean value $\overline{n}(\mathbf{r})$ and the random part $\tilde{n}(\mathbf{r})$ as follows

$$n(\mathbf{r}) = \overline{n}(\mathbf{r}) + \tilde{n}(\mathbf{r}). \tag{7}$$

The average value of the random part $\tilde{n}(\mathbf{r})$ is assumed to be zero $\langle \tilde{n}(\mathbf{r}) \rangle = 0$ since an ensemble of random media is considered, and variance of $\tilde{n}(\mathbf{r})$ is considered to be much smaller in comparison with the background part

$$\sigma_n^2 \equiv \operatorname{Var}[\tilde{n}(\mathbf{r})] = \langle \tilde{n}^2 \rangle \ll \overline{n}^2. \tag{8}$$

Theoretically an ensemble (independent medium realizations) average of a quantity requires infinite number of realizations. Therefore in reality what we can obtain is the statistical average based on the finite number of realizations. Practically each realization corresponds to source and receiver groups, for example, common-shot gathers of a multichannel seismic survey, warranting statistical independence of the wave fields and statistical homogeneity of fluctuations of the medium parameters. Here and hereafter the statistical averaging is denoted by angular brackets$\langle ... \rangle$, or by upper bar $\overline{(...)}$. All other values of interest such as ray trajectory $\mathbf{r}$, eikonal ψ, slowness $\mu(\mathbf{r})$, velocity $v(\mathbf{r})$, and travel-time t can also be presented in the form similar to Eq. (7) and by inequalities similar to Eq. (8).

Substituting $n(\mathbf{r}) = \overline{n}(\mathbf{r}) + \tilde{n}(\mathbf{r})$ and $\psi = \overline{\psi} + \tilde{\psi}$ into eikonal equation (Eq. 2) and taking into account Eq. (8), one can arrive at the well-known first-order approximation for the eikonal (Chernov, 1960; Tatarskii, 1961, 1967, 1971; Ishimaru, 1978, 1997; Rytov *et al.*, 1989b; Kravtsov and Orlov, 1990; Kravtsov 2005),

$$\tilde{\psi} = \int \tilde{n}(\mathbf{r}) d\overline{s}. \tag{9}$$

Correspondingly the first-order travel-time variation $\tilde{t}$ is

$$\tilde{t} = \frac{\tilde{\psi}}{v_0} = \frac{1}{v_0} \int \tilde{n}(\mathbf{r}) d\overline{s} = \int \tilde{\mu}(\mathbf{r}) d\overline{s}. \tag{10}$$

Eqs. (9) and (10) imply the integration of the refractive index fluctuation over the unperturbed ray. This fact was discussed also by Farra and Madariaga (1987), Snieder and Sambridge (1992), and Witte *et al.* (1996). For brevity the upper bar over the regular ray trajectory $\bar{\mathbf{r}}$ and the arc length $\bar{s}$ will be omitted hereafter.

2.2. Model of Quasi-Homogeneous Fluctuations of Medium Parameters

Model of quasi-homogeneous fluctuations (QHF) of refractive index fluctuations, suggested by Rytov *et al.* (1989a,b), deals with covariance function of the form

$$C_n(\mathbf{r}_1,\mathbf{r}_2) \equiv \langle \tilde{n}(\mathbf{r}_1)\, \tilde{n}(\mathbf{r}_2) \rangle = \sigma_n^2(\mathbf{r}_+)K(\mathbf{r}_1 - \mathbf{r}_2; \mathbf{r}_+). \tag{11}$$

Here $\mathbf{r}_+ = (\mathbf{r}_1 + \mathbf{r}_2)/2$ is the center of gravity of the position vectors $\mathbf{r}_1$ and $\mathbf{r}_2$ (Fig. 2), K is a normalized correlation function (correlation coefficient) which is equal to unit at $\mathbf{r}_1 - \mathbf{r}_2 = 0 : K(0;\mathbf{r}_+) = 1$, and tends to zero, when $|\mathbf{r}_1 - \mathbf{r}_2|$ exceeds a characteristic (correlation) length l_c.

In the framework of the QHF model, the correlation coefficient K as well as the variance σ_n^2 may slowly depend on the center of gravity vector $\mathbf{r}_+$. Corresponding characteristic scale l_+ is supposed to be much larger than the correlation length l_c, $l_+ \gg l_c$. The QHF model admits the description of both isotropic and anisometric (statistically anisotropic) fluctuations, which may have different correlation lengths l_x, l_y and l_z in x, y, and z directions, respectively. Moreover, the ratio $l_x:l_y:l_z$ as well as the spatial orientation of the main symmetry axes of the covariance function might slowly change. For statistically homogeneous fluctuations characteristic scale l_+ tends to infinity, so that covariance function (Eq. 11) becomes independent on center of gravity vector $\mathbf{r}_+$. The model of QHF was applied to seismic problems by Kravtsov *et al.* (2003, 2005).

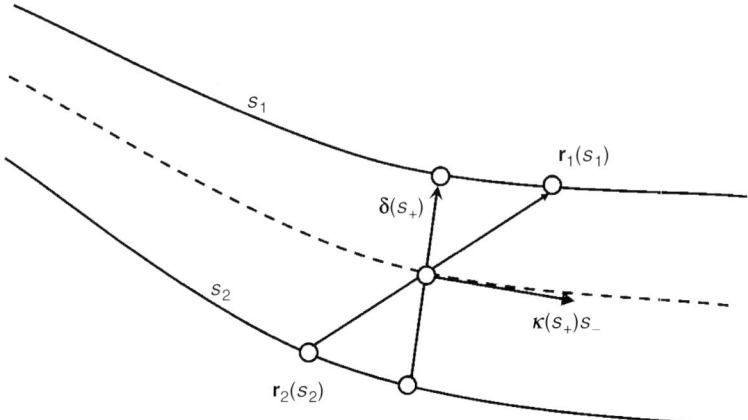

FIG. 2. Representation of the difference of the ray trajectories $\mathbf{r}_1 - \mathbf{r}_2$ in terms of transverse, $\boldsymbol{\delta}(s_+)$, and longitudinal, $\boldsymbol{\kappa}(s_+)s_-$ components (reprinted from Kravtsov *et al.*, 2005).

2.3. Travel-Time Covariance Function in a Medium with Anisometric Fluctuations

According to Eq. (10), covariance function for travel-time fluctuations $\tilde{t}$ can be expressed through the covariance function of medium fluctuations in a following way:

$$C_t(\mathbf{R}_1, \mathbf{R}_2) = \langle \tilde{t}(\mathbf{R}_1)\tilde{t}(\mathbf{R}_2) \rangle = \frac{1}{v_0^2} \int_0^{S(\mathbf{R}_1)} ds_1 \int_0^{S(\mathbf{R}_2)} ds_2 C_n[\mathbf{r}_1(s_1), \mathbf{r}_2(s_2)]. \qquad (12)$$

Here $\mathbf{R}_1$ and $\mathbf{R}_2$ are the locations of the observation points (receivers), $S(\mathbf{R}_1)$ and $S(\mathbf{R}_2)$ are the total arc lengths of the rays arriving to these receivers. The current arc lengths of the rays are given by s_1 and s_2 (Fig. 2). First, let us consider the travel-time fluctuations in conditions of refractive geometry, when regular refraction plays a significant role. Fluctuations in conditions of reflecting geometry will be studied in Section 3.

Let us introduce new variables $s_- = s_1 - s_2$ and $s_+ = (s_1 + s_2)/2$ in Eq. (12) and expand the trajectories $\mathbf{r}_1(s_1)$ and $\mathbf{r}_2(s_2)$ (Fig. 2) into power series in difference variable s_-. Taking into account the zeroth- and first-order terms in $\mathbf{r}_1(s_1) - \mathbf{r}_2(s_2)$ and the zeroth-order term in $\mathbf{r}_+ = (\mathbf{r}_1 + \mathbf{r}_2)/2 = \mathbf{r}_+(s_+)$ the difference term $\mathbf{r}_1 - \mathbf{r}_2$ in Eq. (12) becomes

$$\mathbf{r}_1(s_1) - \mathbf{r}_2(s_2) \cong \boldsymbol{\kappa}(s_+)s_- + \boldsymbol{\delta}(s_+). \qquad (13)$$

Here $\boldsymbol{\kappa}(s_+) = d\mathbf{r}_+/ds$ is a unit vector tangent to the median ray $\mathbf{r} = \mathbf{r}(s_+)$, while the terms $\boldsymbol{\kappa}(s_+)s_-$ and $\boldsymbol{\delta}(s_+)$ are the longitudinal and transverse distances between points $\mathbf{r}_1(s_1)$ and $\mathbf{r}_2(s_2)$ of adjacent rays, respectively (Fig. 2). Using identity $ds_1 ds_2 = ds_- ds_+$, Eq. (12) can be written as follows

$$C_t(\mathbf{R}_1, \mathbf{R}_2) = \frac{2}{v_0^2} \int_0^{S_<} \sigma_n^2[\mathbf{r}_+(s_+)]ds_+ \int_0^\infty ds_- \, K[\boldsymbol{\kappa}(s_+)s_- + \boldsymbol{\delta}(s_+); \mathbf{r}_+(s_+)]. \qquad (14)$$

Here the upper limit of integration over s_- is extended to infinity since K goes to zero at large distances, $|\mathbf{r}_1 - \mathbf{r}_2| \gg l_c$. Besides, $\int_{-\infty}^\infty ds_-$ is transformed into $2 \int_0^\infty ds_-$ and the least value $S_< = \min[S(\mathbf{R}_1), S(\mathbf{R}_2)]$ is taken as the upper limit of integration over s_+. This is commonly accepted in the ray theory of wave propagation through random media (Chernov, 1960; Tatarskii 1961, 1967, 1971; Ishimaru, 1978, 1997; Rytov et al., 1989b).

When the observation points coincide, $\mathbf{R}_1 = \mathbf{R}_2 = \mathbf{R}$, then $\boldsymbol{\delta} = 0$ and the internal integral in Eq. (14) becomes

$$l_{\text{eff}}(\boldsymbol{\kappa}, s_+) = \int_0^\infty K[\boldsymbol{\kappa}(s_+)s_-; \mathbf{r}_+(s_+)]ds_-. \qquad (15)$$

Here $l_{\text{eff}}(\boldsymbol{\kappa}, s_+)$ denotes an effective correlation length in direction $\boldsymbol{\kappa}(s_+) = d\mathbf{r}_+/ds$ at the point $\mathbf{r}_+(s_+)$. In the case of anisometric randomly inhomogeneous medium, the effective correlation length l_{eff} can be calculated by using a normalized correlation function $K(\Delta \mathbf{r}) = \beta[g(\Delta \mathbf{r})]$ of anisometric argument (Iooss, 1998)

$$g(\Delta \mathbf{r}) = \left[\left(\frac{\Delta x}{l_x}\right)^2 + \left(\frac{\Delta y}{l_y}\right)^2 + \left(\frac{\Delta z}{l_z}\right)^2 \right]^{1/2}. \qquad (16)$$

Here $\beta(g)$ stands as a symbol for a model correlation function such as Gaussian, $\beta(g) = \exp(-g^2)$ or exponential, $\beta(g) = \exp(-g)$.

In the case, when the random medium is statistically homogeneous in a horizontal plane (x, y), one can replace l_x and l_y with a horizontal characteristic scale l_h, which typically is larger than vertical scale l_z, $l_h \geq l_z$:

$$g(\Delta \mathbf{r}) = \left[\frac{(\Delta x)^2 + (\Delta y)^2}{l_h^2} + \frac{(\Delta z)^2}{l_z^2} \right]^{1/2}. \tag{17}$$

Then for a ray, incident on a horizontal plane at an angle θ, between the ray and vertical axis z, one has

$$l_{\text{eff}} = \Gamma \left[\left(\frac{\sin \theta}{l_h} \right)^2 + \left(\frac{\cos \theta}{l_z} \right)^2 \right]^{-1/2}, \tag{18}$$

where $\Gamma = 1$ for exponential correlation function $\beta(g) = \exp(-g)$ and $\Gamma = \sqrt{\pi}/2$ for Gaussian correlation function $\beta(g) = \exp(-g^2)$. According to Eq. (18), l_{eff} takes a value Γl_z for a vertical ray, that is for $\theta = 0$, and value Γl_h for a horizontal ray, that is for $\theta = \pi/2$.

In the case $\mathbf{R}_1 = \mathbf{R}_2 = \mathbf{R}$, the covariance expression (Eq. 14) becomes the travel-time variance

$$\sigma_t^2 = C_t(\mathbf{R}) = \frac{2}{v_0^2} \int_0^{S_<} \sigma_n^2[\mathbf{r}_+(s_+)] l_{\text{eff}}(s_+) \mathrm{d}s_+. \tag{19}$$

Note that for an isotropic medium, where all the characteristic scales l_x, l_y, and l_z are equal to each other, $l_x = l_y = l_z = l_c$ and the correlation coefficient does not depend on the ray direction, an effective correlation length l_{eff} is equal to the isotropic characteristic scale l_c:

$$l_{\text{eff}}(s_+) = \int_0^\infty K[s_-; \mathbf{r}_+(s_+)] \mathrm{d}s_- = \Gamma l_c(s_+). \tag{20}$$

2.4. Boundary of GO Applicability

For the validity of the GO method, the inhomogeneity characteristic length l_c should be much greater than the typical wavelength λ of the seismic wave,

$$l_c \gg \lambda, \tag{21}$$

and the cross-section a_f of the Fresnel volume surrounding the ray should be smaller as compared with the transverse scale $l_\perp$ of the inhomogeneities (Kravtsov and Orlov, 1990; Born and Wolf, 1999)

$$a_f \approx \sqrt{L\lambda} < l_\perp. \tag{22}$$

The former is the necessary condition and the latter is the sufficient condition for the applicability of the GO method. Here L stands for the typical distance between the source and the receiver.

According to Eq. (22), the largest distance for GO applicability is of the order of

$$L_{\max} \sim \frac{l_\perp^2}{\lambda}. \tag{23}$$

At larger distances, diffraction phenomena should be taken into account. Some diffraction phenomena for seismic waves in random elastic media have been considered by Shapiro *et al.* (1996), Tong *et al.* (1998), Marquering *et al.* (1999), Nolet and Dahlen (2000), Spetzler and Snieder (2001), and Baig *et al.* (2003). In these papers, they compare the results of their relations by the GO-based results and discuss the limits of the GO method. The most important diffraction phenomenon, arising beyond limiting distance Eq. (23), is the phenomenon of strong fluctuations, which result from random caustics and random foci forming (Rytov *et al.*, 1989b).

However, as it follows from the more powerful methods than GO method, the method of smooth perturbations by Rytov or the method of parabolic equation (Rytov *et al.*, 1989b), some results of GO approximation preserve their applicability even far beyond the distance Eq. (23). It concerns first of all the eikonal variance, which at larger distances acquires a constant multiplier, a value comparable with unit, but preserves its dependence on distance. One can hope therefore that the same is true for travel-time variance Eq. (19). We accept this hypothesis in what follows, though we understand the necessity of its validation by numerical simulation on the basis of the full wave equation. It is worth to notice that beyond the area of the validity of GO the diffraction results should be used for fitting the empirical data. As it was shown by Roth (1997), the parameters, inverted on the basis of diffraction theory, may noticeably differ from the results reconstructed in the frame of GO approach.

3. Travel-Time Fluctuations in Reflection Geometry

3.1. Reflection Geometry

In this section, the travel-time fluctuations will be examined for the reflection geometry, when one layered heterogeneous overburden is placed over a horizontal interface, located at a depth D (Fig. 3a and b). From the following relations of the medium parameters

$$v = \bar{v} + \tilde{v}, \quad n = \bar{n} + \tilde{n}, \quad \mu = \bar{\mu} + \tilde{\mu}, \quad n = v_0/v = v_0\mu, \tag{24}$$

one can derive

$$\bar{n} = v_0/\bar{v} = v_0\bar{\mu}, \quad \tilde{n} = v_0\tilde{\mu} = -v_0\tilde{v}/\bar{v}^2, \quad \sigma_n^2 = \sigma_v^2 v_0^2/\bar{v}^4 = v_0^2 \sigma_\mu^2. \tag{25}$$

Supposing that refraction is weak enough in the layer of depth D, one may accept the mean value $\bar{v}$ as the reference velocity v_0, $\bar{v} \approx v_0$, so that the relations in Eq. (25) becomes as follows,

GEOMETRICAL OPTICS IN ACOUSTIC RANDOM MEDIA 103

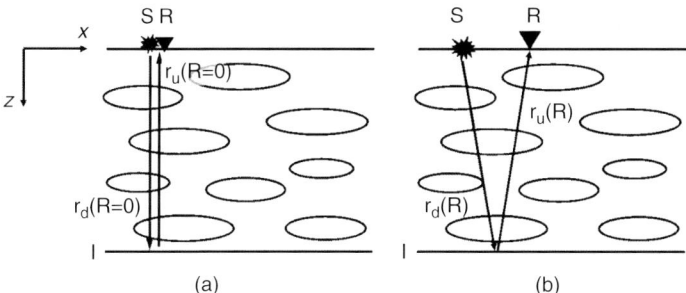

FIG. 3. Geometry of double passage effect (DPE): (a) receiver R is placed close to the source S; (b) receiver R is separated from the source.

$$\bar{n} = 1, \quad \tilde{n} = -\tilde{v}/v_0 = v_0\tilde{\mu}, \quad \sigma_n^2 = \sigma_v^2/v_0^2 = v_0^2\sigma_\mu^2. \tag{26}$$

In the presence of the interface I the unperturbed ray trajectory $\mathbf{r}(s)$ can be composed of down- and up-going parts represented as

$$\mathbf{r}(s) = \begin{cases} \mathbf{r}_d(s_d; \mathbf{R}), & 0 < s_d < S_d(\mathbf{R}), \\ \mathbf{r}_u(s_u; \mathbf{R}), & 0 < s_u < S_u(\mathbf{R}). \end{cases} \tag{27}$$

Here "d" and "u" stands for "down" and "up," respectively. The parameters are the same as they are given in Fig. 3a and b. In case of weak refraction, both descending and ascending parts of the rays look like straight lines [see Eqs. (34) and (35) below] and the total travel-time fluctuations can be obtained as

$$\tilde{t}(\mathbf{R}) = \tilde{t}_d(\mathbf{R}) + \tilde{t}_u(\mathbf{R}), \tag{28}$$

where

$$\tilde{t}_d(\mathbf{R}) = \frac{1}{v_0} \int_0^{S_d(\mathbf{R})} \tilde{n}[\mathbf{r}_d(s_d)] ds_d, \tag{28a}$$

and

$$\tilde{t}_u(\mathbf{R}) = \frac{1}{v_0} \int_0^{S_u(\mathbf{R})} \tilde{n}[\mathbf{r}_u(s_u)] ds_u. \tag{28b}$$

In the next section, these parameters are used in the derivation of the covariance and the variance of travel-time fluctuation for the case of reflection geometry.

3.2. Travel-Time Covariance Function for Small Offsets

In Section 2.2, the general form of the covariance function for a QHF model was given by Eq. (14). In this section, Eq. (14) will be modified for the reflection geometry.

In reflection case, the covariance of the travel-time can be calculated as

$$C_t(\mathbf{R}_1, \mathbf{R}_2) = \langle \tilde{t}(\mathbf{R}_1), \tilde{t}(\mathbf{R}_2) \rangle = \langle [\tilde{t}_d(\mathbf{R}_1) + \tilde{t}_u(\mathbf{R}_1)][\tilde{t}_d(\mathbf{R}_2) + \tilde{t}_u(\mathbf{R}_2)] \rangle$$
$$= C_{dd}(\mathbf{R}_1, \mathbf{R}_2) + C_{du}(\mathbf{R}_1, \mathbf{R}_2) + C_{ud}(\mathbf{R}_1, \mathbf{R}_2) + C_{uu}(\mathbf{R}_1, \mathbf{R}_2). \quad (29)$$

As an example the explicit form of C_{dd} is given as follows,

$$C_{dd}(\mathbf{R}_1, \mathbf{R}_2) = \frac{1}{v_0^2} \int_0^{S_d(\mathbf{R}_1)} ds_{1d} \int_0^{S_d(\mathbf{R}_2)} ds_{2d} C_n[\mathbf{r}_d(s_{1d}; \mathbf{R}_1), \mathbf{r}_d(s_{2d}; \mathbf{R}_2)], \quad (30)$$

or in terms of correlation coefficient as

$$C_{dd}(\mathbf{R}_1, \mathbf{R}_2) = \frac{1}{v_0^2} \int_0^{S_d(\mathbf{R}_1)} ds_{1d} \int_0^{S_d(\mathbf{R}_2)} ds_{2d} \sigma_n^2(\mathbf{r}_{dd+}) K[\mathbf{r}_d(s_{1d}; \mathbf{R}_1) - \mathbf{r}_d(s_{2d}; \mathbf{R}_2); \mathbf{r}_{dd+}], \quad (31)$$

where σ_n^2 is refractive index, K is the correlation coefficient, and the parameter $\mathbf{r}_{dd+}$ in Eq. (31) is defined as

$$\mathbf{r}_{dd+} = [\mathbf{r}_d(s_{1d}; \mathbf{R}_1) + \mathbf{r}_d(s_{2d}; \mathbf{R}_2)]/2. \quad (32)$$

It is assumed that horizontal vectors $\mathbf{R}_1$ and $\mathbf{R}_2$ are small in comparison with the depth D of the interface, $|\mathbf{R}_{1,2}| \ll D$, so that the slightly curved rays might be approximated as straight lines (Fig. 4). In this case, the ray lengths $S_d(\mathbf{R}_1)$ and $S_u(\mathbf{R}_2)$ are practically identical. Since $|\mathbf{R}_{1,2}|/D \ll 1$, the ray length can be calculated as

$$S(\mathbf{R}_{1,2}) = \sqrt{D^2 + (R_{1,2}/2)^2} \approx D\left(1 + R_{1,2}^2/8D^2\right) \approx D, \quad (33)$$

denoting that $S_d(\mathbf{R}_1)$ and $S_u(\mathbf{R}_2)$ differ from depth D only in the second-order term of $|\mathbf{R}_{1,2}|/D$. In this case, the ray trajectories for down- and up-going waves are given as

$$\mathbf{r}_{dj}(s_d; \mathbf{R}_j) = s_d \boldsymbol{\kappa}_{dj}(\mathbf{R}_j), \quad 0 < s_d < D, \quad (34)$$
$$\mathbf{r}_{uj}(s_u; \mathbf{R}_j) = D\boldsymbol{\kappa}_{dj}(\mathbf{R}_j) + s_u \boldsymbol{\kappa}_{uj}(\mathbf{R}_j), \quad 0 < s_u < D, \quad (35)$$

where unit vectors tangent to the rays have the following form:

$$\boldsymbol{\kappa}_{dj} \cong \frac{\mathbf{R}_j}{2D} + \mathbf{i}_z, \quad \boldsymbol{\kappa}_{uj} \cong \frac{\mathbf{R}_j}{2D} - \mathbf{i}_z. \quad (36)$$

Here $\mathbf{i}_z$ is the unit vector in the z direction and $j = 1, 2$.

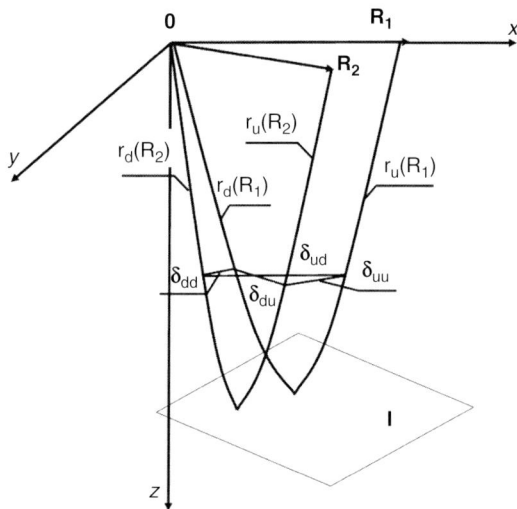

FIG. 4. Geometry for two up- and down-going ray trajectories in 3D space.

For simplicity, considering the statistically homogeneous fluctuations, when correlation function K does not depend on s_+ and $\sigma_n^2 = constant$, Eq. (29) can be expressed in the following form:

$$C_t(\mathbf{R}_1, \mathbf{R}_2) = \frac{2\sigma_n^2}{v_0^2} \int_0^D dz \int_0^\infty ds_- \left\{ \sum_{p=\{d,u\}} \sum_{q=\{d,u\}} K\left[\mathbf{i}_z s_- + \boldsymbol{\delta}_{pq}(z) \right] \right\}, \quad (37)$$

where integral over variables s_{d+} and s_{u+} were transformed into integral along vertical axis z with upper limit of integration $z \approx D$.

In the equation above, $\boldsymbol{\delta} = \mathbf{R}_1 - \mathbf{R}_2$ is the difference vector between observation points and $\boldsymbol{\delta}_{pq}(z)$ represent the horizontal distance (Fig. 4) between the rays in the plane $z = const$. Two down-going and two up-going rays intersect this $z = const$ plane at the points $\mathbf{R}_j^d = \mathbf{R}_j \frac{z}{2D}$, $\mathbf{R}_j^u = \mathbf{R}_j \left(1 - \frac{z}{2D}\right)$, where $j = 1, 2$.

Corresponding vectors, connecting these points are

$$\boldsymbol{\delta}_{dd} = \mathbf{R}_1^d - \mathbf{R}_2^d = \boldsymbol{\delta} \frac{z}{2D}, \tag{38a}$$

$$\boldsymbol{\delta}_{uu} = \mathbf{R}_1^u - \mathbf{R}_2^u = \boldsymbol{\delta}\left(1 - \frac{z}{2D}\right), \tag{38b}$$

$$\boldsymbol{\delta}_{du} = \mathbf{R}_1^d - \mathbf{R}_2^u = \mathbf{R}_1 \frac{z}{2D} - \mathbf{R}_2\left(1 - \frac{z}{2D}\right) = (\mathbf{R}_1 + \mathbf{R}_2)\frac{z}{2D} - \mathbf{R}_2$$
$$= -\mathbf{R}_2 + \left(\frac{\boldsymbol{\delta}}{2} + \mathbf{R}_2\right)\frac{z}{D}, \tag{38c}$$

$$\boldsymbol{\delta}_{ud} = \mathbf{R}_1^u - \mathbf{R}_2^d = \mathbf{R}_1\left(1 - \frac{z}{2D}\right) - \mathbf{R}_2 \frac{z}{2D} = \mathbf{R}_1 - (\mathbf{R}_1 + \mathbf{R}_2)\frac{z}{2D}$$
$$= \mathbf{R}_1 + \left(\frac{\boldsymbol{\delta}}{2} - \mathbf{R}_1\right)\frac{z}{D}. \tag{38d}$$

The covariance function given in Eq. (37) describes the acoustic waves in randomly inhomogeneous media which are reflected from a horizontal interface. The covariance function for midpoint and two-sources-two-receivers geometries, as well as for tilted interfaces, was analyzed by Kravtsov et al. (2003).

The travel-time variance can be obtained from the covariance function as the receiver positions coincide, $\mathbf{R}_1 = \mathbf{R}_2 = \mathbf{R}$. In this case, the parameters $\boldsymbol{\delta}, \boldsymbol{\delta}_{dd}$ and $\boldsymbol{\delta}_{uu}$ in Eq. (37) vanish and the parameters $\boldsymbol{\delta}_{du}, \boldsymbol{\delta}_{ud}$ become

$$\boldsymbol{\delta}_{du}(z) = -\boldsymbol{\delta}_{ud}(z) = \mathbf{R}(1 - z/D). \tag{39}$$

Consequently the variance is obtained as

$$\sigma_t^2(\mathbf{R}) = C_t(\mathbf{R}, \mathbf{R}) = \frac{4}{v_0^2} \int_0^D dz \sigma_n^2(\mathbf{i}_z z) \int_0^\infty ds_- \{K[\mathbf{i}_z s_-; \mathbf{i}_z z] + K[\mathbf{i}_z s_- + \mathbf{R}(1 - z/D); \mathbf{i}_z z]\}, \tag{40}$$

which is used to estimate the medium fluctuations and the horizontal scale of the inhomogeneities from the experimental data.

3.3. Double Passage Effect

The double passage effect (DPE) occurs when the travel-time is measured at zero offset, $\mathbf{R} = 0$, that is, source and receiver are at the same location (Fig. 3a). At zero offset, the wave passes twice through the same randomly inhomogeneous medium. Therefore the fluctuations of the travel-times for down- and up-going waves are equal, $\tilde{t}_d(0) = \tilde{t}_u(0)$, and the total travel-time is $\tilde{t}(0) = \tilde{t}_d(0) + \tilde{t}_u(0) = 2\tilde{t}_d(0)$. For zero offset, the variance is

$$\mathrm{var}[\tilde{t}(\mathbf{R}=0)] = \frac{8}{v_0^2} \int_0^D dz \sigma_n^2(\mathbf{i}_z z) \int_0^\infty ds_-\ K(\mathbf{i}_z s_-; \mathbf{i}_z z). \tag{41}$$

For statistically homogeneous fluctuations, travel-time variance can be written in a shorter form as

$$\mathrm{var}[t(\mathbf{R}=0)] = \frac{8}{v_0^2} D \sigma_n^2 l_z, \tag{42}$$

where

$$l_z = \int_0^\infty ds_-\ K(\mathbf{i}_z s_-). \tag{43}$$

Eq. (41) denotes that the variance of the total travel-time is four times larger than the variance of the one way travel-time

$$\mathrm{var}[\tilde{t}(0)] = 4\,\mathrm{var}[\tilde{t}_d(0)]. \tag{44}$$

The comparison of the travel-time variance at zero (Fig. 3a) and larger offsets (Fig. 3b) can be helpful for estimating the characteristic horizontal length of the inhomogeneities. This idea was suggested by Touati (1996), analyzed by Iooss (1998), Iooss et al. (2000), and generalized by Kravtsov et al. (2003).

At sufficiently large offsets $|\mathbf{R}| = R \gg l_h$, the down- and up-going waves pass through different mediums and therefore the cross-product of $\tilde{t}_d(\mathbf{R})$ and $\tilde{t}_u(\mathbf{R})$ on average is close to zero and the variance is two times larger than the variance of the one way travel-time at large offsets,

$$\text{var}[\tilde{t}(0)] \cong 2\,\text{var}[\tilde{t}_d(|\mathbf{R}| \gg l_h)]. \tag{45}$$

For the case when $R < l_h$ the second cross-covariance term in Eq. (40), which is actually $2C_{du}(\mathbf{R})$, is comparable with the first one, that is, with $2C_{dd}(0)$. In an opposite situation, $R > l_h$, the cross-covariance term $K[\mathbf{i}_z s_- + \mathbf{R}(1 - z/D)]$ in Eq. (40) becomes small enough, because this term is formed by a thick layer $D > z > z_c$, where z_c is estimated from $R(1 - z_c/D) \leq l_h$. As a result, relative thickness of the mentioned layer will be $(D - z_c)/D \sim l_h/R \ll 1$ and

$$C_{du}(\mathbf{R}) \sim C_{dd}(\mathbf{R}) \frac{l_h}{R} \ll C_{dd}(\mathbf{R}). \tag{46}$$

Thus, at $R > l_h$ and $R_{1,2} \geq D$ only auto-covariance remains significant,

$$C_t(\mathbf{R}) \cong C_{dd}(\mathbf{R}) + C_{uu}(\mathbf{R}), \tag{47}$$

whereas contribution of DPE becomes negligible.

4. Travel-Time Fluctuations in Refraction Geometry

4.1. Refracting Medium with a Constant Velocity Gradient

In this section, the travel-time fluctuations of diving-type refracted rays will be examined in a stratified medium with a velocity, linearly increasing with depth. The mean velocity model is written as

$$\bar{v}(z) = v_0 + kz = v_0(1 + z/H), \tag{48}$$

where v_0 is the wave velocity near the surface, k is the constant velocity gradient, and $H = v_0/k$ is the depth, where velocity becomes twice as large as compared with v_0.

For the description of the velocity fluctuations, $\tilde{v}$, it is convenient to introduce the auxiliary dimensionless random field $\tilde{\xi}$, which has a mean value of zero and sufficiently small variance $\sigma_{\tilde{\xi}}^2 \ll 1$. By considering $\tilde{v}$ being proportional to the random field $\tilde{\xi}$ and to the mean velocity $\bar{v}(z)$, the velocity fluctuations can be presented as

$$\tilde{v} = \bar{v}(z)\tilde{\xi}. \tag{49}$$

For statistically homogeneous fluctuations, the variance of the auxiliary random field $\tilde{\xi}$ is a constant value, $\sigma_\xi^2 = const$. According to Eq. (49), the random part $\tilde{n}$ of the refractive index and its variance σ_n^2 take the following form

$$\tilde{n} = -\frac{v_0 \tilde{\xi}}{\bar{v}(z)} = -\bar{n}(z)\tilde{\xi}, \qquad (50)$$

$$\sigma_n^2 = [\bar{n}(z)]^2 \sigma_\xi^2. \qquad (51)$$

Unlike the model, accepted for analysis of the reflection geometry, in this case the variance of the refractive index σ_n^2 depends on the depth z.

In a medium with a mean velocity, linearly increasing with depth, the ray paths are of the curvilinear form. It is assumed that the ray is traveling in the x–z plane and the angle of incidence θ_0 is counted from the vertical z axis. In this case, the ray trajectory satisfies the equation

$$\frac{dz}{dx} = \pm \frac{\sqrt{n^2(z) - \sin^2 \theta_0}}{\sin \theta_0} = \cot \theta, \qquad (52)$$

where θ is the current angle of incidence. According to Eq. (48), the mean value of the refractive index equals to

$$\bar{n}(z) = \frac{v_0}{v(z)} = \frac{1}{1 + z/H}. \qquad (53)$$

Inserting Eq. (53) into Eq. (52), one can find the ray trajectory as

$$z = \sqrt{H^2 + 2xH \cot \theta_0 - x^2} - H = \sqrt{H^2 + xX - x^2} - H > 0, \qquad (54)$$

where $X = 2H \cot \theta_0$ denotes the final point of the ray.

In refraction geometry, there is no necessity to distinguish the down- and up-going branches of the ray, so that travel-time fluctuations might be calculated by direct use of Eq. (10).

4.2. Travel-Time Variance along a Curvilinear Ray

Inserting the right-hand side of Eq. (51) into Eq. (14), we obtain the covariance of travel-times as

$$C_t(\mathbf{R}_1, \mathbf{R}_2) = \frac{2}{v_0^2} \int_0^{S_<} [\bar{n}(z)]^2 \sigma_\xi^2 ds_+ \int_0^\infty ds_- \, K[\boldsymbol{\kappa}(s_+)s_- + \boldsymbol{\delta}(s_+)], \qquad (55)$$

which relates the travel-time fluctuations to the fluctuations of the medium parameters in a refraction geometry. The variance of the travel-time fluctuations can be obtained from Eq. (55) under condition $\mathbf{R}_1 = \mathbf{R}_2 = \mathbf{R}$ as

GEOMETRICAL OPTICS IN ACOUSTIC RANDOM MEDIA   109

$$\sigma_t^2 = \text{var}[\tilde{t}(\mathbf{R})] = \frac{2\sigma_\xi^2}{v_0^2} \int_0^{S(\mathbf{R})} [\bar{n}(z)]^2 ds_+ \, l_{\text{eff}}(\boldsymbol{\kappa}, s_+). \tag{56}$$

It is more convenient to transform the variables in Eq. (56) from the ray coordinates to Cartesian ones. In this case, elementary arc length ds_+ becomes

$$ds_+ = \frac{dx}{\sin\theta}. \tag{57}$$

In turn the refractive index $\bar{n}(z)$ can be rewritten as

$$\bar{n}[z(x)] = \frac{v_0}{\bar{v}(z)} = \frac{1}{1+\dfrac{z}{H}} = \frac{1}{\sqrt{1+\dfrac{\alpha(x)}{H^2}}} = \frac{1}{\sqrt{1+\varepsilon(\gamma-\varepsilon)}}, \tag{58}$$

where $\alpha(x) = xX - x^2$, $\varepsilon = x/H$, and $\gamma = X/H$. By introducing the inhomogeneity scale length ratio $\rho = l_z/l_x$ (Kravtsov et al., 2005), the effective correlation length can be expressed as

$$l_{\text{eff}} = \Gamma l_z \left[(\rho\sin\theta)^2 + (\cos\theta)^2\right]^{-1/2}. \tag{59}$$

Expressing the angle θ and other variables through the new variables we can show that

$$\sin\theta = \sqrt{B^2-\eta^2}/B, \quad \cos\theta = -\eta/B, \quad \bar{n} = \sqrt{B^2-\eta^2} \tag{60}$$

where

$$B = B(\gamma) = \sqrt{1+\left(\frac{\gamma}{2}\right)^2}, \quad \eta = \varepsilon - \frac{\gamma}{2} = \frac{x - X/2}{H}. \tag{61}$$

By using the new variables the travel-time variance can be expressed as

$$\sigma_t^2 = GJ(\rho,\gamma), \tag{62}$$

where

$$G = 2\Gamma\sigma_\xi^2 l_z H/v_0^2, \tag{62a}$$

and

$$J(\rho,\gamma) = B^2 \int_{-\gamma/2}^{\gamma/2} \frac{d\eta}{(B^2-\eta^2)^{3/2}\sqrt{\rho^2(B^2-\eta^2)+\eta^2}}. \tag{62b}$$

Eq. (62) is the fundamental equation for estimating the statistical parameters ($\sigma_\xi^2 l_z$ and ρ) of the random refractive medium.

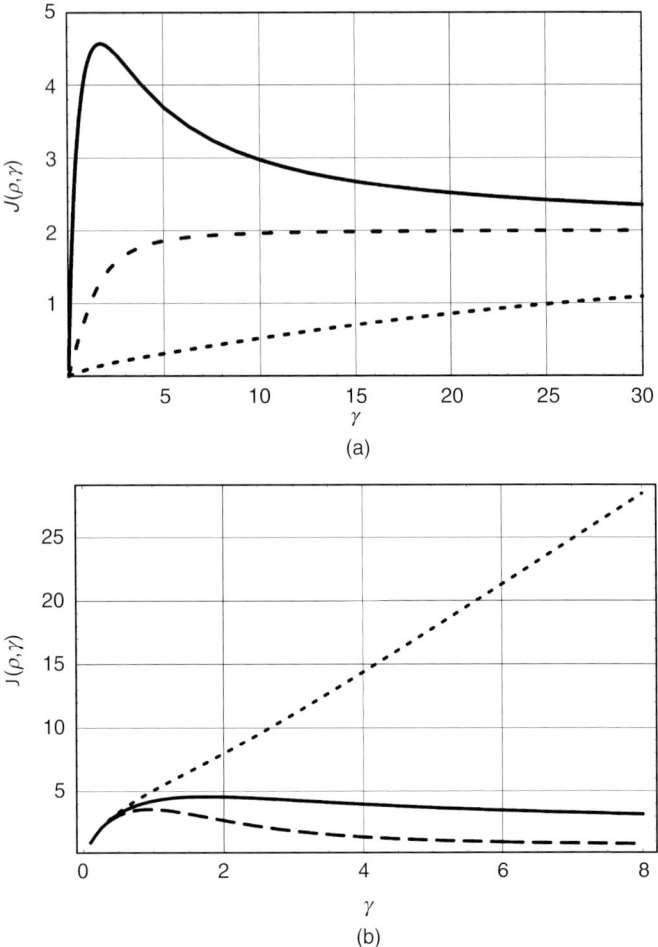

FIG. 5. Dependence of integral $J(\rho, \gamma)$, on dimensionless distance $\gamma = X/H$. (a) corresponds to the model $\sigma_v^2 = [\bar{v}(z)]^2 \sigma_\xi^2$ for $\rho = 0.1$ (continuous curve), $\rho = 1$ (dashed curve), and $\rho = 10$ (dotted curve), respectively. (b) corresponds to the three basic models: (1) $\sigma_v^2 = [\bar{v}(z)]^2 \sigma_\xi^2$, (continuous curve); (2) $\sigma_v^2 = v_0^2 \sigma_\xi^2 = $ const, (dashed curve); and (3) $\sigma_v^2 = [\bar{v}(z)]^4 \sigma_\xi^2 / v_0^2$, (dotted curve). All three curves are calculated for the inhomogeneity scale length ratio $\rho = 0.1$ (reprinted from Kravtsov et al., 2005).

4.3. Dependence of Travel-Time Variance on Offset

Following Kravtsov et al. (2005), let us consider the travel-time variance σ_t^2 dependence on distance X for different models of regular refraction. According to Eq. (62b), travel-time variance σ_t^2 is proportional to integral $J(\rho, \gamma)$. Let the refractive index variance decrease with depth as given by Eq. (51). Fig. 5a presents the dependence of the integral (Eq. 62b) on normalized offset $\gamma = X/H$ for different scale length ratios $\rho = l_z/l_x : \rho = 0.1, \rho = 1$ and $\rho = 10$. In the case, when inhomogeneities are strongly flattened in the horizontal plane ($\rho = 0.1$), the integral $J(\rho, \gamma)$ first increases proportionally to the normalized distance γ, $J(\rho, \gamma) \approx \gamma/\rho$ (continuous curve on Fig. 5a). At $\gamma = 1.9$ the integral $J(\rho, \gamma)$ reaches its maximum value $J_{\max} \approx 4.57$ and then tends to an asymptotic value $J_\infty = 2$ at $\gamma \to \infty$.

The decrease of the travel-time variance at large distances looks quite unusual: the more the distance, the less the travel-time variance. In fact, this is new physical phenomenon in the theory of wave propagation in random media, which has not been observed or discussed before either in acoustics or in adjacent studies in optics and radio physics. There are two factors which could explain this unusual behavior of the travel-time variance: the rapid decrease of the refractive index variance with depth according to Eq. (51) and the strong elongation of the flattened inhomogeneities in the horizontal direction. The decrease of travel-time variance with offset and with decreasing fluctuations of the refractive index in the case of laminated heterogeneities is similar to the Double Passage Effect in the reflection geometry. This decrease is absent if the medium heterogeneities are isometric because the ray propagate down and up in essentially different media. The medium is quickly getting less correlated with increasing offset. However, if the medium is quasi layered, then approximately the same medium will be crossed by both down- and up-going branches of the ray. Also the medium heterogeneity starts to become uncorrelated with increasing offset. Of course, the existence of the phenomenon is due to interplay between the lateral medium correlation and fluctuation decrease with the depth. Thus, for some situations we do not observe any DPE-like effect, however for some situations it can be well observed in the refraction wave geometry as well as in the reflection wave geometry.

When the inhomogeneity scale lengths in horizontal and vertical directions are identical, $\rho = 1.0$ (isometric inhomogeneities), the integral $J(\rho, \gamma)$ increases when $\gamma \leq 2\rho$ and tends to an asymptotic value $J_\infty = 2$ when $\gamma \to \infty$ (Fig. 5a, dashed curve). When $\rho = 10$ (vertically elongated inhomogeneities), the travel-time variance increases regularly (Fig. 5a, dotted curve). As $\gamma \to \infty$, the integral approaches to asymptotic value $J_\infty = 2$ (not shown in the figure). One can see from Fig. 5a that for isometric and vertically elongated inhomogeneities the travel-time variance does not show any decrease with distance, though at $\gamma \to \infty$ in all three cases the integral $J(\rho, \gamma)$ tends to a common asymptotic value $J_\infty = 2$.

When the velocity variance is a constant value,

$$\sigma_v^2 = v_0^2 \sigma_\xi^2 = \text{const}, \tag{63}$$

the refractive index variance $\sigma_n^2 = [\bar{n}(z)]^4 \sigma_\xi^2$ decreases with depth faster than the previous model given in Eq. (51). For the model given in Eq. (63) the travel-time variance σ_t^2 is proportional to the integral

$$J_1(\rho, \gamma) = B^2 \int_{-\gamma/2}^{\gamma/2} \frac{d\eta}{(B^2 - \eta^2)^{5/2} \sqrt{\rho^2(B^2 - \eta^2) + \eta^2}}. \tag{64}$$

The graph of the integral, given in Fig. 5b by dashed curve, shows the faster decrease in comparison with the integral $J(\rho, \gamma)$ of Eq. (62b) (continuous curve in Fig. 5b). On the other hand, in the case of a constant refractive index variance, as it is in the reflection geometry,

$$\sigma_n^2 = \sigma_\xi^2 = \text{const}, \tag{65}$$

the velocity variance increases with depth: $\sigma_v^2 = [\bar{v}(z)]^4 \sigma_n^2/v_0^2 = [\bar{v}(z)]^4 \sigma_\xi^2/v_0^2$. Therefore, the corresponding integral

$$J_2(\rho,\gamma) = B^2 \int_{-\gamma/2}^{\gamma/2} \frac{d\eta}{(B^2 - \eta^2)^{1/2}\sqrt{\rho^2(B^2 - \eta^2) + \eta^2}} \tag{66}$$

increases with distance, as shown by the dotted curve in Fig. 5b. Such a behavior is characteristic for all random media studied before.

Thus, examination of travel-time variance for random medium with different properties promises to be a valuable source of information to relate the observed travel-times to the real media.

4.4. Inverse Problem Solution for Refraction Geometry

This section concerns with the estimation of the medium parameters from the travel-time fluctuations. When dealing with real seismic data, the travel-time fluctuations are calculated as the difference between the observed and the theoretical travel-times. Here we assume that the travel-time fluctuations are in hand and we explain the estimation procedure of the statistical parameters.

To estimate the inhomogeneity scale lengths and the standard deviation of the medium, first the variance of the travel-time fluctuations are calculated and an inversion procedure is employed between the theoretical equation of variance (Eq. 62) and the calculated variance. For this purpose, the Levenberg–Marquard iterative inversion method is used (Press et al., 2001), and the unknown medium quantities ρ and $\sigma_\xi^2 l_z$ are estimated.

The next step is the calculation of the parameters σ_ξ^2, l_z, l_x, and l_y separately. Eq. (55) is the basic formula for this purpose. To estimate the longitudinal and the transverse correlation scales, the observation points $\mathbf{R}_1$ and $\mathbf{R}_2$ should be separated in the longitudinal, X, direction as $\mathbf{R}_1 = (X, 0, 0)$, $\mathbf{R}_2 = (X + \Delta X, 0, 0)$ and in the transverse, Y, direction as $\mathbf{R}_1 = (X, 0, 0)$, $\mathbf{R}_2 = (X, \Delta Y, 0)$ respectively (Fig. 6a and b).

First we show the estimation of the longitudinal correlation scale. In this case, the distance between the rays is $\delta \equiv \sqrt{\delta_z^2 + \delta_x^2} = |\Delta Z|\sin\theta$, where $|\Delta Z|$ is the vertical distance between the neighboring rays, arriving to the points X and $X + \Delta X$:

$$\Delta Z(x) = \sqrt{H^2 + x(X + \Delta X) - x^2} - \sqrt{H^2 + xX - x^2}. \tag{67}$$

When $x\Delta X \ll H^2$, the distance $\Delta Z(x)$ as well as $\sin\theta = \sqrt{H^2 + xX - x^2}/\sqrt{H^2 + (X/2)^2}$ can be expanded into the Taylor series. Keeping only linear term in ΔX for distance between the rays one has

$$\delta(x) = \frac{x\Delta X}{2\sqrt{H^2 + (X/2)^2}}. \tag{68}$$

For a Gaussian correlation function, the inner integral in Eq. (55) can be transformed to the following form by using a similar way given in Kravtsov et al. (2003):

$$\int K_n(\boldsymbol{\kappa} s_- + \boldsymbol{\delta}) ds_- = l_{\text{eff}} \exp\left(-\frac{\delta^2}{l_\delta^2}\right). \tag{69}$$

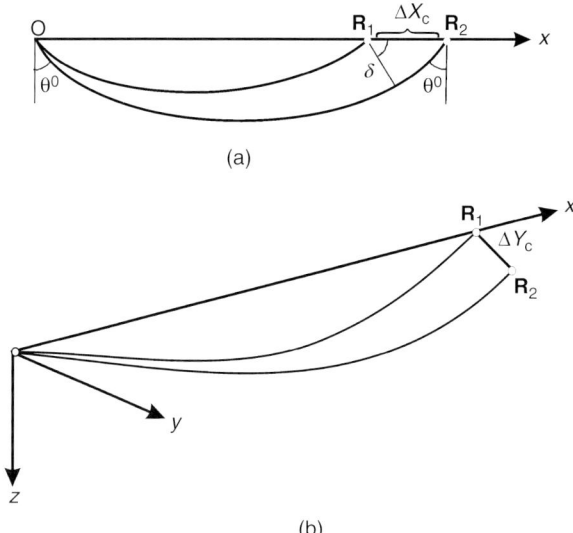

FIG. 6. Longitudinal (a) and transverse (b) location of observation points $\mathbf{R}_1$ and $\mathbf{R}_2$ (reprinted from Kravtsov *et al.*, 2005).

Here the effective correlation length l_{eff} is given by Eq. (59) and the characteristic transverse length l_δ is introduced by

$$l_\delta = \sqrt{(l_z \sin\theta)^2 + (l_x \cos\theta)^2} = l_z\sqrt{\sin^2\theta + (\cos\theta/\rho)^2}. \tag{70}$$

As a result the longitudinal covariance function takes the following form

$$C_l(X, X+\Delta X) = \frac{2\Gamma\sigma_\zeta^2 l_z H}{v_0^2} \int_0^X \frac{\bar{n}^2 d\eta}{\sqrt{(\rho\sin\theta)^2 + \cos^2\theta}} \exp\left[-\frac{(x\Delta X)^2}{(4H^2 + X^2)l_\delta^2}\right]. \tag{71}$$

Using the variables in Eq. (60), Eq. (71) can be rewritten as

$$C_l(X, X+\Delta X) = GP(\rho, \gamma, \Lambda), \tag{72}$$

where G is defined by Eq. (62a) and P is

$$P(\rho, \gamma, \Lambda) = B \int_{-\gamma/2}^{\gamma/2} \frac{\exp(-Q)d\eta}{(B^2 - \eta^2)^{3/2}\sqrt{\rho^2(B^2 - \eta^2) + \eta^2}}. \tag{73}$$

The parameter Q in Eq. (73) is given by

$$Q = \frac{\left(\eta + \frac{\gamma}{2}\right)^2 \Lambda^2}{B^2[\rho^2(B^2 - \eta^2) + \eta^2]}, \tag{74}$$

with $\Lambda = \frac{\Delta X}{l_x}$. For a given normalized distance γ, the dimensionless longitudinal correlation radius $\Lambda_c = \Delta X_c/l_x$ and then l_x can be determined from the following equation:

$$P(\rho, \gamma, \Lambda_c) = \tfrac{1}{2} P(\rho, \gamma, 0) = \tfrac{1}{2} J(\rho, \gamma). \tag{75}$$

Here Λ_c is the particular value of Λ which satisfies Eq. (75) and ΔX_c is the particular longitudinal correlation radius. Inserting the estimated longitudinal scale length l_x in ρ provides the estimation of the vertical scale l_z and subsequently the estimation of the standard deviation σ_ξ from $\sigma_\xi^2 l_z$. The estimation procedure of the longitudinal correlation scale is shown by a flowchart in Fig. 7.

To estimate the transverse correlation scale, we recall the observation points $\mathbf{R}_1 = (X, 0, 0)$ and $\mathbf{R}_2 = (X, \Delta Y, 0)$ given in Fig. 6b. In this case, the distances between the rays in x and z directions are equal to zero, $\delta_z = \delta_x = 0$, while δ_y is proportional to x by $\delta_y = x\Delta Y/X$. We consider $\Delta \mathbf{r} = \boldsymbol{\kappa} s_- + \boldsymbol{\delta}_y$ and write Eq. (17) in the following form

$$g = \left\{ \left[\left(\frac{\kappa_x}{l_x}\right)^2 + \left(\frac{\kappa_z}{l_z}\right)^2 \right] s_-^2 + \left[\frac{(\Delta Y) x}{l_y X} \right]^2 \right\}^{1/2} \tag{76}$$

and derive the transverse travel time covariance $C_t(X, 0, 0; X, \Delta Y, 0)$ in analogy with Eq. (71) as follows

$$C_t(X, 0, 0; X, \Delta Y, 0) = \frac{2\Gamma \sigma_\xi^2 l_z H}{v_0^2} \int_0^X \frac{\bar{n}^2 d\eta}{\sqrt{(\rho \sin \theta)^2 + \cos^2 \theta}} \exp\left[-\frac{(x\Delta Y)^2}{X^2 l_y^2} \right]. \tag{77}$$

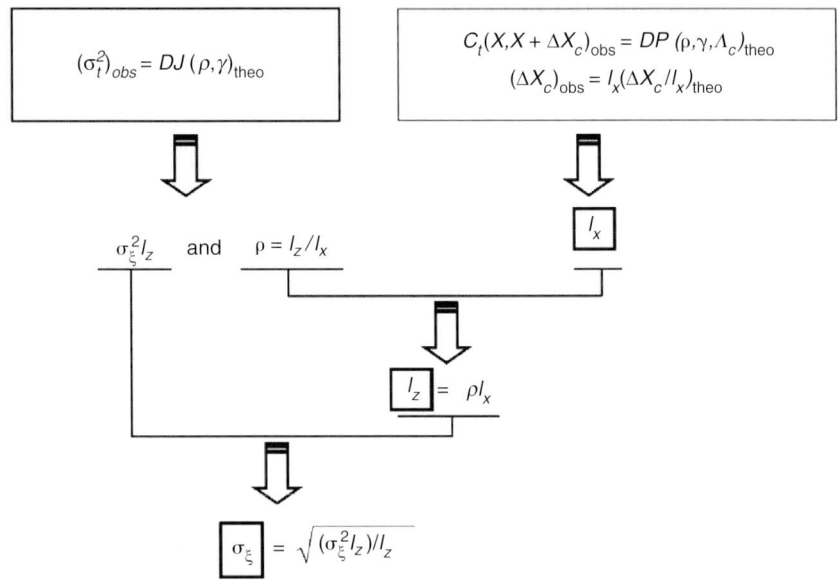

FIG. 7. The flowchart for estimating the statistical parameters of the medium from travel-time fluctuations under longitudinal separation of the observation points like at Fig. 6a.

Instead of longitudinal correlation factor $\exp\left[-(x\Delta X)^2/(4H^2+X^2)l_\delta^2\right]$, which enters in Eq. (71), Eq. (77) contains new factor $\exp\left[-(x\Delta Y)^2/X^2 l_y^2\right]$, which describes the decrease of the correlation along transverse direction. It follows from Eq. (77) that the particular transverse correlation scale ΔY_c is comparable with the horizontal correlation scale $l_y \approx l_x = l_h$ and might be determined from the equation $C_t(X,0,0;X,\Delta Y_c,0) = (1/2)\sigma_t^2 = (1/2)C_t(X,0,0;X,0,0)$. Similar properties have been derived earlier for spherical and cylindrical waves in statistically homogeneous random media (Rytov et al., 1989b).

5. Results of Numerical Simulations

To demonstrate the efficiency of the estimation procedure described above, a numerical simulation was performed for the refraction geometry. A simple ray tracing method was used for a 2D random medium. The model of the random inhomogeneities constituted from 1024 grid points in z direction and from 2048 grid points in x direction, respectively. The grid spacing was chosen 10 m in both directions. The inhomogeneity characteristic lengths were selected as $l_z = 50$ m and $l_x = 500$ m, so that the ratio $\rho = l_z/l_x$ was equal to $\rho = 0.1$.

The background and the random parts of the wave velocity were calculated as given in Eqs. (48) and (49). The near-surface velocity v_0 and the velocity gradient k were considered as $v_0 = 2000$ ms^{-1} and $k = 0.8$ s^{-1}, respectively, which gives $H = v_0/k = 2.5$ km.

The auxiliary random field $\xi(\mathbf{r})$, used in numerical simulations, was supposed to have a Gaussian statistics and anisometric Gaussian covariance function of the form

$$C_\xi(\Delta\mathbf{r}) = \sigma_\xi^2 \exp\left\{-\left[\left(\frac{\Delta x}{l_x}\right)^2 + \left(\frac{\Delta z}{l_z}\right)^2\right]\right\}. \tag{78}$$

The random field $\xi(\mathbf{r})$ with such a covariance function was generated first in the wave number domain and then its spatial representation was obtained by employing an inverse Fourier transform (Frankel and Clayton, 1986). In total, 100 realizations of the random refractive medium were generated in this way.

The travel-time fluctuations were calculated directly by using the basic GO relation given in Eq. (10). The travel-time fluctuations were expressed in the following convenient form:

$$\tilde{t}(X,0) = \frac{1}{v_0}\sqrt{H^2 + \left(\frac{X}{2}\right)^2} \int_0^X \frac{\tilde{n}[x,z(x)]\,dx}{\sqrt{H^2 + xX - x^2}}. \tag{79}$$

The refractive index fluctuations were calculated at each grid point and then the rays were traced by using Eq. (54). Linear interpolation was performed in all cases, when the ray trajectory did not match the grid points (Fig. 8). One hundred travel-time fluctuations were obtained for 100 available medium realizations, and 10 of them are shown in Fig. 9.

The next step was to calculate the travel-time variance by using the following expression

116   KASLILAR ET AL.

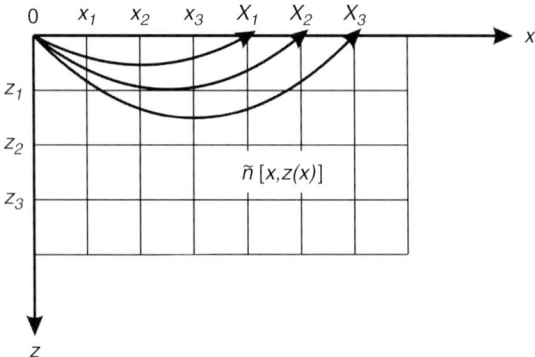

FIG. 8. Rays traveling in a refractive random medium. X_1, X_2, and X_3 represent the final points of the rays while x_1, x_2, x_3 and z_1, z_2, z_3 represents the grid points. Linear interpolation is performed in all cases, when ray trajectories did not match the grid points.

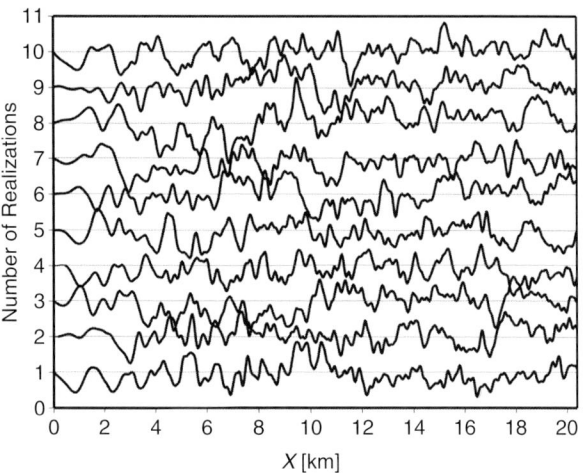

FIG. 9. Examples of travel-time fluctuations along X [km] axis, obtained by numerical simulations (reprinted from Kravtsov et al., 2005). The horizontal lines corresponding to each realization number represent the zero axes of travel-time fluctuations. The vertical distance between them corresponds to 0.005 s.

$$\left[\sigma_t^2(X)\right]_{\text{num}} = \frac{1}{N}\sum_{i=1}^{N}\left[\tilde{t}_i(X)\right]^2, \quad N = 100. \tag{80}$$

The result of the calculation is shown by the thin curve in Fig. 10 as a function of the normalized distance γ. The nonmonotonic variations of the numerical variance

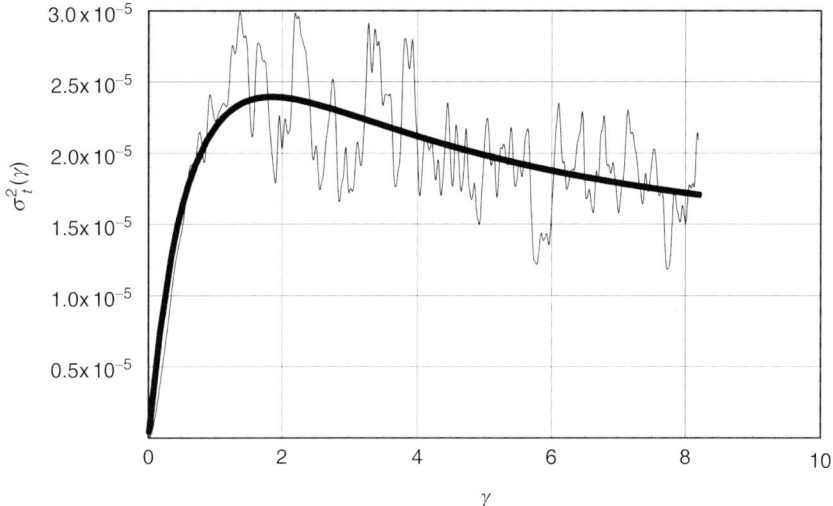

FIG. 10. Variance of the travel-time fluctuations obtained by numerical simulations (thin curve). Thick curve represents the theoretical variance fitted to "numerical" data (reprinted from Kravtsov et al., 2005).

$\left[\sigma_t^2(X)\right]_{num}$ are due to a limited number of realizations. By employing an inversion procedure between the numerical variance Eqs. (80) and (62) (thick curve in Fig. 10), the medium statistical parameters are estimated as $\left(\sigma_\xi^2 l_z\right)_{num} = 0.005$ and $\rho_{num} = l_z/l_x = 0.12$.

Finally the longitudinal correlation radius ΔX_c was estimated from the following equation

$$C_t(X, X + \Delta X_c)_{num} \equiv \frac{1}{N}\sum_{i=1}^{N} \tilde{t}_i(X)\tilde{t}_i(X + \Delta X_c) = \frac{1}{2}\left[\sigma_t^2(X)\right]_{num}. \quad (81)$$

The correlation radius ΔX_c, calculated from Eq. (81), are shown by thick curve in Fig. 11.

Simultaneously the roots Λ_c of the Eq. (75) were calculated for each observation point. Then a fitting procedure was employed between the theoretical curve $[\Delta X_c]_{theo} = l_x \Lambda_c$ (thin curve in Fig. 11) and the numerical curve $[\Delta X_c]_{num}$, thick curve in the same figure. As a result, the horizontal correlation length l_x was estimated as $(l_x)_{num} = 450$ m, which is a good approximation to the model parameter $l_x = 500$ m.

Using the estimated values of ρ, l_x and $\sigma_\xi^2 l_z$, the medium statistical parameters are estimated as follows: $l_z = \rho l_x = 0.12 \times 448 = 54$ m, $\sigma_\xi = \sqrt{0.005/l_z} = \sqrt{0.005/54} \cong 0.01$. Estimated parameters $\sigma_\xi^2 l_z$, ρ, σ_ξ, l_z, and l_x are given in Table 1 together with their actual values, chosen for the simulated random elastic medium. Generally, the inverted parameters are in a good agreement with the actual ones.

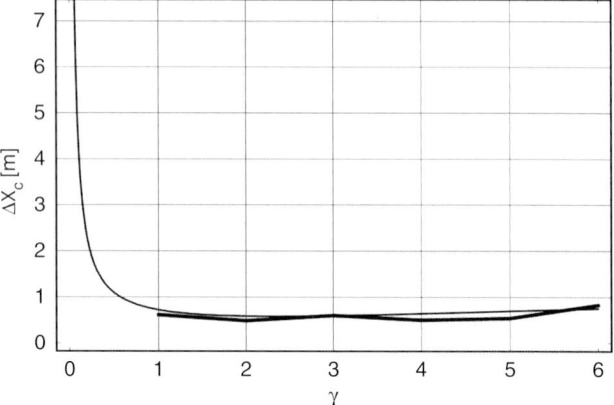

FIG. 11. Longitudinal correlation scale ΔX_c vs. distance $\gamma = X/H$. Numerical calculation (thick curve) and theoretical dependence (thin curve) fitted to the "numerical" data (reprinted from Kravtsov et al., 2005).

TABLE 1. Model and estimated statistical parameters from travel-time fluctuations of refracted waves (reprinted from Kravtsov et al., 2005)

	Model parameters	Estimated parameters
$\sigma_\xi^2 l_z$ (m)	0.005	0.005
ρ	0.1	0.12
σ_ξ	0.01	0.01
l_z (m)	50	54
l_x (m)	500	448

6. Discussion and Conclusion

In this chapter, the travel-time fluctuations are examined in a three-dimensional randomly inhomogeneous and anisometric media for both reflected and refracted seismic waves. It is shown that on the basis of travel-time fluctuations it is possible to estimate the medium statistical parameters such as the standard deviation of the velocity fluctuations and the inhomogeneity scale lengths in vertical and horizontal directions. Besides, a new phenomenon is revealed, concerning the travel-time fluctuations in a stratified refractive random medium: it is shown that the travel-time variance can decrease with offset in a random medium, containing horizontally elongated inhomogeneities. This phenomenon has not been observed before either in acoustics and optics or in radio wave propagation.

The suggested method has been applied to a field seismic data where the properties of the medium were well known by geological observations (Kaslilar et al., 2006). The seismic data were collected along a tunnel wall. The study area was located in the Leventina gneiss complex, which was a part of the Penninic gneiss zone. The zone constitutes from high-grade metamorphic rocks and includes strongly foliated mineral bands like quartz, feldspar, or biotite mica. A high-velocity gradient is obtained around the tunnel wall, which is usually observed in near-surface studies (van Vossen et al., 2005; Wilson and Pavlis, 2000). To estimate the statistical parameters of the medium, the travel-time variance of the real data was calculated from the measured travel-times of 10 shot-receiver group, where each shot-receiver group corresponds to a medium realization. A noticeable deviation between the travel-time variance of the theoretical and real data was observed. This deviation might be ascribed to both insufficiency of the GO method and the limited number of available medium realizations. In spite of this noticeable deviation between the real and theoretical travel-time variances, the statistical parameters were estimated. The obtained results were comparable with the geological observations, however the method needs to be verified by modeling the full wave propagation to understand the applicability and limits of the method for real Earth structures.

For applicability of the method sufficient number of independent source–receiver pairs (medium realizations) are needed. Besides, the characteristic lengths of inhomogeneities should be larger than the wavelength and the width of the first Fresnel zone. When these conditions are not met, the results of GO are still in agreement with the diffraction theory, based on the Rytov's method (Rytov et al., 1989a,b). However detailed numerical analysis, based on the comparison of the GO travel-times with the wave theoretical travel-times is required. The effect of velocity shift and its effect on the estimated parameters (Roth, 1997; Saito, 2006; Samuelides, 1998; Shapiro et al., 1996) need detailed investigation to understand the limits of our method. To obtain more accurate results, one should use the methods which take into account diffraction phenomena, and thus leading to frequency dependences of the phenomenon considered here.

In this chapter, we have concentrated on the simplest—Gaussian—model of autocorrelation function, which is characterized in fact by four parameters: variance $\langle \tilde{n}^2 \rangle$ and three correlation lengths $l_{x,y,z}$. It is obvious that the real auto-correlation function of geological structures differs from Gaussian one. However because of its simplicity the Gaussian auto correlation function (ACF) was used in many works on seismic scattering, for example, by Flatté and Wu (1988), Sato and Fehler (1998), Baig et al. (2003), Petukhin and Gusev (2003), and Saito (2006). Recently, Przybilla (2007) has shown that the Gaussian ACF describes heterogeneities of volcanic regions satisfactorily.

Presented geometrical theory can be modified also for different correlation functions (Klimeš, 2002), for example, for power law type correlation function, which depends on several additional unknown medium parameters. However, extraction of these additional parameters from available travel-time statistics might be accompanied with great technical difficulties. In these conditions, a four parameter Gaussian correlation function presents a kind of practical compromise between multiparameter correlation functions and experimental possibilities.

Information on the statistical parameters, obtained by the statistical methods, is quite important to understand the uncertainties in seismic images. The estimated parameters can be used in seismic inversion as statistical *a priori* models of velocity field. They can be helpful for understanding the seismic attenuation mechanisms, for estimation of the intensity of the heterogeneities in a studied medium and for obtaining valuable information about the dominant scales of the inhomogeneities. Therefore the estimation of the statistical parameters from seismic reflection and refraction data represents an important task to better understand the Earth's internal structure.

Acknowledgements

The authors thank Haruo Sato, Tatsuhiko Saito, and two anonymous reviewers for their valuable comments and suggestions on the manuscript. This work was supported by the sponsors of the Wave Inversion Technology (WIT) Consortium and by the Deutsche Forschungsgemeinschaft (in the framework of the SFB267). A. Kaslilar gratefully acknowledges this financial support and thanks seismic-seismology workgroup of FU-Berlin for friendly collaboration.

References

Aki, K. (1969). Analysis of the seismic coda of local earthquakes as scattered waves. *J. Geophys. Res.* **74,** 615–631.
Aki, K., Chouet, B. (1975). Origin of coda waves: Source, attenuation, and scattering effects. *J. Geophys. Res.* **80,** 3322–3342.
Baig, A.M., Dahlen, F.A., Hung, S.-H. (2003). Traveltimes of waves in three-dimensional random media. *Geophys. J. Int.* **153,** 467–482.
Born, M., Wolf, E. (1999). *Principles of Optics: Electromagnetic Theory of Propagation, Interference and Diffraction of Light.* 7th ed. Cambridge University Press, Cambridge.
Červený, V. (2001). *Seismic Ray Theory.* Cambridge University Press, Cambridge.
Chernov, L.A. (1960). *Wave Propagation in a Random Medium.* Dover Publications, New York.
Farra, V., Madariaga, R. (1987). Seismic waveform modeling in heterogeneous media by ray perturbation theory. *J. Geophys. Res.* **92,** 2697–2712.
Flatté, S.M., Wu, R.S. (1988). Small-scale structure in the lithosphere and asthenosphere deduced from arrival-time and amplitude fluctuations at NORSAR. *J. Geophys. Res.* **93,** 6601–6614.
Frankel, A., Clayton, R.W. (1986). Finite difference simulations of seismic scattering: Implications for the propagation of short-period seismic waves in the crust and models of the crustal heterogeneity. *J. Geophys. Res.* **91,** 6465–6489.
Iooss, B. (1998). Seismic reflection travel times in two-dimensional statistically anisotropic random media. *Geophys. J. Int.* **135,** 999–1010.
Iooss, B., Blanc-Benon, Ph., Lhuillier, C. (2000). Statistical moments of travel times of second order in isotropic and anisotropic random media. *Waves in Random Media* **10,** 381–394.
Ishimaru, A. (1978). *Wave Propagation and Scattering in Random Media.* Academic Press Inc, New York.
Ishimaru, A. (1997). *Wave Propagation and Scattering in Random Media.* SPIE Press, New York; Oxford: Oxford Press..
Kaslilar, A., Kravtsov, Y.A., Shapiro, S.A., Buske, S., Giese, R., Dickmann, Th. (2006). Estimation of the rocks statistical parameters from traveltime measurements. *Stud. Geophys. Geod.* (Special Issue), **50,** 325–336.
Klimeš, L. (2002). Correlation functions of random media. *Pure Appl. Geophys.* **159,** 1811–1831.
Kravtsov, Yu.A. (2005). *Geometrical Optics in Engineering Physics.* Alpha Science International, Harrow, UK.
Kravtsov, Yu.A., Orlov, Yu.I. (1990). *Geometrical Optics of Inhomogeneous Media.* Springer Verlag, Berlin.
Kravtsov, Yu.A., Müller, T.M., Shapiro, S.A., Buske, S. (2003). Statistical properties of reflection travel-times in 3D randomly inhomogeneous and anisomeric media. *Geophys. J. Int.* **154,** 841–851.
Kravtsov, Yu.A., Kaslilar, A., Shapiro, S.A., Buske, S., Müller, T. (2005). Estimating statistical parameters of an elastic random medium from traveltime fluctuations of refracted waves. *Waves in Random and Complex Media* **15** (No. 1), 43–60.

Marquering, H., Dahlen, F.A., Nolet, G. (1999). Three-dimensional sensitivity kernels for finite-frequency traveltimes: The banana-doughnut paradox. *Geophys. J. Int.* **137,** 805–815.

Müller, G., Roth, M., Korn, M. (1992). Seismic-wave traveltimes in random media. *Geophys. J. Int.* **110,** 29–41.

Müller, T.M., Shapiro, S.A., Sick, C. (2002). Most probable ballistic waves in random media: A weak-fluctuation approximation and numerical results. *Waves in Random Media* **12,** 223–245.

Nolet, G., Dahlen, F.A. (2000). Wave front healing and the evolution of seismic delay times. *J. Geophys. Res.* **105,** 19043–19054.

Petukhin, A.G., Gusev, A.A. (2003). The duration-distance relationship and average envelope shapes of small kamchatka earthquakes. *Pure Appl. Geophys.* **160,** 1717–1743.

Press, W.H., Teukolsky, S.A., Vetterling, W.T., Flannery, B.P. (2001). *Numerical recipes in Fortran 77: The Art of Scientific Computing.* Cambridge University Press, USA.

Przybilla, J. (2007). *Akustische und elastische Energietransfertheorie in kontinuierlichen Zufallsmedien und ihre Anwendungen in der Seismologie,* PhD Thesis, Leipzig University, Germany.

Roth, M. (1997). Statistical interpretation of traveltime fluctuations. *Phys. Earth Planet. Inter.* **104,** 213–228.

Rytov, S.M., Kravtsov, Y.A., Tatarskii, V.I. (1989a). *Principles of Statistical Radio Physics, Vol. 3: Elements of Random Fields.* Springer Verlag, Berlin.

Rytov, S.M., Kravtsov, Y.A., Tatarskii, V.I. (1989b). *Principles of Statistical Radio Physics, Vol. 4: Wave Propagation through Random Media.* Springer Verlag, Berlin.

Saito, T. (2006). Velocity shift in two-dimensional anisotropic random media using the Rytov method. *Geophys. J. Int.* **166,** 293–308.

Samuelides, Y. (1998). Velocity shift using Rytov approximation. *J. Acoust. Soc. Am.* **104,** 2596–2603.

Sato, H., Fehler, M. (1998). *Seismic Wave Propagation and Scattering in the Heterogeneous Earth.* Springer-Verlag and American Institute of Physics Press, New York.

Shapiro, S.A., Hubral, P. (1998). *Elastic Waves in Random Media.* Springer, Berlin.

Shapiro, S.A., Kneib, G. (1993). Seismic scattering attenuation: Theory and numerical results. *Geophys. J. Int.* **114,** 373–391.

Shapiro, S.A., Schwarz, R., Gold, N. (1996). The effect of random isotropic inhomogeneities on the phase velocities of seismic waves. *Geophys. J. Int.* **127,** 783–794.

Snieder, R., Sambridge, R. (1992). Ray perturbation theory for travel-times and ray paths in three-dimensional heterogeneous media. *Geophys. J. Int.* **109,** 294–322.

Spetzler, J., Snieder, R. (2001). The effects of small-scale heterogeneity on the arrival time of waves. *Geophys. J. Int.* **145,** 786–796.

Tatarskii, V.I. (1961). *Wave Propagation in a Turbulent Medium.* McGraw-Hill, New York.

Tatarskii, V.I. (1967). *Wave Propagation in a Turbulent Medium.* Dover Publ, New York.

Tatarskii, V.I. (1971). *The Effect of the Turbulent Atmosphere on Wave Propagation.* Nat. Technol. Inform. Service USA, VA, 22151, Springfield.

Tong, J., Dahlen, F.A., Nolet, G., Marquering, H. (1998). Diffraction effects upon finite-frequency travel-times: A simple 2-D example. *Geophys. Res. Lett.* **25,** 1983–1986.

Touati, M. (1996). Contribution géostatistique au traitement des donnéés géophysique, thése de doctorat en géostatistique, ecole des mines de paris, France.

Touati, M., Iooss, B., Galli, A. (1999). Quantitative control of migration: A geostatistical attempt. *Math. Geol.* **31,** 277–295.

van Vossen, R., Curtis, A., Trampert, J. (2005). Subsonic near-surface P-velocity and low S-velocity observations using propagator inversion. *Geophysics* **70,** R15–R23.

Wilson, D.C., Pavlis, G.L. (2000). Near-surface site effects in crystalline bedrock: A comprehensive analysis of spectral amplitudes determined from a dense, three-component seismic array. *Earth Interact.* **4,** 1–31.

Witte, O., Roth, M., Müller, G. (1996). Ray tracing in random media. *Geophys. J. Int.* **124,** 159–169.

ATTENUATION OF SEISMIC WAVES DUE TO WAVE-INDUCED FLOW AND SCATTERING IN RANDOMLY HETEROGENEOUS POROELASTIC CONTINUA

Tobias M. Müller,[1] Boris Gurevich and Serge A. Shapiro

Abstract

Attenuation and dispersion of compressional seismic waves in inhomogeneous, fluid-saturated porous media are modeled in the framework of wave propagation in continuous random media. Two dominant attenuation mechanisms are analyzed in detail. First, attenuation due to wave-induced flow, an intrinsic attenuation mechanism where a passing seismic wave introduces localized movements of the viscous fluid which are accompanied by internal friction. Second, attenuation due to scattering, the so-called apparent attenuation where ordinary elastic scattering is responsible for a redistribution of wavefield energy in space and time. Despite the fact that both attenuation mechanisms have a quite different physical nature, the theory of wave propagation in random media provides a unified framework to model these effects in a consistent manner. In particular, it is shown that the method of statistical smoothing can be applied not only to energy conserving systems (elastic scattering) but also to energy absorbing systems (conversion scattering into diffusion waves). Explicit expressions for attenuation and dispersion for relevant correlation models are presented, and the asymptotic frequency scaling at low- and high frequencies of both attenuation mechanisms are compared and contrasted.

KEY WORDS: Seismic Attenuation, random media, poroelasticity, scattering, wave-induced flow, perturbation theory, Biot's slow wave, self-averaging. © 2008 Elsevier Inc.

1. Introduction

Elastic wave propagation in heterogeneous structures is always accompanied by attenuation and dispersion. It is widely accepted that attenuation is a combination of intrinsic attenuation (absorption) and scattering attenuation (Aki and Chouet, 1975). A key issue in rock characterization is to quantify the magnitude and frequency dependence of these attenuation mechanisms. Closely related is the problem of extracting this information from the recorded wavefield originated by artificial or natural "seismic" sources. In this chapter, we focus on the modeling of the effects of an intrinsic attenuation mechanism, namely, the effect of wave-induced flow, and scattering attenuation in random porous, fluid-saturated media.

---

[1] Author thanks e-mail: tobias.mueller@csiro.au; b.gurevich@curtin.edu.au; shapiro@geophysik.fu-berlin.de

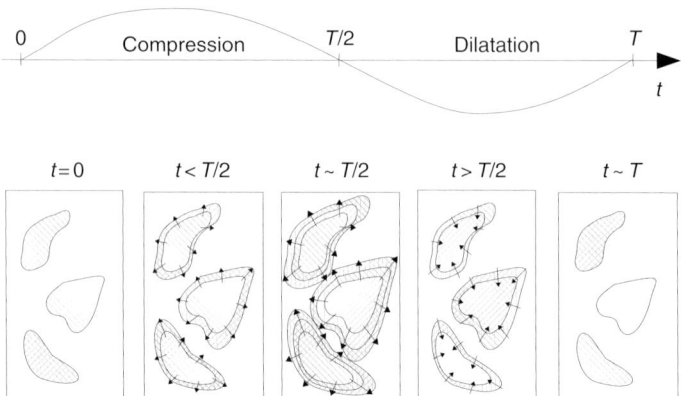

Fig. 1. The mechanism of wave-induced flow. During the compression cycle of a wave, there will be fluid flow from elastically soft inhomogeneities into the background (shown here; the flow direction is indicated by arrows) and flow from the background into elastically stiff inhomogeneities. During the extension cycle of the wave, the fluid flow becomes reversed.

One major cause of elastic wave attenuation in porous fluid-saturated media is viscous dissipation due to the flow of the pore fluid induced by the passing wave. Wave-induced fluid flow occurs as a passing wave creates local pressure gradients within the fluid phase and the resulting fluid flow is accompanied by viscous friction until the pore pressure is equilibrated (this phenomenon is schematically depicted in Fig. 1). The fluid flow can take place on various length scales, for example, from compliant fractures into the equant pores, the so-called squirt flow (Mavko and Nur, 1975; Jones, 1986), or between mesoscopic heterogeneities like fluid patches in partially saturated rocks (White et al., 1975; Murphy, 1982). Theoretical studies of the elastic wave attenuation due to wave-induced flow go back to the 1970s. In such studies, wave propagation in inhomogeneous porous medium is usually analyzed using Biot's equations of poroelasticity with spatially varying coefficients (Biot, 1962). The first models of attenuation, due to wave-induced flow, considered flow caused by a regular assemblage of inhomogeneities of ideal shape such as two concentric spheres or flat slabs (White et al., 1975). A general theory of wave propagation in heterogeneous porous media using a double-porosity approach was recently developed by Pride et al. (2004). A comparative review of a number of models of wave propagation in heterogeneous porous media has been given by Toms et al. (2006).

In real Earth materials, heterogeneities are spatially distributed in a random fashion. The effect of wave-induced flow can be interpreted as conversion scattering from propagating wave modes into the Biot's slow wave mode. At sufficiently low-frequencies, Biot's slow wave is a fluid diffusion wave. Therefore, the theory of wave scattering in random media (e.g., Ishimaru, 1978) applied to Biot's equations of poro-elasticity can be used to study the propagation characteristics of seismic waves in inhomogeneous porous media. Gurevich and Lopatnikov (1995) used the so-called Bourret approximation to quantify attenuation and dispersion of P waves in randomly layered 1-D poroelastic media. Gelinsky et al. (1998) developed a perturbation theory approach based on the Rytov representation of wavefields in 1-D random media illuminating the interplay between elastic scattering and dissipation. For 3-D randomly inhomogeneous porous media, Müller and Gurevich (2005a,b) analyzed the conversion scattering problem using the first-order statistical smoothing approximation and obtained an effective wave number of the coherent field, which accounts for the effect of wave-induced flow. This is different from the usual application of the method of smoothing to energy conserving wave systems, where an apparent dissipation (so-called scattering attenuation) results from the energy transfer from the coherent component of the wavefield into the incoherent component.

Another, so-called "apparent" attenuation mechanism in heterogeneous (porous) media is scattering. Because of elastic scattering by the inhomogeneities, the wavefield becomes distorted and wavefield energy is transferred from the primary signal to the coda. Generally, the wavefield can be described as a sum of coherent and incoherent wavefields. Which part of the wavefield is actually measured in experiments depends on the size of the receiver used. In geophysical applications, the receivers are typically small compared with the wavelength and the size of inhomogeneities, so that the incoherent field will not be averaged out and both parts of the wavefield are present in seismograms. That is, for point-like receivers no aperture averaging, which reduces the fluctuations, takes place. Therefore, there may occur a discrepancy between the recorded wavefield and the coherent wavefield (or equivalently mean field) as shown by Wu (1982a,b) and Sato (1982). Formalisms that take into account these shortcomings and that try to improve the statistical averaging procedure and adopt it for seismology are based on heuristic assumptions like the travel-time-corrected formalism (see for an overview Sato and Fehler, 1998). There is, however, a lack of first principles, that is, wave-equation-based descriptions of scattering attenuation, which go beyond the mean field theory and that are valid for seismograms recorded in single realizations of random media.

Estimates of scattering attenuation in single realizations of 2-D and 3-D random media are obtained by Müller and Shapiro (2001) and Müller et al. (2002) by extending the generalized O'Doherty–Anstey (ODA) formalism of randomly layered 1-D media (Shapiro and Hubral, 1999). It is a dynamic-equivalent medium approach that is applicable to a broad range of frequencies and uses combinations of perturbation approximations including the method of statistical smoothing and the Rytov approximation.

This chapter is organized as follows. In Section 2, we briefly review typical properties of heterogeneities in the Earth reported in literature and introduce the basic random medium descriptors used in the sequel. Then, in Section 3, we present a theory for wave attenuation and dispersion due to wave-induced flow in poroelastic media based on a variant of the method of statistical smoothing. Section 3 is the largest section containing the introduction of wave propagation in poroelastic media, a fairly detailed account of the method of statistical smoothing, and analysis of attenuation and dispersion due to wave-induced flow. In Section 4, a similar formalism is developed for scattering attenuation. We outline the principles of the generalized ODA formalism describing seismic primary wavefields in a single realization of random media. The method of statistical smoothing (describing the coherent part of the wavefield) is combined with the Rytov approximation to obtain expressions for attenuation and dispersion that are applicable over a broad frequency range. Whereas in Sections 3 and 4 the attenuation mechanisms are treated separately, in Section 5 we present results for 1-D poroelastic random media, where the interplay of both attenuation mechanisms can be analyzed in explicit manner again using the generalized ODA formalism. The chapter closes with some concluding remarks (Section 6).

2. Meso- and Macroscopic Heterogeneity in the Earth and Its Description as a Random Medium

Large parts of the Earth's crust in general and hydrocarbon bearing reservoirs, in particular, have a very complex structure including multiscale heterogeneities. This becomes evident from geological surveys, measurements in vertical and horizontal boreholes, and seismic-wavefield analysis (Wu and Flatté, 1990; Mukerji *et al.*, 1995; Sato and Fehler, 1998). Reported scales of heterogeneity range from centimeter up to several kilometers (Murphy *et al.*, 1984; Holliger and Levander, 1992; Wu *et al.*, 1994). Laboratory measurements of rocks reveal that even on the microscopic scale there are heterogeneities due to varying grain sizes and fluid flow channels (Bourbié *et al.*, 1987). An example of centimeter-scale heterogeneities in a porous rock sample is shown in Fig. 2.

Any attempt to describe complex geological structures by deterministic models fails in the sense that the interaction between seismic wavefields and the heterogeneities is not correctly reproduced. In such cases, the concept of random media characterizing the heterogeneities by their statistical moments and their spatial correlations is more suitable. Typical wavelengths in seismology are of the order of hundreds of meters up to a few kilometers (Aki and Richards, 1980). The travel distance can be some hundreds of kilometers. In exploration seismology, the probing pulses have dominant wavelengths of several tens of meters, traveling a few kilometers through the heterogeneous crust (Levander and Gibson, 1991). Altogether, three relevant length scales come into play: the wavelength λ of the probing wave, the characteristic size of heterogeneities a, and the total travel distance L of the wave in the heterogeneous structure. In general, one can expect $max \ \{\lambda, a\} \ll L$. Since there is a huge range of heterogeneity scales, we specify the relevant length scales in the context of the present work. Throughout this chapter, we assume that a is much larger than a typical length scale of the pore space a_{pore}. We refer to *mesoscopic heterogeneity* if

$$a_{\text{pore}} \ll a \ll \lambda. \tag{1}$$

This relation is typically applicable in situations where the effect of wave-induce flow occurs, and these scales are illustrated in Fig. 3. In general, however, there are also *macroscopic heterogeneities* such that

$$a_{\text{pore}} \ll a \lessgtr \lambda. \tag{2}$$

The latter relation is of particular importance for the analysis of scattering attenuation. For instance, if $\lambda \leq a$ diffraction effects need to be accounted for.

In randomly inhomogeneous media, all medium parameters can be presented as random fields $X(\mathbf{r})$. To be more specific, we assume that each of these parameters is the sum of a constant background value, $\bar{X}$ and a fluctuating part, $\tilde{X}(r)$, so that

$$X = \bar{X} + \tilde{X} = \bar{X}(1 + \varepsilon_X), \tag{3}$$

where $\varepsilon_X = \tilde{X}/\bar{X}$ denotes the relative fluctuations. The average over the ensemble of the realizations (denoted by $\langle \cdot \rangle$) of ε_X is assumed to be zero: $\langle \varepsilon_X \rangle = 0$. The spatial autocorrelation function of two random fields is defined as

$$B_{XX}(\delta \mathbf{r}) = \langle \varepsilon_X(\mathbf{r} + \delta \mathbf{r}) \varepsilon_X(\mathbf{r}) \rangle, \tag{4}$$

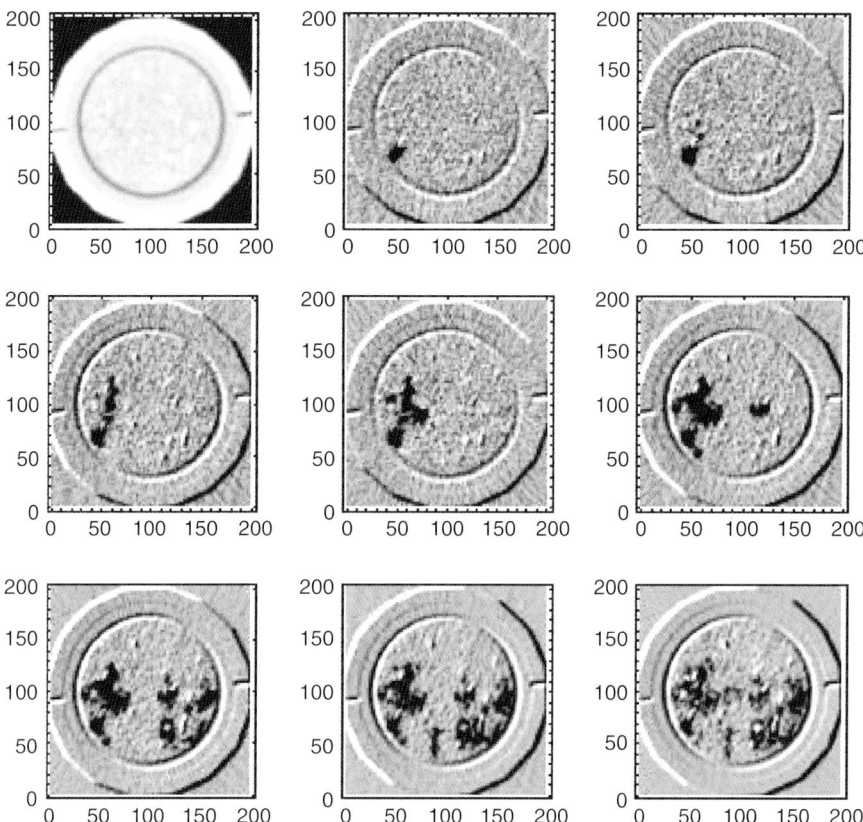

FIG. 2. Experimental evidence of mesoscopic heterogeneities observed during injecting gas into a fully water-saturated limestone sample. The upper left plot shows a 2-D cut through the initially water-saturated rock sample (the sample has a diameter of 5 cm). Gas is injected into the sample at the lower left-hand side of the sample (visualized by black color). During the course of the experiment, the gas phase assumes a complex spatial distribution that evolves with time (from left to right starting from the first row). This experiment illustrates that the two different fluid phases within the porous rock form clusters of mesoscopic sizes. Courtesy Lincoln Patterson, CSIRO (2003).

where the dependence of B on the difference vector δr only is a consequence of the assumption of statistically homogeneous random fields (Rytov *et al.*, 1989), which we use throughout this chapter. The variance of the random process ε_X will be denoted as $B_{XX}(0) = \langle \varepsilon_X^2 \rangle = \sigma_{XX}^2$. In statistically isotropic random media, the fluctuation spectrum $\Phi(\kappa)$ and the correlation function $B(r)$ are related through the 3-D Hankel transform

$$B(r) = \frac{4\pi}{r} \int_0^\infty \kappa \Phi(\kappa) \sin(\kappa r) d\kappa. \tag{5}$$

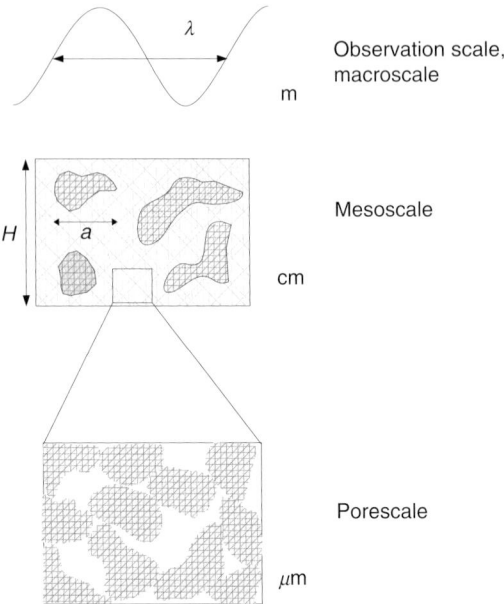

FIG. 3. The scale of heterogeneity that is typically relevant for wave-induced flow to occur at seismic frequencies is on the centimeter scale and is called mesoscopic.

Note that Eq. (5) implies that the integral over the fluctuation spectrum is finite if $B(0)$, that is, the variance is finite. As is typical for statistical wave problems, in the following, we assume that the constant part $\bar{X}$ and statistical properties of the fluctuations ε_X are known.

Ideally, the correlation function should be inferred from experimental data such as X-ray images of rock samples. In many circumstances, the true correlation behavior can be well approximated by simple functions such as an exponential function or combinations of them. A review of frequently used correlation functions in random media is provided by Klimeš (2002). In this chapter, we will make use of the following correlation models. The inhomogeneities are said to be exponentially correlated if variance-normalized $B(r)$ is of the form

$$B(r) = \exp[-|r|/a]. \qquad (6)$$

Here a denotes the correlation length, that is, a characteristic length scale associated with the inhomogeneities. More precisely, the correlation length a is the length scale at which $B(r)$ assumes the value e^{-1}. The Gaussian correlation model is defined as

$$B(r) = \exp[-r^2/a^2]. \qquad (7)$$

Another widely used correlation model is the von Kármán function

$$B(r) = 2^{1-\nu}\Gamma^{-1}(\nu)\left(\frac{r}{a}\right)^{\nu} K_{\nu}(r/a), \qquad (8)$$

ATTENUATION OF SEISMIC WAVES

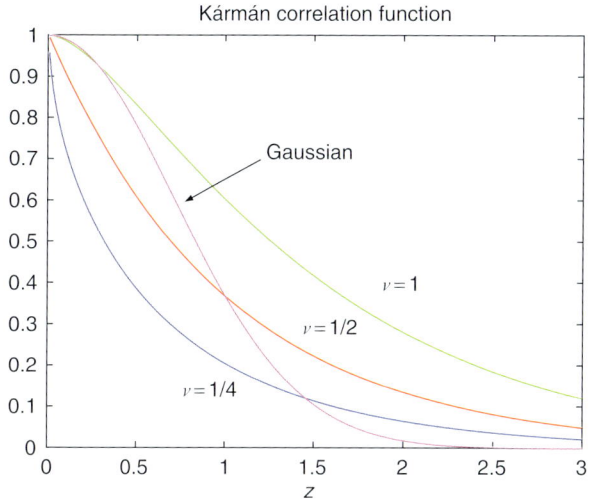

FIG. 4. The von Kármán correlation function for several values of ν with $z = r/a$ and $a = 1$m.

where K_ν is the modified Bessel function of the third kind (MacDonald function) and Γ denotes the Gamma function. The von Kármán correlation function involves an additional parameter, the so-called Hurst coefficient ν, which is assumed to be $0 < \nu \leq 1$. For instance, crustal heterogeneities are best explained using a von Kármán autocorrelation function with a standard deviation of the velocity fluctuations on the order 3–6% as discussed in more detail by Sato and Fehler (1998) and Goff and Holliger (2003). For the case $\nu = 1/2$ the von Kármán function is identical to the exponential correlation function (Eq. 6). Its fluctuation spectrum is given by

$$\Phi(\kappa) = \frac{a^3 \Gamma(\nu + 3/2)}{\pi^{3/2} \Gamma(\nu)(1 + \kappa^2 a^2)^{\nu + 3/2}}. \tag{9}$$

Figure 4 shows the correlation functions used throughout this chapter.

3. Attenuation and Dispersion of Seismic Waves due to Wave-Induced Flow

3.1. Biot's Equations of Dynamic Poroelasticity and Associated Green's Functions

Biot's equations (Biot, 1962) of dynamic poroelasticity provide a general framework for modeling elastic wave propagation through fluid-saturated porous media. One essential feature of Biot's theory is the prediction of a second compressional wave (the Biot slow wave). At frequencies much lower than the so-called characteristic Biot frequency

$$\omega_c = \frac{\phi \eta}{\kappa_0 \rho_f}. \tag{10}$$

(where ϕ is the porosity, κ the permeability, η the fluid viscosity, and ρ_f the fluid mass density), this slow P wave is a diffusion wave (Bourbié et al., 1987). For most fluid-saturated rocks, ω_c is of the order of 10 kHz or larger. In the following, we restrict our analysis to the frequency range $\omega \ll \omega_\text{c}$. Then, the modulus of the wave number of the slow P wave, k_ps, is much larger than that of the propagating (fast) P wave. Therefore, the wave number ratio is a small number:

$$\frac{|k_\text{p}|}{|k_\text{ps}|} \ll 1. \tag{11}$$

We will frequently make use of this relation. Using index notation—summation over repeated indices is assumed and partial derivatives are denoted as i or ∂_i—we can write the equations of motion in the frequency domain (the time-harmonic dependency $\exp(-i\omega t)$ is omitted):

$$\rho\omega^2 u_i + \rho_\text{f}\omega^2 w_i + \tau_{ij,j} = 0 \tag{12}$$

$$\rho_\text{f}\omega^2 u_i + q\omega^2 w_i - p_i = 0, \tag{13}$$

where τ_{ij} is the total stress tensor, p the fluid pressure, u_i and w_i the components of the solid and relative fluid displacement vectors, respectively. The relative fluid displacement is defined as $w_i = \phi(U_i - u_i)$, where U_i is the fluid displacement. The densities of the solid and fluid phase are denoted by ρ_g and ρ_f so that the bulk density is given by $\rho = \phi\rho_\text{f} + (1 - \phi)\rho_\text{g}$. The parameter q is defined as $q = i\eta/(\omega\kappa_0)$. We note that this definition of q is a consequence of the low-frequency assumption (Eq. 11).

To obtain a closed system of wave equations in the displacements u_i and w_i, we complement the equations of motion with the stress–strain relations for an isotropic poroelastic medium (Biot, 1962):

$$\tau_{ij} = G\left[u_{i,j} + u_{j,i} - 2\delta_{ij}u_{j,j}\right] + \delta_{ij}\left[Hu_{j,j} + Cw_{j,j}\right] \tag{14}$$

$$p = -Cu_{j,j} - Mw_{j,j}. \tag{15}$$

Here G is the porous-material shear modulus, and H is the undrained, low-frequency P-wave modulus given by Gassmann's equation (Gassmann, 1951):

$$H = P_\text{d} + \alpha^2 M, \tag{16}$$

where

$$M = \left[(\alpha - \phi)/K_\text{g} + \phi/K_\text{f}\right]^{-1}. \tag{17}$$

In Eqs. (16)–(17) $P_\text{d} = K_\text{d} + 4/3\mu$ is the P-wave modulus of the drained frame, $\alpha = 1 - K_\text{d}/K_\text{g}$ is the Biot–Willis coefficient, $C = \alpha M$, and K_g, K_d, and K_f denote the bulk moduli of the solid phase, the drained frame, and the fluid phase, respectively. The different moduli characterizing a homogeneous poroelastic composite are illustrated schematically in Fig. 5. Symbol δ_{ij} is Kronecker's delta (the identity tensor).

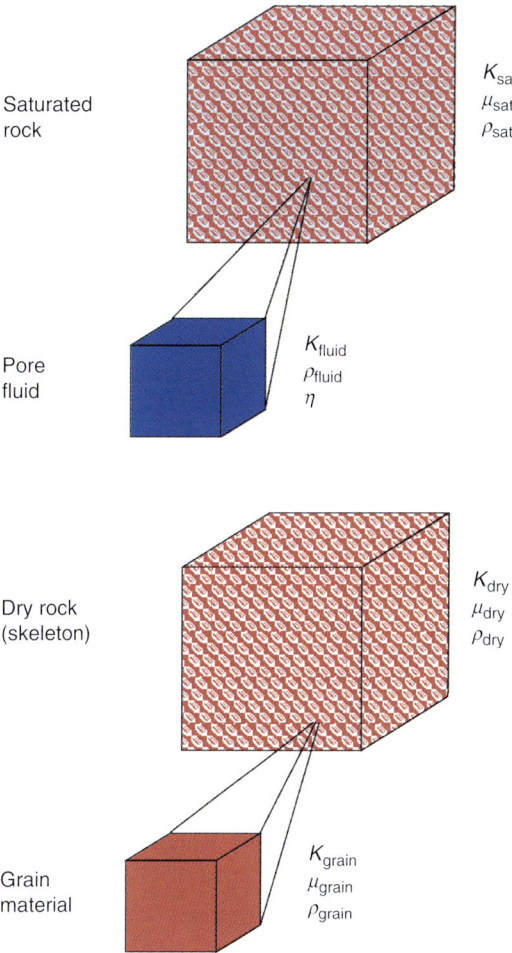

FIG. 5. The different elastic moduli and mass densities characterizing a homogeneous, isotropic poroelastic solid.

It is expedient to write the above system of coupled wave equations in matrix form:

$$\begin{bmatrix} L_{ik}^{(1)} & L_{ik}^{(2)} \\ L_{ik}^{(3)} & L_{ik}^{(4)} \end{bmatrix} \cdot \begin{bmatrix} u_k \\ w_k \end{bmatrix} = \mathbf{0}, \qquad (18)$$

where we defined the linear differential operators as follows:

$$L_{ik}^{(1)} = \rho \omega^2 \delta_{ik} + \partial_j G \left[\delta_{jk} \partial_i + \delta_{ik} \partial_j - 2\delta_{ij} \partial_k \right] + \partial_i H \partial_k \qquad (19)$$

$$L_{ik}^{(2)} = \rho_f \omega^2 \delta_{ik} + \partial_i C \partial_k \qquad (20)$$

$$L_{ik}^{(3)} = L_{ik}^{(2)} \qquad (21)$$

$$L_{ik}^{(4)} = q\omega^2 \delta_{ik} + \partial_i M \partial_k. \qquad (22)$$

In the presence of point sources, Eq. (18) has a nonzero right-hand side that can be written as

$$-\begin{bmatrix} F_i^0 \delta(r_i - r_i') \\ f_i^0 \delta(r_i - r_i') \end{bmatrix}, \qquad (23)$$

where F_k^0 and f_k^0 represent constant forces applied to the bulk and fluid phase, respectively, and $\delta(r_i - r_i')$ denotes the Dirac delta function. The response of system (Eq. 18) to point sources of the form (Eq. 23) can be formulated as

$$\begin{bmatrix} u_i \\ w_i \end{bmatrix} = \begin{bmatrix} G_{ik}^F & G_{ik}^f \\ G_{ik}^f & G_{ik}^W \end{bmatrix} \cdot \begin{bmatrix} F_k^0 \\ f_k^0 \end{bmatrix}, \qquad (24)$$

where G_{ik}^F, G_{ik}^f, and G_{ik}^W denote the Green's tensors. Thus, the point source response of system (Eq. 18) in an isotropic unbounded medium is described by three Green's tensors. The wavefields observed at position **r** due to arbitrary point sources F_i and f_i, applied at position **r**′, can be expressed by a convolution equation of the form

$$\begin{bmatrix} u_i(\mathbf{r}) \\ w_i(\mathbf{r}) \end{bmatrix} = \int_{V'} d^3\mathbf{r}' \begin{bmatrix} G_{ik}^F(\mathbf{r} - \mathbf{r}') & G_{ik}^f(\mathbf{r} - \mathbf{r}') \\ G_{ik}^f(\mathbf{r} - \mathbf{r}') & G_{ik}^W(\mathbf{r} - \mathbf{r}') \end{bmatrix} \cdot \begin{bmatrix} F_k(\mathbf{r}') \\ f_k(\mathbf{r}') \end{bmatrix}. \qquad (25)$$

The complete set of Green's tensors for a homogeneous and isotropic poroelastic continuum was derived by Pride and Haartsen (1996). At low frequencies, that is, if $|k_p|/|k_{ps}| \ll 1$, these tensors read

$$G_{ij}^F(\mathbf{r} - \mathbf{r}_0) = \frac{1}{4\pi\rho\omega^2} \left([k_s^2 \delta_{ij} + \partial_i \partial_j] \frac{e^{ik_s R}}{R} - \partial_i \partial_j \frac{e^{ik_p R}}{R} \right)$$
$$- \frac{C^2}{H^2} \frac{1}{4\pi q \omega^2} \partial_i \partial_j \frac{e^{ik_{ps} R}}{R} \qquad (26)$$

$$G_{ij}^f(\mathbf{r} - \mathbf{r}_0) = \frac{C}{H} \frac{1}{4\pi q \omega^2} \partial_i \partial_j \frac{e^{ik_{ps} R}}{R}, \qquad (27)$$

$$G_{ij}^w(\mathbf{r} - \mathbf{r}_0) = -\frac{1}{4\pi q \omega^2} \partial_i \partial_j \frac{e^{ik_{ps} R}}{R}, \qquad (28)$$

where $R = |\mathbf{r} - \mathbf{r}_0|$. In homogeneous and isotropic media the Green's tensors depend only on R. In the low-frequency version of Biot's equations, the wave numbers of fast P-, S-, and slow P waves are defined as

$$k_{\rm p} = \omega\sqrt{\frac{\rho}{H}} \quad k_{\rm s} = \omega\sqrt{\frac{\rho}{G}} \quad k_{\rm ps} = \sqrt{\frac{i\omega\eta}{\kappa_0 N}} = \omega\sqrt{\frac{q}{N}}, \quad (29)$$

where $N = MP_{\rm d}/H$. Note that the first three terms of G_{ij}^F are formally identical to the elastodynamic Green's tensor (Gubernatis et al., 1977b). In the elastic limit ($K_{\rm d} \to K_{\rm g}$, $\alpha \to 0$ and $\phi \to 0$, $\kappa_0 \to 0$), the set of Green's tensors [(Eq. 26)–(Eq. 28)] reduces to the single elastodynamic Green's tensor (see, e.g., Hudson, 1980):

$$G_{ij}^{\rm elast}(\mathbf{r} - \mathbf{r}_0) = \frac{1}{4\pi\rho\omega^2}\left([k_{\rm s}^2\delta_{ij} + \partial_i\partial_j]\frac{e^{ik_{\rm s}R}}{R} - \partial_i\partial_j\frac{e^{ik_{\rm p}R}}{R}\right), \quad (30)$$

where the P-wave number is now given by $k_{\rm p} = \omega\sqrt{\rho/(K_{\rm d} + 4/3G)}$.

Based on Eq. (25), we now derive a wavefield representation in a randomly inhomogeneous medium.

3.2. The Basic Poroelastic Scattering Equation

Using the decomposition (Eq. 3), the differential operators L_{ik} can be also decomposed as (Karal and Keller, 1964)

$$L_{ik} = \bar{L}_{ik} + \tilde{L}_{ik}, \quad (31)$$

where the perturbing operator $\tilde{L}_{ik}$ satisfies $\langle \tilde{L}_{ik}\rangle = 0$. Substitution of Eq. (31) into matrix Eq. (18) yields

$$\begin{bmatrix}\bar{L}_{ik}^{(1)} & \bar{L}_{ik}^{(2)} \\ \bar{L}_{ik}^{(2)} & \bar{L}_{ik}^{(4)}\end{bmatrix}\cdot\begin{bmatrix}u_k \\ w_k\end{bmatrix} = -\begin{bmatrix}\tilde{L}_{ik}^{(1)} & \tilde{L}_{ik}^{(2)} \\ \tilde{L}_{ik}^{(2)} & \tilde{L}_{ik}^{(4)}\end{bmatrix}\cdot\begin{bmatrix}u_k \\ w_k\end{bmatrix}. \quad (32)$$

In the most general case, the perturbing operators $\tilde{L}_{ik}$ contain fluctuations of all poroelastic moduli and densities. The right-hand side of Eq. (32) can be thought of as a source term in the homogeneous system (Eq. 18) due to the presence of inhomogeneities (so-called secondary sources). Thus, Eq. (32) can be understood as an inhomogeneous equation with constant coefficients, whose formal solution can be written by substituting the source term into Eq. (25):

$$\begin{bmatrix}u_i \\ w_i\end{bmatrix} = \begin{bmatrix}u_i^0 \\ w_i^0\end{bmatrix} + \int_V dV \begin{bmatrix}G_{ij}^F & G_{ij}^f \\ G_{ij}^f & G_{ij}^w\end{bmatrix}\cdot\begin{bmatrix}\tilde{L}_{jk}^{(1)} & \tilde{L}_{jk}^{(2)} \\ \tilde{L}_{jk}^{(2)} & \tilde{L}_{jk}^{(4)}\end{bmatrix}\cdot\begin{bmatrix}u_k \\ w_k\end{bmatrix}. \quad (33)$$

Equation (33) is the basic poroelastic scattering equation. The total wavefields u_i and w_i are composed of wavefields propagating in the homogeneous background medium, u_i^0 and w_i^0, and scattered wavefields (the second term). By definition, u_i^0 and w_i^0 satisfy the homogeneous Eq. (18). The scattered wavefields are represented by volume integrals whose kernels involve the Green's tensors and the secondary sources. The scattered wavefields vanish if there are no fluctuations in the medium parameters. The integration volume encompasses the inhomogeneous part of the medium. According to Eq. (25), the wavefields can be represented as a convolution of Green's tensors with the source function. Let us denote Green's tensors for the homogeneous background medium by $^0G_{ik}^{F,f,w}$ and for the inhomogeneous medium by $G_{ik}^{F,f,w}$. Substituting these wavefield representations into Eq. (33), we obtain an equation for the Green's tensors of the inhomogeneous medium:

$$\begin{bmatrix} G_{im}^F & G_{im}^f \\ G_{im}^f & G_{im}^w \end{bmatrix} = \begin{bmatrix} ^0G_{im}^F & ^0G_{im}^f \\ ^0G_{im}^f & ^0G_{im}^w \end{bmatrix} + \\ + \int dV \begin{bmatrix} ^0G_{ij}^F & ^0G_{ij}^f \\ ^0G_{ij}^f & ^0G_{ij}^w \end{bmatrix} \cdot \begin{bmatrix} \tilde{L}_{jk}^{(1)} & \tilde{L}_{jk}^{(2)} \\ \tilde{L}_{jk}^{(2)} & \tilde{L}_{jk}^{(4)} \end{bmatrix} \cdot \begin{bmatrix} G_{km}^F & G_{km}^f \\ G_{km}^f & G_{km}^w \end{bmatrix}. \quad (34)$$

To simplify the equations that follow, we introduce a shorthand notation. The latter equation can be symbolically rewritten as

$$\mathbf{G} = \mathbf{G}^0 + \int \mathbf{G}^0 \tilde{\mathbf{L}} \mathbf{G}, \quad (35)$$

where $\mathbf{G}$, $\mathbf{G}^0$, and $\tilde{\mathbf{L}}$ represent matrices, whose elements are tensors of rank two.

3.3. First-Order Statistical Smoothing Approximation

We will now analyze Eq. (35) using a statistical approach. Since the matrix of perturbing operators $\tilde{\mathbf{L}}$ in Eq. (35) contains fluctuating medium parameters, the resulting matrix of Green's tensors also contains randomly fluctuating elements. Because individual realizations of the random wavefields are never known, it is natural to analyze the statistical moments of $\mathbf{G}$. Solving Eq. (35) by iteration, we obtain the scattering series:

$$\mathbf{G} = \mathbf{G}^0 + \int \mathbf{G}^0 \tilde{\mathbf{L}} \mathbf{G}^0 + \int\int \mathbf{G}^0 \tilde{\mathbf{L}} \mathbf{G}^0 + \int\int\int \ldots \quad (36)$$

Averaging this equation by the ensemble of realizations and regrouping the scattering terms yields

$$\bar{\mathbf{G}} = \mathbf{G}^0 + \int\int \mathbf{G}^0 \mathbf{Q} \bar{\mathbf{G}}, \quad (37)$$

where $\bar{\mathbf{G}} = \langle \mathbf{G} \rangle$ is the matrix of mean Green's tensors, and $\mathbf{Q}$ is the matrix operator defined as

$$\mathbf{Q} = \begin{bmatrix} Q_{ik}^{(1)} & Q_{ik}^{(2)} \\ Q_{ik}^{(3)} & Q_{ik}^{(4)} \end{bmatrix} \\ = \left\langle \tilde{\mathbf{L}} \mathbf{G}^0 \tilde{\mathbf{L}} + \int \tilde{\mathbf{L}} \mathbf{G}^0 \tilde{\mathbf{L}} \mathbf{G}^0 \tilde{\mathbf{L}} + \int \ldots \right\rangle. \quad (38)$$

Operator $\mathbf{Q}$ given by Eq. (38) corresponds to the kernel-of-mass operator in the acoustic formulation (Rytov *et al.*, 1989). The linear integral equation in $\bar{\mathbf{G}}$ [Eq. (37)] is the poroelastic analog of the Dyson equation. It is not possible to obtain an exact solution of Eq. (37). A first-order statistical smoothing consists in the first-order truncation of the infinite series expression for the operator **Q**. Then, we obtain the following approximation for the mean Green's tensor:

$$\tilde{\mathbf{G}} = \mathbf{G}^0 + \iint \mathbf{G}^0 \left\langle \tilde{\mathbf{L}} \mathbf{G}^0 \tilde{\mathbf{L}} \right\rangle \bar{\mathbf{G}} \tag{39}$$

$$= \mathbf{G}^0 + \iint \mathbf{G}^0 \mathbf{Q}^B \bar{\mathbf{G}}. \tag{40}$$

The truncation of the series (Eq. 38) implies that the first-order statistical smoothing is valid when $|\varepsilon_X| \ll 1$, that is, when the absolute value of the relative fluctuations of X is a small parameter. Note also that the elements of matrix operator $\mathbf{Q}^B$ only contain terms involving the second statistical moment of the fluctuating parts of the $\tilde{L}_{ik}$'s, that is, they are of the order $O(\varepsilon^2)$. Higher-order correlations are neglected within the accuracy of the first-order statistical smoothing approximation.

Since Eq. (40) contains a double volume convolution, it is expedient to work with its spatial Fourier transform:

$$\bar{\mathbf{g}} = \mathbf{g}^0 + \left(8\pi^3\right)^2 \mathbf{g}^0 \mathbf{q} \bar{\mathbf{g}}, \tag{41}$$

where $\bar{\mathbf{g}}$, $\mathbf{g}^0$, and $\mathbf{q}$ denote the spatial Fourier transforms of $\bar{\mathbf{G}}$, $\mathbf{G}^0$, and $\mathbf{Q}^B$, respectively. Carrying out the necessary matrix multiplications in Eq. (41), we find that this system splits up into two pairs of coupled equations. Since we are only interested in the characteristics of the fast P wave, which are exclusively contained in the Green's tensor $\bar{g}^F$ [see also Eqs. (24) and (26)], we analyze only those two equations that involve $\bar{g}^F_{ik}$. We obtain

$$\bar{g}^F = g^F + \left(8\pi^3\right)^2 \left[g^F q^{(1)} \bar{g}^F + g^F q^{(2)} \bar{g}^f + g^f q^{(3)} \bar{g}^F + g^f q^{(4)} \bar{g}^f \right] \tag{42}$$

$$\bar{g}^f = g^f + \left(8\pi^3\right)^2 \left[g^f q^{(1)} \bar{g}^F + g^f q^{(2)} \bar{g}^f + g^w q^{(3)} \bar{g}^F + g^w q^{(4)} \bar{g}^f \right], \tag{43}$$

where we omitted subscripts for brevity. The quantities g without upper bar denote the background space Green's tensors. Since all quantities $q^{(i)}$ ($i = 1, \ldots, 4$) are of the order $O(\varepsilon^2)$, $\bar{g}^f$ is also of the order $O(\varepsilon^2)$. Inserting the expression for $\bar{g}^f$ [Eq. (43)] into Eq. (42) and neglecting terms of higher order than $O(\varepsilon^2)$ we obtain

$$\bar{g}^F = g^F + \left(8\pi^3\right)^2 \left[g^F q^{(1)} \bar{g}^F + g^F q^{(2)} g^f + g^f q^{(3)} \bar{g}^F + g^f q^{(4)} g^f \right]. \tag{44}$$

Equation (44) is an implicit equation for the mean Green's tensor $\bar{g}^F$. Because of its tensorial character, an explicit solution for $\bar{g}^F$ is still difficult to construct. Note, however, that we are not interested in mean Green's tensor itself but only in a *mean wave number* of the P wave (for brevity denoted as wave number in the following) contained in $\bar{g}^F$.

3.4. Effective Fast Wave Number Accounting for Conversion Scattering into Slow P Waves

To extract an effective wave number from Eq. (44), we have to introduce further simplifications. Because of the assumption of small fluctuations in the medium parameters

($\varepsilon \ll 1$), we can assume that mean Green's tensor $\bar{g}_{ik}^F(K)$ is of the same functional form as a background Green's tensor $g_{ik}^F(K)$ given by the spatial Fourier transform of Eq. (26), however, involving some effective wave number. Further, we consider an incoming, plane P wave propagating in x_3-direction (i.e., only the displacement component u_3 is nonzero).

The resulting coherent P wave in the inhomogeneous medium will also propagate in x_3-direction [if condition (1) is fulfilled]. Therefore, only the tensor components $i = j = 3$ of g_{ij}^F need to be analyzed. Noting that in this case the Green's tensor $g_{ik}^F(K)$ yields the largest contribution for the spatial wave number $K = k_p$, we can approximate the spatial Fourier transform of the full Green's tensor (Eq. 26) by

$$g_{33}^F \approx \frac{-1}{8\pi^3 \rho \omega^2} \left(1 + \frac{K^2}{k_p^2 - K^2} \right). \tag{45}$$

We assume that the mean Green's tensor component is given by

$$\bar{g}_{33}^F \approx \frac{-1}{8\pi^3 \bar{\rho} \omega^2} \left(1 + \frac{K^2}{\bar{k}_p^2 - K^2} \right), \tag{46}$$

where $\bar{k}_p$ is the yet unknown effective wave number. Substituting Eqs. (45) for g^F and Eq. (46) for $\bar{g}^F$ into Eq. (44), we obtain after algebraic manipulations

$$\bar{k}_p \approx k_p \left(1 + \frac{4\pi^3}{\rho \omega^2} q_{33}^{(1)} \right). \tag{47}$$

Here, we neglected terms that contain combinations of the tensor components $q_{33}^{(i)}$. This introduces no additional inaccuracy because higher-order correlations are neglected within the accuracy of the first-order statistical smoothing $O(\varepsilon^2)$.

The remaining problem is the evaluation of $q_{33}^{(1)}$ in Eq. (47), or equivalently, of $Q_{33}^{(1)}$ in space domain. In explicit form, from the first term in the expansion of Q as given by Eq. (38) we obtain

$$Q_{ik}^{(1)}(\mathbf{r}' - \mathbf{r}'') = \Big\langle \tilde{L}_{ij}^{(1)}(\mathbf{r}') G_{jl}^F(\mathbf{r}' - \mathbf{r}'') \tilde{L}_{lk}^{(1)}(\mathbf{r}'')$$
$$+ 2\tilde{L}_{ij}^{(1)}(\mathbf{r}') G_{jl}^f(\mathbf{r}' - \mathbf{r}'') \tilde{L}_{lk}^{(2)}(\mathbf{r}'')$$
$$+ \tilde{L}_{ij}^{(2)}(\mathbf{r}') G_{jl}^w(\mathbf{r}' - \mathbf{r}'') \tilde{L}_{lk}^{(2)}(\mathbf{r}'') \Big\rangle, \tag{48}$$

where for statistically homogeneous random media both Q_{ik} and G_{ik} depend only on the difference vector $\mathbf{r}' - \mathbf{r}''$. It is interesting to note that in the elastic limit, only the first term of $Q_{ik}^{(1)}$ is nonzero. In the poroelastic case, we have to analyze all three terms. Equation (48) involves the perturbing operators $\tilde{L}_{ij}^{(1)}$ and $\tilde{L}_{ij}^{(2)}$ (but not $\tilde{L}_{ij}^{(4)}$). Let us now specify the perturbing operators resulting for a particular case where the density fluctuations are neglected [it can be shown that incorporation of density fluctuations yields a correction to the background wave number (the second term in Eq. (47)] which scales with ω^3, whereas the other fluctuations result in a ω^2 scaling. Then from Eqs. (19)–(20), we obtain

$$\tilde{L}_{ik}^{(1)} = \partial_k \tilde{G} \partial_i + \partial_j \delta_{ik} \tilde{G} \partial_j - 2\partial_i \tilde{G} \partial_k + \partial_i \tilde{H} \partial_k \tag{49}$$

$$\tilde{L}_{ik}^{(2)} = \partial_i \tilde{C} \partial_k. \tag{50}$$

The computation of the three $Q_{ik}^{(1)}$ terms in Eq. (48) in the wave number domain using the perturbing operators [Eqs. (49)–(50)] results into

$$q_{33}^{(1)} = q^{HH} + q^{HG} + q^{HC} + q^{GG} + q^{GC} + q^{CC}, \tag{51}$$

where

$$q^{HH} = \frac{1}{8\pi^3} k_p^2 \left(\frac{H^2}{P_d} B_{HH}(0) + \frac{C^2}{N} k_{ps}^2 \int_0^\infty r B_{HH}(r) \exp[ik_{ps}r] dr \right) \tag{52}$$

$$q^{HG} = -\frac{1}{3\pi^3} k_p^2 \left(\frac{GH}{P_d} B_{HG}(0) + \frac{\alpha^2 MG}{P_d} k_{ps}^2 \int_0^\infty r B_{PM}(r) \exp[ik_{ps}r] dr \right) \tag{53}$$

$$q^{HC} = -\frac{1}{4\pi^3} \frac{C^2}{N} k_p^2 \left(B_{HC}(0) + k_{ps}^2 \int_0^\infty r B_{HC}(r) \exp[ik_{ps}r] dr \right) \tag{54}$$

$$q^{GG} = \frac{1}{15\pi^3} \frac{G}{NH^2} k_p^2 \Bigg([4C^2G + 4NHG + NH^2] B_{GG}(0)$$

$$+ 4C^2 G k_{ps}^2 \int_0^\infty r B_{GG}(r) \exp[ik_{ps}r] dr \Bigg) \tag{55}$$

$$q^{GC} = \frac{1}{3\pi^3} \frac{\alpha^2 MG}{P_d} k_p^2 \left(B_{GC}(0) + k_{ps}^2 \int_0^\infty r B_{GC}(r) \exp[ik_{ps}r] dr \right) \tag{56}$$

$$q^{CC} = -\frac{1}{8\pi^3} \frac{C^2}{N} k_p^2 \left(B_{CC}(0) + k_{ps}^2 \int_0^\infty r B_{CC}(r) \exp[ik_{ps}r] dr \right). \tag{57}$$

Here, B_{HH}, B_{HC}, B_{HG}, B_{GG}, B_{GC}, and B_{CC} denote the (cross-) correlation functions of the random fields $\tilde{H}$, $\tilde{G}$, and $\tilde{C}$ defined by Eq. (4). The upper bar denoting the background properties is omitted.

We will now assume that all correlation functions are of the same functional form and only differ by theirs variances, that is, $B_{XY} = \sigma_{XY}^2 B(r)$ with $B(0) = 1$ and $B(\infty) = 0$. Substituting then expressions (52)–(57) into Eq. (47), we obtain the final result for the effective wave number of the fast P wave

$$\bar{k}_p = k_p \left(1 + \Delta_2 + \Delta_1 k_{ps}^2 \int_0^\infty r B(r) \exp[ik_{ps}r] dr \right), \tag{58}$$

with the dimensionless coefficients

$$\Delta_1 = \frac{\alpha^2 M}{2P_d} \left(\sigma_{HH}^2 - 2\sigma_{HC}^2 + \sigma_{CC}^2 + \frac{32}{15} \frac{G^2}{H^2} \sigma_{GG}^2 - \frac{8}{3} \frac{G}{H} \sigma_{HG}^2 + \frac{8}{3} \frac{G}{H} \sigma_{GC}^2 \right) \tag{59}$$

$$= \frac{\alpha^2 M}{2P_d} \left(\left\langle \left(\varepsilon_H - \frac{4}{3} \frac{G}{H} \varepsilon_G - \varepsilon_C \right)^2 \right\rangle + \frac{16}{45} \frac{G^2}{H^2} \sigma_{GG}^2 \right) \tag{60}$$

$$\Delta_2 = \Delta_1 + \frac{1}{2}\sigma_{HH}^2 - \frac{4G}{3H}\sigma_{HG}^2 + \left(\frac{4G}{H} + 1\right)\frac{4\,G}{15H}\sigma_{GG}^2. \tag{61}$$

The structure of the effective wave number can be explained as follows. Because of the presence of random inhomogeneities, there are two terms added to the background wave number k_p. The first term, Δ_2, is frequency independent and consists in a weighted sum of the variances of the random fields $\tilde{H}$, $\tilde{G}$, and $\tilde{C}$. The second term is frequency dependent and contains an integral over the correlation function multiplied by a weighted sum of the variances, Δ_1. It is important to note that the expression for $\bar{k}_P$ describes only the process of conversion scattering from fast into slow P waves. The contribution of purely elastic scattering is left out. The corresponding result would include additional terms involving the correlation functions B_{HH}, B_{GG}, and B_{HG} which describe the elastic scattering (P to P and S waves) and produces the typical Rayleigh frequency dependence for scattering attenuation as analyzed in the next section. Therefore, analysis of the properties of $\bar{k}_p$ gives insight into the relationship between the properties of elastic waves and wave-induced flow.

The corresponding 1-D effective wave number can be obtained by the following procedure. We consider the limiting situation:

(a) To degenerate the 3-D random medium into a 1-D random medium, we stretch the correlation lengths perpendicular to the direction of wave propagation, $a_\perp$, to infinity so that the correlation function becomes only a function of z with parameter $a_\|$, that is, the correlation length parallel to the direction of wave propagation. Obviously, if the wave propagates mainly in z direction we can also write the spatial wave vector as $\mathbf{K} \approx (0, 0, k_p)^T$.

(b) Since in such a 1-D random medium there are only two directions of wave propagation ($\pm z$), we can use the small-angle approximation (Rytov et al., 1989) (or Fresnel approximation) of the propagator-like term $(\exp[ik_{ps}R])/R$:

$$\frac{\exp[ik_{ps}R]}{R} \approx \frac{\exp[ik_{ps}z]}{z}\exp\left[\frac{ik_{ps}r_t^2}{2z}\right], \tag{62}$$

where r_t denotes the absolute value of the transverse coordinate vector $\mathbf{r}_t = (x, y)^T$.

Steps (a) and (b) are illustrated in Fig. 6. The detailed derivation of the 1-D wave number from the 3-D wave number can be found in Müller and Gurevich (2005a) and coincides with the earlier reported results of Gurevich and Lopatnikov (1995) and Gelinsky et al. (1998).

3.5. Attenuation and Dispersion due to Wave-Induced Flow

Equation (58) for the effective wave number enables us to derive expressions for the attenuation and dispersion due to the presence of mesoscopic inhomogeneities. By definition, the real part of $\bar{k}_p$ yields the phase velocity

$$v(\omega) = \omega/\Re\{\bar{k}_p\} = v_0\left[1 - \Delta_2 + 2\Delta_1 \bar{k}^2 \int_0^\infty rB(r)\exp[-\bar{k}r]\sin(\bar{k}r)\mathrm{d}r\right], \tag{63}$$

ATTENUATION OF SEISMIC WAVES

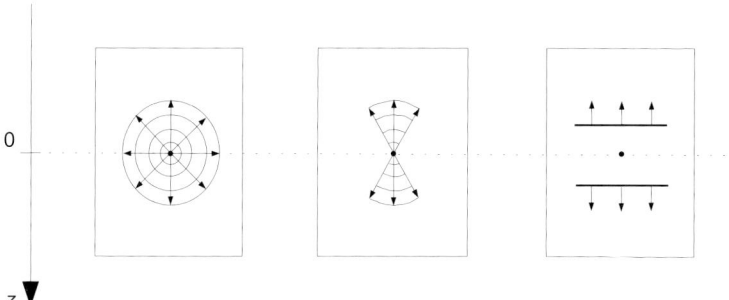

Fig. 6. From 3-D point source to 1-D line source results.

where v_0 is the constant background P-wave velocity defined as $v_0 = \sqrt{H/\rho}$ (ρ is the bulk density) and $\bar{k}$ denotes the real part of the wave number of the slow P-wave k_{ps}

$$\bar{k}(\omega) = \sqrt{\frac{\eta}{2\kappa_0 N}\omega}. \tag{64}$$

The imaginary part of the wave number yields the attenuation coefficient γ and the reciprocal quality factor Q^{-1}, which for low-loss media can be written as

$$Q^{-1} = 2\gamma/\Re\{\bar{k}_p\} = 2\Im\{\bar{k}_p\}/\Re\{\bar{k}_p\}. \tag{65}$$

Then, from Eq. (58) we find

$$Q^{-1}(\omega) = 4\Delta_1 \bar{k}^2 \int_0^\infty rB(r)\exp[-\bar{k}r]\cos(\bar{k}r)dr. \tag{66}$$

From Eqs. (63) and (66), the meaning of the coefficients [Eqs. (60)–(61)] becomes clear. The attenuation Q^{-1} and the frequency-dependent part of v are proportional to Δ_1. Thus Δ_1 is the measure of the magnitude of attenuation and velocity dispersion, that is, the dynamic effects. In contrast, Δ_2 produces a frequency-independent velocity shift in Eq. (63).

To gain further insight into the general properties of the results for attenuation and velocity dispersion, it is useful to express the Eqs. (63) and (66) in terms of the fluctuation spectrum (power spectrum), that is, the spatial Fourier transform of the correlation function. Substituting Eq. (5) into Eqs. (63) and (66), changing the order of integration and integrating over r we obtain

$$v(\omega) = v_0\left[1 - \Delta_2 + 16\pi\Delta_1 \int_0^\infty \frac{\bar{k}^4 \kappa^2}{4\bar{k}^4 + \kappa^4}\Phi(\kappa)d\kappa\right], \tag{67}$$

and

$$Q^{-1}(\omega) = 16\pi\Delta_1 \int_0^\infty \frac{\bar{k}^2 \kappa^4}{4\bar{k}^4 + \kappa^4}\Phi(\kappa)d\kappa. \tag{68}$$

From Eq. (68), we observe that the dynamic behavior of attenuation is controlled by the integrand, that is, by the product of fluctuation spectrum $\Phi(\kappa)$ and the function

$$\Theta(\kappa, \bar{k}) = \frac{\bar{k}^2 \kappa^4}{4\bar{k}^4 + \kappa^4}. \tag{69}$$

The function $\Theta(\kappa, \bar{k})$ acts like a filter and controls which part of the fluctuation spectrum yields a relevant contribution to attenuation. A similar filter function can be deduced from Eq. (67). In analogy to the acoustic scattering problem (Ishimaru, 1978), we refer to Θ as the spectral filter function. Analyzing the product $\Phi\Theta$ in terms of the dimensionless, spatial wave number κa (a is the characteristic length scale of the inhomogeneities), we identify three different regimes for different values of $\bar{k}a$.

If $\bar{k}a \ll 1$ then $\Theta(\kappa a, \bar{k}a)$ behaves like $\bar{k}a$ and, therefore, the product $\Phi\Theta$ and hence the attenuation becomes small. Since $\bar{k}$ is inversely proportional to the diffusion wavelength $\lambda_d = \sqrt{\kappa_0 N/\omega\eta}$, this case corresponds to the relaxed or low-frequency regime where the induced pore pressure is equilibrated during one wave cycle. In the opposite case, if $\bar{k}a \gg 1$ then $\Theta(\kappa a, \bar{k}a) \approx (\kappa a)^4/(\bar{k}a)^2$. This means that the contribution of Φ at small spatial wave numbers is suppressed, but its contribution at large wave numbers is amplified. However, since $\Phi(\kappa)$ becomes very small for large κ, the product of Φ and Θ becomes small again. In other words, in the high-frequency (unrelaxed) regime only the behavior of Φ at large $\bar{k}a$ is important.

There is an intermediate regime with $\bar{k}a \approx 1$ where $\Phi\Theta$ (and Q^{-1}) attains its maximum. Since in our approximation attenuation, due to wave-induced flow and the process of conversion scattering from fast into slow P waves are equivalent, maximum attenuation is observed at the "resonance" condition $\lambda_d = a$. The interplay between Φ and Θ is illustrated in Fig. 7.

We now give explicit results for Q^{-1} and v for several correlation functions of practical interest. The choice of a single correlation function $B(r)$ implies that the correlation length is the same for the three random functions $H(r)$, $G(r)$, and $C(r)$. Substituting the exponential correlation function (Eq. 6) into Eqs. (63) and (66) we find

$$Q^{-1}(\omega) = \Delta_1 \frac{4(a\bar{k})^2(2\bar{k}a + 1)}{(1 + 2\bar{k}a + 2\bar{k}^2 a^2)^2}, \tag{70}$$

and

$$v(\omega) = v_0 \left[1 + \Delta_1 \frac{4(a\bar{k})^3(1 + \bar{k}a)}{(1 + 2\bar{k}a + 2\bar{k}^2 a^2)^2} - \Delta_2 \right]. \tag{71}$$

For the Gaussian correlation function we obtain

$$Q^{-1}(\omega) = 2\Delta_1(a\bar{k})^2 \left[1 - \frac{\sqrt{\pi}}{4} \sum_{z=z_-}^{z_+} a\bar{k}z \exp[(a\bar{k}z)^2/4] \operatorname{erfc}[a\bar{k}z/2] \right], \tag{72}$$

$$v(\omega) = v_0 \left[1 + \Delta_1 (a\bar{k})^2 \frac{\sqrt{\pi}}{4} \sum_{z=z_-}^{z_+} a\bar{k}z^* \exp[(a\bar{k}z)^2/4] \operatorname{erfc}[a\bar{k}z/2] - \Delta_2 \right], \tag{73}$$

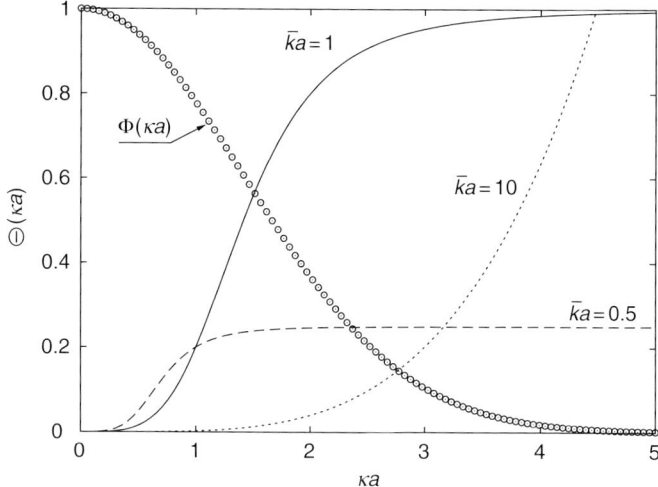

FIG. 7. The spectral filter Θ as a function of dimensionless spatial wave number κa for varying $\bar{k}a$. The general behavior of the fluctuation spectrum Φ is also shown (circles). Elastic wave attenuation due to fluid flow is proportional to the product of Φ and Θ. Largest attenuation occurs at $\bar{k}a \approx 1$.

where $z_+ = 1 + i$, $z_- = 1 - i$, z^* denotes complex conjugation, and erfc is the complementary error function. Finally, for the von Kármán correlation function (Eq. 9) we obtain

$$Q^{-1}(\omega) = c_1 \Delta_1 \left[c_{23} F_2 \left(1, \frac{1}{2} + \frac{\nu}{2}, 1 + \frac{\nu}{2}; \frac{1}{4}, \frac{3}{4}; -4(a\bar{k})^4 \right) \right.$$

$$- \frac{1}{2} \Gamma \left(\nu + \frac{3}{2} \right) (2\nu + 3) B^{-\frac{3}{4} - \frac{\nu}{2}} \cos \left[\left(\frac{3}{4} + \frac{\nu}{2} \right) A \right]$$

$$\left. + \Gamma \left(\nu + \frac{5}{2} \right) B^{-\frac{5}{4} - \frac{\nu}{2}} \left\{ 2(a\bar{k})^2 \cos[c_3 A] + \sin[c_3 A] \right\} \right], \quad (74)$$

$$\nu(\omega) = \nu_0 \left[1 - \Delta_2 - \frac{c_1}{2} \Delta_1 \right.$$

$$\left(-4c_2(1+\nu)(a\bar{k})^2 {}_3F_2 \left(1, 1 + \frac{\nu}{2}, \frac{3}{2} + \frac{\nu}{2}; \frac{3}{4}, \frac{5}{4}; -4(a\bar{k})^4 \right) \right.$$

$$+ \frac{1}{2} \Gamma \left(\nu + \frac{1}{2} \right) (2\nu + 3) c_3 B^{-\frac{3}{4} - \frac{\nu}{2}} \cos \left[\left(\frac{3}{4} + \frac{\nu}{2} \right) A \right]$$

$$\left. \left. + \Gamma \left(\nu + \frac{3}{2} \right) \left(\frac{3}{2} + \nu \right) B^{-\frac{5}{4} - \frac{\nu}{2}} \left\{ 2(a\bar{k})^2 \cos[c_3 A] + \sin[c_3 A] \right\} \right) \right], \quad (75)$$

where we used $c_1 = 16\sqrt{\pi}(a\bar{k})^3/(\Gamma(\nu)(2\nu+3))$, $c_2 = \Gamma(\nu+1)(2\nu+3)/(2\sqrt{\pi}a\bar{k})$, $c_3 = (1/4 + \nu/2)$, $A = 2\arctan\left(2a^2\bar{k}^{-2}\right)$, and $B = 1 + 4(a\bar{k})^4$. $_3F_2$ is the generalized hypergeometric function.

Let us consider various scenarios how mesoscopic inhomogeneities can affect attenuation and dispersion of P waves. In all numerical examples, we assume that the background material is a porous sandstone with parameters specified in Table 1. In the first example, we assume that the correlation function is of exponential type (Eq. 6) with varying correlation length a. Further, we assume that there are fluctuations of all bulk moduli and the shear modulus specified through their variances: $\sigma^2_{K_d K_d} = 0.12$, $\sigma^2_{K_g K_g} = 0.02$, $\sigma^2_{GG} = 0.1$, and $\sigma^2_{K_f K_f} = 0.14$. The fluctuations of K_d, K_g, and G are fully correlated so that the coefficient of correlation for two different random fields $R = \sigma^2_{XY}/\sqrt{\sigma^2_{XX}\sigma^2_{YY}}$ is equal to one. In our case, the cross-variances become $\sigma^2_{K_d K_g} = 0.049$, $\sigma^2_{K_d G} = 0.110$, and $\sigma^2_{GK_g} = 0.048$. The fluctuations of porous-material parameters and fluid bulk modulus are uncorrelated. Using these variances, we compute the variances of the poroelastic parameters H, C, and G: $\sigma^2_{HH} = 0.051$, $\sigma^2_{CC} = 0.081$, $\sigma^2_{HG} = 0.098$, $\sigma^2_{HC} = 0.025$, and $\sigma^2_{GC} = 0.098$. The frequency dependence of attenuation and phase velocity for this model according to Eqs. (70) and (71) is shown in Fig. 8. The frequency is normalized by Biot's critical frequency $f_c = 100$ kHz. From Fig. 8, we can observe that even weak fluctuations of the bulk moduli can produce significant attenuation of the fast P wave ($Q^{-1} \gtrsim 0.01$).

Next, we consider the influence of the cross-correlations of the fluctuations. Obviously, if there is an inhomogeneity with low-P-wave modulus but relatively high fluid bulk modulus (that is, negatively correlated fluctuations in K_d and K_f), we expect an increased wave-induced fluid flow during the compression cycle of the wave. This means that both the dispersion and attenuation characteristics should be more pronounced than in the case of uncorrelated fluctuations. Such a behavior can be observed in Fig. 9, where

TABLE 1. Parameters of the background solid and fluid phases used for the computation of the numerical examples

Porous material	
K_g (GPa)	40
K_d (GPa)	4.5
G (GPa)	9
$\rho_g \left(\frac{kg}{m^3}\right)$	2650
ϕ	0.17
κ_0 (mD)	250
Pore fluid	
K_f (GPa)	2.17
η (Pa s)	0.001
$\rho_f \left(\frac{kg}{m^3}\right)$	1000
Poroelastic parameters	
P_d (GPa)	16.5
α	0.89
M (GPa)	10.4
H (GPa)	24.7
N (GPa)	6.9
ω_B (kHz)	680

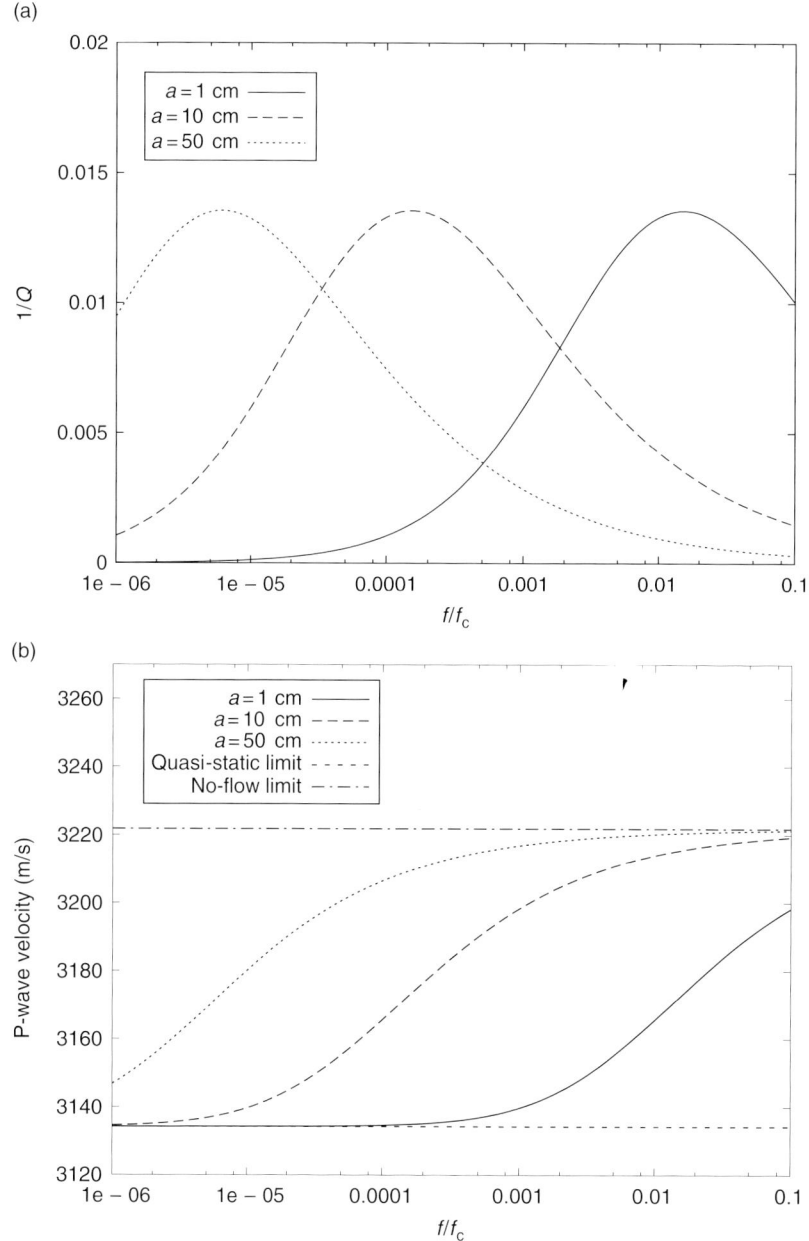

FIG. 8. (a) Reciprocal quality factor as a function of frequency (normalized by Biot's critical frequency $f_c \approx 100$ kHz) for different correlation lengths. (b) P-wave velocity versus frequency for the same models. It can be observed that for larger correlation lengths the dispersion curves are shifted toward lower frequencies. The horizontal curves denote the quasi-static and no-flow limits, respectively.

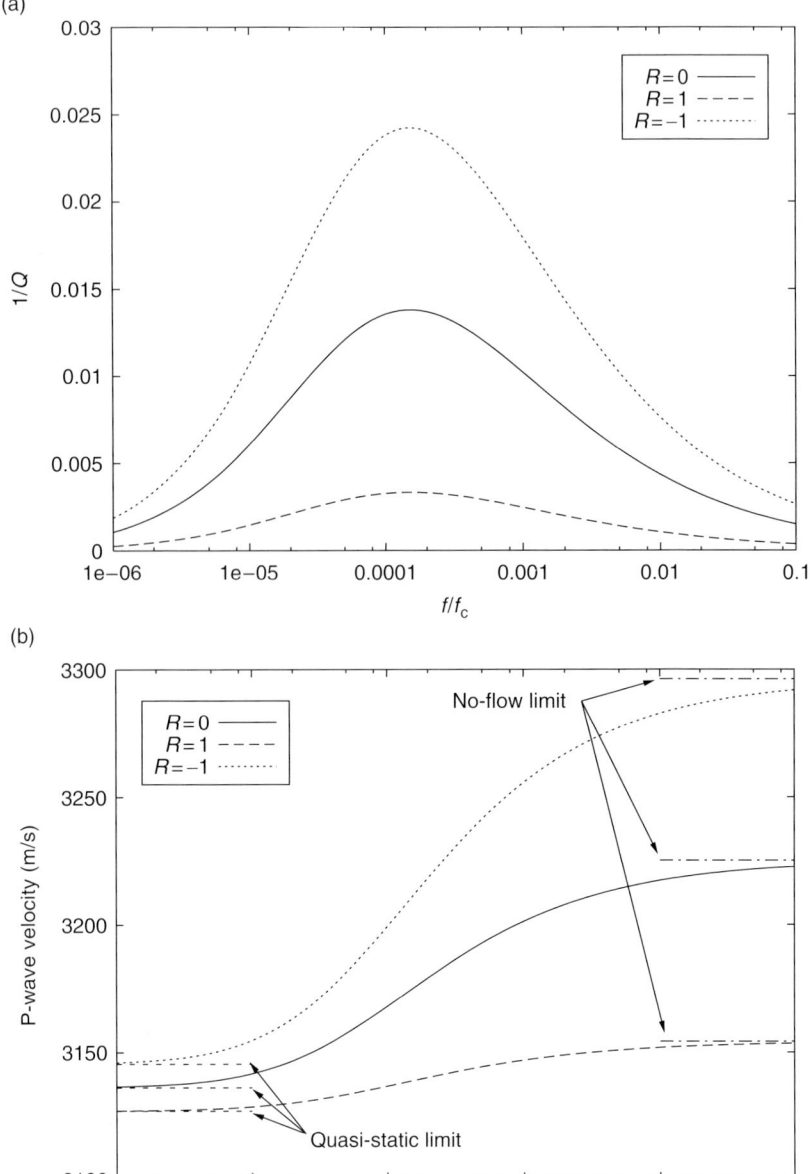

FIG. 9. Q^{-1} versus normalized frequency for differently correlated fluctuations. A significant amount of attenuation ($Q^{-1} > 0.01$) can be observed if the fluctuation of the solid and frame material are negatively correlated with the fluctuations of the fluid bulk modulus (that means a soft frame inhomogeneity contains a fluid with increased bulk modulus). The correlation length is constant ($a = 10$ cm). (b) P-wave velocity for the same model. It can be observed that for negative cross-correlation the dispersion effect is most pronounced.

P-wave velocity and attenuation are computed for the above sandstone model with $\sigma^2_{K_d K_d} = 0.10$, $\sigma^2_{K_s K_s} = 0.02$, $\sigma^2_{GG} = 0.08$, and $\sigma^2_{K_f K_f} = 0.16$. The fluctuations of K_d, K_g, and G are positively correlated. The fluctuations of the fluid bulk modulus and those of all other fluctuating parameters are either positively (coefficient of correlation $R = 1$) or negatively ($R = -1$) correlated. The case of uncorrelated fluctuations ($R = 0$) between K_f and all other moduli is also displayed in Fig. 9. We note that such scenarios may produce a significant amount of P-wave attenuation ($0.01 < Q^{-1} < 0.1$) even if the relative fluctuations in the medium parameters are small.

To demonstrate the influence of the correlation function on the frequency dependence of attenuation and velocity dispersion in Fig. 10, we show Q^{-1} and P-wave velocity for exponential, Gaussian, and von Kármán correlation function. Note that all curves are generated using the same medium parameters (those from Fig. 8). The resulting differences in magnitude and frequency dependence of attenuation are only due to the use of a different correlation model [see Eqs. (6)–(8)]. Largest attenuation is obtained for the Gaussian correlation model. Whereas at low frequencies the frequency dependence is the same for all correlation models, one can observe that at high frequencies different asymptotes are obtained. Only the Gaussian correlation model is symmetric about its maximum. The variability of both attenuation and velocity dispersion for different correlation models indicates the importance of the geometrical shape of mesoscopic inhomogeneities for the wave-induced flow.

3.6. Asymptotic Behavior at Low and High Frequencies

At low frequencies, we can assume $B(r) \approx B(0)$ and replace the exponential in Eq. (58) by 1 because $B(r)$ will vanish before the exponential term changes noticeably from its value at small arguments. Obviously, an asymptote exists only if the resulting expression $\int_0^\infty rB(r)dr$ has a finite positive value. This is the case for a large class of correlation functions. Then the attenuation coefficient γ scales like $\gamma \propto \omega^2$ or, in terms of the quality factor,

$$Q^{-1} \propto \omega. \tag{76}$$

It is important to note that the same low-frequency behavior is reported for 1-D and 3-D periodic structures (Norris, 1993; Johnson, 2001; Pride *et al.*, 2004). In contrast, in 1-D random media the following asymptotic scaling is found (see also Section 5):

$$Q_{1D}^{-1} \propto \sqrt{\omega}. \tag{77}$$

The implications of these scalings for 1-D and 3-D disorder are discussed by Müller and Gurevich (2004). Moreover, the physical origin of this different scaling is elucidated by Müller and Rothert (2006).

At high frequencies only the behavior of $B(r)$ at small arguments is important. Assuming that the correlation function can be expanded in power series around the origin

$$B(r/a) = 1 - r/a + O((r/a)^2), \tag{78}$$

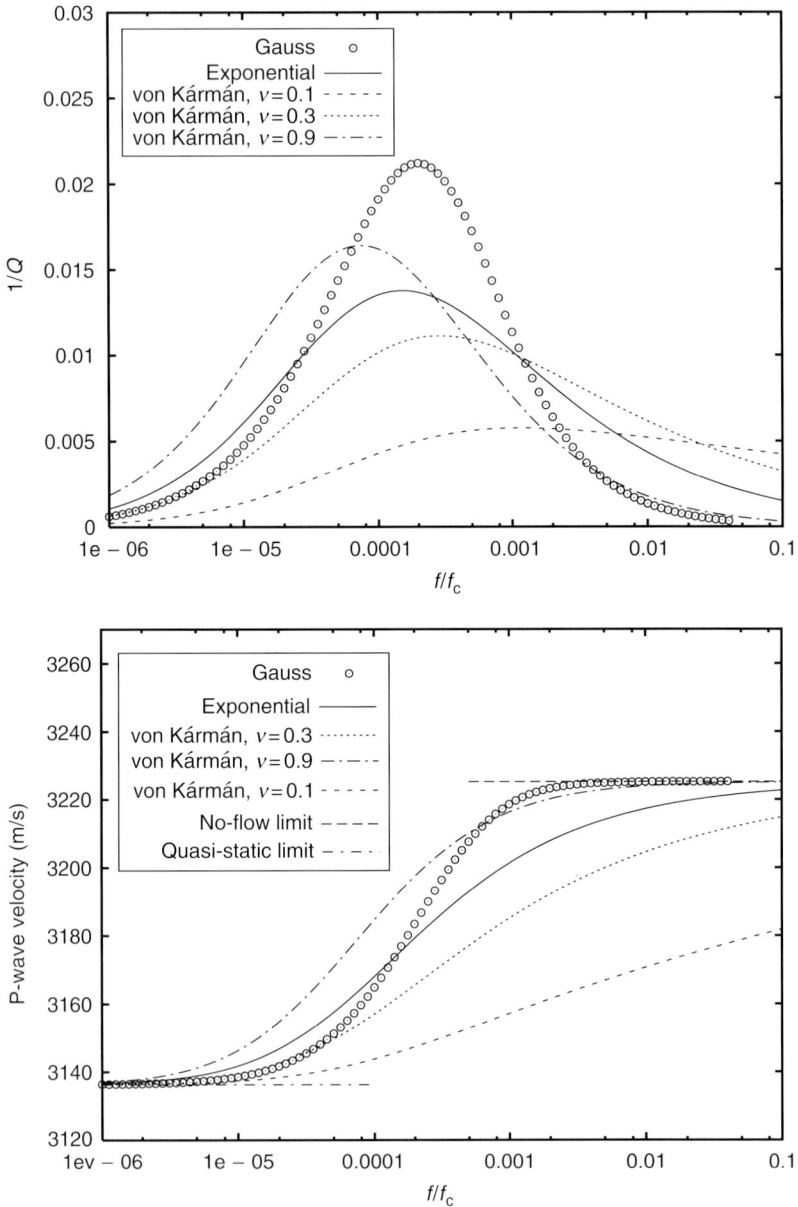

FIG. 10. Q^{-1} and P-wave velocity as a function of normalized frequency for Gaussian, exponential, and von Kármán (the Hurst coefficient ν is denoted in the legend) correlation functions. The model parameters are the same as in Fig. 9, where the fluctuations of K_f are uncorrelated with those of K_d, G, and K_g.

we can evaluate the integral in Eq. (58) and obtain

$$Q^{-1} \propto \frac{1}{\sqrt{\omega}}. \tag{79}$$

The same asymptote has been found in 1-D/3-D periodic and 1-D random structures. It is, however, important to note that the scaling (Eq. 79) is not universal for any kind of disorder (see also Fig. 4). For example, the Gaussian correlation function (Eq. 7) at high frequencies yields the proportionality $Q^{-1} \propto 1/\omega$, a much faster decrease of attenuation with frequency as compared to Eq. (79). The reason for this discrepancy is that the Gaussian correlation function behaves differently at small argument. Instead of Eq. (78), we have $B^{\text{Gauss}}(r/a) = 1 + O(r^2/a^2)$ which means that on small scale $r \ll a$ the medium is almost homogeneous. As a consequence, at high frequencies the passing wave will create less fluid flow as compared with a medium characterized by Eq. (78) and, therefore, the decrease of Q^{-1} with frequency is stronger.

We now analyze the asymptotic behavior of phase velocity in the cases of low- and high frequencies. In both cases, the phase velocity has a finite limit. The physical situation, however, is different for these two limits: in the low-frequency limit, there is enough time during the wave cycle to equilibrate the induced pore pressure. We refer to this relaxed limit (Mavko and Jizba, 1991) as "quasi-static" and denote the corresponding phase velocity as v_{qs}. In the high-frequency limit, there is no time to develop a wave-induced fluid flow. This situation is called no-flow (unrelaxed) limit, and we denote the phase velocity as v_{nf}. From Fig. (2), it can be observed that

$$v_{\text{nf}} \geq v_{\text{qs}}. \tag{80}$$

Physically, this relation can be explained by the additional stiffening of the porous frame in the no-flow limit.

From Eq. (63), it is straightforward to deduce v_{qs}. The low-frequency limit is obtained by neglecting the third term in Eq. (63) and obtain

$$v_{\text{qs}} = v_0(1 - \Delta_2). \tag{81}$$

To determine the no-flow velocity, we need to compute the limit $\omega \to \infty$ in Eq. (63). Since in the limit $\omega \to \infty$ only the value of B at zero correlation lag yields a contribution [see Eq. (78)] we can replace $B(r)$ by $B(0) = 1$. Thus, the third term in Eq. (63) gives

$$2\Delta_1 \bar{k}^2 \int_0^\infty r \exp[-\bar{k}r]\sin(\bar{k}r)dr = \Delta_1, \tag{82}$$

so that

$$v_{\text{nf}} = v_0(1 + \Delta_1 - \Delta_2). \tag{83}$$

From Eqs. (81) and (83), it follows that the relative magnitude of the dispersion effect is

$$\frac{v_{\text{nf}} - v_{\text{qs}}}{v_0} = \Delta_1 \tag{84}$$

with Δ_1 defined in Eq. (60). It is interesting to note that the limiting velocities do not depend either on the correlation function or on the transport properties of the porous material. In other words, v_{qs} and v_{nf} are independent of the geometry of the inhomogeneities.

4. ATTENUATION OF SEISMIC WAVES IN RANDOM POROUS MEDIA DUE TO SCATTERING

4.1. The Generalized ODA Formalism

In the previous section, we showed how attenuation and dispersion due to wave-induced flow in a poroelastic medium can be treated using the method of statistical smoothing. The same method can in principle be used to model attenuation and dispersion due to scattering. However, since statistical smoothing describes signatures of the ensemble-averaged field, its application to the evaluation of scattering attenuation in heterogeneous media including macroscopic heterogeneity is limited (Sato, 1982; Wu, 1982a,b). Instead, a wavefield approximation valid in single realizations of the random medium (Shapiro and Hubral, 1999; Müller and Shapiro, 2001; Müller et al., 2002) should be employed.

The problem of multiple scattering and pulse propagation in randomly layered media has been studied in great mathematical detail (Papanicolaou, 1971; Burridge et al., 1988; Asch et al., 1991). For the case of elastic waves propagating in 1-D random media, Shapiro and Hubral (1999) characterized the transmitted wavefield with the help of an approach similar to a second-order Rytov approximation and showed that the scattering attenuation coefficient and phase increment are self-averaged quantities. This wavefield description is known as the generalized ODA theory. It accounts for back- and forward multiple scattering contributions to the primary wavefield. There are also results generalizing the ODA theory for so-called locally layered media, that is, media with a 1-D microstructure but with a 3-D heterogeneous macroscopic (on a scale much larger than the wavelength) background (Solna and Papanicolaou, 2000). Only slightly more recently, Müller and Shapiro (2001) and Müller et al. (2002) adopted the strategy of the generalized ODA theory to 2-D and 3-D random media resulting into quantification of scattering attenuation and dispersion of the ballistic wavefield (so-called seismic primaries) in a broad frequency range (accounting for back- and forward scattering) and to describe pulse propagation in single realizations of random media.

To outline the principle ideas of the generalized ODA formalism, we concentrate on the analysis of a scalar wavefield u. Let us consider a time-harmonic plane wave propagating in a random medium and describe the wavefield inside the random medium with help of the Rytov transformation, using the complex exponent $\Psi = \chi + i\phi$, where the real part χ represents the fluctuations of the logarithm of the amplitude (so-called log–amplitude fluctuations) and the imaginary part ϕ represents phase fluctuations. Omitting the time dependence $\exp(-i\omega t)$, we write

$$\tilde{u}(\omega, \mathbf{r}) = u_0(\omega, \mathbf{r}) e^{\chi(\omega, \mathbf{r}) + i\phi(\omega, \mathbf{r})}, \tag{85}$$

where $u_0 = A_0 \exp(i\phi_0)$ is the wavefield in the homogeneous reference medium (zero fluctuations) with the amplitude $A_0 = 1$ and the unwrapped phase ϕ_0. To be more specific, we assume that the initially plane wave propagates vertically along the z axis. Equation (85) can then be written as

$$\tilde{u}(\omega, z' = L, \mathbf{r}_\perp) = e^{iK(\omega, L, \mathbf{r}_\perp)L}, \tag{86}$$

where L is the travel distance and $\mathbf{r}_\perp$ denotes the transverse coordinates relative to the z axis and we introduced the complex wave number

$$K(\omega, L, \mathbf{r}_\perp) = \left(\frac{\phi(\omega, L, \mathbf{r}_\perp)}{L} + \frac{\phi_0(\omega, L)}{L} \right) - i \frac{\chi(\omega, L, \mathbf{r}_\perp)}{L} \tag{87}$$

$$= \varphi(\omega, L, \mathbf{r}_\perp) + i\gamma(\omega, L, \mathbf{r}_\perp), \tag{88}$$

with real functions ϕ and γ denoting the phase increment and attenuation coefficient, respectively.

An essential feature of the ODA approach is its ability to describe the wavefield in a single realization of the random medium. That is to say, it is an approximation of nonaveraged wavefields. This is possible due to the use of self-averaged wavefield attributes. The latter are quantities that assume their ensemble-averaged values when propagating in an infinitely long single realization of the random medium. Hence, we use the following wavefield approximation in all space dimensions:

$$\tilde{u}(\omega, L, r_\perp) \approx e^{i(\langle \varphi \rangle + i \langle \gamma \rangle)L} \tag{89}$$

Note that Eq. (89) is aimed to describe the ballistic wavefield in single realizations of random media. Its left-hand side is not subjected to statistical averaging. The right-hand side contains, however, the ensemble-averaged wavefield attributes $\langle \phi \rangle$ and $\langle \gamma \rangle$. To keep Eq. (89) physically meaningful, we require self-averaging of the quantities γ and ϕ. It can be shown that the attenuation coefficient and the phase increment are self-averaged quantities in the sense that they are Gaussian random variables and their relative fluctuations decrease with increasing travel distances. However, the weak-wavefield-fluctuation regime permits only finite travel distances. Therefore, we expect that only a *partial* self-averaging can be observed. The self-averaging property of these wavefield attributes can be numerically demonstrated (Fig. 11). In the following, we consider the approximations and properties of the wavefield attributes $\langle \gamma \rangle$ and $\langle \phi \rangle$ leading to estimates of scattering attenuation and dispersion.

4.2. Effective Wave Number in 3-D Random Media

Simple approximations for the wavefield attributes $\langle \chi \rangle$ and $\langle \phi \rangle$ in the weak-wavefield-fluctuation regime are obtained by the following procedure. First, we note that the wavefield can be separated into a coherent and fluctuating (incoherent) part: $u = \langle u \rangle + u_f$ where $\langle u \rangle$ denotes the ensemble-averaged wavefield. A measure of the wavefield fluctuations is the ratio

$$\varepsilon = \left| \frac{u_f}{\langle u \rangle} \right|, \tag{90}$$

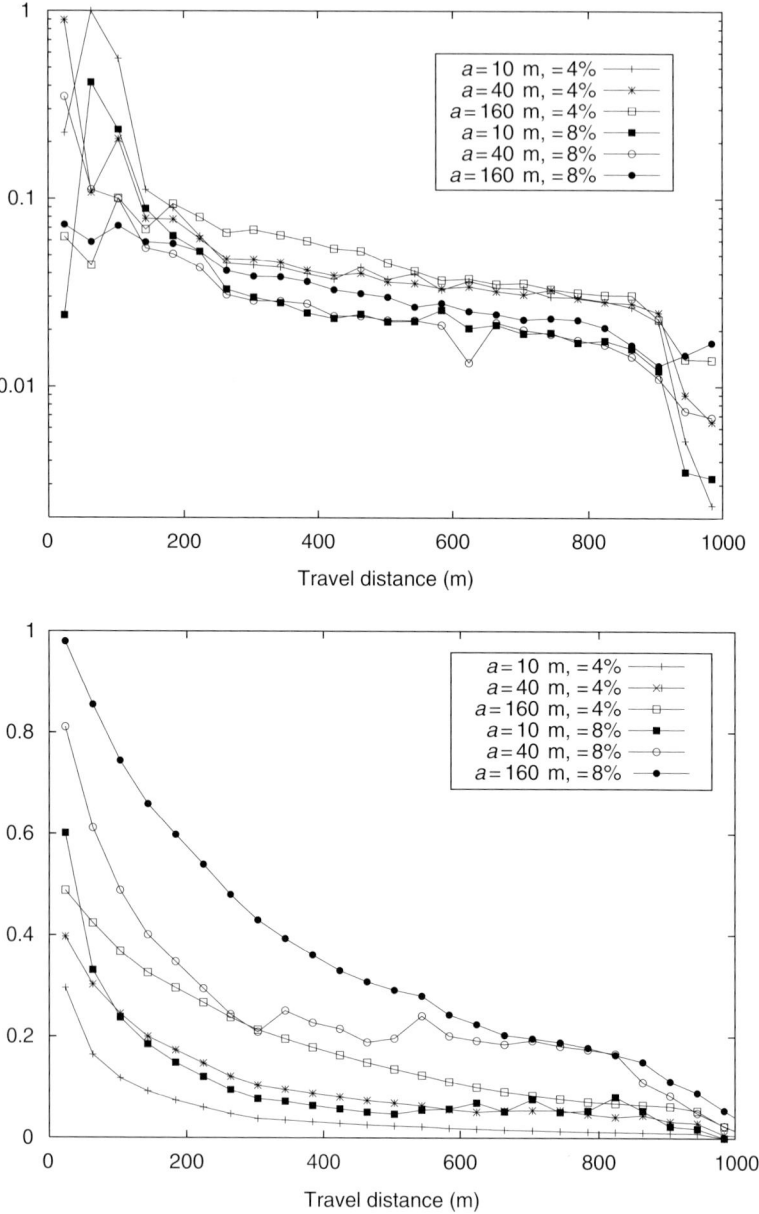

FIG. 11. Top: The relative standard deviations of the attenuation coefficient $\sigma_\alpha/\langle\alpha\rangle$ as a function of travel distance for a 2-D exponentially correlated random medium evaluated for the six plane wave transmission experiments. For all investigated correlation lengths and standard deviations of the velocity fluctuations (denoted by a and σ in the legend), we can observe a decrease of $\sigma_\alpha/\langle\alpha\rangle$ with increasing travel distances (note that the ordinate axis is logarithmically spaced). The

and the statistical average $\langle \varepsilon^2 \rangle = \frac{I_t}{I_c} - 1$, where $I_t = \langle |u|^2 \rangle$ is the total intensity and $I_c = |\langle u \rangle|^2$ is the coherent intensity. Then, the range of weak fluctuation is defined by

$$\langle \varepsilon^2 \rangle \ll 1 \tag{91}$$

and means that the coherent intensity is of the order of the total intensity. In terms of the medium parameters, the latter inequality corresponds to

$$\sigma_n^2 (ka)^2 L/a < 1, \tag{92}$$

where σ_n^2 denotes the variance of the medium parameter fluctuations. As shown in Shapiro and Kneib (1993) and Shapiro et al. (1996), the averaged fluctuations of log–amplitude and phase due to the presence of inhomogeneities in 2-D and 3-D random media can be described within the weak-wavefield-fluctuation regime [i.e., in the regime, where Eqs. (91) and (92) are satisfied]

$$\langle \chi \rangle = -\sigma_{\chi\chi}^2 + O(\varepsilon^3), \tag{93}$$
$$\langle \phi \rangle = \phi_c - \phi_0 - \sigma_{\chi\phi}^2 + O(\varepsilon^3). \tag{94}$$

It is important to note that the derivation of Eq. (93) is based on the assumption that the total intensity remains constant and reflects the property of energy conservation within the parabolic approximation (Rytov et al., 1989). Conversely, to derive Eq. (94) no assumption for I_t has to be made. Thus, in Eq. (93) the backscattering is neglected but it is still present in Eq. (94). The log-amplitude variance $\sigma_{\chi\chi}^2$ and the log-amplitude phase cross-variance $\sigma_{\chi\phi}^2$ in Eqs. (93)–(94) can be calculated with help of the 2-D and 3-D Rytov approximation (Ishimaru, 1978). The coherent phase ϕ_c follows from the method of statistical smoothing.

Simple expressions for the quantities $\sigma_{\chi\chi}^2, \sigma_{\chi\phi}^2$ and ϕ_c are known (Ishimaru, 1978). The results in 3-D media are

$$\sigma_{\chi\chi}^2 = 2\pi^2 k^2 L \int_0^\infty d\kappa \kappa \left(1 - \frac{\sin(\kappa^2 L/k)}{\kappa^2 L/k}\right) \Phi^{3D}(\kappa) \tag{95}$$

$$\sigma_{\chi\phi}^2 = 4\pi^2 k^3 \int_0^\infty d\kappa \kappa \left(\frac{\sin^2(\kappa^2 L/2k)}{\kappa^2}\right) \Phi^{3D}(\kappa) \tag{96}$$

$$\sigma_{\phi\phi}^2 = 2\pi^2 k^2 L \int_0^\infty d\kappa \kappa \left(1 + \frac{\sin(\kappa^2 L/k)}{\kappa^2 L/k}\right) \Phi^{3D}(\kappa). \tag{97}$$

The terms in brackets are the so-called spectral filter functions or Fresnel filters (since they act on the fluctuation spectra like filters). For the 2-D case, the results can be

---

strong decrease of $\sigma_\alpha/\langle \alpha \rangle$ for $L > 950$ m is caused by numerical problems in determining the attenuation coefficient when the wavefield fluctuations become strong. Bottom: The relative standard deviations of the travel time fluctuations $\sigma_t/\langle t \rangle$ as a function of travel distance for the same numerical experiments. The decrease of $\sigma_t/\langle t \rangle$ with travel distance can be interpreted as a decrease of the relative phase increment fluctuations $(\varphi - \langle \varphi \rangle)/\langle \varphi \rangle$ since $\varphi = \omega t$.

obtained by skipping κ in the integral over dκ, dividing by π and using the 2-D fluctuation spectra. Explicit expressions for the coherent phase are obtained with help of the method of statistical smoothing. The result in 3-D media is

$$\phi_c - \phi_0 = \pi k^2 L \int_0^\infty d\kappa \kappa \ln\left(\frac{2k+\kappa}{2k-\kappa}\right)^2 \Phi^{3D}(\kappa), \tag{98}$$

whereas in 2-D media one obtains

$$\phi_c - \phi_0 = 4\pi k^3 L \int_{2k}^\infty d\kappa \frac{\Phi^{2D}(\kappa)}{\sqrt{\kappa^2 - 4k^2}}. \tag{99}$$

Stipulating that the wavefield attributes must satisfy a causality condition, we can extend the range of applicability of Eqs. (93)–(94) using of the Kramers–Kronig relations. For a comprehensive review of this topic we refer to (Weaver and Pao, 1981; Beltzer, 1989). The Kramers–Kronig equations allow one to reconstruct the attenuation from the dispersion behavior and vice versa since both quantities are related by a pair of Hilbert transforms. To this end it is sufficient to derive a Kramers–Kronig relationship for the complex wave number $K(\omega)$ of the plane wave response. It is expedient to use the formulation of the Kramers–Kronig relationship of Weaver and Pao in (Weaver and Pao, 1981) corresponding to twice-subtracted dispersion relations [see their Eqs. (71) and (72)]:

$$\varphi(\omega') = B\omega' + \frac{2\omega'}{\pi} \int_0^\infty \frac{\gamma(\omega) - \gamma(\omega')}{\omega^2 - \omega'^2} d\omega + \varphi(0) \tag{100}$$

$$\gamma(\omega') = \frac{-2\omega'^2}{\pi} \int_0^\infty \left[\frac{\varphi(\omega)}{\omega} - \frac{\varphi(\omega')}{\omega'}\right] \frac{d\omega}{\omega^2 - \omega'^2} + \gamma(0), \tag{101}$$

where $B = \lim_{\omega \to \infty} \gamma(\omega)/\omega$.

In the following, we derive the scattering attenuation coefficient $\langle \gamma \rangle$ by applying Eq. (101) to the phase increment $\langle \phi \rangle = \langle \phi \rangle / L$ with $\langle \phi \rangle$ taken from Eq. (94). Let's begin with the $-\sigma_{\chi\phi}^2$ part of the phase increment which is in 3-D random media given by Eq. (96). Inserting Eq. (96) into Eq. (101) we get, after integration,

$$\gamma R(\omega') = 2\pi^2 \frac{\omega'^2}{v_0^2} \int_0^\infty d\kappa \kappa \Phi(\kappa) \left[1 - \frac{\sin(\kappa^2 L v_0/\omega')}{\kappa^2 L v_0/\omega'}\right] \tag{102}$$

$$= \frac{\sigma_{\chi\chi}^2}{L}, \tag{103}$$

which corresponds to the log-amplitude variance in the Rytov approximation [see Eq. (95)]. Thus, the contributions resulting from the Rytov approximation in Eqs. (93)–(94) satisfy the Kramers–Kronig relationship (Eq. 101).

Now the question arises, what happens with the ϕ_c part of Eq. (94) when subjected to the Kramers–Kronig relation (Eq. 101)? We show that this results in an additional term γ_B. To do so, we note that the phase increment resulting from the method of smoothing [see Eq. (98)] can be written [for instance, the real part of equation 4.59 in Rytov et al. (1989)]

as

$$\varphi_c(\omega') = \phi_c(\omega')/L$$
$$= \frac{\omega}{v_0} + \frac{\omega'^2}{v_0^2} \int_0^\infty dr\, \sin(2\omega'/v_0 r) B(r). \tag{104}$$

Inserting Eq. (104) into Eq. (101) and performing the integrations we obtain

$$\gamma_B(\omega') = 2\frac{\omega'^2}{v_0^2} \int_0^\infty B(r)\sin^2(\omega'/cr)dr - \frac{\omega'^2}{v_0^2} \int_0^\infty B(r)dr. \tag{105}$$

Since correlation function and fluctuation spectrum are related by Eq. (5) we obtain

$$\gamma_B(\omega) = 2\pi^2 \frac{\omega^2}{v_0^2} \int_0^{2\omega/v_0} d\kappa \Phi(\kappa)\kappa - 2\pi^2 \frac{\omega^2}{v_0^2} \int_0^\infty d\kappa \Phi(\kappa)\kappa$$
$$= -2\pi^2 k^2 \int_{2k}^\infty d\kappa \Phi(\kappa)\kappa. \tag{106}$$

That means that we derived two pairs of wavefield attributes, $\{\sigma_{\chi\chi}^2, -\sigma_{\chi\phi}^2\}$ and $\{\gamma_B L, \phi_c - \phi_0\}$, each of them related by Eq. (101). Therefore, with help of the Kramers–Kronig relations we obtained the following logarithmic wavefield attributes in 2-D as well as 3-D random media:

$$\langle \chi \rangle = -(\gamma_R + \gamma_B)L = -\sigma_{\chi\chi}^2 - \gamma_B L + O(\varepsilon^3), \tag{107}$$

$$\langle \phi \rangle = \phi_c - \phi_0 - \sigma_{\chi\phi}^2 + O(\varepsilon^3). \tag{108}$$

Thus, accepting the existence of the Kramers–Kronig relations as a physical constraint, we arrived at Eqs. (107) and (108). The hybrid character of Eqs. (107) and (108) concerning the combination of Rytov approximation and the method of statistical smoothing extends the wavefield approximation (Eq. 89) in a sense that now at least a part of the backscattering is taken into account. But accounting for backscattering means also that the wavelength can exceed the size of inhomogeneities. Therefore, we can expect Eq. (107) to be valid in a broader frequency range than Eq. (93). We used the 3-D wavefield attributes in Eqs. (102)–(106). We note that Eqs. (106)–(108) are valid for 2-D random media if we divide in Eq. (106) the expression inside the integral over κ by $\pi\kappa$ and use the 2-D fluctuation spectrum and use the corresponding 2-D approximations for the quantities in Eqs. (107)–(108).

It is known that the scattering of seismic waves in media with a large characteristic size of heterogeneity compared with the wavelength is confined within small angles around the forward direction. This means the conversion between P- and S waves in elastic media can be neglected. In fact, following Ishimaru (1978) it is not difficult to show that the complex exponent $\Psi = \chi + i\phi$ in elastic random media is exactly the same as in acoustic random media. Therefore, under the assumptions of weak-wavefield fluctuations and $\lambda < a$, the propagation of elastic waves shows the same behavior as acoustic waves and approximations [Eqs. (107)–(108)] are valid. In the regime $\lambda \geq a$, however, the coherent phase differs in acoustic and (poro)elastic media. Gold et al. (2000) applied the method of statistical smoothing in elastic random media to compute the coherent wave number in 2-D and 3-D random media. The real part of these wave numbers multiplied by the travel distance defines the coherent phase. In 3-D random media the results are:

$$\phi_c^P = k_p L \Re \left\{ \left(1 - \frac{1}{\rho_0 \omega^2} \int_{-\infty}^{\infty} d^2 r e^{-ik_p z} \left[\omega^4 B_{\rho\rho} G_{33} - k_p^2 B_{\lambda\lambda} G_{jl,jl} - 4k_p^2 B_{\mu\mu} G_{33,33} \right. \right. \right.$$
$$\left. \left. \left. + 4i\omega^2 k_p B_{\rho\mu} G_{33,3} + 2i\rho\omega^2 k_p B_{\lambda\rho} G_{m3,m} - 4k_p^2 B_{\lambda\mu} G_{3m,3m} \right] \right)^{-1/2} \right\}. \quad (109)$$

G_{ij} means the 3-D elastodynamic Green's function (Eq. 30). The subscripts λ and μ denote the Lamé parameters. Therefore, the results obtained for acoustic waves can also be applied with slight modification to elastic media. That is, in elastic media Eqs. (89), (93), and (94) can be used exactly as in the acoustic case with the only difference: instead of Eqs. (99) and (98) for ϕ_c Eq. (109) must be used.

To summarize, analytic expression for the phase and the attenuation coefficients of a plane wave in the weak-wavefield-fluctuation regime are derived by linking the Rytov approximation with the method of statistical smoothing using the Kramers–Kronig relations. The resulting expressions have practically no restriction in the frequency domain. We note that this approach is not based on a rigorous mathematical treatment but rather a combination of perturbation approximations guided by physical arguments. Numerical experiments yield quite good support for this approach (details can be found in Müller et al., 2002).

4.3. Scattering Attenuation and Asymptotic Behavior

The derived logarithmic wave field attributes enable us to analyze the frequency behavior of the scattering attenuation coefficient in a broad frequency range. The attenuation coefficient in 3D is given by

$$\gamma^{3D} = \gamma_R(\omega) + \gamma_B(\omega)$$
$$= 2\pi^2 k^2 \int_0^\infty d\kappa \kappa \Phi^{3D}(\kappa) \left[1 - \frac{\sin(\kappa^2 L/k)}{\kappa^2 L/k} \right] - 2\pi^2 k^2 \int_{2k}^\infty d\kappa \kappa \Phi^{3D}(\kappa)$$
$$= 2\pi^2 k^2 \int_0^\infty d\kappa \kappa \Phi^{3D}(\kappa) \left[\vartheta(\kappa - 2k) - \frac{\sin(\kappa^2 L/k)}{\kappa^2 L/k} \right], \quad (110)$$

and analogously in 2-D random media

$$\gamma^{2D} = 2\pi k^2 \int_0^\infty d\kappa \Phi^{2D}(\kappa) \left[\vartheta(\kappa - 2k) - \frac{\sin(\kappa^2 L/k)}{\kappa^2 L/k} \right], \quad (111)$$

where ϑ is the Heaviside step function. The structure of the attenuation coefficient is similar to that found in the previous section and involves an integral over the fluctuation spectrum multiplied by the so-called Fresnel filter. This product is shown in Fig. 12. The frequency dependence of the reciprocal quality factor $Q^{-1} = 2\gamma/k$ is shown in Fig. 13.

The lower part of Fig. 13 depicts Q^{-1} as a function of ka in a 3-D Gaussian-correlated random medium for various L/a. The upper part of Fig. 13 shows the reciprocal quality factor for waves propagating in 2-D, 3-D exponentially and Gaussian-correlated random media according to Eqs. (110)–(111). In the same figure, Q^{-1} for 1-D random media is displayed. We used the corresponding expressions of the ODA theory of Shapiro and Hubral (1999) (their formulas 4.18). All curves are normalized by $\sqrt{\pi}\sigma_n^2$. Whereas in 1D the attenuation is independent of the parameter L/a, in 2D and 3D we observe a dependence in L/a. In Gaussian random media the maximum attenuation occurs roughly at

$$\omega_{\max} \approx \frac{v_0 L}{a^2} \tag{112}$$

which corresponds to the so-called Fresnel length. Note that in 2D and 3D, attenuation in Gaussian random media is slightly larger than in exponentially correlated ones for $ka < 1$. This relation becomes reversed for $ka > 1$.

In the low-frequency limit ($ka \to 0$), we can approximate the spectral filter function $1 - (\sin(\kappa^2 L/k))/(\kappa^2 L/k)$ in Eq. (95) by 1. Then Eq. (110) can be rewritten as

$$\gamma_{\text{low}}^{3D} = 2\pi^2 k^2 \int_0^\infty d\kappa\, \kappa \Phi^{3D}(\kappa) - 2\pi^2 k^2 \int_{2k}^\infty d\kappa\, \kappa \Phi^{3D}(\kappa) = 2\pi^2 k^2 \int_0^{2k} d\kappa\, \kappa \Phi^{3D}(\kappa). \tag{113}$$

Equation (113) coincides with the mean field attenuation coefficient as derived by various authors (e.g., Rytov *et al.*, 1989). Inserting, for example, the fluctuation spectrum for 3-D exponential media $\Phi^{3D}(\kappa) = \dfrac{\sigma_n^2 a^3}{\pi^2 (1 + \kappa^2 a^2)^2}$, we get

$$\gamma_{\text{low}}^{3D} = 4\sigma_n^2 \frac{k^2 a^3}{1 + 4k^2 a^2} \approx 4\sigma_n^2 a^3 k^4, \tag{114}$$

yielding the expected Rayleigh frequency dependency ω^{d+1}, where d denotes the spatial dimension.

In general, the high-frequency behavior of the attenuation coefficient depends on the used fluctuation spectrum. In the case of Gaussian-correlated fluctuations, the high-frequency limit of α is obtained as follows. Applying a Taylor expansion of the spectral filter function in Eq. (95) for small argument L/k (or equivalently $D \ll 1$) yields

$$\left(1 - \frac{\sin(\kappa^2 L/k)}{\kappa^2 L/k}\right) = \frac{1}{6}\left(\frac{\kappa^2 L}{k}\right)^2 + O\left(\left(\frac{\kappa^2 L}{k}\right)^4\right). \tag{115}$$

This results in the frequency-independent equation:

$$\gamma_{\text{high}}^{3D} = \frac{\pi^2}{3} L^3 \int_0^\infty d\kappa\, \kappa^5 \Phi^{3D}(\kappa). \tag{116}$$

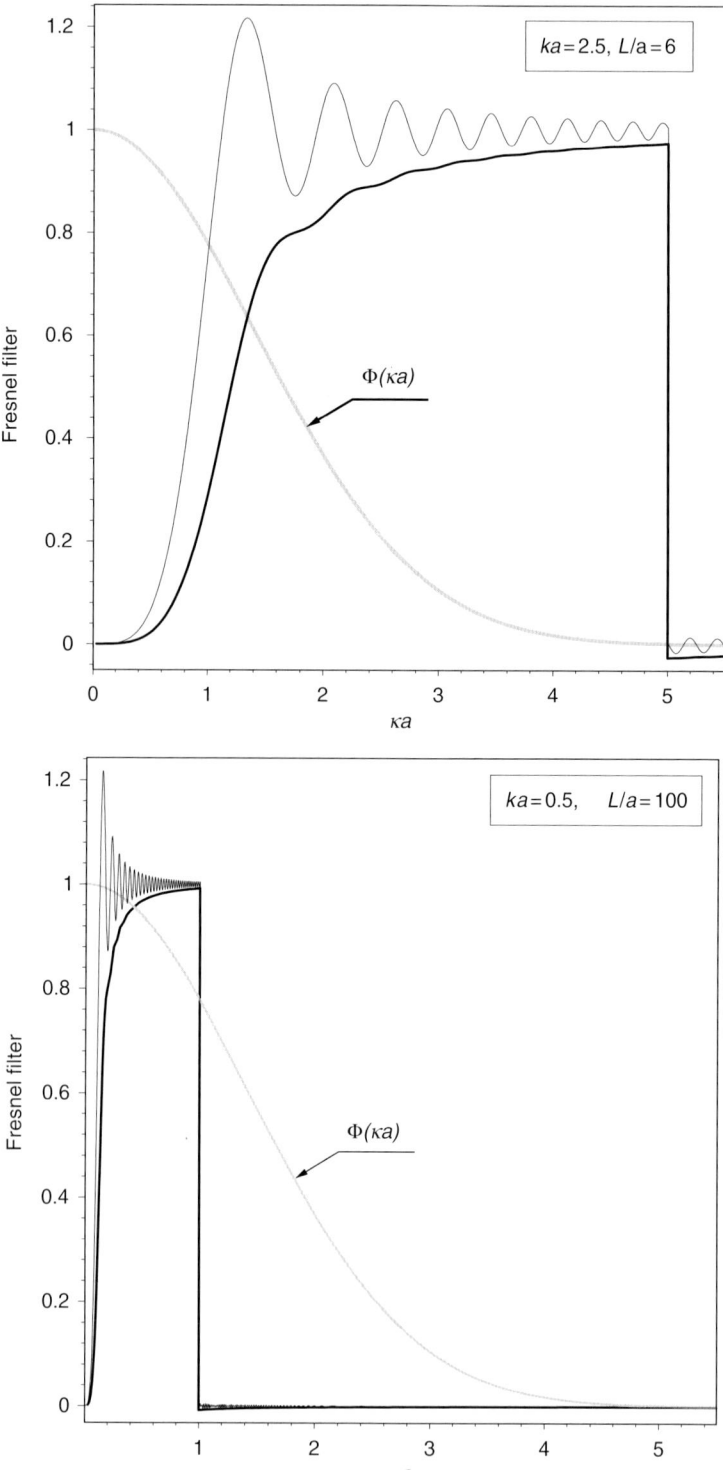

This corresponds to the log–amplitude variance divided by the travel distance in the geometrical optics limit.

5. The Interplay Between Attenuation Due Interlayer Flow and Scattering

5.1. 1-D Poroelastic Random Media

In the preceding sections, we analyzed the two attenuation mechanisms under consideration separately and assumed the validity of the superposition principle. In the case of randomly layered media, it is possible to analyze the combined effects of wave-induced flow and scattering within a single theoretical framework: the poroelastic extension of the generalized ODA formalism. In 1-D random media, the wave-induced flow degenerates to the so-called *interlayer flow*.

The total-frequency-range results for the attenuation coefficient and the vertical-phase increment of the P wave have been found by Gelinsky *et al.* (1998). The further analysis will be concentrated on the attenuation of the P wave. The reciprocal quality factor Q^{-1} is (Shapiro and Müller, 1999)

$$Q^{-1} = \frac{\alpha^2 N \Delta_1}{H} F_{\text{flow}}(x_{\text{flow}}) + \frac{N^2 \Delta_2}{M^2} F_{\text{scat}}(x_{\text{scat}}), \tag{117}$$

where the notation of the poroelastic parameters is the same as in Section 3. The quantities $\Delta_{1,2}$ are measures of heterogeneity of the medium. There are two different linear combinations of the normalized variances and covariances of the poroelastic parameters:

$$\Delta_1 = \left\langle \left(\varepsilon_P - \varepsilon_M - 2\frac{P - \alpha^2 M}{P} \varepsilon_\alpha \right)^2 \right\rangle, \quad \Delta_2 = \sigma_{HH}^2. \tag{118}$$

The factors before the functions F do not contain any dynamic information. Accordingly, they do not contain any information about the permeability. The dynamic dependence of the attenuation is contained in the quantities $F_{\text{flow}}(x)$ and $F_{\text{scat}}(x)$. These are positive functions of the following form:

$$\begin{aligned}F_{\text{flow}}(x_{\text{flow}}) &= \sqrt{2} x_{\text{flow}} \int_0^\infty d\zeta B(\zeta) \exp(-\zeta x_{\text{flow}}) \cos(\zeta x_{\text{flow}} + \pi/4) \\ F_{\text{scat}}(x_{\text{scat}}) &= \frac{1}{2} x'_{\text{scat}} \int_0^\infty d\zeta B(\zeta) \cos(2\zeta x_{\text{scat}}),\end{aligned} \tag{119}$$

Fig. 12. Behavior of the Fresnel filter, the term in brackets of Eq. (110) (denoted by the thin black curve), as a function of the normalized wave number κa. For large frequencies and small travel distances (small wave parameter D), the Fresnel filter excludes only the low wave number components of the fluctuation spectrum $\Phi(\kappa a)$ indicated by the grey curve (top). The opposite behavior can be observed for low frequencies and large travel distances (large D) (bottom).

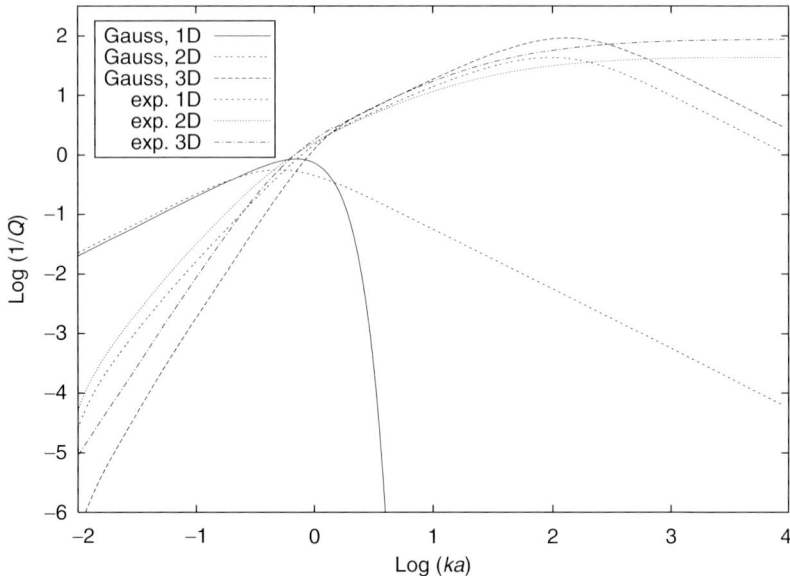

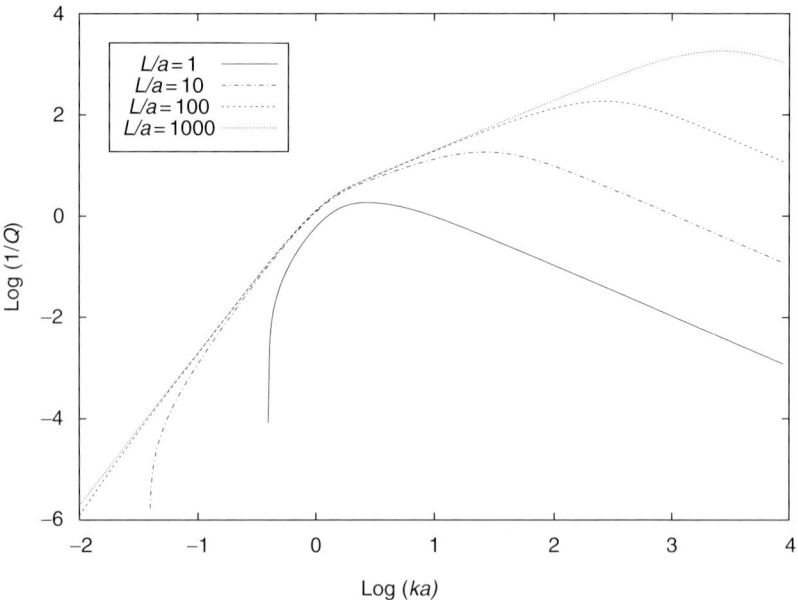

Fig. 13. Top: Scattering attenuation ($1/Q$) as a function of κa for fixed $L/a = 50$ for Gaussian and exponentially correlated random media in 1D, 2D, and 3D. Bottom: The L/a dependency of the reciprocal quality factor is shown for a 3-D Gaussian random medium. Note that all curves are normalized by $\sqrt{\pi}\sigma_n^2$.

where $\zeta = z/a$. Their arguments depend on frequency: $x_{\text{flow}} = \sqrt{\omega a^2 r/2N}$; $x_{\text{scat}} = \omega a/v_0$, where $v_0 = \sqrt{H/\rho}$ is the P-wave velocity in the Gassmann limit. Here, $r = \eta/\kappa_0$. The second term in Eq. (117) describes the elastic scattering.

The values of Δ_1 and Δ_2 in Eq. (117) are usually of the order of 0.01–0.1. Under the assumptions $K_{\text{f}} \ll K_{\text{g}}$, $\phi \ll 1$ and accepting that fluctuations of the K_{g} are smaller than the fluctuations of other parameters, the following rough approximation is obtained:

$$\Delta_1 \approx \langle (\varepsilon_{\text{P}} - \varepsilon_{K_{\text{f}}} + \varepsilon_\phi)^2 \rangle. \tag{120}$$

This relation shows the role of cross-correlations between the fluctuations. For instance, if the fluctuations of the skeleton elastic moduli and the fluctuations of the bulk modulus of the pore fluid are anticorrelated, then a maximum interlayer-flow contribution in the attenuation of the P wave will be observed. Such a situation is possible in partially saturated reservoirs. In realistic situations, the low-frequency range attenuation due to the interlayer flow can be of the order of $n \times 100$ in terms of the Q factor. The contribution of the elastic scattering is also of the same order. In extreme cases (of partial gas saturation) the resulting Q can reach the order of $n \times 10$.

We also compare the result for the attenuation with those obtained in the 3-D case. For the same sandstone model as above, we compute the interlayer-flow contribution of Q^{-1} [the first term in Eq. (117)] for the case that only parameter K_{f} exhibits fluctuations with $\sigma_{K_{\text{f}}}^2 = 0.2$ and $a = 10$ cm (Fig. 14). It can be observed that the magnitude of attenuation in 1-D and 3-D random media is of the same order. However, note that the attenuation in 3D is slightly larger. Maximal attenuation in the 3-D case is observed at

$$\omega_{\text{max}}^{\text{3D}} = 2\kappa_0 N/a^2 \eta, \tag{121}$$

whereas in the 1-D case at

$$\omega_{\text{max}}^{\text{1D}} = \kappa_0 N/a^2 \eta. \tag{122}$$

Thus, maximum of attenuation in 3D occurs at a frequency twice as large as compared with the 1-D case. In our example, this difference has important implications for the observability of the attenuation mechanism. For typical seismic frequencies (10–100 Hz), the attenuation due to wave-induced fluid flow is larger in 3-D inhomogeneous structures (this is indicated by the arrow in Fig. 14). It can be also observed that the low- and high-frequency velocities coincide for the two cases. This is, however, a consequence of the constant shear modulus in this example. Numerical validation of the proposed model can be found in Gelinsky et al. (1998). Using a poroelastic reflectivity code attenuation and dispersion are computed in a broad frequency range. The numerical results show good agreement with the theoretical predictions (Fig. 15).

5.2. Asymptotic Scaling of Attenuation

The frequency and permeability dependencies of the attenuation are defined by functions $F_{\text{flow}}(x)$ and $F_{\text{scat}}(x)$. Therefore, they are controlled by the correlation properties of the medium heterogeneities (i.e., by the disorder of rock structures).

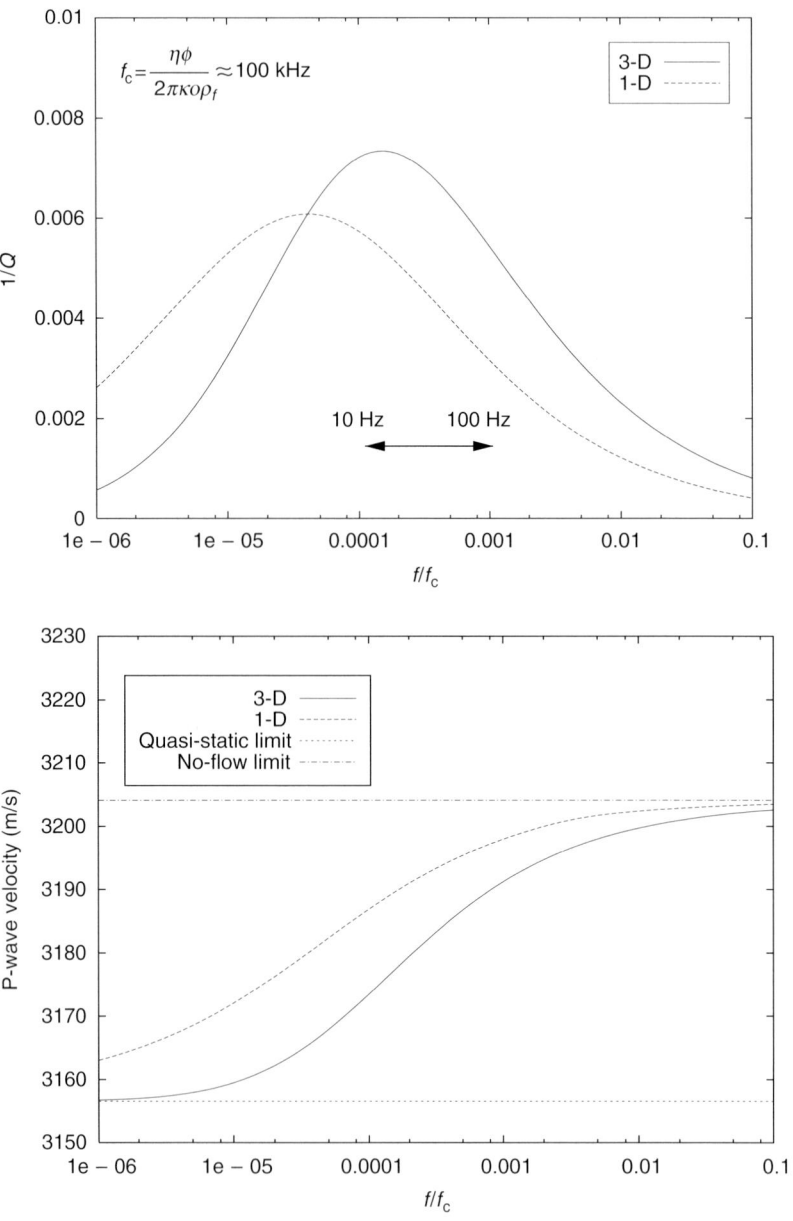

FIG. 14. Wave-induced flow attenuation in terms of Q^{-1} (top) and velocity dispersion (bottom) as a function of normalized frequency for a model with fluctuations in the parameter K_f with $\sigma^2_{K_f K_f} = 0.2$ and $a = 10$ cm. For the same parameters, the result of the 1-D poroelastic extension of the O'Doherty–Anstey theory is also shown (dashed curve).

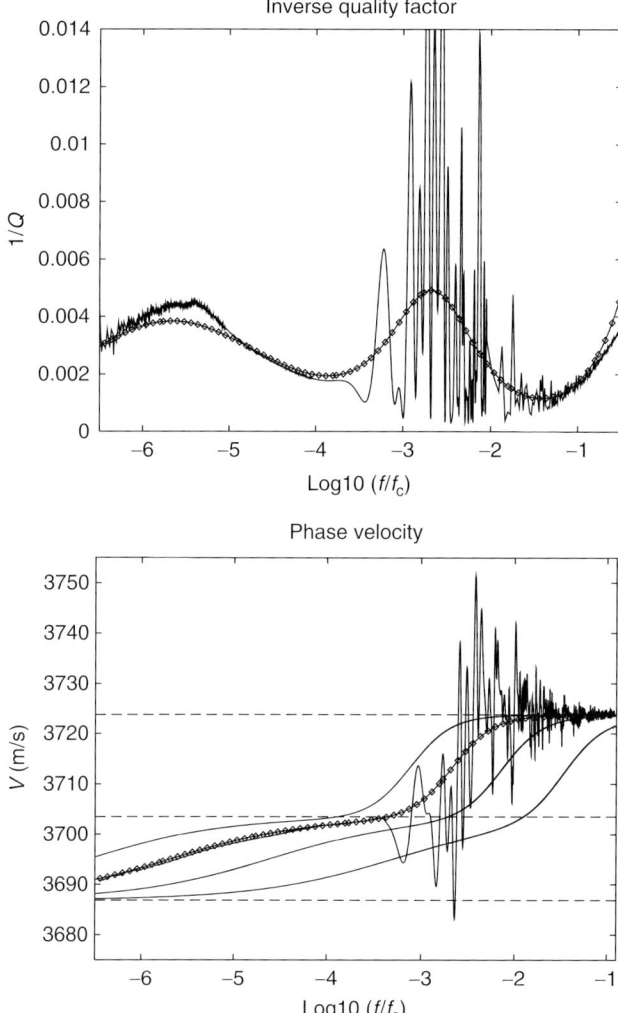

FIG. 15. Comparison of theory (diamonds) and numerical experiment (solid line) for 1/Q (top) and phase velocity (bottom). The correlation length is $a = 1.25$ m. In the lower plot, the quasistatic, no-flow, ray velocities (from bottom to top) are indicated by dashed lines.

Let us now consider the general case of randomly heterogeneous media. In this case, the correlation between any two points decreases with an increasing distance between them. Therefore, the function $B(\zeta)$ is a rapidly decreasing function for $\zeta > 1$. Taking this into account and considering the limit $x_{\text{flow}} \to 0$ one obtains

$$F_{\text{flow}}(x_{\text{flow}}) \approx x_{\text{flow}} \int_0^1 d\zeta \Phi(\zeta), F_{\text{scat}}(x_{\text{scat}}) \approx \frac{1}{2} x_{\text{scat}} \int_0^1 d\zeta \Phi(\zeta). \tag{123}$$

Therefore, we observe that in any medium with disorder, in the low-frequency range the Q^{-1} contribution of the interlayer flow is proportional to x_{flow}, that is, to $\omega^{1/2}$. The contribution of the elastic scattering is proportional to ω. Thus, the interlayer-flow-associated attenuation coefficient is proportional to $\omega^{3/2}$. This frequency dependence is analogous to the Rayleigh-scattering frequency dependence, and it is universal for all poroelastic media with disorder. Moreover, the low-frequency-range attenuation coefficient is proportional to $\kappa_0^{-1/2}$. There is no universal behavior in the case of $x_{\text{flow}} \gg 1$, where the frequency dependence of the interlayer-flow attenuation is controlled by the statistics of the heterogeneities.

Substituting the von Kármán correlation function (Eq. 8) into Eq. (119) we obtain

$$F_{\text{flow}}(x_{\text{flow}}) = 2^{\nu-1/2}\nu\Gamma(\nu+1/2)z\left[\left(1+2ix_{\text{flow}}^2\right)^{\frac{-2\nu-1}{4}}P_{-1/2+\nu}^{-1/2-\nu}(\tilde{z})\right.$$

$$\left. - i\left(1 - 2ix_{\text{flow}}^2\right)^{\frac{-2\nu-1}{4}} P_{-1/2+\nu}^{-1/2-\nu}(z)\right] \qquad (124)$$

$$F_{\text{scat}}(x_{\text{scat}}) = \sqrt{\pi}\Gamma(\nu+1/2)\Gamma^{-1}(\nu)x_{\text{scat}}\left(1+x_{\text{scat}}^2\right)^{-1/2-\nu}, \qquad (125)$$

where the function $P_{-1/2+\nu}^{-1/2-\nu}$ is the associated Legendre function of the first kind, $z = x_{\text{flow}} + ix_{\text{flow}}$, and $\tilde{z}$ its complex conjugated. To derive the first result, the tables of integrals (Prudnikov et al., 1988, p. 349, eq. 2.16.6.3) has been used.

The high-frequency limit ($x_{\text{flow, scat}} \to \infty$) can now be obtained by an asymptotic expansion. Therefore, the relation for large arguments of the Legendre function is used (Gradshteyn and Ryzhik, 1980, p. 1011, eq. 8.776) and provides the following proportionalities:

$$F_{\text{flow}} \propto x_{\text{flow}}^{-2\nu}, \; F_{\text{scat}} \propto x_{\text{scat}}^{-2\nu}. \qquad (126)$$

In spite of their similarity, these relations result in different frequency dependencies of the fluid-flow and scattering contributions to the attenuation:

$$Q_{\text{flow}}^{-1} \propto \omega^{-\nu}, \; Q_{\text{scat}}^{-1} \propto \omega^{-2\nu}. \qquad (127)$$

In addition, the permeability dependence is

$$Q_{\text{flow}}^{-1} \propto \kappa^{\nu}. \qquad (128)$$

As a particular example, we consider the case $\nu = 1/2$. Then the correlation function assumes the exponential form $B(\zeta) = \exp(-\zeta)$ and the functions F are

$$F_{\text{flow}}(x_{\text{flow}}) = \frac{x_{\text{flow}}}{1+2x_{\text{flow}}+2x_{\text{flow}}^2}, \; F_{\text{scat}}(x_{\text{scat}}) = \frac{x_{\text{scat}}/2}{1+4x_{\text{scat}}^2}. \qquad (129)$$

This example shows that functions $F_{\text{flow}}(x)$ and $F_{\text{scat}}(x)$ are positive functions with magnitudes of the order $O(1)$. These functions reach their maxima at $x = O(1)$. Therefore, the permeability controls the location of the maximum of the interlayer-flow part in the frequency range. This maximum is reached at frequencies of the order $O(2Nk/a^2\eta)$.

In realistic situations, if a is in the range of $10^{-2} - 10$ m the maximum of the Q^{-1} factor of the interlayer-flow attenuation can be reached at any frequency lower than 10^3 Hz. Therefore, this attenuation will be significant at least in a part of the seismic frequency range.

6. Concluding Remarks

The models proposed in this chapter provide expressions for frequency-dependent attenuation and dispersion in 3-D randomly inhomogeneous porous media accounting for the effect of wave-induced fluid flow and scattering. Our results are based on perturbation theory and, therefore, are restricted to weak-contrast media. However, we think that this approximate solution reveals the exact solution's essential dependence on frequency and medium parameters. In our approach, the dynamic characteristics depend on the correlation properties of the medium fluctuations. Closed-form expressions for $Q^{-1}(\omega)$ and $v(\omega)$ are obtained for several correlation functions. In this chapter, we focused the analysis on wave propagation in statistically isotropic random media. However, the results can be probably generalized to the case of statistically anisotropic random media. An advantage of the statistical approach is its flexibility to handle complex geometrical distributions of the inhomogeneities. Only the spatial correlation of the fluctuations needs to be known to compute the dynamic wavefield attributes.

We analyzed the properties of the coherent wave propagating in poroelastic random media. Neglecting the ordinary elastic scattering, we only accounted for conversion scattering from fast P into Biot's slow P wave. This process of conversion scattering is equivalent to the mechanism of pore pressure relaxation due to wave-induced perturbations. Thus, our results describe the relationship between the dynamic properties of the coherent wavefield and the mechanism of wave-induced fluid flow. In particular, we have derived an explicit expression for the effective wave number of the fast compressional wave [Eq. (58)] by applying a first-order statistical smoothing of Biot's equations of poroelasticity with randomly varying coefficients. Attenuation and dispersion depend on linear combinations of the spatial correlations of the fluctuating poroelastic parameters. The observed frequency dependence is typical for a relaxation phenomenon. The low- and high-frequency asymptotes of the attenuation coefficient of a plane compressional wave in 1D and 3D are analyzed. The low-frequency behavior of attenuation is found to be $Q^{-1} \propto \omega$ in 3D and $Q^{-1} \propto \sqrt{\omega}$ in 1D, whereas at high frequencies $Q^{-1} \propto \omega^{-1/2}$ in all space dimensions. It is interesting to note that these asymptotes coincide with those predicted by the periodicity-based approaches (Johnson, 2001; Pride et al., 2004). Consequently, in 3-D space the observed frequency dependency of attenuation due to fluid flow has universal character independent of the type of disorder (periodic or random). This result is somewhat unexpected if we remember that in 1-D space the attenuation asymptotes are different for periodic and random structures. Several modeling choices of the approach including the effect of cross-correlations between fluid and solid phase properties are demonstrated.

To account properly for scattering attenuation in general random (porous) media, we extended the generalized ODA formalism to 2-D and 3-D random media. Using the fact that the attenuation coefficient and phase increment are self-averaged quantities, we derive approximations for their ensemble-averaged counterparts. In particular, we use a combination of the method of statistical smoothing and the Rytov approximation to obtain expressions for attenuation and dispersion that is valid in the entire frequency range. At low-frequencies attenuation scales according to the Rayleigh dependence $Q^{-1} \propto \omega^D$, where D denotes the spatial dimension. At high frequencies, we recover the geometrical optic limit.

ACKNOWLEDGMENTS

This work was kindly supported by the Deutsche Forschungsgemeinschaft (contract MU 1725/1–3), CSIRO Petroleum, and Centre of Excellence for Exploration and Production Geophysics.

REFERENCES

Aki, K., Chouet, B. (1975). Origin of coda waves: Source, attenuation and scattering effects. *J. Geophys. Res.* **80**, 3322–3342.
Aki, K., Richards, P.G. (1980). *Quantitative Seismology: Theory and Methods*, vol. 1. W.H. Freeman and Company, San Francisco, CA.
Asch, M., Kohler, W., Papanicolaou, G., Postel, M., White, B. (1991). Frequency content of randomly scattered signals. *SIAM Rev.* **33**, 519–625.
Beltzer, A.I. (1989). The effective dynamic response of random composites and polycrystals— A survey of the causal approach. *Wave Motion* **11**, 211–229.
Biot, M.A. (1962). Mechanics of deformation and acoustic propagation in porous media. *J. Appl. Phys.* **33**, 1482–1498.
Bourbié, T., Coussy, O., Zinszner, B. (1987). *Acoustics of Porous Media*. Editions Technip, Paris.
Burridge, R., Papanicolaou, G., White, B.S. (1988). One-dimensional wave propagation in a highly discontinuous medium. *Wave Motion* **10**, 19–44.
Gassmann, F. (1951). Über die Elastizität poröser Medien. *Viertel. Naturforsch. Ges. Zürich* **96**, 1–23.
Gelinsky, S., Shapiro, S.A., Müller, T.M., Gurevich, B. (1998). Dynamic poroelasticity of thinly layered structures. *Int. J. Solids Struct.* **35**, 4739–4752.
Goff, J.A., Holliger, K. (2003). *Heterogeneity in the Crust and Upper Mantle*. Kluwer Academic/Plenum Publishers, New York.
Gold, N., Shapiro, S.A., Bojinski, S., Müller, T.M. (2000). An approach to upscaling for seismic waves in statistically isotropic heterogeneous elastic media. *Geophysics*. **65**, 1837–1850.
Gradshteyn, I.S., Ryzhik, J.M. (1980). *Table of Integrals, Series and Products*. Academic Press, London.
Gubernatis, J.E., Domany, E., Krumhansl, J.A. (1977b). Formal aspects of the theory of the scattering of ultrasound by flaws in elastic materials. *J. Appl. Phys.* **48**, 2804–2811.
Gurevich, B., Lopatnikov, S.L. (1995). Velocity and attenuation of elastic waves in finely layered porous rocks. *Geophys. J. Int.* **121**, 933–947.
Holliger, K., Levander, A.R. (1992). A stochastic view of lower crustal fabric based on evidence from the Ivrea zone. *Geophys. Res. Lett.* **19**, 1153–1156.
Hudson, J.A. (1980). *The Excitation and Propagation of Elastic Waves*. Cambridge University Press, Cambridge.
Ishimaru, A. (1978). *Wave Propagation and Scattering in Random Media*. Academic Press Inc., NY.
Johnson, D.L. (2001). Theory of frequency dependent acoustics in patchy-saturated porous media. *J. Acoust. Soc. Am.* **110**, 682–694.
Jones, T. (1986). Pore fluids and frequency dependent wave propagation in rocks. *Geophysics* **51**, 1939–1953.

Karal, F.C., Keller, J.B. (1964). Elastic, electromagnetic and other waves in random media. *J. Math. Phys.* **5**, 537–547.
Klimes, L. (2002). Correlation functions of random media. *Pure Appl. Geophys.* **159**, 1811–1831.
Levander, A.R., Gibson, B.S. (1991). Wide-angle seismic reflections from two-dimensional random target zones. *J. Geophys. Res.* **96**, 251–260.
Mavko, G., Jizba, D. (1991). Estimating grain-scale fluid effects on velocity dispersion in rocks. *Geophysics* **56**, 1940–1949.
Mavko, G., Nur, A. (1975). Melt squirt in aesthenosphere. *J. Geophys. Res.* **80**, 1444–1448.
Mukerji, T., Mavko, G., Mujica, D., Lucet, N. (1995). Scale-dependent seismic velocity in heterogeneous media. *Geophysics* **60**, 1222–1233.
Müller, T.M., Gurevich, B. (2004). 1-D random patchy saturation model for velocity and attenuation in porous rocks. *Geophysics* **69**, 1166–1172.
Müller, T.M., Gurevich, B. (2005a). A first-order statistical smoothing approximation for the coherent wave field in random porous media. *J. Acoust. Soc. Am.* **117**(4), 1796–1805.
Müller, T.M., Gurevich, B. (2005b). Wave-induced fluid flow in random porous media: Attenuation and dispersion of elastic waves. *J. Acoust. Soc. Am.* **117**(5), 2732–2741.
Müller, T.M., Rothert, E. (2006). Seismic attenuation due to wave-induced flow: Why Q in random structures scales differently. *Geophys. Res. Lett.* **33**, L16305.
Müller, T.M., Shapiro, S.A. (2001). Most probable seismic pulses in single realizations of 2-d and 3-d random media. *Geophys. J. Int.* **144**, 83–95.
Müller, T.M., Shapiro, S.A., Sick, C.M.A. (2002). Most probable ballistic waves in random media: A weak-fluctuation approximation and numerical results. *Waves Random Media* **12**, 223–246.
Murphy, W.F. (1982). Effects of partial water saturation on attenuation in Massilon sandstone and Vycor porous glass. *J. Acoust. Soc. Am.* **71**, 1458–1468.
Murphy, W.F.I., Roberts, J.N., Yale, D., Winkler, K.W. (1984). Centimeter scale heterogeneities and microstratification in sedimentary rocks. *Geophys. Res. Lett.* **11**, 697–700.
Norris, A.N. (1993). Low-frequency dispersion and attenuation in partially saturated rocks. *J. Acoust. Soc. Am.* **94**, 359–370.
Papanicolaou, G.C. (1971). Wave propagation in a one-dimensional random medium. *SIAM Rev.* **21**(1), 13–18.
Pride, S.R., Haartsen, M.W. (1996). Electroseismic wave properties. *J. Acoust. Soc. Am.* **100**, 1301–1315.
Pride, S.R., Berryman, J.G., Harris, J.M. (2004). Seismic attenuation due to wave-induced flow. *J. Geophys. Res.* **109**(B1), B01201.
Prudnikov, A.P., Brychkov, Y.A., Marichev, O.I. (1988). *Integrals and Series*. Gordon and Breach Science Publ., NY.
Rytov, S.M., Kravtsov, Y.A., Tatarskii, V.I. (1989). *Wave Propagation Through Random Media: Volume 4 of Principles of Statistical Radiophysics*. Springer Verlag, Berlin.
Sato, H. (1982). Amplitude attenuation of impulsive waves in random media based on travel time corrected mean wave formalsim. *J. Acoust. Soc. Am.* **71**, 559–564.
Sato, H., Fehler, M. (1998). *Wave Propagation and Scattering in the Heterogenous Earth*. AIP-press, NY.
Shapiro, S.A., Hubral, P. (1999). *Elastic Waves in Random Media*. Springer, Heidelberg.
Shapiro, S.A., Kneib, G. (1993). Seismic attenuation by scattering: Theory and numerical results. *Geophys. J. Int.* **114**, 373–391.
Shapiro, S.A., Müller, T.M. (1999). Seismic signatures of permeability in heterogeneous porous media. *Geophysics* **64**, 99–103.
Shapiro, S.A., Schwarz, R., Gold, N. (1996). The effect of random isotropic inhomogeneities on the phase velocity of seismic waves. *Geophys. J. Int.* **127**, 783–794.
Solna, K., Papanicolaou, G. (2000). Ray theory for a locally layered random medium. *Waves Random Media* **10**, 151–198.

Toms, J., Müller, T.M., Ciz, R., Gurevich, B. (2006). Comparative review of theoretical models for elastic wave attenuation and dispersion in partially saturated rocks. *Soil Dyn. Earthquake Eng.* **26**, 548–565.

Weaver, R.L., Pao, Y. (1981). Dispersion relations for linear wave propagation in homogeneous and inhomogeneous media. *J. Math. Phys.* **22**, 1909–1918.

White, J.E., Mikhaylova, N.G., Lyakhovitsky, F.M. (1975). Low-frequency seismic waves in fluid saturated layered rocks. *Izvestija Academy of Sciences USSR, Phys. Solid Earth* **11**(10), 654–659.

Wu, R.-S. (1982a). Attenuation of short period seismic waves due to scattering. *Geophys. Res. Lett.* **9**(1), 9–12.

Wu, R.-S. (1982b). Mean field attenuation and amplitude attenuation due to wave scattering. *Wave Motion* **4**, 305–326.

Wu, R.-S., Flatté, S.M. (1990). Transmission fluctuations across an array and heterogeneities in the crust and upper mantle. *PAGEOPHYSICS* **132**(1/2), 175–196.

Wu, R.-S., Xu, Z., Li, X.-P. (1994). Heterogeneity spectrum and scale-anisotropy in the upper crust revealed by the German continental deep-drilling (ktb) holes. *Geophys. Res. Lett.* **21**, 911–914.

OBSERVING AND MODELING ELASTIC SCATTERING IN THE DEEP EARTH

PETER M. SHEARER AND PAUL S. EARLE

ABSTRACT

Seismic scattering in the deep Earth below the mantle transition zone is observed from precursors and codas to a number of body-wave arrivals, including P, P_{diff}, PKP, $PKiKP$, $PKKP$, and $P'P'$. Envelope-stacking methods applied to large teleseismic databases are useful for resolving the globally averaged time and amplitude dependence of these arrivals. Stacks of P coda near 1 Hz from shallow earthquakes exhibit significant variations among different source and receiver locations, indicating lateral variations in scattering strength. At least some deep-mantle, core-mantle boundary, and inner-core scattering is indicated by the observations, but the strength and scale length of the random velocity heterogeneity required to explain the data are not yet firmly established. Monte Carlo seismic "particle" algorithms, based on numerical evaluation of radiative transfer theory with Born scattering amplitudes for random elastic heterogeneity, provide a powerful tool for computer modeling of scattering in the whole Earth because they preserve energy and can handle multiple scattering through depth-varying heterogeneity models. Efficient implementation of these algorithms can be achieved by precomputing ray tracing tables and discretized scattering probability functions.

KEY WORDS: Seismic scattering, coda waves, Monte Carlo algorithms, deep Earth heterogeneity   © 2008 Elsevier Inc.

1. INTRODUCTION

Observing and modeling seismic scattering in the mantle and core is important because of the constraints these studies provide on small-scale heterogeneity. However, investigating seismic scattering in the deep Earth is challenging because strong lithospheric scattering can mask scattered arrivals from deeper in the mantle. By correctly identifying the scattering origin of PKP precursors, Cleary and Haddon (1972) found the first definitive evidence for deep-Earth scattering. PKP precursors have an unusual ray geometry that provides a unique window into scattering within the lowermost mantle and at the core-mantle boundary (CMB). Early modeling of PKP precursors focused on the (CMB) region as their likely source and used single-scattering theory applied to random media models to provide a first-order fit to precursor amplitudes (e.g., Haddon and Cleary, 1974; Doornbos, 1978; Bataille and Flatté, 1988). However, more recent work (Hedlin et al., 1997; Cormier, 1999; Margerin and Nolet, 2003a,b) showed that the scattering must extend at least 600 km into the mantle above the core, and it seems likely that some amount of scattering is present throughout the mantle.

In addition to PKP precursors, there are a number of other seismic observations that suggest deep scattering, including P_{diff} coda (Bataille et al., 1990; Tono and Yomogida, 1996; Bataille and Lund, 1996; Earle and Shearer, 2001). $PKKP$ precursors (Chang and Cleary, 1978, 1981; Doornbos, 1980; Earle and Shearer, 1997), and $PKiKP$ coda (Vidale and Earle, 2000; Vidale et al., 2000; Poupinet and Kennett, 2004; Koper et al., 2004). In principle, the coda of deep-turning P and S waves is sensitive to lower-mantle scattering, but it is tricky to separate this scattering from the much stronger scattering that occurs in the shallow mantle and crust. Nonetheless, several recent studies have found evidence in these phases for a deep-scattering contribution (Cormier, 2000; Lee et al., 2003; Shearer and Earle, 2004).

All of these results are valuable because they provide estimates on the strength of heterogeneity in the deep mantle and inner core at scale lengths (e.g., ~10 km) much smaller than those that can be imaged using tomographic methods. These velocity anomalies are almost certainly compositional in origin because small-scale thermal perturbations would quickly diffuse away, and thus they provide insight regarding the degree of mixing in mantle convection models. It is therefore important to further develop seismic observations to resolve additional details regarding the heterogeneity, including its strength, scale length, and depth dependence.

Accurate modeling of the seismic observations requires a more complete theory than methods based on the Born approximation, which do not conserve energy and ignore the effects of multiple scattering. For example, Margerin and Nolet (2003a,b) found that Born theory is accurate for whole-mantle scattering models only when the deep-velocity heterogeneity is less than 0.5%. Finite difference/element methods can handle velocity models of arbitrary complexity but are not yet numerically feasible for global scattering problems at high frequencies (1 Hz). Faster algorithms are possible through use of the parabolic and Markov approximations (e.g., Sato and Fehler, 2006), but three-dimensional global calculations remain difficult. Here, we focus on Monte Carlo methods based on radiative transfer theory that simulate the random walk of millions of seismic energy "particles." Although these methods discard phase information, they are a powerful and practical approach to modeling whole-Earth, high-frequency scattering.

We begin by describing the processing and stacking methods that are suited for global seismic observations and then present some specifics regarding how the Monte Carlo method can be efficiently implemented. Results from the Shearer and Earle (2004) analysis of P coda will be highlighted, but we describe our algorithms in more detail and present some new results concerning lateral variability in teleseismic P coda.

2. Data Stacking

Waveform stacking has several advantages over analysis of individual seismograms:

1. It generally increases the signal-to-noise ratio, making it possible to identify and characterize weak seismic arrivals that are hard to resolve on single records.
2. It reduces the volume of data to be modeled to a more manageable level. For example, the information in thousands of global seismograms can be reduced to a single time-versus-epicentral-distance image of the average wave field.
3. It can provide a spatially averaged measure of the wave field that is less biased than results from small numbers of seismograms. By processing all of the data, it reduces the selection bias problem that may affect studies that focus on the most visible or anomalous phases in individual records.

Conventional seismogram-stacking methods do not work well for imaging scattered seismic arrivals because coda waves are generally incoherent among the different recording stations, especially at the high frequencies where the scattered wave field is typically observed. In other words, the timing of the peaks and troughs in the seismograms varies randomly among the records. Small-scale arrays are an important exception to this and have provided valuable constraints on the slowness and back-azimuth of scattered arrivals. However, our focus in this chapter concerns the use of single stations that are too far apart for standard array processing techniques to work. In this case, it is necessary to develop stacking methods that work for incoherent data.

One approach is to discard the phase information in the seismograms and consider only their energy content as defined by their envelope functions. This method has been used successfully to image *PKP* and *PKKP* precursors (Hedlin *et al.*, 1997; Earle and Shearer, 1997; Shearer *et al.*, 1998), P_{diff} coda (Earle and Shearer, 2001), and P coda (Shearer and Earle, 2004).

Figure 1 illustrates the envelope-function stacking technique applied to P and its coda (Shearer and Earle, 2004). Before stacking, we manually review the data and remove seismograms with dropouts, data glitches, or contaminating arrivals from aftershocks or local earthquakes. Once the data are cleaned, they are processed and stacked using the following steps shown in Fig. 1:

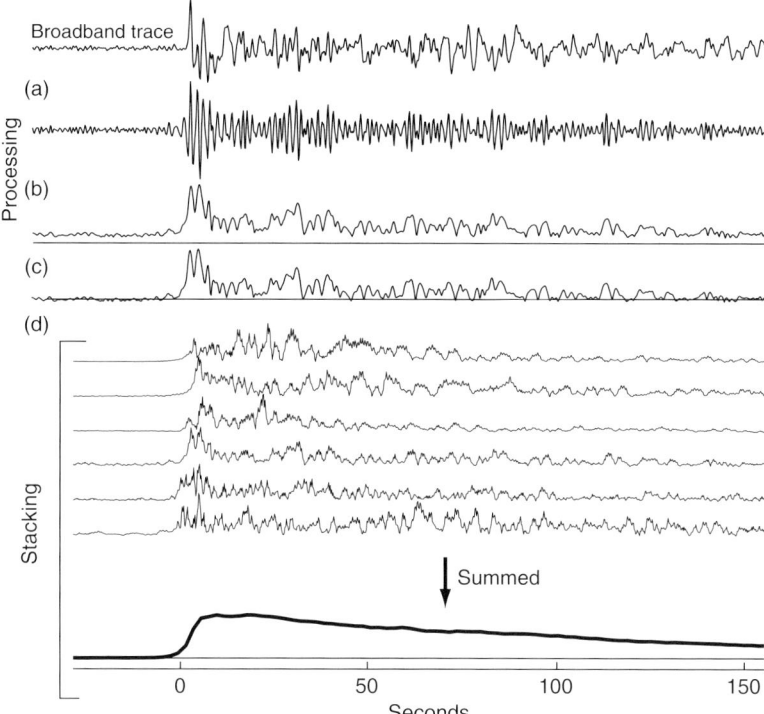

FIG. 1. Illustration of the different processing steps in the envelope-stacking technique. Starting with each original broadband trace (a) band-pass filtering is applied, (b) the envelope function is computed, (c) the power in the preevent noise is removed. The processed traces are then stacked (d). Notice the large variability seen in the individual seismograms compared to the smooth stacked trace at the bottom.

(a) The traces are band-pass filtered between 0.5 Hz and 2.5 Hz. This frequency band falls in a low noise region of the spectrum and provides the greatest sensitivity to deep-Earth structure with scale length of about 10 km.
(b) The envelope function (e.g., Kanasewich, 1981) is calculated for each seismogram.
(c) We assume that the noise and signal are uncorrelated so that, upon averaging, their energies will sum. Thus, to account for varying noise levels between traces, the envelope functions are squared and the average noise in a time window preceding the reference phase (in this example the P wave) is subtracted from the entire trace. Then, the square root of the noise corrected trace is taken. The squared envelope is used because the recorded signal is the square root of the sum of the squared noise plus the squared signal.
(d) The final stack is made by normalizing the traces to their maximum amplitudes, aligning them on the reference phase arrival time, and averaging all traces in a target distance window as a function of time.

We stack the traces in amplitude rather than power because we have found that this produces slightly more robust results than power stacks. However, in practice the differences between amplitude and power stacks are usually fairly small.

2.1. Shallow- Versus Deep-Earthquake Teleseismic P Coda

Results of this stacking method applied to teleseismic P coda are plotted in Fig. 2 (originally in Shearer and Earle, 2004), which compares stacks for over 7500 records from shallow earthquakes (depth $\leq$ 50 km) and 650 records from deep earthquakes (depth $\geq$ 400 km). Data are taken from vertical- component seismograms from $M_W = 6$ to 7 events in the IRIS FARM archive from 1990 to 1999. The stacked envelopes are binned at 5° distance intervals and 2-s time intervals. This figure shows the striking difference in teleseismic coda strength between shallow and deep earthquakes. The shallow-event coda is much stronger and longer-lasting than the deep-event coda. At 50 s following the P arrival, it is 2–5 times larger in amplitude (4–25 times larger in energy). This difference indicates that teleseismic P coda from shallow events is dominated by near-source scattering above 600-km depth. Note that both stacks should have equal coda contributions from near-receiver scattering, but the energy difference between the shallow and the deep coda is much more than a factor of two. However, as discussed in Shearer and Earle (2004), this does not necessarily imply lateral variations in scattering strength with stronger scattering in active earthquake areas. In fact, with Monte Carlo modeling, it is possible to achieve a reasonable fit to both the shallow- and deep-earthquake coda amplitudes with a single model in which scattering strength varies only with depth.

2.2. Regional Variations in Teleseismic P Coda Amplitude

The Shearer and Earle (2004) teleseismic P coda study considered only spherically averaged coda amplitudes. Here, we analyze data from this study in more detail to identify variations in coda levels among different source regions and station locations. Fig 3 illustrates such a difference between two Asian stations as seen in coda stacks of 120 quakes recorded at YAK and 163 quakes recorded by AAK (all quake depths are less than 50 km). Station AAK has consistently higher P coda amplitudes than station YAK.

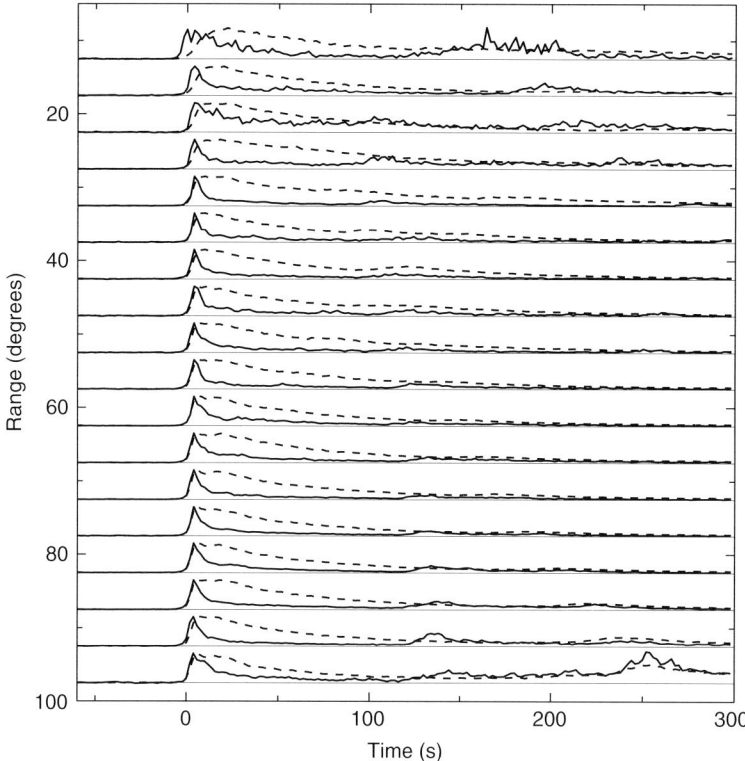

Fig. 2. A comparison between the shallow-event, envelope-function stack (dashed), and the deep-event stack (solid line). Time is relative to P and the stacks have been scaled to the same P-wave maximum amplitude. Note the much more extended coda from the shallow events. Figure taken from Shearer and Earle (2004).

To characterize these variations systematically, we measure the average ratio of the P coda amplitude to the P amplitude at source-receiver distances between 60° and 95° (the observed ratio changes little over this distance interval). The P amplitude is measured as the average absolute value in the demeaned trace in a 40-s window starting at the predicted P arrival time. The coda amplitude is measured similarly in a 60-s window starting 60 s after the P arrival time. To account for possible biasing effects related to the specific subset of events recorded by each station, we assume that the logarithm of this relative coda level, c, for the ith source and the jth receiver can be approximated as the sum of a source term, q, and a receiver term, r,

$$\log(c_{ij}) = \log(q_i) + \log(r_j). \tag{1}$$

This equation does not have a direct physical basis; it is an empirical approach to test how much of the variation in the coda amplitudes can be explained with a simple decomposition into source- and receiver-side contributions. Because we have many receivers for

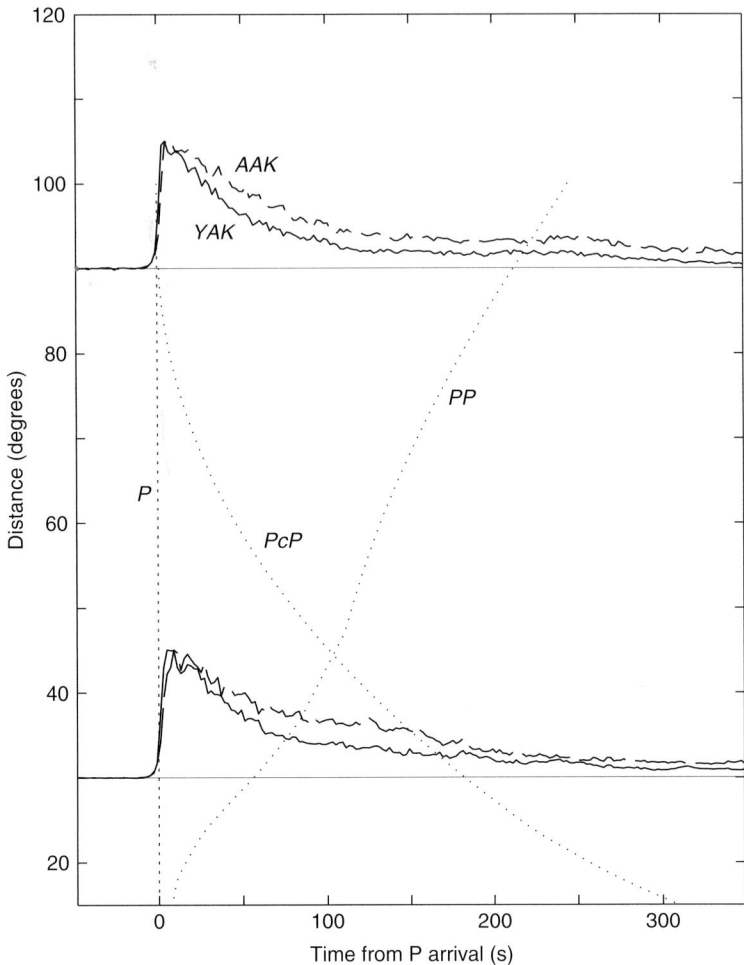

FIG. 3. A comparison of stacked P coda amplitudes for shallow earthquakes (≤50 km) between stations AAK (dashed) and YAK (solid) at 30 and 90. Note the higher coda amplitudes for AAK.

each source and many sources for each receiver, this is an over-determined problem, which we solve using a robust, iterative least squares approach. To remove the nonuniqueness in this equation (a constant could be added to the log q terms and subtracted from the log r terms), we constrain the average log r to be zero.

Fig. 4 plots the resulting individual source and receiver terms q and r. Before plotting, we scale the receiver terms to have the same median value as the source terms. Symbol size is proportional to the log deviation from the median amplitude ratio of 0.432. Plus symbols have relatively high coda levels and diamonds have relatively low coda levels. Because results can be quite variable for small numbers of traces, we only include terms constrained by at least 10 traces. Overall, the quake coda levels have about twice the

OBSERVING AND MODELING ELASTIC SCATTERING 173

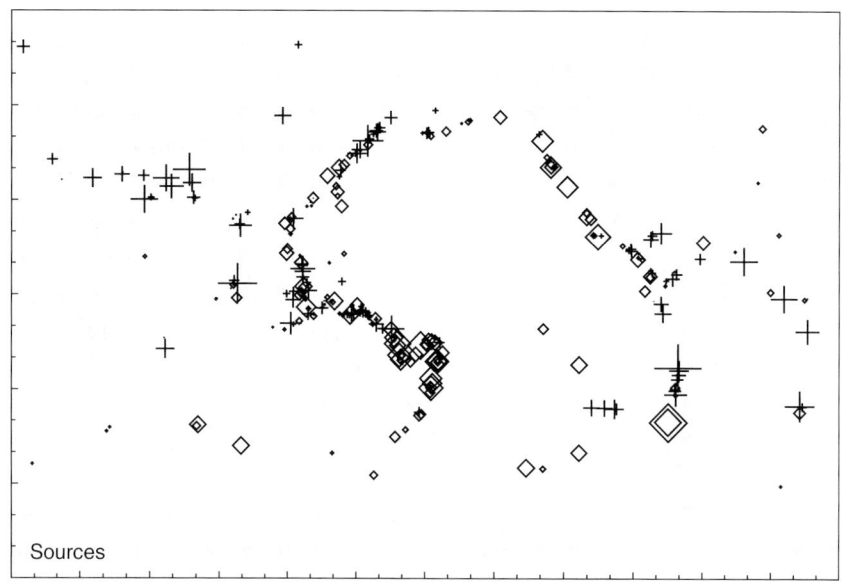

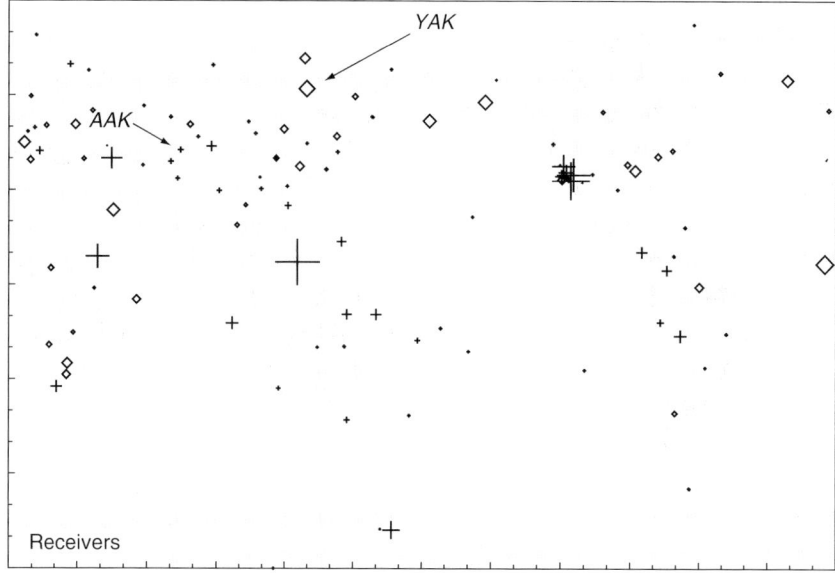

FIG. 4. Differences in teleseismic P coda amplitudes at source locations (top) and station locations (bottom). Coda levels higher than the global median of 0.43 are plotted as + symbols and weaker levels are shown as diamonds, with the size of the symbol scaling as the logarithm of the perturbation. Stations AAK and YAK (see Fig. 3) are labeled in the bottom figure.

variability of the station coda levels. The median variations in quake coda levels are about ±16% while the median variations in station coda levels are about ±8%. This is consistent with near-source rather than near-receiver scattering being the dominant contributor to teleseismic P coda, as is also implied by the differences in the shallow- and deep-earthquake stacks plotted in Fig. 2. The relatively sharp changes in coda levels that can be seen in these plots over short distance intervals also indicates that the bulk of the coda originates from near-surface scattering, rather than from deeper in the mantle.

Both the quake and the station terms exhibit some degree of spatial coherence. Stronger coda is seen from events in central Asia, the Kurils and South America and weaker coda is observed in Japan, western North America, and the southwest Pacific. The station terms are less coherent but generally have weaker coda in northeast Asia, parts of Africa, and North America outside of California. There is little, if any, spatial correlation between the source and the receiver terms. This may indicate that the near-source and near-receiver scattering processes are different (i.e., more S-to-P near the source, more P-to-P near the stations), or could reflect strong variations in heterogeneity very close to earthquake source regions that do not always extend far enough to produce correlated variations in coda strength for teleseismic arrivals at nearby seismic stations.

Our station term results have only limited correlation with the scattering Q (Q_{Sc}) estimates at 1 Hz of Korn (1990, 1993) from P coda for 9 stations in Australia and around the Pacific. We observe stronger than average P coda for stations GUMO and TATO, which have lower than average Q_{Sc} at 1 Hz in Korn (1993). In contrast, we also observe relatively strong P coda for NWAO and weak P coda for AFI, opposite to Korn's results. Because of differences between our simple station term inversion method and the energy flux modeling approach of Korn (1990, 1993), it is not clear how well correlated results should be between the methods.

3. Monte Carlo Methods

Monte Carlo methods have been used in physics since the 1950s to model radiation transport by using a computer to simulate the random scattering of large numbers of individual particles [see Dupree and Fraley (2002) for a recent introduction to many of these techniques]. The Monte Carlo approach uses computer-generated random numbers to sample the different possible variables in a problem. For example, neutron scattering can be simulated by tracking the behavior of individual neutrons, radiated in random directions from a source and scattered in random directions during their propagation, thus in effect simulating the results of an actual experiment. In general, the accuracy of the solution grows with the number of particle trajectories that are computed and thus Monte Carlo methods have become increasingly useful as faster computers have become available. Typically, the algorithms converge such that the variance of the results decreases as $1/\sqrt{n}$, where n is the number of particles.

The concept of seismic "particles" may not seem useful upon initial consideration because there is no wave-particle duality for seismic waves, as exists for electromagnetic waves. However, if one is willing to consider energy transport alone and discard phase information in seismic records, then a particle-based, Monte Carlo approach can be very valuable. It is particularly suited to studying scattering at high frequencies, where the waveforms are incoherent and typical modeling efforts consider only the envelope function. It bridges the gap between Born theory for weak scattering and computationally intensive finite difference/element calculations for complicated models.

3.1. Seismology Applications

The first use of the Monte Carlo approach in seismology was by Gusev and Abubakirov (1987) who modeled acoustic wave scattering in a uniform whole space using particles randomly radiated from an isotropic source. They assumed a constant probability of scattering per unit volume, resulting in an exponential distribution of path lengths. They did not explicitly include intrinsic attenuation but noted that it could easily be modeled in the constant Q case by multiplying the energy of each particle by $e^{-2\pi ft/Q}$. They considered both isotropic scattering and forward scattering with a Gaussian angle distribution and showed that their results agree with the diffusion model at large lapse times, but that only the forward scattering model produces realistic pulse broadening and coda envelopes at short distances. Abubakirov and Gusev (1990) presented a more detailed account of the Monte Carlo technique and used a forward scattering model to compute master curves describing the relationships between the mean free path and both pulse broadening and the intensity ratio of the direct and the scattered waves. Applying these results to S-coda observations in Kamchatka, they obtained S-wave mean free paths of 100–150 km over a 1.5–6 Hz frequency range. Gusev and Abubakirov (1996) expanded their Monte Carlo method to include scattering angles predicted by specific models of random velocity heterogeneity, including Gaussian and power-law media, and argued that a power-law exponent of 3.5–4 is in qualitative agreement with the features of observed S-wave envelopes.

Hoshiba (1991) used a Monte Carlo method to model isotropic S-wave scattering in a uniform whole space, and demonstrated that the results agree with Born theory for weak scattering and with the radiative transfer theory of Wu (1985) and the diffusion model for strong scattering. Hoshiba (1994, 1997) extended his method to include depth-dependent scattering strength, layered velocity models, and intrinsic attenuation. He simulated SH-wave reflection and transmission coefficients at layer interfaces as probabilities of reflection or transmission, which is a practical way to handle the energy partitioning at interfaces without the complexities of generating additional particles.

Margerin *et al.* (1998) modeled isotropic S-wave scattering using a Monte Carlo method for a layer over half-space model (i.e., the crust and the upper mantle) and included both surface and crust-mantle boundary (Moho) reflected/transmitted phases, using probabilities to handle reflection and transmission coefficients, but did not model S polarity, phase conversions, and intrinsic attenuation. Margerin *et al.* (2000) extended their method to fully elastic waves, including S polarity and phase conversions, and modeled scattering off randomly distributed spherical inclusions within a uniform whole space. They explored the dependence of their results of the relative size of the spheres compared to the seismic wavelengths. Margerin and Nolet (2003a,b) applied a Monte Carlo method to model *PKP* precursors with whole Earth P-to-P scattering in the mantle. They showed that their results were in good agreement with geometrical ray theory for the main *PKP* arrivals and that the scattered arrivals agreed with Born theory for weak random velocity heterogeneity.

Bal and Moscoso (2000) included S-wave polarizations in Monte Carlo simulations of randomly heterogeneous lithosphere and showed that S waves can become depolarized after multiple scattering. Yoshimoto (2000) used a finite difference ray tracing method to implement a Monte Carlo method for a complex velocity profile for the lithosphere and found that ray bending caused by velocity gradients and the Moho can have large effects on the shape of the S coda envelope.

Wegler et al. (2006) and Przybilla et al. (2006) performed a series of tests of Monte Carlo simulations based on radiative transfer theory, for both acoustic and fully elastic P- and S scattering, and compared their results to those predicted by various analytical solutions in 3D and finite difference solutions to the full wave equation in 3D. In general, they found good agreement between the Monte Carlo approach and other methods, except in the case of extreme velocity perturbations, such as can occur in volcano seismology.

Although all of these methods work by computing trajectories for a large number of particles, they differ in some important details. A fundamental distinction can be made between two different approaches: (1) algorithms that simply count the number of particles that hit different cells in the model (e.g., Gusev and Abubakirov, 1987, 1996; Yoshimoto, 2000; Shearer and Earle, 2004) and (2) those that compute the probability of particles at a series of discrete receivers (e.g., Hoshiba, 1991, 1994, 1997; Margerin et al., 1998, 2000). The former provide the energy density at every point in the model and thus can be termed global methods in contrast to the local nature of the calculations in (2). Although for some models there are computational advantages to (2), the simplicity and flexibility of (1) have made it a more popular choice, given the speed and storage capabilities of modern computers.

3.2. Monte Carlo Implementation

The theoretical basis for the Monte Carlo approach is provided by radiative transfer theory (e.g., Wu, 1985; Sato, 1995; Ryshik et al., 1996; Bal and Moscoso, 2000; Margerin, 2005). We will not review this theory here. Instead, we will focus on the practical aspects of writing a computer program to perform seismic Monte Carlo calculations in an efficient manner for radially symmetric Earth models. As an example, we will give details of the global elastic algorithm of Shearer and Earle (2004).

The fundamental principle involved is that each seismic "particle" represents an energy packet and that our treatment of the particles (i.e., propagation, reflection/transmission, phase conversions, and scattering) should be designed to conserve energy in a logically consistent manner. We will use geometrical ray theory to compute particle trajectories. A nice aspect of the particle approach is that geometrical spreading terms are not required because the energy reduction with distance is naturally included as the decrease in particle density as the particles spread out from the source. The energy partitioning that occurs at interfaces or scattering points is handled not by splitting the energy into different particles but by assigning appropriate probabilities to the changes in particle directions and using computer-generated random numbers to sample these probability distributions. Thus, we track only one particle at a time, making the code straightforward to parallelize and run on multiple processors, if desired.

The output of a Monte Carlo simulation will typically appear noisy at first (i.e., producing spiky and irregular envelopes) when only a small number of particles are computed, but will become increasingly smooth as more particles are included in the calculation. The number of particles required to give adequate results will vary, depending upon the details of the model, the portion of the output wave field that is of greatest interest, and how much resolution in time and space is desired.

3.3. The Monte Carlo Source

Despite the fact that earthquake radiation patterns are not uniform, most Monte Carlo simulations assume isotropic radiation from the source. This can be justified for two reasons: (1) At high frequencies, observed P and S amplitudes show considerable scatter compared to that predicted by double-couple sources (e.g., Nakamura et al., 1999; Hardebeck and Shearer, 2003) presumably caused by strong crustal and near-surface focusing and scattering effects. Thus, details of the radiation pattern tend to be lost during high-frequency wave propagation. (2) Results are often averaged over many earthquakes and stations at different azimuths from the source. This will lessen the bias caused for individual ray paths by neglecting radiation pattern effects. However, it is important to recognize that some bias may still be present. For example, if most earthquakes in a region are strike-slip, the expected P-wave fraction of energy radiated in the near-vertical direction is much less than that predicted by assuming an isotropic source. This bias cannot be removed by averaging over azimuth. Isotropic average radiation will only occur for a truly random distribution of focal mechanisms, which is unlikely to be the case for real Earth observations.

If the background Earth model is radially symmetric, then the expected average energy observed on Earth's surface from an isotropic source will vary only as a function of distance. It thus makes sense to combine the energy from all of the particles hitting the surface at a particular distance (regardless of azimuth) to compute the average predicted wave field. It follows that because of the symmetry of the problem and the randomness of individual scattering events, it is only necessary to shoot the particles at a single azimuth from the source. In this case, however, the number or the energy of the particles must be weighted as $\sin \theta$, where θ is the takeoff angle from the vertical, to account for the greater number of particles expected at more horizontal takeoff angles for a spherically isotropic source.

For efficiency it is usually desirable for ray tracing and other information to be computed and stored at certain discrete values, in which case the sampling will be limited to these values. It is not necessary for the takeoff angle sampling to be uniform, provided suitable weights are assigned to the particles. This technique is termed event biasing in the Monte Carlo literature. For example, the rays could be evenly sampled in ray parameter rather than angle or proportionally more rays could be fired at steeper angles for better sampling of core phases compared to crustal phases. Both upgoing and downgoing rays from the source should be included unless the source is assumed to be exactly at the surface. In Shearer and Earle (2004), we spray rays evenly spaced in 10,000 values of ray parameter (from $p = 0$ to $p = 1/c$, where c is the P or S velocity at the source), and set the energy of the ith particle proportional to $(\theta_i - \theta_{i-1}) \sin \theta_i$, where the takeoff angle $\theta_i = \sin^{-1}(cp_i)$. Note that $c\, dp = \cos \theta\, d\theta$, $\theta_i - \theta_{i-1} \approx d\theta_i$, and thus $(\theta_i - \theta_{i-1}) \sin \theta_i$ is proportional to $\tan \theta$ when dp is constant.

For S waves, it is simplest to assume random polarizations for the particles leaving the source, again making the assumption that radiation pattern differences will tend to average out when results are combined from many different sources and receivers. In addition, multiple scattering will at some point remove the information about the original source polarization. In complete modeling of both P and S waves, an S-to-P total initial energy scaling factor, $q = E_i^S / E_i^P$, must be assumed. This can be done by radiating q times more S particles or by radiating equal numbers of P and S particles but assigning q times more energy to the S particles. The latter approach is more efficient for resolving both P and S phases. Theoretical results for a double-couple source suggest that $q = 23.4$ for a Poisson solid (e.g., Sato, 1984) and this factor was used by Shearer and Earle (2004). However, this assumes that the S and P corner frequencies are identical and earthquake source studies indicate that the P-wave corner frequency is typically higher than the S-wave corner frequency, with observations ranging from $q = 9$ to 14 (Boatwright and Fletcher, 1984; Abercrombie, 1995; Prieto et al., 2004). These values are for total radiated energy integrated over the entire frequency band and are not necessarily appropriate at the single fixed frequency used in each Monte Carlo calculation. The question of the relative sizes of P and S radiation from earthquakes is an active area of research to which scattering studies may be able to contribute by better separation of intrinsic attenuation, scattering attenuation, and source effects in observed earthquake spectra.

Because the number of radiated particles will vary depending upon how long the program is kept running, it is simplest to initially consider only relative energy at the source and perform the normalization to absolute energy at a later stage. This may be accomplished by keeping track of the total initial energy of the radiated particles and multiplying the observed energy by the ratio of the desired radiated energy to the total initial particle energy. In many applications, only the relative time versus distance behavior of the wave field is important, in which case the calibration to absolute amplitude is not required.

3.4. Particle Trajectories

Most seismic applications of the Monte Carlo approach have assumed acoustic or elastic body-wave propagation and used ray theory to track the particle trajectories. For simple whole-space or homogeneous layer models, the ray paths are straight lines. However, for more realistic models containing velocity gradients the ray paths are curved. To save computer time when computing results for millions of particles, it is advantageous in this case to precompute ray tracing results (time and distance) at discrete values of ray parameter, p, and depth in the model. The Shearer and Earle (2004) algorithm computed dt and dx values within 10-km-thick layers in the model for 10,000 values of p, saving the results in separate arrays for P and S waves. Another array records whether the ray passes through, reflects off, or turns (changes direction) within each layer. Of course, a specific velocity versus depth model must be assumed. Standard models such as PREM (Dziewonski and Anderson, 1981) and IASP91 (Kennett, 1991) predict body-wave travel times that generally agree within a few seconds, but differences in the velocity gradients among the models can produce significant differences in ray theoretical amplitudes.

Modeling of S-wave polarizations is complicated by the fact that the polarization will rotate along curved ray paths. In radiative transfer theory, S-wave polarizations can be handled using the Stokes parameters and this is the approach described in Margerin et al. (2000) and Bal and Moscoso (2000). Shearer and Earle (2004) adopt the simpler scheme of assigning S polarity as an angle in a local SV versus SH coordinate system, an angle that will remain constant along curved ray paths in radially symmetric models, although it will of course change following reflection/transmission or scattering events. This approach is more limited than the Stokes method because it assumes that the S polarization is always linear, whereas phase shifts between the SV and SH components can occur for some reflections at interfaces. Thus, the Shearer and Earle (2004) algorithm should be considered only approximate for S polarizations.

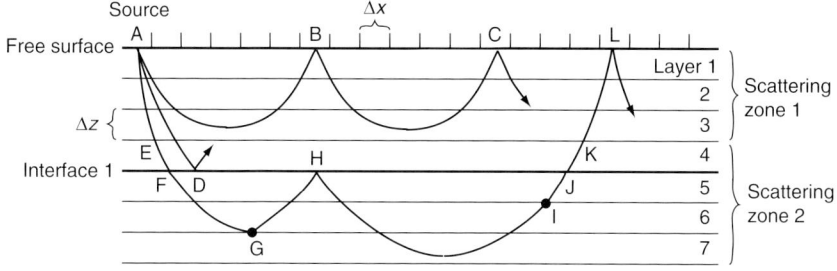

FIG. 5. Example particle trajectories for a Monte Carlo simulation method. The one-dimensional background velocity model is defined as velocity points at a depth spacing of Δz. In this example, there is a velocity jump between layers 4 and 5 and the mean free path is constant within 2 different scattering zones. See text for a description of the particle behavior.

Figure 5 shows some examples of particle trajectories. Here, the model is discretized at depth intervals of Δz. Consider the simple case of ray path ABC, which does not experience any scattering. The ray is radiated downward from the source at a specific ray parameter. The program marches downward through the layers, adding the precomputed time and distance increments for each ray segment. In layer 3, the ray experiences a turning point and its vertical direction changes. The program then goes up through the layers and continues to accumulate time and distance increments. The direction changes again when the ray reflects off the free surface at B.

The free surface is important because it is normally where we output results of Monte Carlo simulations to compare with observations. For isotropic sources and Earth models that are radially symmetric in their bulk properties (i.e., excluding random small-scale perturbations), the observed wave field is a function of time and epicentral distance only. It is therefore convenient to discretize the output into small bins in a time versus distance array (the distance increment Δx is shown in Fig. 5), into which the accumulated energy from each arriving seismic particle is summed. At this point, separate results can be saved for P- and S energy (derived from the wave type) as well as the vertical, radial, and tangential components (derived from the local ray angle, wave type, and S polarization). Later, these energies are normalized by the surface area in each bin (e.g., to account for the greater surface area between 90° and 91° from the source compared to between 10° and 11° from the source). In fully elastic calculations, P-to-S and S-to-P conversions occur at the free surface and must be included in the modeling (see Section 3.4.1 below).

Following its reflection at point B, the ray again travels downward until it turns and reflects again at surface point C, where the ray energy is added to a different part of the time–distance array. The calculation for each particle continues until a maximum time limit is reached, at which point the algorithm starts over with a new particle from the source.

3.4.1. Interfaces

All standard Earth models contain significant velocity changes at the surface, the Moho, the CMB, and the inner-core boundary (ICB) and minor velocity jumps near 410- and 660-km depth. Thus, it is important for Monte Carlo simulations to correctly model the reflection and the transmission behavior at interfaces. As described in Hoshiba (1997) and Margerin and Nolet (2003a), this is handled by computing energy-normalized reflection and transmission coefficients and converting them into probabilities that are used to pick (based on a computer-generated random number) the wave type (P or S), direction (up or down), and polarization (in the case of S waves) for a single particle that leaves the interface. In this way the energy partitioning at each interface is modeled stochastically as the average response of thousands of individual particles. Because these are spherical interfaces, Snell's law is obeyed and the ray parameter does not change.

In Fig. 5, there are two interfaces, one at the free surface and one between layers 4 and 5, where the velocity jumps discontinuously. For program efficiency, the reflection and the transmission coefficients are precomputed for all of the discrete values of ray parameter used to calculate the ray paths. Assuming the ray path ABC represents a P wave, there is some energy converted to SV upon each free surface reflection. Thus, there was a random chance that the particle might have changed to an S wave (with the probability determined by the energy normalized reflection coefficient), but in this example we assume that it did not. Ray path AD hits interface 2 and is reflected. However, note that other particles traveling along the exact same path may be transmitted though the interface. Whether a particle is reflected or transmitted will depend upon the value of a random number generated by the computer whenever the particle hits an interface.

The interface energy partitioning described above will result in the majority of particles going into the seismic phases with the largest reflection and transmission coefficients and thus into the highest energy parts of the wave field. This may not always be desirable if a target phase of interest is of relatively low amplitude because of a small reflection coefficient along its ray path, in which case most of the particles are "wasted" and comparatively few particles will illuminate the phase. Examples of such phases include *PcP*, *PKiKP*, and *PKKP* along much of their travel time curves. To improve the performance of Monte Carlo algorithms in these cases, the appropriate reflection coefficients can be artificially increased, provided that the energy of the reflected particles is decreased and the energy of transmitted particles is increased, such that average energy over many particles is preserved even if energy conservation is violated for individual particles. This is an example of a particle biasing technique, which is a common approach in many Monte Carlo analyses in physics, although to our knowledge it has not yet been applied in seismology.

3.4.2. Scattering Events

Scattering strength may be described either in terms of the probability of scattering as the particle passes through a given volume or as the mean free path between scattering events. The scattering coefficient, g, is defined as the scattering power per unit volume per unit solid angle (e.g., Sato, 1977) and has units of reciprocal length. The total scattering coefficient, g_0, is defined as the average of g over all directions and can also be expressed as

$$g_0 = \ell^{-1}, \tag{2}$$

where ℓ is the mean free path. In general, these parameters will vary with depth in the Earth for physically based random heterogeneity models because they are dependent on the seismic wave number k. Thus, one approach would be to test for a scattering event by generating random numbers at short intervals along the ray path. This is the best method in some respects because it can accurately handle depth-dependent scattering of arbitrary complexity and is straightforward to code. However, it requires generation of a large number of random numbers along the ray paths. Thus, it is more efficient to approximate the scattering probability as constant within large depth intervals, in which case the path length r to a scattering event is given by an exponentially distributed random number with mean value ℓ^P or ℓ^S for P waves and S waves, respectively. Thus, individual values of r for P and S waves are computed as

$$r_P = -\ell^P \ln x$$
$$r_S = -\ell^S \ln x, \qquad (3)$$

where x is a random number between 0 and 1. This is the approach taken by Margerin and Nolet (2003a,b) and Shearer and Earle (2004) and is accurate assuming that the mean free path is much larger than the ray path segments in the model.

Note that the depths separating the intervals of different scattering probabilities need not coincide with the velocity interfaces in the model. Whenever a particle enters a layer with a different scattering probability, a random number determines the ray path length to the next scattering event. As the computer tracks the ensuing particle trajectory, if the accumulated path length within the layer exceeds this number, then a scattering event occurs. If the particle leaves the layer and enters a layer with a different scattering probability, then a new random number is generated for the new layer.

Consider ray AEFGHIJKL in Fig. 5. In this case, there is a uniform scattering probability in layers 1–3 and a different scattering probability in layers 4–7. When the ray leaves the source, a random number determines the path length to the next scattering point according to Eq. (3) and the mean free path in the top scattering zone. For this example, this length exceeds the distance AE and the particle is not scattered. At point E, a new random path length is computed from the mean free path for the lower scattering zone. This path length is exceeded by the downgoing ray somewhere in layer 6 and a scattering event occurs at point G (for coding simplicity, scattering events are forced to occur at boundaries between layers). The random orientation of the scattered ray is then computed (see Section 3.5), which in general will involve a change in ray parameter, ray vertical direction (upgoing or downgoing), ray azimuth and may also involve a change in wave type (P or S). A new path length to the next scattering event is computed for the scattered ray. The particle is reflected at H and another scattering event occurs at I. The scattered particle is then transmitted at J and leaves the lower scattering zone at K, at which point a path length is computed for the distance to the next scattering event in the upper scattering zone. The particle hits the free surface at L and the energy of the particle is added to the appropriate bin in the time versus distance array.

3.5. Scattering Angles

Once a scattered event occurs, the next step is to assign a new particle trajectory, and in the case of fully elastic simulations to assign the new wave type (P or S) and S-wave polarization. The simplest approach is to assume that the scattering is isotropic, that is, the probability is equal in all directions. However, Gusev and Abubakirov (1987) and Abubakirov and Gusev (1990) showed that this approach does not lead to realistic early coda and pulse broadening of the direct arrival. To fit most seismic observations, some form of anisotropic scattering is required with more scattering in the forward direction. This can be achieved with empirical scattering functions, such as the Gaussian function analyzed by Gusev and Abubakirov. However, this approach does not provide a connection to the physical properties of the velocity and density perturbations that are causing the scattering.

Thus, most recent Monte Carlo simulations in seismology compute scattering probabilities based on a physical model of the scattering medium, which is generally done using Born scattering coefficients for random heterogeneity models. Although Born theory is for single scattering, it can be used to model multiple scattering when the distance between scattering events is much longer than the seismic wavelength and the scale length of the random heterogeneity, which is generally the case for elastic scattering in the Earth. The required conditions can be expressed as (e.g., Wegler et al., 2006)

$$\ell \gg \lambda/2\pi \tag{4}$$

and

$$\ell \gg a, \tag{5}$$

where ℓ is the mean free path, λ is the seismic wavelength, and a is the heterogeneity correlation distance. Comparisons to finite difference calculations have confirmed the validity of Born theory to compute scattering probability in radiative transfer theory (Wegler et al., 2006), except in cases of extreme scattering from very heterogeneous media (such as may occur in volcano seismology). For the P coda simulations described here, the smallest mean free path is 82 km, which is much longer than the wavelength of mantle P waves at 1 Hz and the 8-km correlation length of the random heterogeneity models.

We now summarize the Born results that are necessary to implement a fully elastic Monte Carlo method that includes P-to-S and S-to-P scattering, using the appropriate results from Sato and Fehler (1998; hereafter referred to as S&F; see also Wu, 1985, and Wu and Aki, 1985a,b, for more details on Born theory in seismology). The simplest equations are obtained when the P velocity α and S velocity β are assumed to have the same fractional velocity fluctuations (S&F, 4.47):

$$\xi(\mathbf{x}) = \frac{\delta\alpha(\mathbf{x})}{\alpha_0} = \frac{\delta\beta(\mathbf{x})}{\beta_0}, \tag{6}$$

where α_0 and β_0 are the mean P and S velocities of the medium. We further assume that the fractional density fluctuations are proportional to the velocity variations (S&F, 4.48):

$$\frac{\delta\rho(\mathbf{x})}{\rho_0} = \nu\xi(\mathbf{x}), \tag{7}$$

where ν is the density/velocity fluctuation scaling factor.

The basic scattering patterns are given by (S&F, 4.50)

$$X_r^{PP}(\psi,\zeta) = \frac{1}{\gamma_0^2}\left[\nu\left(-1+\cos\psi+\frac{2}{\gamma_0^2}\sin^2\psi\right)-2+\frac{4}{\gamma_0^2}\sin^2\psi\right]$$

$$X_\psi^{PS}(\psi,\zeta) = -\sin\psi\left[\nu\left(1-\frac{2}{\gamma_0}\cos\psi\right)-\frac{4}{\gamma_0}\cos\psi\right]$$

$$X_r^{SP}(\psi,\zeta) = \frac{1}{\gamma_0^2}\sin\psi\cos\zeta\left[\nu\left(1-\frac{2}{\gamma_0}\cos\psi\right)-\frac{4}{\gamma_0}\cos\psi\right]$$

$$X_\psi^{SS}(\psi,\zeta) = \cos\zeta[\nu(\cos\psi-\cos 2\psi)-2\cos 2\psi]$$

$$X_\zeta^{SS}(\psi,\zeta) = \sin\zeta[\nu(\cos\psi-1)+2\cos\psi], \tag{8}$$

where X_r^{PP} is the radial component of P-to-P scattering, X_ψ^{PS} is the ψ component of P-to-S scattering, and so on. The angles ψ and ζ are defined as in Fig. 6 and the velocity ratio $\gamma_0 = \alpha_0/\beta_0$.

Assuming a random media model, the scattered power per unit volume is given by the scattering coefficients for the various types of scattering (P to P, P to S, etc.) (S&F, 4.52):

$$g^{PP}(\psi,\zeta;\omega) = \frac{l^4}{4\pi}|X_r^{PP}(\psi,\zeta)|^2 P\left(\frac{2l}{\gamma_0}\sin\frac{\psi}{2}\right)$$

$$g^{PS}(\psi,\zeta;\omega) = \frac{1}{\gamma_0}\frac{l^4}{4\pi}|X_\psi^{PS}(\psi,\zeta)|^2 P\left(\frac{l}{\gamma_0}\sqrt{1+\gamma_0^2-2\gamma_0\cos\psi}\right)$$

$$g^{SP}(\psi,\zeta;\omega) = \gamma_0\frac{l^4}{4\pi}|X_r^{SP}(\psi,\zeta)|^2 P\left(\frac{l}{\gamma_0}\sqrt{1+\gamma_0^2-2\gamma_0\cos\psi}\right)$$

$$g^{SS}(\psi,\zeta;\omega) = \frac{l^4}{4\pi}(|X_\psi^{SS}(\psi,\zeta)|^2+|X_\zeta^{SS}(\psi,\zeta)|^2) P\left(2l\sin\frac{\psi}{2}\right), \tag{9}$$

where $l = \omega/\beta_0$ is the S wave number for angular frequency ω, P is the power spectral density function (PSDF) for the random media model (see S&F, pp. 14–17). A popular choice for P is the exponential autocorrelation function, in which case we have (S&F, 2.10)

$$P(m) = \frac{8\pi\varepsilon^2 a^3}{(1+a^2 m^2)^2}, \tag{10}$$

where a is the correlation distance, ε is the root mean square (RMS) fractional fluctuation ($\varepsilon^2 = \langle\xi(\mathbf{x})^2\rangle$), and m is the wave number [i.e., the argument of the P functions in Eq. (9) above]. For example,

$$P\left(2l\sin\frac{\psi}{2}\right) = \frac{8\pi\varepsilon^2 a^3}{\left(1+4a^2 l^2\sin^2\frac{\psi}{2}\right)^2}. \tag{11}$$

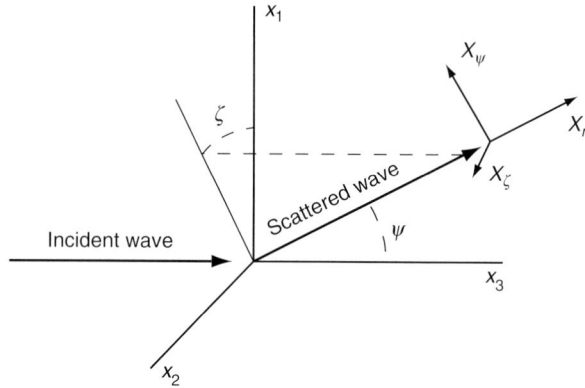

FIG. 6. The ray-centered coordinate system used in the scattering equations. The incident ray is in the x_3 direction. For S waves, the initial polarization is in the x_1 direction. The scattered ray direction is defined by the angles ψ and ζ. The scattered ray polarization is defined by X_r, X_ψ, and X_ζ. Figure taken from Shearer and Earle (2004).

The PSDF defines the strength of the heterogeneity as a function of its scale length and controls how the amplitude of the scattering varies with seismic wavelength. Observations at a single frequency, such as the P coda results presented in this chapter, mainly constrain the heterogeneity at scale lengths near the seismic wavelength and cannot determine the PSDF very completely. Analysis of broadband data and consideration of scattered arrival amplitudes as a function of frequency will be necessary to make quantitative estimates of the PSDF.

The total scattering coefficients, g_0^{PP}, and so on, are given by the averages of these coefficients over the unit sphere. The mean free path ℓ for a ray between scattering events is given by the reciprocals of these coefficients:

$$\ell^P = \frac{1}{g_0^{PP} + g_0^{PS}}$$

$$\ell^S = \frac{1}{g_0^{SP} + g_0^{SS}} \qquad (12)$$

and these values are used to assign path lengths using random numbers as described above. When a scattering event occurs, a second random number is used to decide whether the scattered wave is P or S, according to the relative sizes of g_0^{PP} and g_0^{PS} for an incident P wave or g_0^{SP} and g_0^{SS} for an incident S wave. A third random number (see below) is then used to determine the scattering angle (ψ and ζ) and the S polarization (if required). The particle then travels along its new ray direction until the next scattering event.

Despite the apparent complexity of the scattering equations, there are only three free parameters used to describe this model: the RMS fractional fluctuation ε, the correlation distance a, and the velocity density scaling factor ν. Of course, a more general PSDF than the exponential model would require more parameters. Larger values of ν will generally increase the amount of backward scattering. Shearer and Earle (2004) use $v = 0.8$, an estimate for the lithosphere obtained using Birch's law (S&F p. 101). Simpler equations can be obtained for the case of velocity perturbations alone or for purely acoustic waves. The random heterogeneity described by Eq. (11) is isotropic so that the scattering properties do not depend upon the angle of the incident wave. However, anisotropic PSDFs may be important in some regions of the Earth, in which case more free parameters will be required to define the model, which may also affect the relative strength of forward versus backscattering (e.g., Hong and Wu, 2005).

An efficient way to implement these scattering kernels in a computer program is to precompute the scattered power and S-wave polarizations at a series of small intervals of solid angle. The probability of scattering at each discrete angle is then given by its relative power and a single random number can be used to assign the scattered ray path. For example, consider the scattering pattern plotted in Fig. 7 [the S-to-P coefficient g^{SP} in Eq. (9), computed for $\gamma_0 = \sqrt{3}, v = 0.8, \beta_0 = 6/\sqrt{3}$ km/s, $\omega = 2\pi, \varepsilon = 0.01$, and $a = 1$ km], which is plotted at $6°$ increments in ψ and ζ. Assign a unique cell number to each of the n cells in the scattering surface and save the normalized scattering probabilities in a one-dimensional array, P, of dimension n. Define a second array, S, with the cumulative probabilities in P, that is, $S(1) = P(1), S(i + 1) = S(i) + P(i + 1), S(n) = 1$. The scattered ray angle is then defined by the smallest value of S that is larger than a computed random number between 0 and 1.

This approach has the computational advantage that the scattering probability arrays are computed only once and then angles are obtained for individual scattering events during the Monte Carlo simulation from a single random number, without the need to recompute any of the terms in Eq. (9). The accuracy of this approach depends upon how finely the scattering angles are sampled. Shearer and Earle (2004) used an angle spacing of $1.8°$. If desired, additional random numbers can be used to add a small amount of scatter to the ray angle so that the scattered ray angles are not restricted to the exact angles of the precomputed cells. The final step is to convert from the ray-based coordinate system used to define the scattering angles (i.e., as plotted in Fig. 7) to the absolute ray parameters needed to continue propagating the particle in the model. These include the ray azimuth (degrees from north), the ray parameter (approximated as the closest ray parameter in the precomputed table), and the ray vertical direction (upgoing or downgoing).

The scattering depends on the wave number, which is a function both of the wave frequency and the local background seismic velocity. Thus, Monte Carlo calculations that include scattering must be performed for a specific frequency. In addition, Earth's changing velocity with depth results in scattering kernels that vary continuously with depth. For the most accurate whole-Earth calculations, the kernels could be computed and stored at 10-km depth intervals. But this would require random numbers to be generated every 10 km along each ray path, greatly slowing the code. Thus, for practical purposes, it is useful to approximate the scattering properties, including the mean free path, as constant within fairly coarse depth intervals. Margerin and Nolet (2003a) assume that the mean free path is constant within the entire mantle. Shearer and Earle (2004) use four mantle layers, separated at depths of 200, 600, and 1700 km.

3.6. Intrinsic Attenuation

Energy converted to heat or crystal dislocations during wave propagation is termed intrinsic attenuation (as opposed to scattering attenuation in which some energy in the main

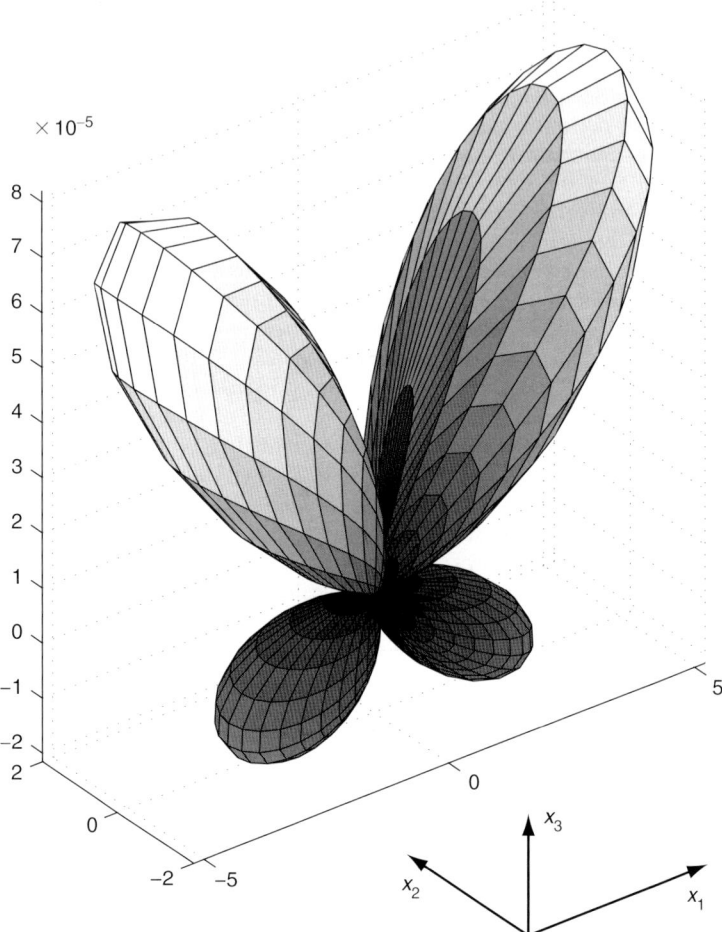

FIG. 7. An example of an S-to-P scattering pattern as a function of ray angle, computed using the Born equations for a random media model. The distance from the origin gives the value of the coefficient g^{SP} in Eq. (9). The incident S wave is traveling in the x_3 direction and is polarized in the x_1 direction.

pulse is scattered into other seismic waves). Intrinsic attenuation defined by a quality factor Q will reduce the wave energy by $e^{-\omega t/Q}$, where $\omega = 2\pi f$ is the angular frequency and t is the travel time along the ray. For global Earth models, Q varies strongly with depth so this correction is most easily performed in Monte Carlo methods by accumulating a value of $t^* = \int dt/Q$ along each particle. The energy at any desired point (such as when the particle hits the surface) is then computed using the reduction factor $e^{-\omega t^*}$.

For fully elastic calculations, both P- and S-wave attenuation must be defined. The PREM Earth model (Dziewonski and Anderson, 1981) contains depth-dependent values for both bulk and shear attenuation, $^\kappa Q$ and $^\mu Q$, from which can be computed P and S factors, $^\alpha Q$ and $^\beta Q$. However, the PREM values are accurate only at frequencies below about 0.2 Hz because of the frequency dependence of Q at higher frequencies (e.g., Sipkin and Jordan, 1979; Lundquist and Cormier, 1980; Anderson and Given, 1982; Der et al., 1986; Warren and Shearer, 2000). At 1 Hz, there is much less attenuation than predicted by the PREM model. Warren and Shearer (2000) analyzed high-frequency P and PP spectra and proposed a frequency-dependent mantle P-attenuation model that is generally consistent with prior work. At 1 Hz, $^{\alpha}Q = 227$ from 0- to 220-km depth and $^{\alpha}Q = 1383$ from 220 km to the CMB. Corresponding S attenuation can be computed using $^{\beta}Q = (4/9)^{\alpha}Q$, a commonly used approximation that assumes a Poisson solid and that all attenuation is in shear. The outer core is generally assumed to have zero attenuation, but the inner core is observed to be strongly attenuating (e.g., Bhattacharyya et al., 1993; Yu and Wen, 2006), with $^{\alpha}Q$ values varying between about 200 and 600 and some evidence of depth dependence.

It is likely that most high-frequency estimates of Q derived from teleseismic body waves contain a mixture of both intrinsic and scattering attenuation. Thus, these published values are only useful as a starting point for intrinsic Q in whole-Earth Monte Carlo simulations; the true intrinsic Q values are likely to be higher once scattering effects are included. Shearer and Earle (2004) found this to be the case in their Monte Carlo modeling of teleseismic P amplitudes and coda at 1 Hz, for which they obtained $^{\alpha}Q = 450$ from 0 km to 200 km and $^{\alpha}Q = 2500$ between 200 km and the CMB, values significantly higher than the Warren and Shearer (2000) Q values derived from P and PP spectra.

4. Fit to Teleseismic P Coda

Figure 8 shows the fit achieved to stacks of teleseismic P coda at 1 Hz by Shearer and Earle (2004) using their Monte Carlo method. The bottom plots show stacks of P coda amplitudes (obtained using the method described in Section 2) relative to the maximum P-wave amplitude for both shallow and deep earthquakes. These plots discard absolute P amplitude information, which is shown separately in the top plots. P amplitude versus distance is particularly sensitive to the intrinsic attenuation in the mantle. To model these observations, Shearer and Earle (2004) found that most scattering occurs in the lithosphere and upper mantle, but that a small amount of lower-mantle scattering was also required. Their preferred exponential autocorrelation random heterogeneity model contained 4% RMS velocity heterogeneity at 4-km scale length from the surface to 200 km depth, 3% heterogeneity at 4-km scale between 200 km and 600 km, and 0.5% heterogeneity at 8-km scale length between 600 km and the CMB. They assumed equal and correlated P and S fractional velocity perturbations and a density/velocity scaling ratio of 0.8. Intrinsic attenuation was $^{\alpha}Q_I = 450$ above 200 km and $^{\alpha}Q_I = 2500$ below 200 km, with $^{\beta}Q_I = (4/9)\,^{\alpha}Q_I$ (an approximation that assumes all the attenuation is in shear). This model produced a reasonable overall fit, for both the shallow- and deep-event observations, of the amplitude decay with epicentral distance of the peak P amplitude and the P coda amplitude and duration (see Fig. 8). These numbers imply that at 1 Hz, the total attenuation is dominated by scattering in the upper mantle and by intrinsic energy loss in the lower mantle.

To show the sensitivity of coda amplitudes to changes in the strength of the heterogeneity, Fig. 9 plots Monte Carlo predictions for a model with 30% less RMS heterogeneity (at all mantle depths) and a model with 30% more RMS heterogeneity. The resulting mean free paths in the upper 200 km are 82 km, 140 km, and 283 km for the three

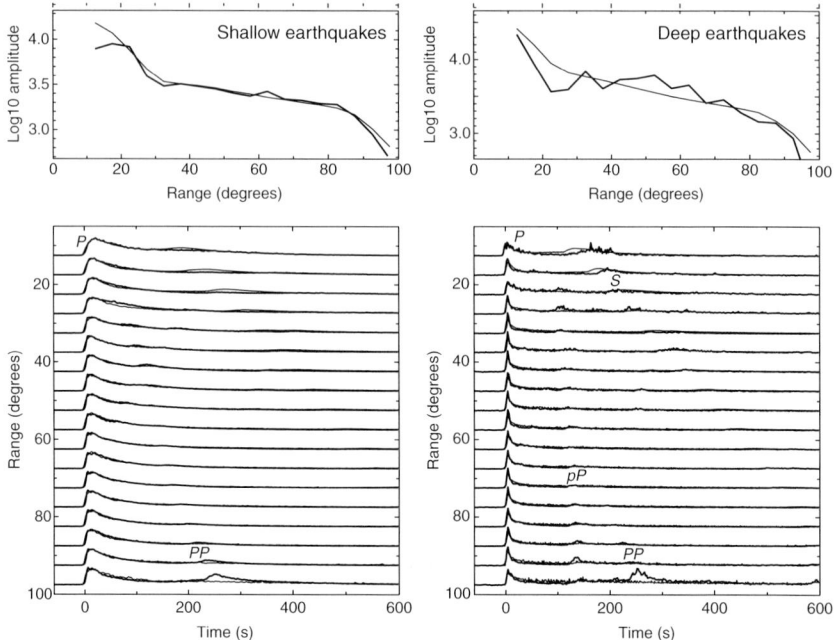

FIG. 8. Comparisons between envelope function stacks of teleseismic P-wave arrivals (solid lines) with predictions of a Monte Carlo simulation for a whole-Earth scattering model (thin lines) as obtained by Shearer and Earle (2004). The left panels show results for shallow earthquakes (≤50 km); the right panels are for deep earthquakes (≥400 km). The top panels show peak P-wave amplitude versus epicentral distance. The bottom panels show coda envelopes in 5° distance bins plotted as a function of time from the direct P arrivals, with amplitudes normalized to the same energy in the first 30 s. Figure adapted from Shearer and Earle (2004).

models. As expected, the coda amplitudes are very sensitive to the strength of the heterogeneity. At distances between 40 and 100 degrees, the differences in coda level are difficult to see at times later than about 150 s. However, at closer distances the differences persist to much longer times.

All of the P-coda observations and modeling presented here are for the vertical component. However, it is also possible to constrain mantle heterogeneity and scattering by studying the transverse component of teleseismic P coda (Nishimura et al., 2002; Kubanza et al., 2006). Our Monte Carlo code computes and saves all three components of output, but we have not yet analyzed the transverse component results.

5. Conclusions

Envelope-stacking methods and Monte Carlo modeling provide a powerful set of tools for analyzing whole-Earth scattering. Detailed applications of these approaches to a variety of seismic phases have only begun, but promise to provide reliable constraints on the average strength of small-scale heterogeneity as a function of depth in the mantle

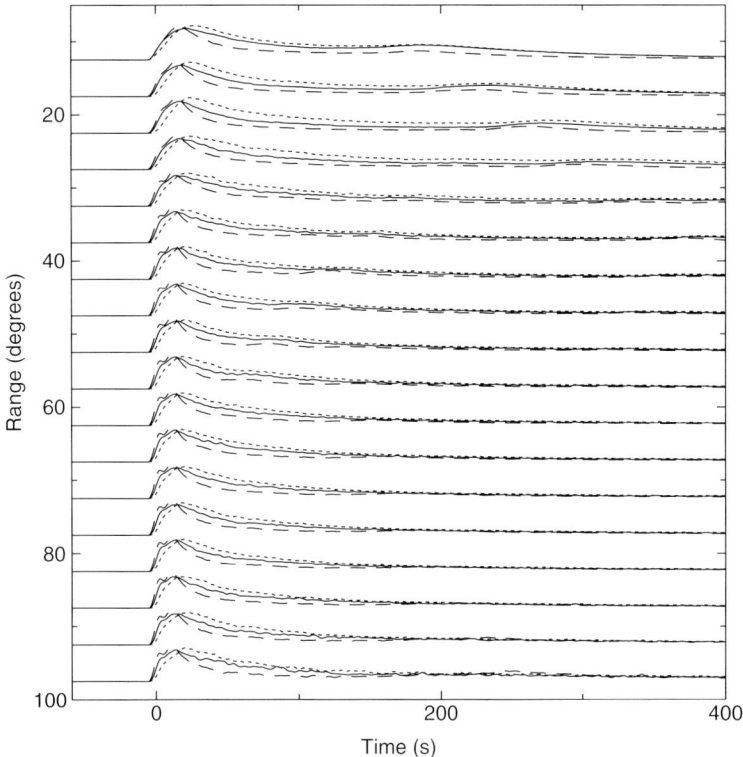

FIG. 9. A comparison of Monte Carlo-predicted P-coda amplitudes for the Shearer and Earle (2004) heterogeneity model (solid line), a model with 30% more root mean square (RMS) heterogeneity (dotted line), and a model with 30% less RMS heterogeneity (dashed line). The Shearer and Earle exponential autocorrelation random heterogeneity model contains 4% RMS velocity heterogeneity at 4-km scale length from the surface to 200-km depth, 3% heterogeneity at 4-km scale between 200 km and 600 km, and 0.5% heterogeneity at 8-km scale length between 600 km and the core-mantle boundary. It contains equal and correlated P and S fractional velocity perturbations and a density/velocity scaling ratio of 0.8. Intrinsic attenuation is $^{\alpha}Q_I = 450$ above 200 km and $^{\alpha}Q_I = 450$ above 200 km and $^{\alpha}Q_I = 2500$ below 200 km, with $^{\beta}Q_I = (4/9)\, ^{\alpha}Q_I$.

and inner core and the relative strength of intrinsic and scattering attenuation mechanisms. In addition to their value in resolving details of Earth structure, these results should help in making better estimates of earthquake source spectra, including the relative sizes of P and S corner frequencies, high-frequency spectral decay rates, and the ratio of radiated S energy to radiated P energy. Whole-Earth scattering studies and Monte Carlo simulations will become increasingly practical as global seismic data become more readily available and computer speeds continue to improve. In addition, it is clear that significant lateral variations in scattering strength exist in many regions, as can be seen in simple comparisons of teleseismic P coda levels among different stations and sources.

Acknowledgments

We thank George Choy, Stu Sipkin, Vernon Cormier, Ludovic Margerin, and Haruo Sato for detailed and constructive reviews. In particular, Hauro Sato noted the relevance of the Korn studies and Ludovic Margerin pointed out the differences between the global and the local Monte Carlo methods and the value of the Stokes parameters for handling general S-wave polarizations. This research was supported by National Science Foundation Grant EAR0229323.

References

Abercrombie, R.E. (1995). Earthquake source scaling relationships from −1 to 5 M_L using seismograms recorded at 2.5-km depth. *J. Geophys. Res.* **100**, 24,015–24,036.
Abubakirov, I.R., Gusev, A.A. (1990). Estimation of scattering properties of lithosphere of Kamchatka based on Monte-Carlo simulation of record envelope of a near earthquake. *Phys. Earth Planet. Inter.* **64**, 52–67.
Anderson, D.L., Given, J.W. (1982). Absorption band Q model for the Earth. *J. Geophys. Res.* **87**, 3893–3904.
Bal, G., Moscoso, M. (2000). Polarization effects of seismic waves on the basis of radiative transport theory. *Geophys. J. Int.* **142**, 571–585.
Bataille, K., Flatté, S.M. (1988). Inhomogeneities near the core-mantle boundary inferred from short-period scattered PKP waves recorded at the global digital seismograph network. *J. Geophys. Res.* **93**, 15,057–15,064.
Bataille, K., Lund, F. (1996). Strong scattering of short-period seismic waves by the core-mantle boundary and the P-diffracted wave. *Geophys. Res. Lett.* **23**, 2413–2416.
Bataille, K., Wu, R.S., Flatté, S.M. (1990). Inhomogeneities near the core-mantle boundary evidenced from scattered waves: A review. *Pure Appl. Geophys.* **132**, 151–173.
Bhattacharyya, J., Shearer, P., Masters, G. (1993). Inner core attenuation from short-period *PKP* (*BC*) versus *PKP(DF)* waveforms. *Geophys. J. Int.* **114**, 1–11.
Boatwright, J., Fletcher, J.B. (1984). The partition of radiated seismic energy between *P* and *S* waves. *Bull. Seismol. Soc. Am.* **74**, 361–376.
Chang, A.C., Cleary, J.R. (1978). Precursors to PKKP. *Bull. Seismol. Soc. Am.* **68**, 1059–1079.
Chang, A.C., Cleary, J.R. (1981). Scattered PKKP: Further evidence for scattering at a rough coremantle boundary. *Phys. Earth Planet. Inter.* **24**, 15–29.
Cleary, J.R., Haddon, R.A.W. (1972). Seismic wave scattering near the core-mantle boundary: A new interpretation of precursors to PKP. *Nature* **240**, 549–551.
Cormier, V.F. (1999). Anisotropy of heterogeneity scale lengths in the lower mantle from PKIKP precursors. *Geophys. J. Int.* **136**, 373–384.
Cormier, V.F. (2000). D'' as a transition in the heterogeneity spectrum of the lowermost mantle. *J. Geophys. Res.* **105**, 16,193–16,205.
Der, Z.A., Lees, A.C., Cormier, V.F. (1986). Frequency dependence of Q in the mantle underlying the shield areas of Eurasia, part III, the Q model. *Geophys. J.R. Astron. Soc.* **87**, 1103–1112.
Doornbos, D.J. (1978). On seismic wave scattering by a rough core-mantle boundary. *Geophys. J.R. Astron. Soc.* **53**, 643–662.
Doornbos, D.J. (1980). The effect of a rough core-mantle boundary on PKKP. *Phys. Earth Planet. Inter.* **21**, 351–358.
Dupree, S.A., Fraley, S.K. (2002). A Monte Carlo Primer, A Practical Approach to Radiation Transport. Kluwer Academic/Plenum, New York.
Dziewonski, A.M., Anderson, D.L. (1981). Preliminary reference Earth model (PREM). *Phys. Earth Planet. Inter.* **25**, 297–365.
Earle, P.S., Shearer, P.M. (1997). Observations of *PKKP* precursors used to estimate small-scale topography on the core-mantle boundary. *Science* **277**, 667–670.

Earle, P.S., Shearer, P.M. (2001). Distribution of fine-scale mantle heterogeneity from observations of Pdiff coda. *Bull. Seismol. Soc. Am.* **91**, 1875–1881.
Gusev, A.A., Abubakirov, I.R. (1987). Monte-Carlo simulation of record envelope of a near earthquake. *Phys. Earth Planet. Inter.* **49**, 30–36.
Gusev, A.A., Abubakirov, I.R. (1996). Simulated envelopes of non-isotropically scattered body waves as compared to observed ones: Another manifestation of fractal heterogeneity. *Geophys. J. Int.* **127**, 49–60.
Haddon, R.A.W., Cleary, J.R. (1974). Evidence for scattering of seismic PKP waves near the mantle-core boundary. *Phys. Earth Planet. Inter.* **8**, 211–234.
Hardebeck, J.L., Shearer, P.M. (2003). Using S/P amplitude rations to constrain the focal mechanisms of small earthquakes. *Bull. Seismol. Soc. Am.* **93**, 2434–2444.
Hedlin, M.A.H., Shearer, P.M., Earle, P.S. (1997). Seismic evidence for small-scale heterogeneity throughout Earth's mantle. *Nature* **387**, 145–150.
Hong, T.-K., Wu, R.-S. (2005). Scattering of elastic waves in geometrically anisotropic media and its implication to sounding of heterogeneity in the Earth's deep interior. *Geophys. J. Int.* **163**, 324–338.
Hoshiba, M. (1991). Simulation of multiple-scattered coda wave excitation based on the energy conservation law. *Phys. Earth Planet. Inter.* **67**, 123–136.
Hoshiba, M. (1994). Simulation of coda wave envelope in depth dependent scattering and absorption structure. *Geophys. Res. Lett.* **21**, 2853–2856.
Hoshiba, M. (1997). Seismic coda wave envelope in depth-dependent S wave velocity structure. *Phys. Earth Planet. Inter.* **104**, 15–22.
Kanasewich, E.R. (1981). Time Sequence Analysis in Geophysics. University of Alberta Press, Edmonton, Alberta, Canada.
Kennett, B.L.N. (1991). IASPEI 1991 Seismological Tables. Research School of Earth Sciences, Australian National University, Canberra, ACT.
Koper, K.D., Franks, J.M., Dombrovskaya, M. (2004). Evidence for small-scale heterogeneity in Earth's inner core from a global study of PKiKP coda waves. *Earth Planet. Sci. Lett.* **228**, 227–241.
Korn, M. (1990). A modified energy flux model for lithospheric scattering of teleseismic body waves. *Geophys. J. Int.* **102**, 165–175.
Korn, M. (1993). Determination of site-dependent scattering Q from P-wave coda analysis with an energy flux model. *Geophys. J. Int.* **113**, 54–72.
Kubanza, M., Nishimura, T., Sato, H. (2006). Spatial variation of lithospheric heterogeneity on the globe as revealed from transverse amplitudes of short-period teleseismic P-waves. *Earth Planets Space* **58**, e45–e48.
Lee, W.S., Sato, H., Lee, K. (2003). Estimation of S-wave scattering coefficient in the mantle from envelope characteristics before and after the ScS arrival. *Geophys. Res. Lett.* **30**, 24, 2248, doi:10.1029/2003GL018413.
Lundquist, G.M., Cormier, V.F. (1980). Constraints on the absorption band model of Q. *J. Geophys. Res.* **85**, 5244–5256.
Margerin, L. (2005). Introduction to radiative transfer of seismic waves. In: Seinsmic Data Analysis and Imaging with Global and Local Arrays, AGU Monograph Series American Geophysical Union.
Margerin, L., Nolet, G. (2003a). Multiple scattering of high-frequency seismic waves in the deep Earth: Modeling and numerical examples. *J. Geophys. Res.* **108**, B5, 2234, doi:10.1029/2002JB001974.
Margerin, L., Nolet, G. (2003b). Multiple scattering of high-frequency seismic waves in the deep Earth: PKP precursor analysis and inversion for mantle granularity. *J. Geophys. Res.* **108**, B11, 2514, doi:10.1029/2003JB002455.
Margerin, L., Campillo, M., van Tiggelen, B.A. (1998). Radiative transfer and diffusion of waves in a layered medium: New insight into coda Q. *Geophys. J. Int.* **134**, 596–612.

Margerin, L., Campillo, M., Tiggelen, B.V. (2000). Monte Carlo simulation of multiple scattering of elastic waves. *J. Geophys. Res.* **105**, 7873–7892.

Nakamura, A., Horiuchi, S., Hasegawa, A. (1999). Joint focal mechanism determination with source-region station corrections using short-period body-wave amplitude data. *Bull. Seismol. Soc. Am.* **89**, 373–383.

Nishimura, T., Yoshimoto, K., Ohtaki, T., Kanjo, K., Purwana, I. (2002). Spatial distribution of lateral heterogeneity in the upper mantle around the western Pacific region as inferred from analysis of transverse components of teleseismic P-coda. *Geophys. Res. Lett.* **29**, 23, 10.1029/2002GL015606.

Poupinet, G., Kennett, B.L.N. (2004). On the observation of high frequency PKiKP and its coda in Australia. *Phys. Earth Planet. Inter.* **146**, 497–511.

Prieto, G.A., Shearer, P.M., Vernon, F.L., Kilb, D. (2004). Earthquake source scaling and self-similarity estimation from stacking P and S waves. *J. Geophys. Res.* **109**, B08310, doi: 10.1029/2004JB003084.

Przybilla, J., Korn, M., Wegler, U. (2006). Radiative transfer of elastic waves versus finite difference simulations in two-dimensional random media. *J. Geophys. Res.* **111**, doi: 10.1029/2005JB003952.

Ryshik, L.V., Papanicolaou, G.C., Keller, J.B. (1996). Transport equations for elastic and other waves in random media. *Wave Motion* **24**, 327–370.

Sato, H. (1977). Energy propagation including scattering effects, single isotropic scattering approximation. *J. Phys. Earth* **25**, 27–41.

Sato, H. (1984). Attenuation and envelope formation of three-component seismograms of small local earthquakes in randomly inhomogeneous lithosphere. *J. Geophys. Res.* **89**, 1221–1241.

Sato, H. (1995). Formulation of the multiple non-isotropic scattering process in 3-D space on the basis of energy transport theory. *Geophys. J. Int.* **121**, 523–531.

Sato, H., Fehler, M.C. (1998). Seismic Wave Propagation and Scattering in the Heterogeneous Earth. Springer-Verlag, New York.

Sato, H., Fehler, M.C. (2006). Synthesis of seismogram envelopes in heterogeneous media. In: Advances in Geophysics, Vol. 48, Elsevier, doi: 10.1016/S0065-2687(06)48010-9.

Shearer, P.M., Earle, P.S. (2004). The global short-period wavefield modelled with a Monte Carlo seismic phonon method. *Geophys. J. Int.* **158**, 1103–1117.

Shearer, P.M., Hedlin, M.A.H., Earle, P.S. (1998). PKP and PKKP precursor observations: Implications for the small-scale structure of the deep mantle and core. In: The Carlo-Mantle Boundary Region, Geodynamics 28, American Geophysical Union, pp. 37–55.

Sipkin, S.A., Jordan, T.H. (1979). Frequency dependence of Q_{ScS}. *Bull. Seismol. Soc. Am.* **69**, 1055–1079.

Tono, Y., Yomogida, K. (1996). Complex scattering at the core-mantle boundary observed in short-period diffracted P-waves. *J. Phys. Earth* **44**, 729–744.

Vidale, J.E., Earle, P.S. (2000). Fine-scale heterogeneity in the Earths inner core. *Nature* **404**, 273–275.

Vidale, J.E., Dodge, D.A., Earle, P.S. (2000). Slow differential rotation of the Earth's inner core indicated by temporal changes in scattering. *Nature* **405**, 445–448.

Warren, L.M., Shearer, P.M. (2000). Investigating the frequency dependence of mantle Q by stacking P and PP spectra. *J. Geophys. Res.* **105**, 25,391–25,402.

Wegler, U., Korn, M., Przybilla, J. (2006). Modeling full seismograms envelopes using radiative transfer theory with Born scattering coefficients. *Pure Appl. Geophys.* **163**, 503–531, doi: 10.1007/s00024-005-0027-5.

Wu, R.-S. (1985). Multiple scattering and energy transfer of seismic waves–separation of scattering effect from intrinsic attenuation–I. Theoretical modelling. *Geophys. J.R. Astron. Soc.* **82**, 57–80.

Wu, R., Aki, K. (1985a). Scattering characteristics of elastic waves by an elastic heterogeneity. *Geophysics* **50**, 582–595.

Wu, R., Aki, K. (1985b). Elastic wave scattering by a random medium and the small-scale inhomogeneities in the lithosphere. *J. Geophys. Res.* **90**, 10,261–10,273.
Yoshimoto, K. (2000). Monte Carlo simulation of seismogram envelopes in scattering media. *J. Geophys. Res.* **105**, 6153–6161.
Yu, W.Y., Wen, L. (2006). Seismic velocity and attenuation structures in the top 400 km of the Earth's inner core along equatorial paths. *J. Geophys. Res.* **111**, B07308, doi: 10.1029/2005JB003995.

A SCATTERING WAVEGUIDE IN THE HETEROGENEOUS SUBDUCTING PLATE

TAKASHI FURUMURA AND BRIAN L.N. KENNETT

Abstract

The subducting plate is an efficient waveguide for high-frequency seismic waves. Such effects are often observed in Japan as anomalously large ground acceleration and distorted pattern of seismic intensity extending along the eastern seaboard of the Pacific Ocean from deep earthquakes in the Pacific plate, and Kyushu to Shikoku region from deep events in the Philippine Sea plate.

Seismograms in these high intensity zones show low-frequency ($f < 0.25$ Hz) onset for both P and S waves, followed by large, high-frequency ($f > 2$ Hz) later arrivals with a very long coda. Such observations are not explained by a traditional plate model comprising just high wave speed and low attenuation material in the slab.

A new plate model that can produce such guided high-frequency waves is characterized by multiple forward scattering of seismic waves due to small-scale heterogeneities within the plate. The preferred model requires anisotropic heterogeneity of elongated properties in the subduction slab with longer correlation distance (10 km) in the plate downdip direction and much shorter correlation distance (0.5 km) across the plate thickness. The standard deviation of P- and S-wave velocities and density from average is 3%. Such a quasi-laminated structure in the plate, which is equivalent to random distribution of anisotropic heterogeneities of elongated properties in parallel to the plate surface, can guide high-frequency signals with wavelengths shorter than the correlation distances along the plate. In contrasts, low-frequency signals with longer wavelength are not affected by the small scale heterogeneities and travel through the heterogeneous plate as a forerunner of the scattering signals.

The high wave speed property of the plate and a strong velocity gradient from the center to the outer part of the plate due to the thermal regime allows low-frequency ($f = 0.3$–0.5 Hz) seismic waves to escape into the surrounding, low wave-speed mantle by refraction of seismic waves. The net result is a frequency-dependent waveguide in the subducting plate with efficient guiding of high-frequency ($f > 2$ Hz) signals by multiple forward scattering and loss of intermediate frequency ($f = 0.3$–0.5 Hz) signals due to internal velocity gradients. Very low frequency signals ($f < 0.15$ Hz) with wavelength larger than the plate thickness are not significantly affected by the presence of the plate.

We demonstrate the presence of the frequency selective wave propagation effect from comparisons of observations from deep earthquakes that occurred recently in the Philippine Sea plate and in the Pacific plate. A good representation of the behavior of scattering waveguide is provided by 2D finite-difference calculations for seismic waves using heterogeneous slab models. The results of the simulations demonstrate that the frequency dependency of the models is quite sensitive to the thickness of the plate, and also depends on the scale lengths of heterogeneity distribution in the plate.

KEY WORDS: Seismic intensity, numerical simulation, scattering, subducting plate, waveguide.
© 2008 Elsevier Inc.

1. INTRODUCTION

Large number of earthquakes occurs in the area around Japan due to the simultaneous subduction of the Pacific plate in northern Japan and the Philippine Sea plate in western Japan (Fig. 1). Since the subducting plate acts as an efficient waveguide of

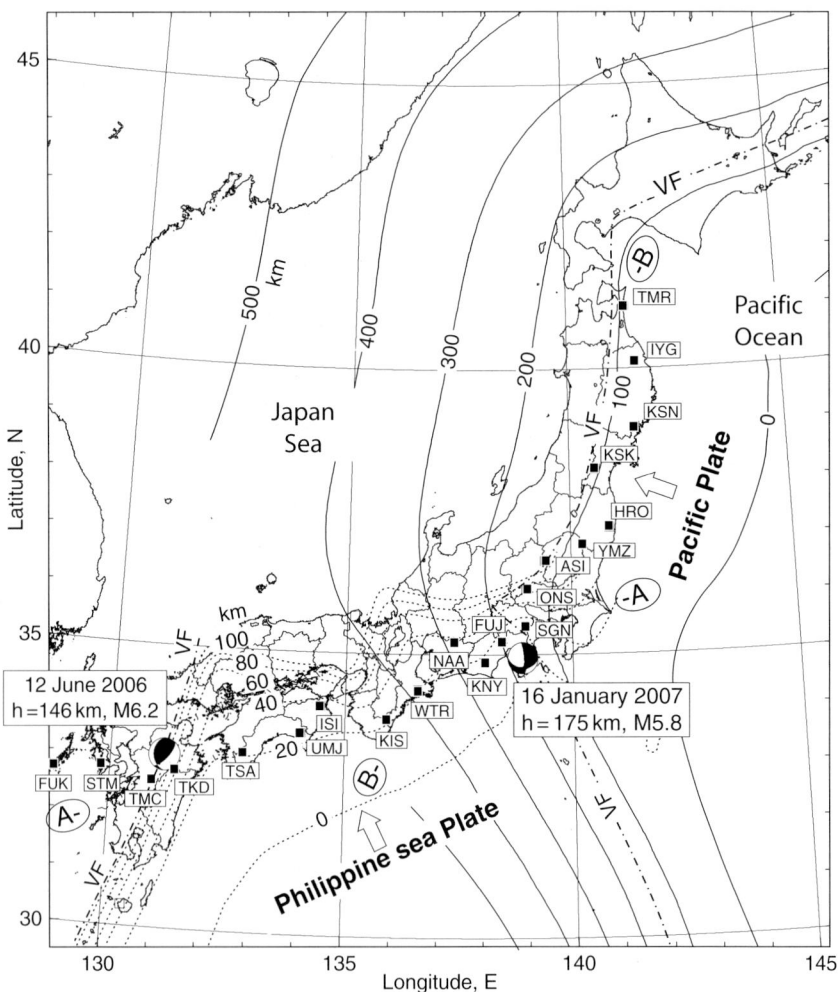

FIG. 1. The structure of the Japanese subduction zones with isodepth contours of the upper surface for the Pacific plate in northern Japan (solid curves) and for the Philippine Sea plate in western Japan (dashed curves). The associated volcanic fronts are shown with dashed lines and denoted VF. The epicenters of two events in the Philippine Sea plate on 12 June 2006 (the PHS event) and in the Pacific plate on 16 January 2007 (the PAC event) are indicated. The solid squares show the F-net broadband stations used in the analysis. A–A′ and B–B′ indicate the orientations of seismogram record sections in Fig. 3 for the PHS and PAC events, respectively.

high-frequency signals anomalously large ground motions are often produced in the fore-arc of northern Japan from intermediate to deep earthquakes occurring in the Pacific plate, and a similar effect occurs in western Japan from the earthquakes in the Philippine Sea plate. Sometimes the deep earthquakes are not felt near the epicenter, but many people are surprised by large intensities occurring in an area several hundred kilometers away from the epicenter.

Such observations of anomalously large ground motions and distorted intensity patterns for intraplate earthquakes in Japan have been recognized since modern seismic observations were started in Japan in the early 1900s (e.g. Hasegawa, 1918; Ishikawa, 1926a,b), but no satisfactory explanation for the cause of such anomalous intensity was made for over 30 years.

Utsu (1966) was the first to give an explanation. He proposed that the attenuation structure of the subduction zone in northern Japan, with a high Q and high wavespeed (high-V) dipping plate descending through a low-Q and low wavespeed (low-V) upper mantle, is the main reason for efficient propagation in the plate and so gives strong seismic signals on the fore-arc side of the subduction zone. Since then the high-V and high-Q models have been widely accepted as the expression of the subducting plate, and as a reason for the presence of high-frequency seismic waves.

Another important feature of the ground motions associated by intra-plate events are that a very long duration of high-frequency ($f > 2$ Hz) ground shaking are observed in the area of larger intensity. Such long duration and intense shaking is why people felt strong shocks from deep intra-plate earthquakes, but this type of behavior is not explained by the traditional subduction zone model comprising just a high-V and high-Q plate.

Several studies have considered the waveguide effect of high-frequency signals in the thin, low velocity zone of former oceanic crust at the top of the plate as an explanation of the time separation of low frequency arrivals from higher frequency waves, with apparent dispersion (e.g. Abers, 2000; Abers et al., 2003; Martin et al., 2003). The low velocity waveguide effect for deep ($h > 200$–500 km) Pacific plate earthquakes would be difficult to sustain, since the former oceanic crust is unlikely able to survive much beyond a depth of about 110 km without a transformation to a higher velocity eclogite. Moreover, even if the low velocity zone extended to greater depth it is still difficult to reproduce the observed large amplitudes and long duration of high frequency waves with trapped signals traveling within a thin low-Q region.

Furumura and Kennett (2005) proposed an alternative explanation of the waveguide effect for high-frequency signals in the Pacific plate by assuming strong heterogeneities in the plate. With the aid of 2D and 3D finite-difference method (FDM) simulations of seismic wave propagation they demonstrated that multiple forward scattering of high-frequency signals in the heterogeneous plate is the main cause of the guiding of high-frequency signals with very long coda. They used anisotropic heterogeneity model of von Karmann type function with a correlation length of about 10 km in downdip direction, and much shorter correlation length of about 0.5 km across the thickness, and a standard deviation of wave speed fluctuation from the average background model of 2%. This model proves to be very suitable to describe the observed characteristics of the broadband waveforms for the Pacific plate subduction zone.

The aim of this study is to review the scattering waveguide effect in the heterogeneous subduction plate through comparison between observations from dense seismic arrays of the F-net broadband and the K-NET and KiK-net strong motion instruments across Japan for two recent earthquakes; the first occurred on 12 June 2006 in the subducting Philippine Sea plate at depth of $h = 146$ km below western Oita prefecture (M6.2), southwestern Japan (hereafter we call the PHS event) and the second on 16 January 2007 in the Pacific plate at depth of $h = 175$ km beneath the Izu Peninsula (M5.8), central Japan (hereafter the PAC event). The frequency-dependent propagation properties of the P and S waves in heterogeneous plate is demonstrated by finite-difference method (FDM) simulation of seismic waves using different classes of plate structure and internal heterogeneities in the subducting plate. Through comparisons between the observations and the computer simulations for the two events, we will examine the similarity and dissimilarity in the heterogeneous structure in the Philippine Sea plate and the Pacific plate, through such parameters as the scale length of internal heterogeneities within the plate and the thickness of heterogeneities that characterize the guiding properties of the high-frequency signals.

2. Anomalous Intensity Patterns from Two Deep Events in the Subducted Philippine Sea Plate and in the Subducted Pacific Plate

Figure 2 illustrates the distribution of peak ground acceleration (PGA) for the PHS and PAC events.

The seismic intensity scale, which was originally defined by the strength of felt shaking and the damage rate of low-rise buildings, is very sensitive to higher-frequency signals in the 0.5–2 Hz range. Thus, the anomalous pattern of intensities from intermediate and deep events is more pronounced when we map PGA rather than the peak ground velocity or the displacement. Thus, we illustrate distributions of PGA contours in Fig. 2 using the records of 3-component K-NET and KiK-net accelerograms (squares and triangles in Fig. 2).

The PGA pattern from the PHS event shows a substantial northeast-southwest elongation of isoseismic contours with the largest PGA (>150 cm/s/s) values at the stations in the Shikoku and Chugoku regions over 100–200 km away from the epicenter. The area of larger PGA (>50 cm/s/s) occurs on the fore-arc side of the volcanic front (dashed line in Figs. 1 and 2), and the PGA contours appear to be correlated with the configuration of the isodepth contours of the Philippine Sea plate (Yamazaki and Ooida, 1985). The attenuation of PGA on the back-arc side, behind the volcanic front, is very dramatic in the Kyushu and Chugoku areas. At similar epicentral distances, the PGA value in the area of raised intensity is more than 10 times larger than that behind the volcanic front.

A similar observation of an anomalous pattern of the PGA distribution is found in northern Japan associated with the PAC event as is shown in Fig. 2b. There is a significant extension of PGA contours in the northeast direction along the strike of subducting Pacific plate with largest PGA (>20 cm/s/s) occurring in the southern Tohoku region more than 100–200 km away from the epicenter. The extension of PGA contours along the subducting Pacific plate is more striking than for the Philippine Sea plate event, and the area where the strong motion instruments were triggered extends more than 700 km from the epicenter along the eastern seaboard of the Pacific Ocean. In contrast, only very small PGA is observed at the stations to the west of the volcanic front even at the vicinity of the epicenter.

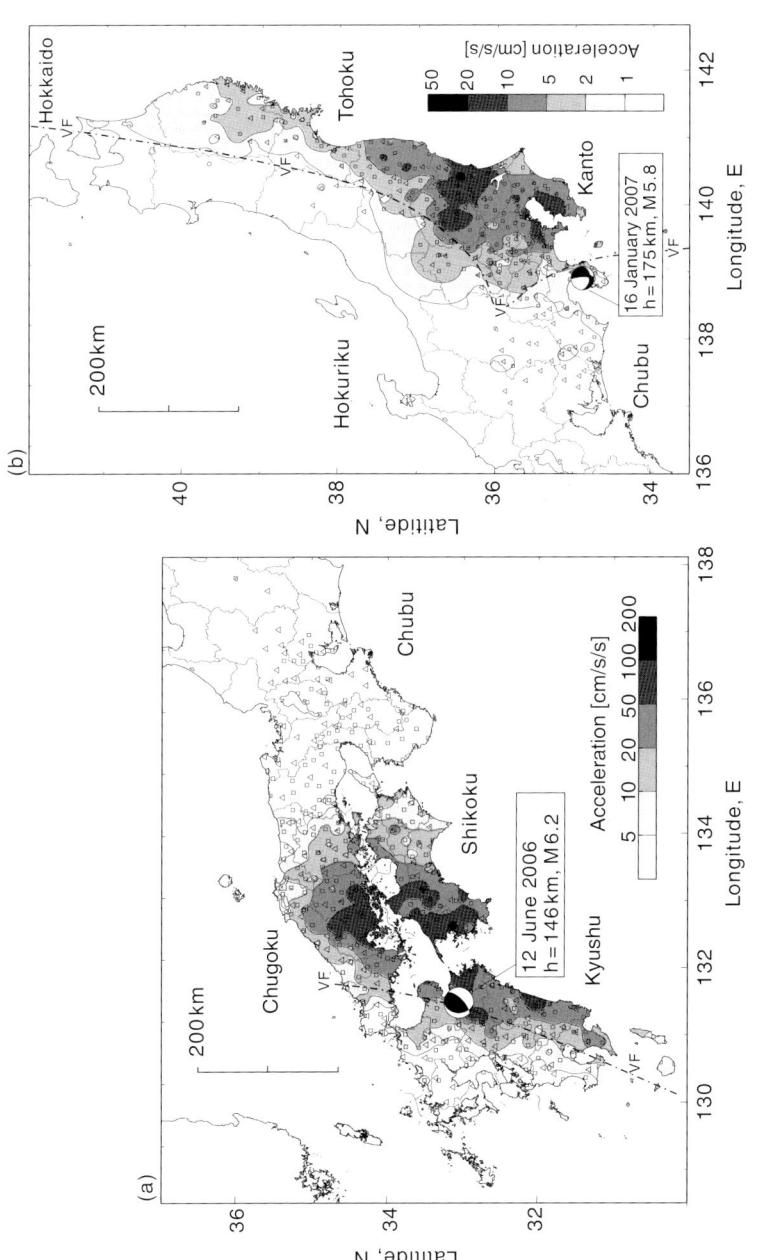

FIG. 2. The pattern of peak ground acceleration (PGA) for (a) the PHS event and (b) the PAC event. Triangles (KiK-net) and squares (K-NET) indicate the triggered stations induced by the earthquake. The location of the volcanic front (VF) is marked by dashed lines.

Figure 3 displays radial-component broadband records of ground velocity from the PHS event along a profile from Kyushu to the Chubu region (A–A' in Fig. 1), and from the PAC event along a profile from the Chubu to Tohoku regions crossing the volcanic front (B–B', Fig. 1). The seismic traces are multiplied by the epicentral distance to

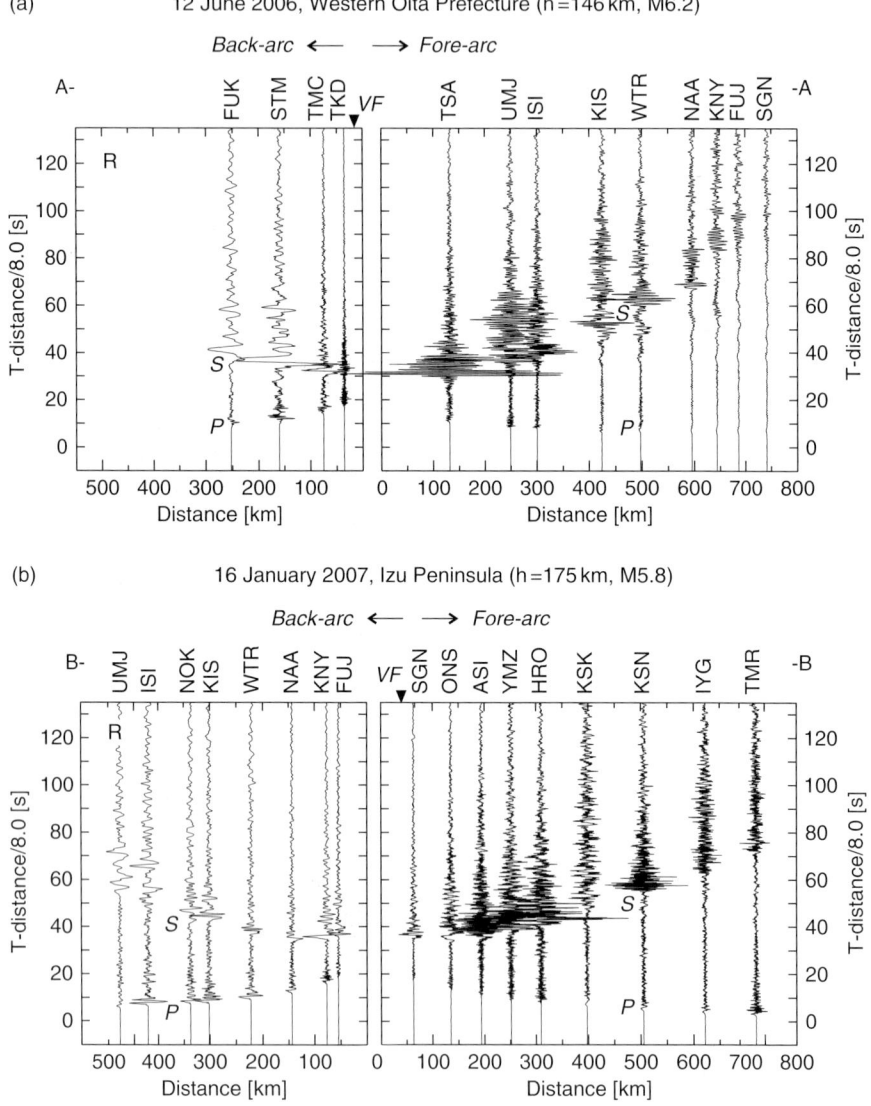

FIG. 3. Radial-component record section of ground velocity from (a) the PHS event along the profile A–A' in Fig. 1 and (b) from the PAC event along the profile B–B' in Fig. 1. The seismograms are recorded the labeled stations of the F-net broadband network (see Fig. 1 for locations). VF denotes the position of the volcanic front for each case.

roughly compensate the geometrical spreading of P and S body waves. The contrast in the appearance of the waveforms across the volcanic front is very striking, especially for S waves. A very large and high-frequency ($f > $ 1–2 Hz) P- and S-wave signal with very long coda is observed at fore-arc stations as the wave traveling along the plate, but the attenuation of high-frequency signals is very significant in the records from back-arc stations, especially for S waves.

2.1. Separation of Low-Frequency Precursors and High-Frequency Coda

An interesting feature in the waveforms from deeper subduction zone earthquakes that cause large shocks for stations in the fore-arc region is that the seismograms show a low-frequency ($f < $ 0.25 Hz) P-and S-wave onset followed by delayed high-frequency ($f > $ 2 Hz) signals with a long coda (e.g., see Fig. 7 in Furumura and Kennett, 2005). This pattern seems to be a common characteristics of the waveform from major inslab earthquakes such as those occurring in the Cocos plate of Nicaragua (Abers *et al.*, 2003) and the Nazca plate at the Chili-Peru subduction zone (e.g., Martin *et al.*, 2003).

Figure 4 illustrates the three-component broadband waveforms at WTR for the PHS event (Fig. 4a) and at KSN from the PAC event (Fig. 4b). Both stations lie in the fore-arc for the different seduction zones. The long-period P wave precursor from these events is very clearly recognized in the radial and vertical motions with an offset of about 2 s prior to the high frequency arrivals. There is no clear precursor in tangential motion. The precursor therefore represents the direct propagation of P waves from the source to the station. The following high-frequency waves are sustained for a very long time with large amplitude, which is quite similar for all three components. This character suggests strong internal scattering of high-frequency signals due to heterogeneities in the plate. There is also large P-wave energy on the transverse component in the higher frequency part, indicating a complicated pattern of wave propagation along the heterogeneous plate.

The scattering and guiding effects for the high-frequency signals along the plate are very striking in the records from the Pacific plate event, but are not so much clear for the Philippine Sea plate. The difference may indicate stronger heterogeneities in the Pacific plate.

2.2. Frequency Selective Propagation Properties in the Subducting Plate

The contrast in the high-frequency content of the waveforms across the volcanic front of the Philippine Sea plate and of the Pacific plate are examined by taking the spectral ratio of broadband records between fore-arc and back-arc stations at comparable epicentral distances. Since the F-net stations are all placed in hard rock sites, we expect that local site amplification effects at each station can be ignored.

Figure 5a illustrates the spectral ratio of the S wave for a 24.6 s time window at the fore-arc stations NSK and TSA relative to the back-arc stations STM and SBR respectively for the PHS event. Figure 5b is a similar spectral ratio for the PAC event between the fore-arc stations ASI and ONS and the back-arc stations KNM and TGA respectively.

For the PAC event (Fig. 5a), both sets of spectral ratios show a smooth rise from about 0.15–30 as the frequency increases from 0.3 to 1 Hz, and then the rate of increase diminishes in the higher frequency band (due to the loss of higher-frequency signals at back-arc stations). The lower values in the spectral ratio ($<$1) in the low-frequency band below about 0.15 Hz arise from the radiation pattern of the S waves.

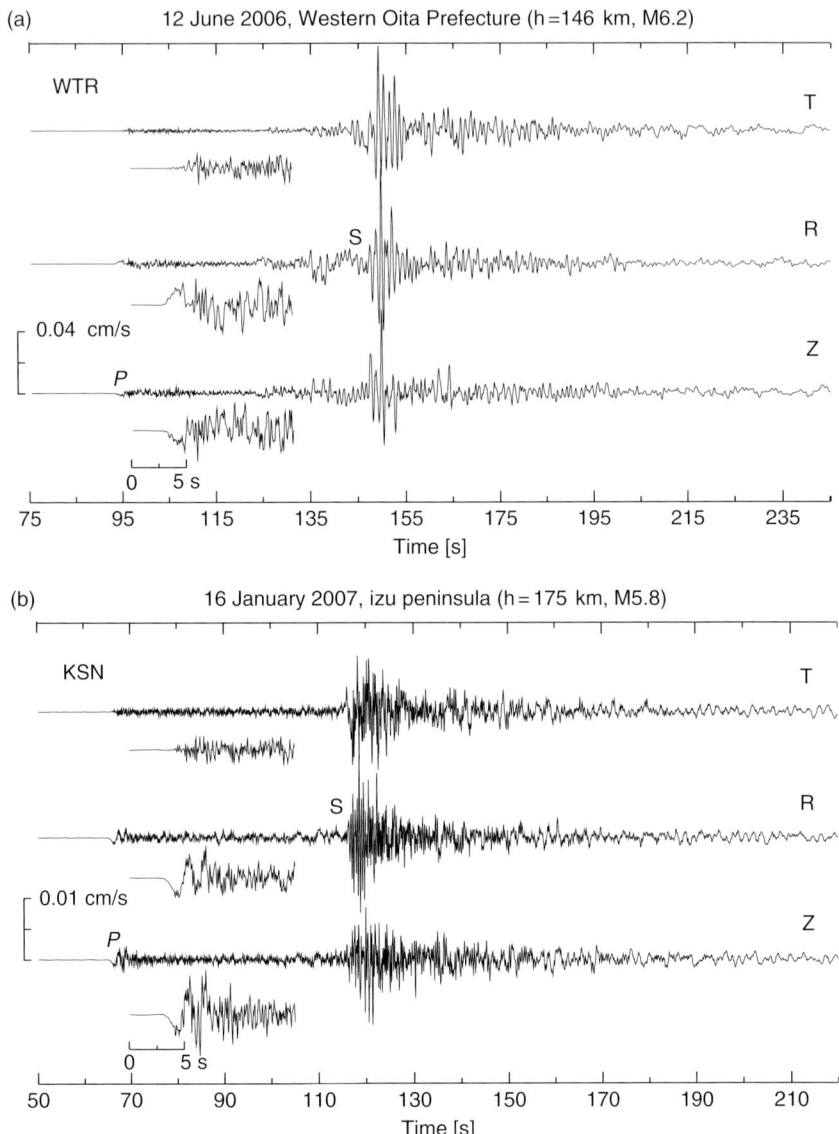

FIG. 4. Three-component broadband records of transverse (T), radial (R), and vertical (Z) motion at station (a) WTR from the PHS event and (b) at station KSN from the PAC event. Both stations are located on the fore-arc side of the subducting plate (see Fig. 1). An expanded P-wave segment for the three-component records is displayed at the left.

The guiding of high-frequency signals within the subducting Philippine Sea plate from source to fore-arc stations is clearly confirmed by Fig. 5a for frequencies above about 0.3 Hz. We have strongly frequency-dependent anelastic attenuation (Q) properties, with

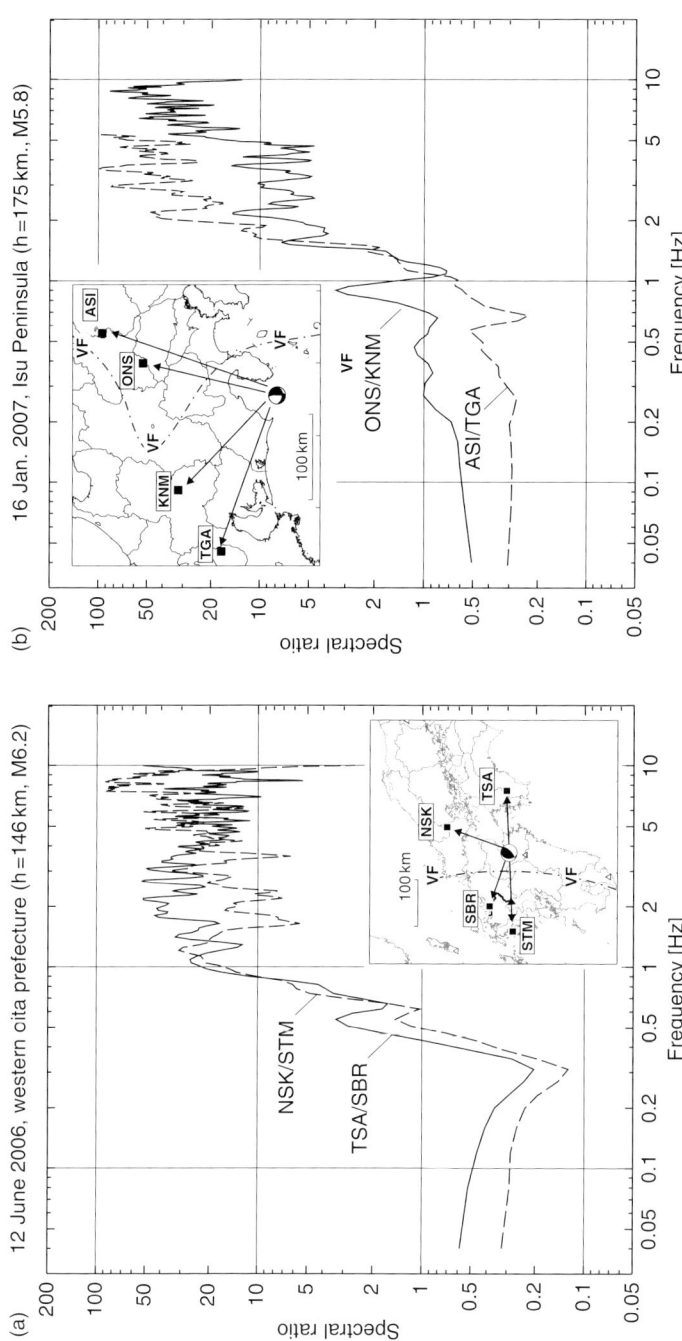

FIG. 5. Spectral ratio of broadband waveforms at fore-arc stations relative to back-arc stations for (a) the PHS event (NSK/STM and TSA/SBR) and (b) the PAC event (ONS/KNM and ASI/TGA). The locations of the F-net broadband stations are shown in the insert maps as solid squares.

an increase in Q values in the slab relative to that in the wedge mantle with increasing frequency above 0.3 Hz.

The significant drop in the spectral ratio around 0.3 Hz for the PHS event indicates that the seismic waves in this frequency band are dramatically attenuated whilst traveling up the Philippine Sea plate. The results demonstrate a complicated, frequency selective, waveguide property for the subducting plate due to strong scattering of high-frequency signals with wavelengths shorter than the heterogeneities in the plate. They also indicate escape of seismic waves from the plate in the lower frequency band. The low-frequency signals with wavelengths longer than the dominant scale of heterogeneities cannot be captured by such small-scale heterogeneities in the plate and thus easily escapes from the high-V plate to the surrounding low-V mantle.

Such a defocusing effect from the subducting plate in the low-frequency band is not so clearly observed for the PAC event (Fig. 5b). In this case there is an almost linear increase in the guiding properties of the high-frequency signals over a wide frequency range above 0.5 Hz, and no clear drop in the spectral ratio can be found in the low-frequency band below 0.5 Hz.

The difference in the frequency selective properties for the guiding of seismic signals in the Philippine Sea plate compared with the Pacific plate are likely to be related to the specific characteristics of each plate, such as plate thickness and the shape and scale lengths of heterogeneities in the plate.

3. 2D FDM Modeling of Scattering Wavefield

We illustrate that the guiding of high-frequency signals is produced by heterogeneities in the plate leading to the time separation of low- and high-frequency signals, by FDM simulation of high-frequency seismic waves in 2D heterogeneous structures.

The 2D model covers a region of 204.8 km × 204.8 km, which is discretized with a uniform grid interval of 0.1 km. The seismic wavefield is calculated explicitly by solving the equation of motions and the constitutive equations using a 16th-order staggered-grid FDM in space and 2nd-order scheme in time (Furumura and Chen, 2004).

The simulation model has P- and S-wave velocity of $V_P = 8$ km/s, $V_S = 4.6$ km/s, and density of $\rho = 2.6$ t/m^3. Frequency independent intrinsic anelastic properties for P- and S-wave with $Q_P = 2400$ and $Q_S = 1200$ are introduced in the FDM simulation using the memory variable technique of Robertsson et al. (1994). An absorbing boundary of Cerjan et al. (1985) is applied to the 20-grid points surrounding the model in order to reduce artificial reflections from model boundaries. An explosive line source with a pulse width of 10 Hz is introduced in the model, which radiates P wave isotropically.

The first simulation (Fig. 6a) employs a random media with an isotropic distribution of random fluctuation in the average P- and S-wave velocities and density following a von Karmann distribution function with a correlation distance of $a = 3$ km, Hurst number of $v = 0.5$, and a standard deviation of fluctuation from average background model of $\sigma = 4\%$. The stochastic fluctuations of elastic parameters were first produced in the wave number domain by applying a wave number filter to a sequence of random numbers, following the procedure of Frankel (1989), and then the result is transformed back into the physical domain using a Fast Fourier Transform.

Snapshots of seismic wave propagation at times $T = 8$ and 16 s from source initiation are illustrated in Fig. 6a. The scattering of seismic wavefield produced by the presence of

SCATTERING WAVEGUIDE IN HETEROGENEOUS PLATE 205

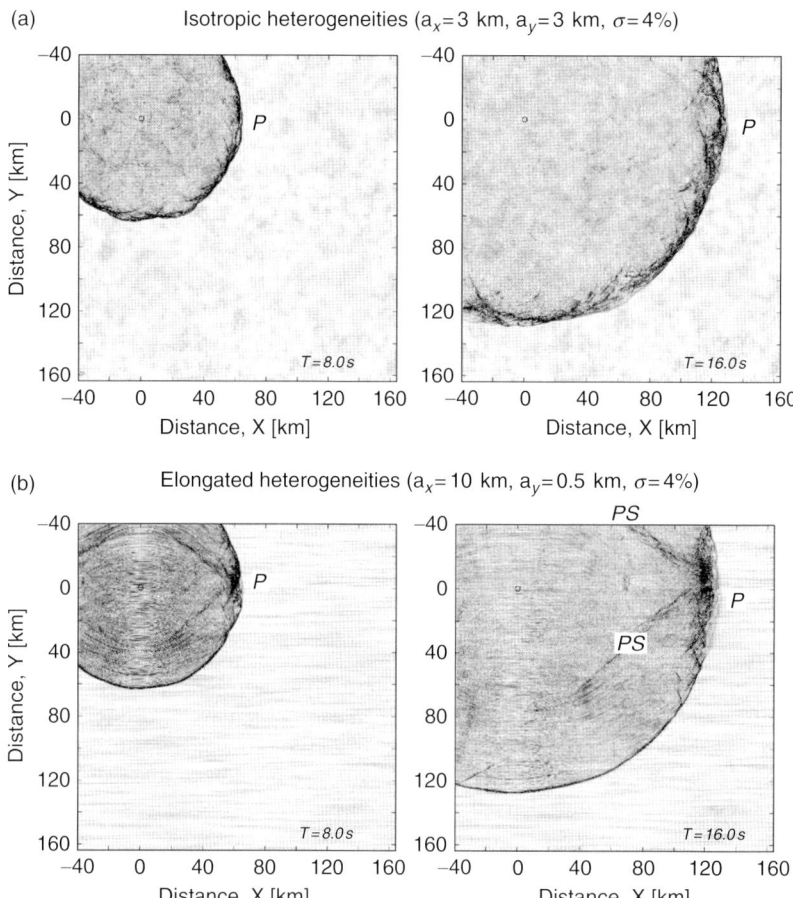

FIG. 6. Snapshots of seismic wave propagation (vector mean of two horizontal motions) at 8 and 16 s after initiation of an explosive source, derived from the 2D FDM simulation of seismic wave for models of (a) isotropic velocity fluctuation of von Karmann type function with a correlation length of $a = 3$ km and standard deviation of $\sigma = 4\%$, and (b) anisotropic, elongated heterogeneities with a correlation length of $a_x = 10$ km/$a_y = 0.5$ km, and a standard deviation of $\sigma = 4\%$. The wavefronts of P and PS converted waves are marked.

small-scale heterogeneities in the model is evident; the P wavespeed is distorted, and a strong coda is developed behind the wavefront. Such behavior resembles the patterns that are seen in the actual seismic waves observed at fore-arc stations, with an incoherent pattern of P and S phases between even close stations and an associated large coda. However, the simulation results for isotropic distribution of random heterogeneity model cannot explain the separation of the low-and high-frequency components of the initial P wave which is another important feature of the observations for intraplate earthquakes.

The second simulation assumes an elongated heterogeneity with a longer correlation distance of $a_x = 10$ km in x direction and much shorter correlation distance of

$a_y = 0.5$ km in y direction, retaining the areal scale of heterogeneity at $a_x \times a_y = 5$ km^2 with a standard deviation of fluctuations $\sigma = 4\%$.

A similar style of quasi-lamina structure in the lithosphere has been considered for the interpretation of the result of a long-range refraction experiment in the former Soviet Union using nuclear sources (e.g., Morozova et al., 1999) and also in modeling the oceanic crust of the subducting Nazca plate based on array observations of natural earthquakes (Buske et al., 2002; Patizig et al., 2002). These observational data and corresponding computer simulations demonstrate that high-frequency Pn and Sn signals with long tails observed in continental structure can be produced by horizontally elongated heterogeneities in the crust or uppermost mantle (e.g., Morozova et al., 1999; Tittgemeyer et al., 1999, 2000; Ryberg et al., 2000; Nielsen et al., 2003; Nielsen and Thybo, 2003).

Figure 6b shows an example of such quasi-laminated structure that produces a strong waveguide effect for high-frequency waves. The enhanced forward scattering within the quasi-lamina structure arises from successive post-critical reflections at the surfaces of the lamellae. In the snapshot of the later time frame (16 s), we find a strong concentration of P-wave energy as the wave traveling in the direction parallel to the lamellae and associated coherent P-to-S reflections in the shape of a laid "V". The two features together form a characteristic shape like a sliced mushroom. Thus, the seismic waves can travel longer distances in the direction parallel to the lamellae with only limited loss compared with the homogeneous structure and the heterogeneous model with isotropic heterogeneities as Fig. 6a.

The low-frequency precursor to the main P wavefront propagating parallel to the lamellae is clearly seen in the snapshot for the later time frame ($T = 16$ s). Multiple post-critical reflections of seismic waves in the quasi-lamina structure modify the effective wave speed for high-frequency signals, but low-frequency waves with wavelength much longer than the correlation distance ($a_y = 0.5$ km in normal to the scatterers) can easily travel through such small heterogeneities at faster propagation speed. Such generation of the low-frequency forerunner is explained as an inhomogeneous wave (Ikelle et al., 1993) or as a tunneling wave (Fuchs and Schulz, 1976) traveling in quasi-lamina structure.

We have also examined the development of the long-period precursors in the wavefront of the P wave using different classes of heterogeneity models with varying standard deviation of fluctuation, $\sigma = 2\%$, 4%, 6%, and 8% (Fig. 7), while retaining the same correlation length for the heterogeneities as in the previous model (Fig. 6b).

The results of the simulations demonstrate that larger velocity fluctuations with an enhanced velocity contrast across the flat interfaces of the lamellae can easily produce post-critical reflections with a smaller incident angle, and can trap high-frequency energy within the quasi-lamina structure by multiple forward scattering (Fig. 7b and c). The low frequency precursors in front of the trapped high-frequency signals are largely concentrated within the critical angle defined by the contrast for the low-to-high wave speed at the interface of lamellae. For a typical velocity contrast of $(1 \sim \sigma)/(1+\sigma)$, the critical angles calculated by $i_c = \sin^{-1}((1 \sim \sigma)/(1+\sigma))$ are $i_c = 73.8°, 67.4°, 62.5°$, and $57.4°$ for the case of velocity fluctuations of $\sigma = 2\%$, 4%, 6%, and 8% respectively (Fig. 7).

Figure 7d compares synthetic seismograms of radial and transverse components for the common source (open circle) to receiver (separation 150 km; solid square) configuration for different fluctuations of $\sigma = 2\%$, 4%, 6%, and 8%. The concentration of large seismic signals following the long-period P precursor produced by the stochastic

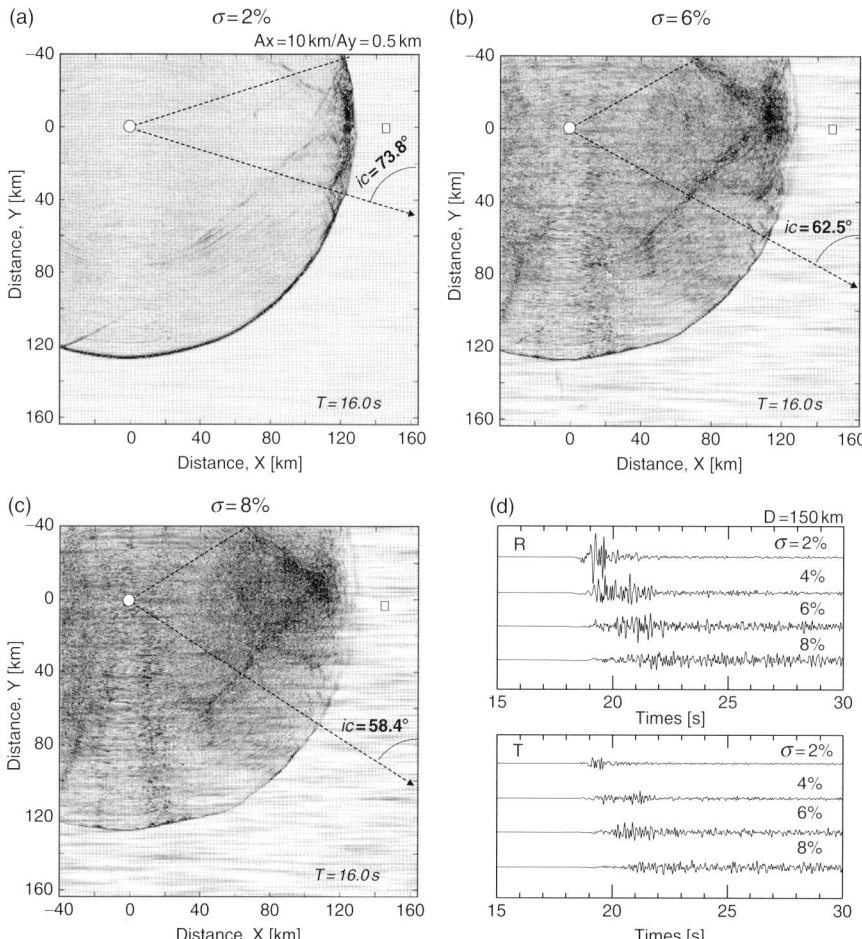

FIG. 7. Snapshots of seismic wave propagation for elongated heterogeneities ($a_x = 10$ km/$a_y = 0.5$ km) with varying standard deviation of the fluctuations, (a) $\sigma = 2\%$, (b) 6%, and (c) 8%. (d) Synthetic seismograms for radial (R) and transverse (T) motions at common stations (squares) at a distance $D = 150$ km from the source (open circle). ic indicates critical angle defined by the standard deviation of the velocity fluctuations across the low/high-wavespeed interface of the heterogeneities.

waveguide effect of the quasi-lamina structure is clearly demonstrated in this simulation. As the random fluctuation in elastic parameter increases, the relative amplitude of the low-frequency P wave and the duration of the coda increase very dramatically. With a large fluctuation in random heterogeneity (6–8%) the amplitude of the coda in tangential motion is almost as large as that for the P wave in radial motion. Thus, the strong concentration of P wave energy in the tangential motion from the in-slab earthquakes observed in Fig. 3 indicates the existence of strong heterogeneities in the plate.

4. 2D FDM Modeling of Slab Guided Waves

We now present 2D simulations of seismic wave propagation from in-slab earthquakes to illustrate the stochastic waveguide effect of the subducting heterogeneous plate.

The 2D model is taken along a profile cutting across Kyushu to Kanto region through the hypocenter of the PHS event which is nearly perpendicular to the subducting Philippine Sea plate (line A–A′ in Fig. 1). The physical parameters for the crust and upper mantle structure for western Japan are based on the $ak135$ reference Earth model (Kennett et al., 1995). The shape of upper boundary of the Philippine Sea plate is based on the model of Yamazaki and Ooida (1985). The 2D numerical model covers a region of 720 km horizontally and 307 km in depth, which is discretized into 12,000 by 5120 grid points with a uniform grid interval of 0.06 km.

The simulation employs a double-couple line source with a 45° reverse fault mechanism which is placed at the depth of $h = 160$ and 5 km below the plate surface. The source radiates seismic P and S waves with a maximum frequency of 16 Hz. We conducted parallel FDM simulations using 16 nodes of the Earth Simulator which required CPU time of 65 min to simulate 240 s seismic wave propagation with 160,000 time steps in the computations.

4.1. Base Model: High-Q and High-V Subduction Zone

The first simulation employs a simple plate model with the geometry illustrated in Fig. 8, assuming a uniform velocity increase (+3%) in the descending plate relative to the $ak135$ reference Earth model, while lowered (−5%) P- and S-wave velocities are assigned to the wedge mantle. We consider the thermal regime of the subducting zone with increasing temperature from the interior to the exterior of the plate. Such thermal effects in the plate can be approximated by 2% larger velocity and density at the center of the plate with a decrease to normal values toward either side of the plate, using a cosine function as shown in Fig. 8.

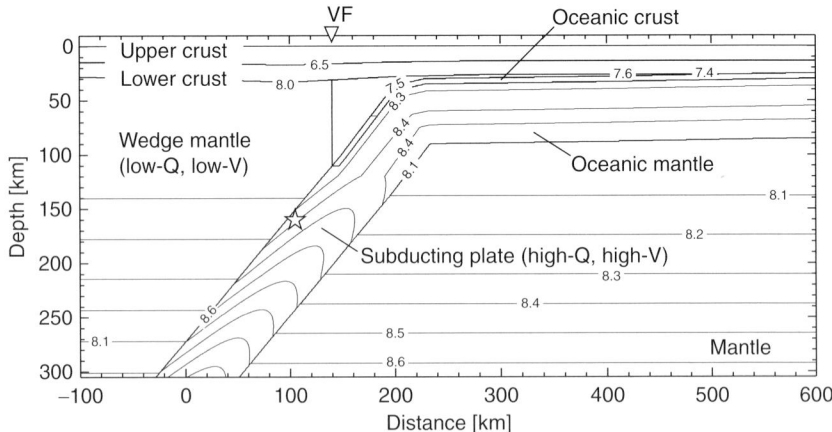

FIG. 8. 2D model of depth variation of the P wave structure used in the FDM simulation of seismic wave propagation along profile A–A′ shown in Fig. 1. Triangle and star denote the position of the volcanic front and seismic source, respectively.

SCATTERING WAVEGUIDE IN HETEROGENEOUS PLATE

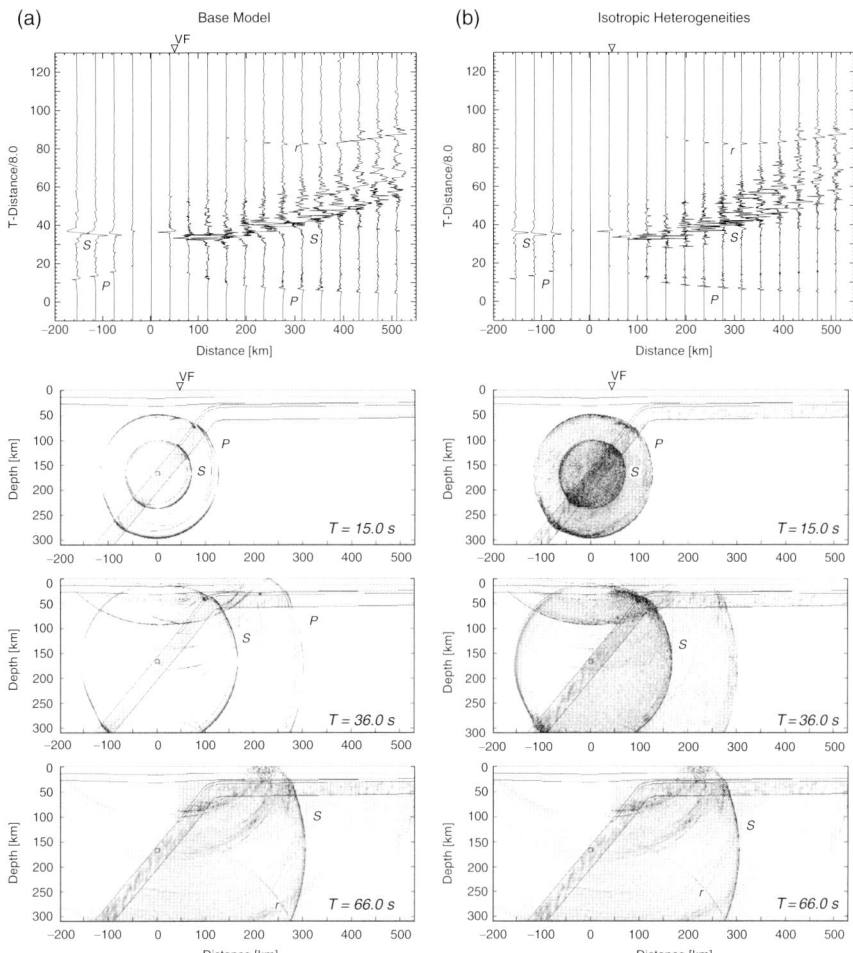

FIG. 9. Snapshots of seismic wave propagation at $T = 15$, 36, and 66 s after source initiation, and waveforms for radial motion from 2D FDM simulation. A double couple M6 event source is set in the slab at $h = 160$ km (small circle). (a) The case of base model with high-Q and high-V plate model. V_P distribution is shown in Fig. 8. (b) The case of a homogeneous, where isotropic velocity fluctuation of von Karmann type is superimposed to the base model. The correlation length is $a = 2.2$ km and standard deviation is $\sigma = 3\%$. Wavefronts of P and S waves are marked, and r indicates artificial reflections from the bottom of the model. VF denotes the position of the volcanic front.

High-Q values are assigned for P and S waves in the subducting plate ($Q_P = 2400$, $Q_S = 1200$) and somewhat smaller values for the oceanic crust ($Q_P = 400$, $Q_S = 200$). Stronger attenuation ($Q_P = 120$, $Q_S = 60$) is placed in the uppermost mantle above the plate on the back-arc side of the subduction zone, compared with that for the surrounding mantle ($Q_P = 400$, $Q_S = 200$).

The results of the FDM simulation as snapshots of seismic wavefield at $T = 15$, 36, and 66 s from the earthquake initiation are shown in Fig. 9a together with the waveforms of radial component velocity ground motion. The synthetic seismograms are obtained by convolution with a triangular slip-velocity function assuming a rise time of $T = 0.6$ s for a M6 in-slab event.

At $T = 15$ s, the P and S waves radiated from the 160 km deep source in the plate shows a clear and nearly spherical wavefront traveling upward in the subducting plate and through the surrounding mantle. The high-Q and high-V plate brings about very large seismic signals traveling upward within the plate compared with the overlying low-Q and low-V wedge mantle.

As the P wave enters the low-V oceanic crust, trapped P waves with larger energy are built up due to the superposition of multiple reflections in the thinner low-V layer ($T = 15$ s). The P-to-SV wave conversion is striking at the top of the oceanic crust, and reduces the energy in the trapped P wave signals in the oceanic crust.

The record section of synthesized ground motions demonstrates a remarkable contrast in the shape of waveform as the volcanic front is crossed at distance of about 40 km. The waveforms at stations on the fore-arc side show large amplitude of P and S waves with dominantly higher-frequency signals. The synthetic seismograms at back-arc stations show very simple P and S pulses with a dramatic loss of high-frequency signals for waves traveling through the low-Q and low-V mantle wedges. These features are consistent with the observed broadband records (Fig. 3).

However, the guiding of high-frequency seismic signals within the plate is not strong enough in this simple model without heterogeneous plate structure. Moreover, the snapshot at $T = 15$ and 36 s show some refraction of seismic waves from plate interior to the low-V mantle outside, which gradually attenuates the seismic waves traveling in the plate.

The strong velocity gradient in the plate, caused by the thermal regime, causes significant defocusing of seismic signals from the plate center towards the surrounding mantle, resulting in very weak P- and S-wave signals at stations near the volcanic front and on the fore-arc side of the subduction zone.

The results of these computer simulations demonstrate that the traditional plate model comprised of just a high-Q and high-V plate structure cannot trap sufficient high-frequency seismic energy within the plate. Furthermore, the simulated waveforms shown in Fig. 9a seem to be too simple to explain the complex properties of the observed waveforms in the broadband records at fore-arc stations (Fig. 3), which are characterized by a very long duration of high-frequency coda.

4.2. Heterogeneous Plate Model: Isotropic Heterogeneities in the Plate

Furumura and Kennett (2005) proposed that multiple scattering of seismic wave due to heterogeneities in the plate is the main cause of large amplitude and long-duration of high-frequency P and S waves associated by in-slab zone earthquakes.

The heterogeneous plate model assumes a random media with an isotropic random fluctuation in P- and S-wave velocity and density following a von Karmann distribution function with a correlation distance of $a = 2.2$ km in all direction, Hurst number of $v = 0.5$ and a standard deviation of fluctuations of $\sigma = 3\%$. This random heterogeneity is imposed on the whole plate including the oceanic crust and mantle. The scale of heterogeneities are selected to produce strong seismic scattering for P waves for frequencies over $f > 2$ Hz, with average wave speed in the oceanic mantle of about $V_P = 8$ km/s and corresponding slownesses for P and S waves of 1.6 s/km and 2.7 s/km, respectively.

As is shown in Fig. 9b the high-frequency seismic wavefield is changed produced by introducing isotropic small-scale heterogeneity in the plate, especially for stations on the fore-arc side. The distortion of the P wavefront traveling though the heterogeneous plate is clearly seen in the snapshots as a result of multiple scattering of P and S waves in the heterogeneous plate. The scattering wavefield is mainly composed of the P and S multiple reflections, and P-to-SV and SV-to-P conversions are very weak in this model.

The heterogeneous plate model leads to a significant improvement in the duration of high-frequency P- and S-wave coda in the waveforms at fore-arc stations. However, the scattering of P and S waves isotropically within the plate largely dissipates seismic energy into the surrounding mantle, as is clearly seen in the snapshots at $T = 15$ and 36 s in Fig. 9b. The result is relatively weak P and S wavefronts traveling upward in the plate as compared with the homogeneous plate model of Fig. 9a. Thus, the model of internal isotropic heterogeneities in the plate produces large and long duration of P- and S-wave coda by multiple scattering, but is insufficient to keep high-frequency seismic signal within the plate.

4.3. Anisotropic Heterogeneities in the Plate

Following the study of Furumura and Kennett (2005), we improve the heterogeneous plate model by introducing a stochastic distribution of anisotropic correlation properties for the random heterogeneities in the plate. We elongate the correlation distance; $a_x = 10$ km in the downdip direction, and much shorter correlation distance of $a_z = 0.5$ km in the direction of thickness. The heterogeneity is placed in the whole plate through the oceanic crust and the oceanic mantle. Such a quasi-laminated structure produces a strong waveguide effect on high-frequency seismic signals with enhanced forward multiple P- and S-wave scattering within the plate by wide-angle reflections. This leads to a large concentration of seismic energy in the center of the plate and thus the seismic waves travel in the high-V plate with less attenuation than the previous model with isotropic correlations (Fig. 9b). The synthetic seismograms for this model with anisotropic correlations show very large P- and S-wave coda with a long duration similar to those seen in the broadband waveform of fore-arc stations (Fig. 3).

Figure 10a shows the results of FDM simulation. The scattering wavefield produced by the quasi-lamina in the plate shows a strong P-to-SV conversion following the P wave front, which is much stronger than for the earlier models with a homogeneous plate and with isotropic scatters within the plate (Fig. 9a and b). Such P-to-SV conversion associated with multiple P-wave reflections at the lamina interface removes some energy from the trapped P-wave signals within the plate, and so the trapped P-wave energy gradually leaks into the surrounding mantle by SV conversions.

For the S wave, some SV-to-P conversions associated with multiple SV-wave reflections at the interface of lamellae arise near the source region. However, such conversions to P waves disappear as the distance from the source increases, and the multiple post-critical S-wave reflections between lamellae begin to trap S-wave energy within the plate. Thus, the heterogeneous plate with a quasi-laminated plate structure acts as a perfect waveguide for S wave, but it is not so efficient for P wave. The trapped S-wave signal propagating upward in the plate is clearly seen in the snapshots in Fig. 10a. This leads to a very large and long tail for the S-wave coda in fore-arc sites.

A weak separation of a low-frequency precursor and following high-frequency later coda is found in the records at distant stations but the results of computer simulation

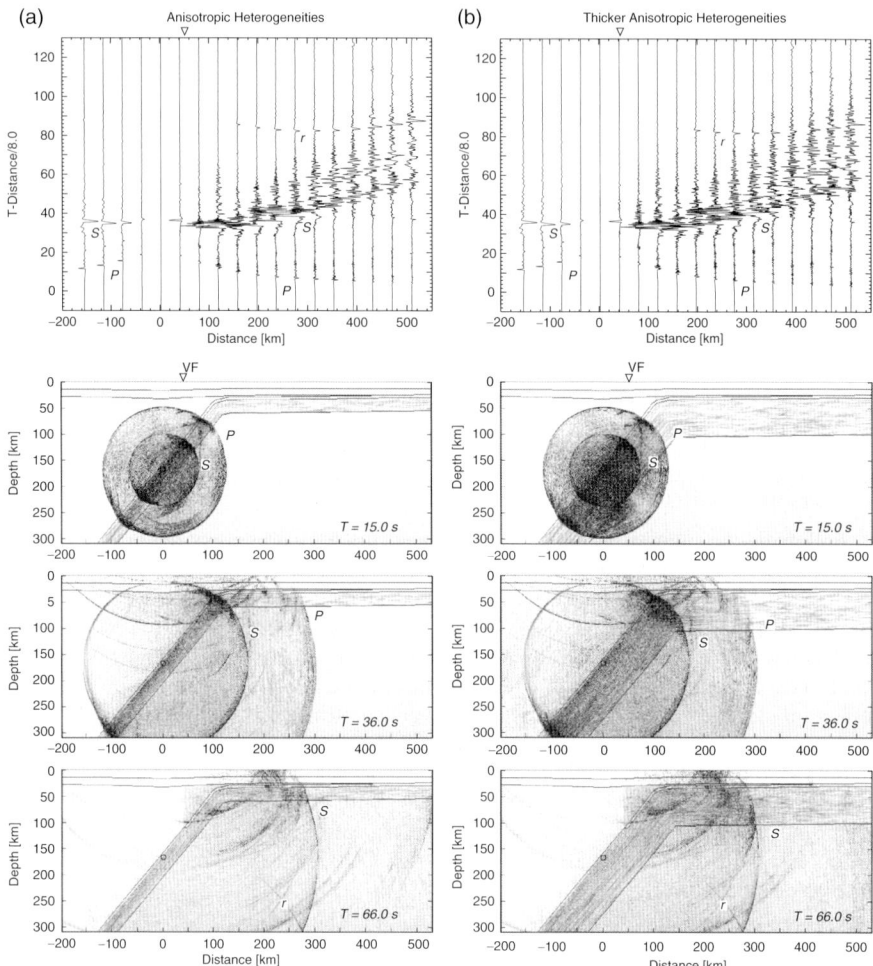

FIG. 10. Same as Fig. 9 but for a random plate model with (a) superimposed horizontally elongated anisotripic heterogeneities with a longer correlation length of $a_x = 10$ km in the downdip direction, $a_z = 0.5$ km across the thickness, and standard deviation from the average background model is $\sigma = 3\%$. (b) Thicker ($H = 85$ km) plate model with same elongated anisotropic heterogeneities.

using the present 2D subduction zone model is not as clear as the observations for Philippine Sea plate event (Fig. 3a) and as compared with the observations and corresponding simulations for the Pacific plate subduction zone event in northern Japan (Furumura and Kennett, 2005, Fig. 13). The present simulation model for the subduction zone structure in Kyushu could reproduce main feather in the broadband waveform with large and long coda in the fore-arc side stations and significant drop of high-frequency signals in the back-arc stations. However, in order for improving the match between simulation and observed seismograms in terms of the separation of low-frequency precursor with high-frequency later code, we may need future effort to constructing proper 2D model by projection of actual 3D heterogeneous structure in the subducting zone.

4.4. Effect of Plate Thickness

The Philippine Sea plate descending below western Japan is young (30 Ma) and relatively thin (about 35–45 km), and the guiding effect of high-frequency waves within the plate tends to less effective as compared with the old (120 Ma) and thicker Pacific plate (about 80–120 km) descending below northern Japan. Moreover, the enhanced internal velocity gradient in the thinner Philippine Sea plate caused by the thermal regime, as shown in Fig. 8, tends to cause strong refraction of seismic waves towards the mantle outside.

We examine the effect of plate thickness on the guiding of high-frequency S-wave signals using a thicker ($H = 85$ km) plate model. The new plate model with a thick quasi-laminated structure leads to a strong bundle of S-wave coda trapped in the heterogeneous plate (Fig. 10b, $T = 36$ s) as compared with the thin plate model (Fig. 10a). The result is a very large S-wave with long coda traveling to fore-arc stations as we observed in the broadband seismograms for the PAC event occurring in the subducting Pacific plate (Fig. 3b). These computer simulations demonstrate clearly that the trapping properties of the high-frequency seismic waves in the heterogeneous plate are very sensitive to the shape of heterogeneities and the thickness of the waveguide.

The separation of low- and high-frequency signals in the initial P wave is slightly improved in the present simulation of thick plate model but it is still not so clear as clear as observations for the PAC event (see Figs. 3b and 4b). This may because the bending structure of the Pacific plate near the trench in Kanto region does not match to the present model for the Philippine Sea plate subduction zone in Kyushu. Note that the separation of low- to high-frequency signals in the initial P wave is very sensitive to the source and receiver configurations with the bending structure of the subducting plate. Our previous simulation of Furumura and Kennett (2005) for the model of the Pacific plate using the same elastic properties and heterogeneity distributions as the present simulation model but different configuration of the subduction zone structure demonstrated the low-frequency precursors very clearly.

As we have seen in the previous set of simulations (Figs. 6 and 7) the low-frequency P-wave precursor appears in a limited area of P wavefront. The detection of this phase may depend on the bending structure of the subducting plate and the geometry of source and receiver locations.

In order to clarify the effect of plate thickness on the guiding of high-frequency S waves within the plate, we evaluate the Fourier spectral ratio of S waves at a fore-arc station at distance $D = 100$ km, and a back arc station at the same epicentral distance ($D = -100$ km). Figure 11a compares Fourier spectral ratio of the S wave for different plate thickness of $H = 35$, 50, and 85 km.

The simulated S waves guided by the thick ($D = 85$ km) plate shows a gentle drop in the spectral ratio in lower-frequency band of about 0.2 Hz. This means that the plate is an efficient waveguide for higher-frequency signals above 0.2 Hz. The high-frequency seismic signals with wavelengths shorter than the correlation length of the heterogeneity are easily trapped in the quasi-laminated plate due to multiple post-critical S wave

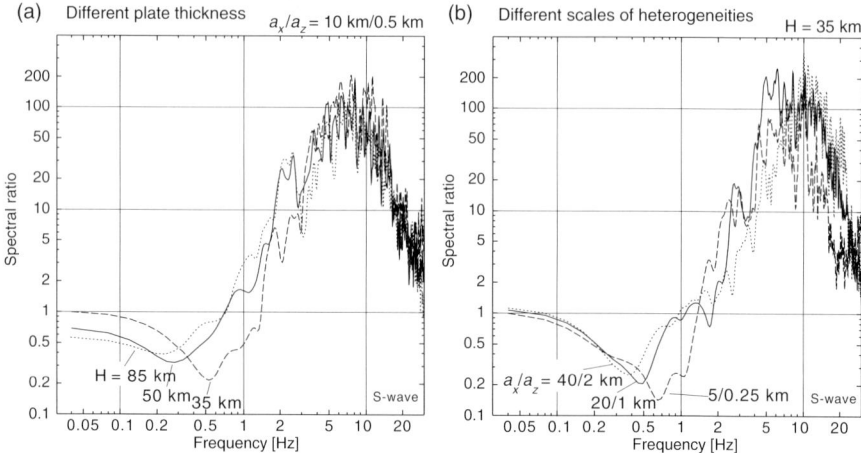

FIG. 11. Fourier spectral ratio of the synthetic seismograms of S wave at a fore-arc station ($D = 100$ km) relative to a back-arc station ($D = -100$ km) (a) for three models assuming varying plate thickness of $H = 35$, 50, and 85 km and (b) for different scales of random heterogeneities in the plate and a fixed plate thickness of $H = 35$ km.

reflections within the plate, but very low-frequency signals with longer wavelength are not sensitive to the internal structure of the plate.

A trough in the spectral ratio at frequencies between 0.15 and 0.5 Hz for the different plate thickness models arises from the anti-waveguide effect of the High-V plate with a large internal velocity gradient induced by the thermal regime. The simulation results demonstrate that the peak frequency for the anti-waveguide effect of the plate moves gradually to lower frequencies from 0.5 to 0.2 Hz as the thickness of the plate increases from $H = 35$ to 85 km.

When comparing the simulation results with observation, the spectral ratio of S waves for the PHS event (Fig. 4a) roughly corresponds to the characteristics for the plate thickness of $H = 50$ km. The case of the PAC event (Fig. 4b) roughly correlates with the simulation result for thick ($H = 85$ km) plate model, inconsistent to the thick Pacific plate descending below Kanto region.

4.5. Effect of Heterogeneity Scale in the Plate

In the simulations with varying plate thickness we have so far fixed the scale of random heterogeneity in the plate; correlation distance of $a_x = 10$ km in the plate downdip direction and $a_z = 0.5$ km in the direction of thickness. However, the frequency-selective propagation characteristics in the subducting plate are also influenced by the distribution properties of the heterogeneities within the plate.

We have conducted a set of additional simulation experiments to examine how the correlation scales of the randomness in the plate modify the propagation of high-frequency waves by multiple internal scattering within the plate. We consider three stochastic random media with varying correlation distances of nonisotropic heterogeneities with an elongated correlation distances in the plate downdip direction (a_x) and in the plate thickness direction (a_z); (a) $a_x/a_z = 5$ km/0.25 km, (b) $a_x/a_z = 20$ km/1 km, and (c) $a_x/a_z = 40$ km/2 km. All models have the same aspect ratio of 20 and a standard deviation from the background of $\sigma = 3\%$.

The Fourier spectral radio of S waveforms in the fore-arc relative to that for the back-arc station at same epicentral distance ($D = \pm 100$ km) are compared in Fig. 11b. Simulation results demonstrate that shift of the minimum frequency of S waves, that can be trapped within the plate, varies with changes in heterogeneity scale.

As we can see from Fig. 11b, there is strong tradeoff between the plate thickness and the heterogeneity scale on the frequency-dependent waveguide properties of the subducting plate wave. The present configurations provide a reasonable representation of the observed behavior, but it is difficult to pin down the heterogeneity distribution closely from limited observational waveform data.

5. Discussion and Conclusion

Since the extensive studies of Utsu (1966, 1967, 1969) and Utsu and Okada (1968), it has been recognized that the anomalous patterns of seismic intensity from in-slab earthquakes are caused by the guiding of high-frequency waves though the high-Q and high-V plate. However, a simple plate model comprising just a high-Q and high-V plate does not explain the frequency-dependent propagation characteristics of observed seismic waves and the long coda following the P- and S-waves. In particular, the high-V property of the plate tends to shed waves from the plate because strong internal velocity gradient refract seismic energy to the mantle outside. Such anti-waveguide effects are more significant for the thin Philippine Sea plate than for the thick Pacific plate.

To solve the difficulties of past models, Furumura and Kennett (2005) proposed a new heterogeneous, quasi-laminated plate model, which is described by a non-isotropic heterogeneity in the plate structure. This heterogeneity generates trapped high-frequency ($f > 2$ Hz) waves that propagate along the plate due to multiple forward scattering within the plate. Low-frequency ($f < 0.25$ Hz) waves, with longer wave length can easily tunnel through the lamella features without affecting small scale heterogeneities or as the inhomogeneous waves with the incidence angle larger than the critical angle or as the tunneling waves (Fuchs and Schulz, 1976). The result is a separation of a low-frequency precursor and following high-frequency later signals with a long coda.

Such a non-isotropic heterogeneity structure in the subducting plate appears to be an intrinsic property of the oceanic lithosphere. Such heterogeneity may most likely be produced at the mid-ocean ridge with injection and underplating as the lithosphere is formed, and as the oceanic plate travel from the ridge to the subduction zone.

The trapping effect of the high-frequency signals for the thin ($H = 35$ km) Philippine Sea plate is somewhat weaker than for the thick ($H = 85$ km) Pacific plate, and this may be the reason for the less clear pattern of stretched intensity contours along the subducting Philippine Sea plate in western Japan compared with the clear pattern of anomalous intensity along the Pacific coast from the Pacific plate events in northern Japan.

Numerical 2D simulations of seismic P and S waves for the heterogeneous quasi-lamina plate structure demonstrate clearly how such frequency-dependent properties can arise, and how they bring about strong dependence on the velocity gradient in the plate, the thickness of the plate heterogeneities, and the character of the heterogeneity distribution in the plate.

The present simulations are for a 2D model assuming that the source and structure are invariant in the out-of-plane direction, and thus they can account only for the scattering effects in the in-plane motion. The exclusion of out-of-plane scattering that will occur in the actual full 3D scattering situation may underestimate the amplitude and duration of P- and S-wave coda. We can reproduce the main features of the observed seismic wavefield in the present 2D modeling using suitable stochastic random heterogeneities in the plate, but before reaching firm conclusions we may need to improve the understanding of the nature of actual 3D heterogeneous wavefield. The further study should include analysis of dense seismic array data, and corresponding high-resolution 3D simulations with the aid of high-performance computing technologies.

ACKNOWLEDGMENTS

The computations were conducted as a joint science research project "Seismic wave propagation and strong ground motion in heterogeneous 3D structure" between the Earthquake Research Institute, University of Tokyo and the Earth Simulator Center. A part of CPU time was supported by the JST-CREST Project on "Multi-scale and multi-physics simulation". We acknowledge the National Research Institute for Earth Science and Disaster Prevention (NIED) for the use of K-NET and KiK-net strong motion records and F-net broadband waveform data. We thank for careful review and constructive comments from Profs. M. Ohtake, H. Sato, and S.G. Shapiro were very helpful for revision of the manuscript.

REFERENCES

Abers, G.A. (2000). Hydrated subducted crust at 100–250 km depth. *Earth Planet. Sci. Lett.* **176**, 323–330.
Abers, G.A., Plank, T., Hacker, B.R. (2003). The wet Nicaraguan slab. *Geophys. Res. Lett.* **30**, doi: 10.1029/2002GL015649.
Buske, S., Luth, S., Meyer, H., Patizig, R., Reichert, C., Shapiro, S., Wigger, P., Yoon, M. (2002). Broad depth range seismic imaging of the subducted Nazca Slab, North Chile. *Tectonophysics* **350**, 273–282.
Cerjan, C., Kosloff, D., Kosloff, R., Reshef, M. (1985). A nonreflecting boundary condition for discrete acoustic and elastic wave equations. *Geophysics* **50**, 705–708.
Frankel, A. (1989). A review of numerical experiments on seismic wave scattering. *Pure Appl. Geophys.* **131**, 639–885.
Fuchs, K., Schulz, K. (1976). Tunneling of low-frequency waves through the subcrustal lithosphere. *J. Geophys.* **42**, 175–190.
Furumura, T., Chen, L. (2004). Large scale parallel simulation and visualization of 3-D seismic wavefield using the Earth Simulator. *Comput. Model. Eng. Sci.* **6**, 153–168.
Furumura, T., Kennett, B.L.N. (2005). Subduction zone guided waves and the heterogeneity structure of the subducted plate: Intensity anomaly in northern Japan. *Geophys. Res.* **110**, doi: 10.1029/2004JB003486.
Hasegawa, K. (1918). An earthquake under the Sea of Japan. *J. Meteorol. Soc. Jap.* **37**, 203–207 (in Japanese).
Ikelle, L., Yung, S.K., Daube, F. (1993). 2D random media with ellipsoidal autocorrelation functions. *Geophysics* **58**, 1359–1372.

Ishikawa, T. (1926a). On the abnormal distribution of felt areas of an earthquake. *J. Meteorol. Soc. Jpn. Ser. 2* **4**, 137–146 (in Japanese).
Ishikawa, T. (1926b). On seismograms of earthquakes exhibiting abnormal distribution of seismic intensities. *Quart. J. Seismol.* **2**, 7–15 (in Japanese).
Kennett, B.L.N., Engdahl, E.R., Buland, R. (1995). Constraints on the velocity structure in the Earth from travel times. *Geophys. J. Int.* **122**, 108–124.
Martin, S., Rietbrock, A., Haberland, C., Asch, G. (2003). Guided waves propagating in subduced oceanic crust. *J. Geophys. Res.* **108**, doi: 10.1029/2003JB002450.
Morozova, E.A., Morozov, I.B., Smithson, S.B., Solodilov, L.N. (1999). Heterogeneity of the uppermost mantle beneath Russian Eurasia from the ultra-long range profile QUARTZ. *Geophys. Res. Lett.* **104**, 20329–20348.
Nielsen, L., Thybo, H. (2003). The origin of teleseismic Pn waves: Multiple scattering of upper mantle whispering gallery phases. *J. Geophys. Res.* **108**, 2460, doi: 10.1029/2003JB002487.
Nielsen, L., Thybo, H., Levander, A., Solodilov, N. (2003). Origin of upper-mantle seismic scattering—evidence from Russian peaceful nuclear explosion data. *Geophys. J. Int.* **154**, 196–204.
Patizig, R., Shapiro, S., Asch, G., Giese, P., Wigger, P. (2002). Seismogenic plane of the northern Andean Subduction Zone from aftershocks of the Autofagasta (Chili) 1955 earthquake. *Geophys. Res. Lett.* **29**, doi: 10.1029/2001GL013244.
Robertsson, J.O.A., Blanch, J.O., Symes, W.W. (1994). Viscoelastic finite-difference modeling. *Geophysics* **59**, 1444–1456.
Ryberg, T., Tittgemeyer, M., Wenzel, F. (2000). Finite-difference modelling of P-wave scattering in the upper mantle. *Geophys. J. Int.* **141**, 787–800.
Tittgemeyer, M., Wenzel, F., Ryberg, T., Fuchs, K. (1999). Scales of heterogeneities in the continental crust and upper mantle. *Pure Appl. Geophys.* **156**, 29–52.
Tittgemeyer, M., Wenzel, F., Fuchs, K. (2000). On the nature of Pn. *J. Geophys. Res.* **105**, 16173–16180.
Utsu, T. (1966). Regional difference in absorption of seismic waves in the upper mantle as inferred from abnormal distribution of seismic intensities. *J. Fac. Sci. Hokkaido Univ. Ser. VII* **2**, 359–374.
Utsu, T. (1967). Anomalies in seismic wave velocity and attenuation associated with a deep earthquake zone (I). *J. Fac. Sci. Hokkaido Univ. Ser. VII* **3**, 1–25.
Utsu, T. (1969). Anomalous seismic intensity distributions in western Japan. *Geophys. Bull. Hokkaido Univ.* **21**, 45–51 (in Japanese).
Utsu, T., Okada, H. (1968). Anomalies in seismic wave velocity and attenuation associated with a deep earthquake zone (II). *J. Fac. Sci. Hokkaido Univ. Ser. VII* **3**, 65–84.
Yamazaki, F., Ooida, T. (1985). Configuration of subducted Philippine Sea plate beneath the Chubu district, central Japan. *Zishin* **38**, 193–201 (in Japanese).

LABORATORY EXPERIMENTS OF SEISMIC WAVE PROPAGATION IN RANDOM HETEROGENEOUS MEDIA

OSAMU NISHIZAWA AND YO FUKUSHIMA

ABSTRACT

Subsurface structures contain small-scale random heterogeneities which generate scattered waves. Considerable fluctuations appear in seismic waveform when heterogeneity scales are comparable or a little smaller than the dominant seismic wavelengths. Waveform fluctuation by scattering can be examined in laboratory experiments by using random heterogeneous materials as scale models. A laser Doppler vibrometer is used to accurately record waveforms propagating through a model specimen. By taking advantages of laboratory experiments, one can reveal some quantitative relationships between wave fluctuations and the intensity or the characteristic scale length of random heterogeneity. Variations in travel times, fluctuations of amplitude, phase, and particle-motion, as well as envelope formation are examined with respect to the statistical properties of random heterogeneities. The variations can be characterized in terms of the scale-invariant values, ka and kL, the wavelength-normalized values for the characteristic scale length of heterogeneity and the wave travel distance, respectively. On the basis of experimental results, we obtain the boundary between equivalent homogeneous media approach and scattering random media approach. The two different approaches come from two different properties of the same medium in wave propagation problems, depending on the values of ka and kL. The boundary is critical in seismic imaging techniques, because strong scattered waves degrade the seismic signals used for imaging and deteriorate the image quality.

KEY WORDS: Laser Doppler vibrometer, scattering, scale-model experiment, seismic waves in random media, equivalent homogeneous media © 2008 Elsevier Inc.

1. INTRODUCTION

Laboratory scale-model experiments have been conducted to study seismic wave propagation in random heterogeneous media. Wave frequencies of laboratory experiments are mostly from 0.1 MHz to 1 MHz. Piezoelectric transducers are commonly used to study elastic waves as both sources and receivers. However, when piezoelectric transducers are used as receivers, they have the following limitations for accurate waveform recording (Nishizawa et al., 1998):

1. Their frequency response is non-flat.
2. The size of the transducer is comparable to the wavelength.
3. Wave field becomes complicated because of a surface topographic change due to the attached transducers.

The first problem can be solved by employing calibrated transducers, but the second and the third problems are inherent to piezoelectric transducers. For example, waves with P-wave velocities in the 3–7 km/s range (a typical range for laboratory model specimens) correspond to the wavelength 3–70 mm for 0.1–1 MHz frequency range, which are comparable to the size of the transducer. In this situation, the transducer response is an average response over the surface area of the transducer. The phase differences between two points in short distances less than a transducer size cannot be detected. The third limitation is associated with perturbation of wave field including the scattered/diffracted waves produced by topographic bumps of sample surface that are resulted from attached transducers. The second and third limitations are serious because they disable to detect accurate vibrations produced by elastic waves. Because of those reasons, piezoelectric transducers are not optimally suited for laboratory scale-model experiments (Nishizawa et al., 1997; Scales and Malcolm, 2003).

The laser Doppler vibrometer (abbreviated as LDV) is an easy-to-use high-precision vibration sensor that overcomes the limitations of the piezoelectric transducers (Yamamoto et al., 1992; Nishizawa et al., 1997; Scales and van Wijk, 1999). LDV irradiates the surface of an object with a laser beam. Elastic wave vibrations are detected as Doppler shifts in laser frequency (f_D) in the reflected beam. The particle velocity V of the beam-reflecting object is given by (Nishizawa et al., 1997, 1998):

$$f_D = \frac{2V}{\lambda_L}, \qquad (1)$$

where λ_L is the laser wavelength (633 nm for He–Ne laser). The laser-beam irradiation area is less than 50 μm in diameter, which assures the response area much smaller than the elastic wavelength. Scale-model experiments offer a promising approach to study seismic wave propagation in heterogeneous media, by using LDV as a wave sensor.

In this study, we focus on irregularly shaped heterogeneities, not the heterogeneities made up of layers or blocks. Irregularly shaped structures are found almost everywhere in the earth's interior as shown by the three-dimensional images revealed by seismic methods. Those images are mostly large-scale heterogeneity of which characteristic scale lengths are larger than the seismic wavelengths used for surveys. However, well-log surveys reveal that subsurface structures contain irregularly shaped small-scale heterogeneities (Shiomi et al., 1997; Goff and Holliger, 1999, 2003; Wu et al., 1994) that are below the resolution limits associated with the seismic wavelength.

In the following sections, "random media" refers to materials with irregularly shaped structures of which characteristic scale lengths extend from long wavelengths to short wavelengths compared to the seismic wavelength. The terms "small-scale" and "large-scale" are defined by the relationship of seismic wavelength to the characteristic length of heterogeneity through which waves propagate. Scale-invariant values should be used when interpreting experimental results and applying the results to the real scale problems of field observations.

We present experimental studies of elastic wave propagation in random media. Wave fluctuations in random media are shown by the wavelength-normalized relationships between the characteristic parameters of random medium and the travel distance.

We begin with a technique for characterizing random media and an illustration of LDV system for waveform measurements. We then present observations that show how random heterogeneities affect travel-time fluctuations, amplitude and phase fluctuations, distortions in shear-wave particle velocity, and wave envelope formations. We also show effects of cracks on wave fluctuations. Crack effects become more pronounced when rocks contain oriented cracks. In the final section of the chapter, we suggest the idea for applying laboratory results to seismic data analyses.

2. LABORATORY EXPERIMENTS

2.1. Statistical Description of Heterogeneity

Rocks are typical and easily available random media consisting of random mixtures of irregularly shaped mineral grains. We first present a method to determine the randomness parameters of rock specimens. The randomness parameters allow us to obtain quantitative relationships between waveform fluctuations and random heterogeneities, on the basis of laboratory scale-model experiments (Nishizawa et al., 1997; Scales and Malcolm, 2003).

Figure 1 shows a three-value image of mineral distribution in granite converted from a surface photo image. The white, gray, and black colors represent feldspar, quartz, and biotite grains, respectively. Assigning seismic velocity (the P-wave velocity, V_P) of mineral for each color, we obtain one-dimensional seismic velocity distributions along traverses across the sample. Because biotite shows strong velocity anisotropy (Aleksandrov and Ryzhova, 1961), we give directional variations of seismic velocity in biotite by assuming a random lattice orientation. One-dimensional fractional velocity fluctuations are then obtained by subtracting each averaged velocity along the traverse. Spectra of the twenty different traverse lines are calculated by the fractional velocity fluctuations, and the averaged spectra and auto-correlation functions are obtained (Fig. 2). Figure 2(a)–(c), (a′)–(c′), (a″)–(c″) shows, respectively, the image, the spectra, and the auto-correlation

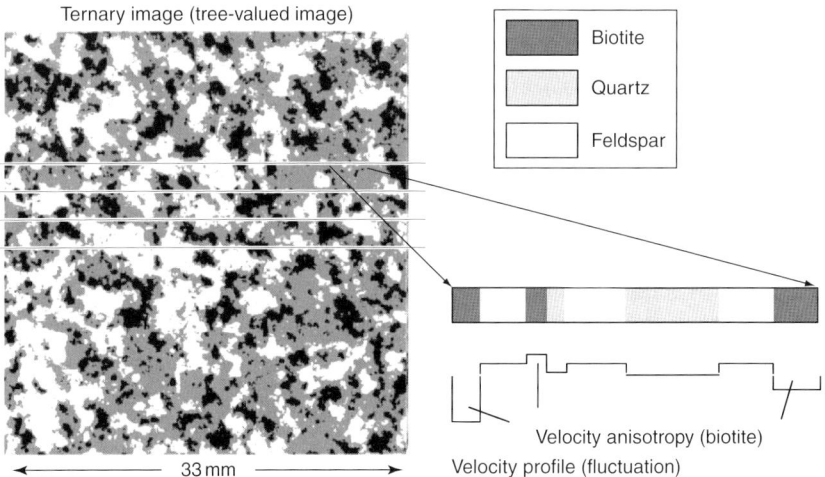

FIG. 1. Mineral grain distribution in the Oshima granite as a three-valued surface image. Velocity fluctuation is obtained by assigning velocity for each mineral by incorporating velocity anisotropy in biotite (Spetzler et al., 2002).

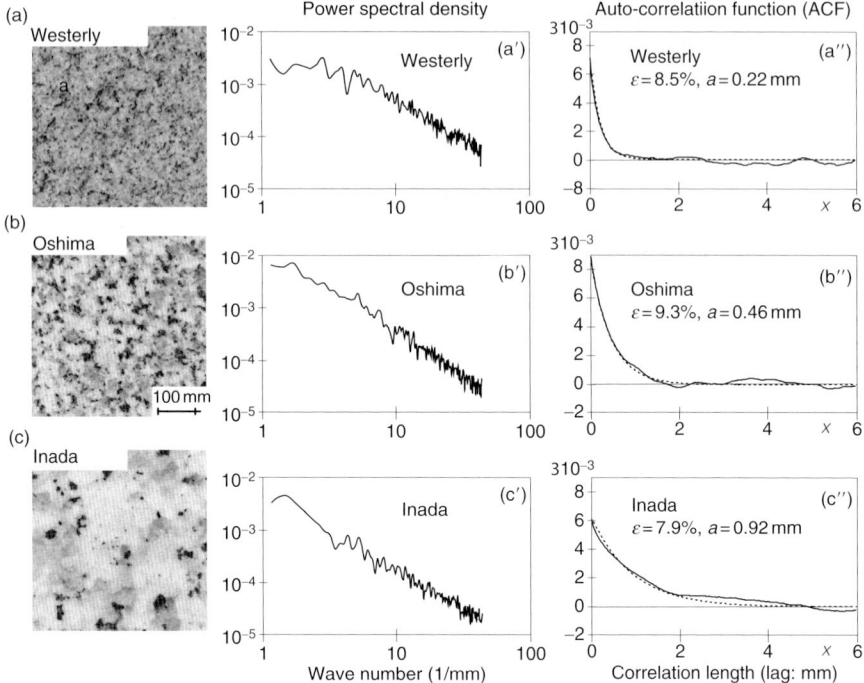

FIG. 2. Surface images, power spectral density, and auto-correlation functions (ACF) in the Westerly, Oshima, and Inada granites. Power spectral densities and ACF are calculated from the V_P (P-wave velocity) fractional fluctuations in the granite rocks. The dashed curve shown on the observed ACF indicates the best-fit exponential function (Sivaji et al., 2002).

functions of the fractional velocity fluctuations for the (a) Westerly, (b) Oshima, and (c) Inada granites.

The averaged auto-correlation is fitted to the exponential function with respect to the one-dimensional distance x:

$$C(x) = \varepsilon^2 \exp\left(-\frac{|x|}{a}\right), \qquad (2)$$

where ε is the root mean square (RMS) of the fractional velocity fluctuation and a is the correlation length. We characterize the randomness of three granites by two parameters: ε and a.

We may have to characterize the heterogeneity associated with cracks because the rocks contain very thin cracks. However, most of the cracks in intact crystalline rock samples are inter-grain or intra-grain cracks having very small apertures (10^{-2} μm to several 10th-μm) and similar lengths as the grain sizes (Simmons et al., 1975; Wong et al., 1989; Sano et al., 1992). We cannot detect cracks from the surface image because crack apertures are too small to be imaged. We therefore consider that the cracks in granite will not affect estimation of the characteristic scale length of heterogeneity, but may cause a slight increase of the fluctuation intensity ε^2.

The role of cracks is more pronounced in the shear wave. This will be discussed later in the context of the three-component vibration measurements made in the Oshima granite. Cracks in the Oshima granite have a preferred orientation (Subsections 4.3 and 4.4). In the discussion about P-wave fluctuations, we assume that the random media is isotropic or nearly isotropic, and the medium can be characterized by a and ε. The three granites can be differentiated only by the parameters a since there is only a little variation in ε. Thus, the differences observed in wave fields in the three granites can be attributed to differences in a.

Most of subsurface structures are primarily approximated by layered structures. Layered structure is also quite common for sedimentary and metamorphic rocks which exhibit planar microstructures such as bedding in sedimentary rocks and foliation in metamorphic rocks. It is therefore reasonable to consider different values of a and ε in the horizontal and the vertical directions for modeling subsurface structures (Kravtsov et al., 2003; Saito, 2006a,b). However, for simplification, we consider only the isotropic case where a and ε are equal for all directions. Later in the Subsections 4.3 and 4.4, we focus on the effects of oriented cracks in the Oshima granite where the cracks behave as oriented discontinuous scatterers embedded into an isotropic random medium.

2.2. Wave Fields in Random Media

Figure 3 shows an experimental setup. A one- or two-cycle sine-wave pulse is produced by a wave synthesizer and is fed to a piezoelectric transducer after power amplification. A wave-recording system captures transmitted wave before the onset, triggered by a pulse from the synthesizer. A laser beam is irradiated on the opposite side of the transducer's surface to measure the surface vibration.

The LDV generates high-frequency electronic circuit noise, which masks the elastic waves in ultrasonic frequency range, 0.1–1 MHz. When waveforms are perfectly reproducible, the electronic noise is reduced by waveform stacking. We repeat measurements about 1000–2000 times and stack waveforms to improve signal-to-noise ratio (Nishizawa et al., 1997).

Vibrations are measured through an optical lens unit which is used for both beam irradiation and detection. Measured surface vibration is the vertical component in the normal irradiation. An autoregressive (AR) reflection sheet (Fig. 4) attached to the surface reflects the beam parallel to the incident direction. This enables to obtain waveforms in the three orthogonal directions by irradiating laser beam in more than three directions. As will be shown later, we obtain three-component waveforms (Nishizawa et al., 1998; Fukushima et al., 2003).

Figure 5 (b) and (c) illustrates snapshots of wave fields in the three-granite rocks. The measured vibration is vertical to the surface. Samples are square prisms 300 mm × 300 mm on a side. These surfaces are referred to as the major surfaces. For all the specimens, the distances between the major surfaces range 90–100 mm [Fig. 5(a)]. The early part of the wave measured near the center of a major surface is free from diffracted waves generated from edges and vertices of the sample since the traveling distances of such diffracted waves are much larger than the distances of transmitted direct P and S waves and the twice reflected P waves (denoted as $PP'P''$, hereinafter). This allows an assumption that the waveform variations in the transmitted and reflected waves are due to the fluctuations caused by random heterogeneity in the specimens.

Waves are generated by a piezoelectric transducer attached at the center of one of the major surfaces. Waves are observed at 2 mm intervals within a 50 mm × 50 mm lattice

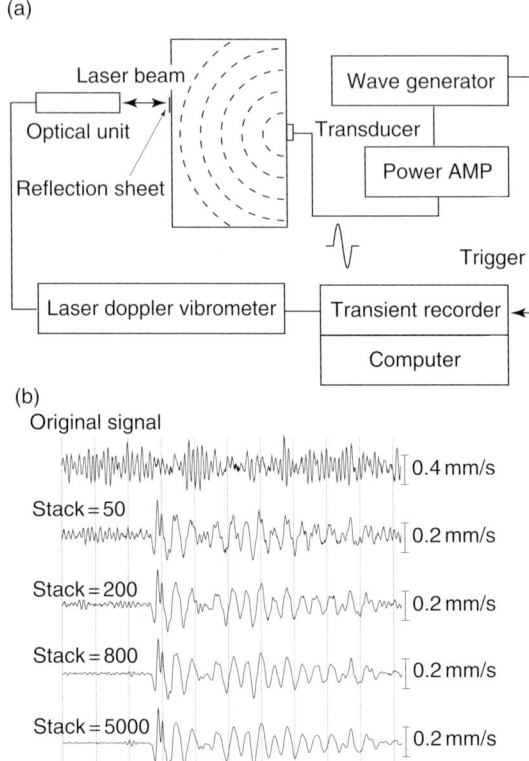

FIG. 3. (a) A schematic illustration of LDV used for measuring transmitted waveforms. (b) Original signal and stacked signals, which illustrate improvement of signal-to-noise ratio by increasing the number of stacking of repeated observations (Nishizawa et al., 1997).

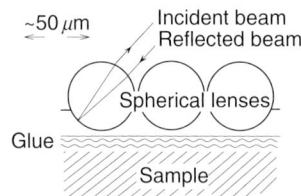

FIG. 4. Autoregressive reflection sheet. The sheet is covered with sorted glass spheres of about 50 μm in diameter. When the laser beam passing through a glass sphere, the directions of the incident and reflected beams coincide with each other (Nishizawa et al., 1997).

on the opposite surface of the transducer. One of the corners (O) of the 50 mm × 50 mm observation area is located at the intersection of the elongated axis of the disk-shaped transducer with another major surface. A disk-shaped transducer with a thickness-dilation mode produces axially symmetric energy radiation patterns for the radial and

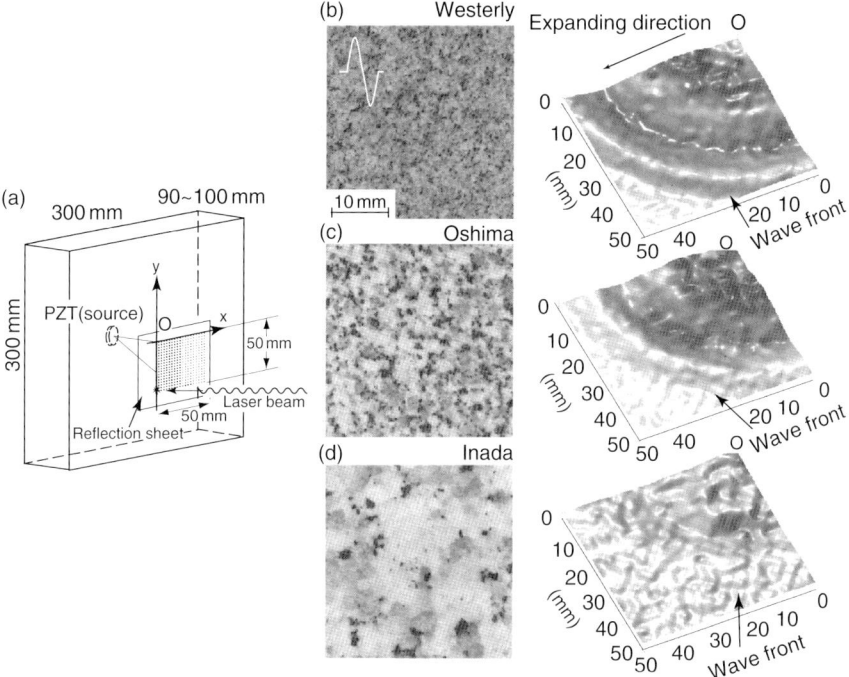

FIG. 5. (a) Observation of wave field on the sample surface. The sample is a rectangular prism with 300 mm × 300 mm major surfaces and 90–100 mm thickness. Wave fields are observed in an area 50 mm × 50 mm with the corner O located opposite the source. Surface images of the (b) Westerly, (c) Oshima, and (d) Inada granites are shown in the left column, together with snapshots of observed wave field for the transmitted P wave in the right column. Source signal is a 0.5-MHz sine-wave burst. The emitted wave first arrives at the corner O and spreads radially outward from O. The length scale and wavelength of P wave are shown on the image of Westerly granite.

the transverse (in the plane including the axis) vibrations as shown in Fig. 6. After arrival of the transmitted P wave at the corner O that is the first arrival point of the transmitted wave, wave spreads in a concentric manner from the point O.

Wave fluctuation is recognized as collapse of the concentric waveform. Waveform collapsing is most pronounced in the Inada granite, followed by the Oshima granite and then the Westerly granite that shows a concentric circle pattern. The order is associated with the correlation length of the random heterogeneity a [Fig. 5(b)–(d)]. The wave front is more poorly resolved with increasing of a.

Waveform fluctuation can be studied by comparing waveforms that propagate along identical ray paths but different locations in a random medium. In this case, waveforms are expected to be identical if the medium is homogeneous, but the waveforms may fluctuate if the medium contains random heterogeneity. One of the simplest cases is to use a source having an axially symmetric radiation pattern and observe waveforms by a circular array of which center is located on the symmetric axis.

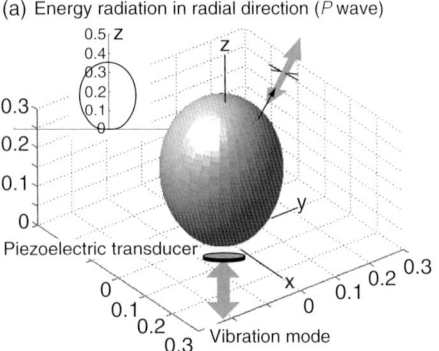
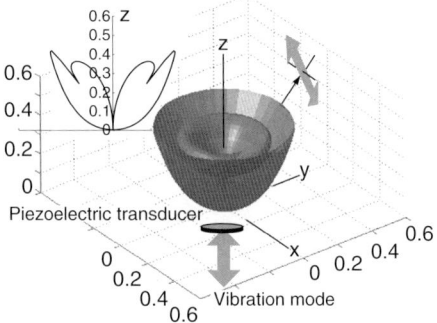

FIG. 6. Amplitude radiation patterns from a disc-shaped transducer with a thickness vibration mode mounted on steel for (a) radial component and (b) transverse component. The disc diameter is 5 mm, and the frequency is 0.5 MHz. An arbitrary scale is used to depict relative directional variations of amplitude. A cross-sectional view is presented in each 3D perspective view.

Figure 7(a) illustrates a transducer and a circular array located on a specimen. Figure 7(b) illustrates a sectional view of ray paths in a homogeneous medium for the direct P and S waves and the reflected P wave ($PP'P''$). The source–receiver geometry assures an identical waveform between receivers for an axially symmetric source if the medium is homogeneous. To compare waveforms in rocks with the waveform in a homogeneous medium, we first observed waveforms in steel, which is shown in Fig. 8. The circular array in steel and rocks has a radius of 10 mm. One hundred eighty waveforms were observed at two degree intervals. Waves show high coherence between each observation point: P, S, and $PP'P''$ are very clear. Then the waveforms can be used as a control data set for comparing the waveforms in rocks. Slight variations in waveforms represent the perturbations associated with the beam positioning error and electronic circuit noise. Figure 9 shows circular-array waveforms in the three granites. Considerable variations in waveforms are apparent. Coherence diminishes as the correlation distance a increases. In the following analysis, we use the waveforms presented in Fig. 9 in addition to other circular-array observations.

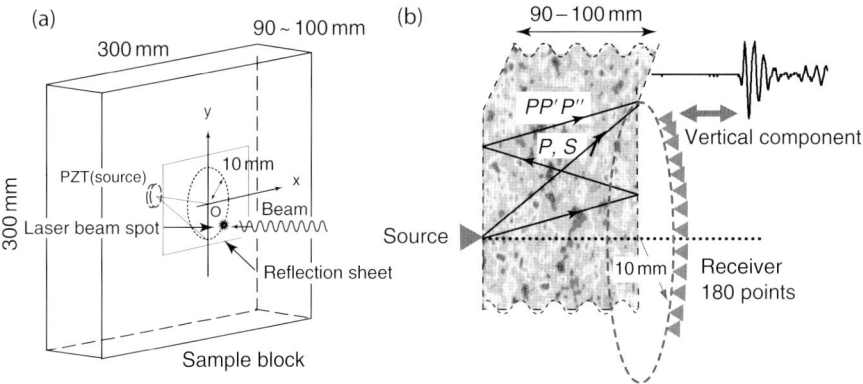

FIG. 7. Illustration of the layout of a circular array and source on the major surfaces of a sample: (a) sample block and (b) partial cross-sectional view. The circular array consists of 180 observation points. In (b), ray paths of the direct P, and the multiple of P wave ($PP'P''$) are shown. Measured vibrations are the component normal to the major surface.

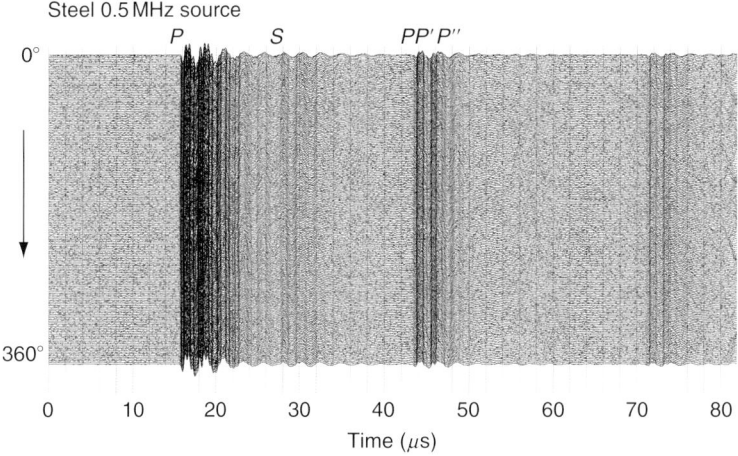

FIG. 8. A wave gather along a circular array for the steel sample. The array has a radius 10 mm and 180 observation points. Source signal is a 0.5 MHz single shot sine-wave burst (Nishizawa and Kitagawa, 2007).

3. SCALE-INVARIANT EXPRESSION

Wave frequencies in laboratory experiments are much higher than those in field seismic surveys. Laboratory results are interpreted in terms of scale-invariant values that allow one to extrapolate the high-frequency laboratory observations to the low frequency observations encountered in field seismic explorations or earthquakes. Aki and Richards (1980) classified the approaches to wave propagation in random media

228 NISHIZAWA AND FUKUSHIMA

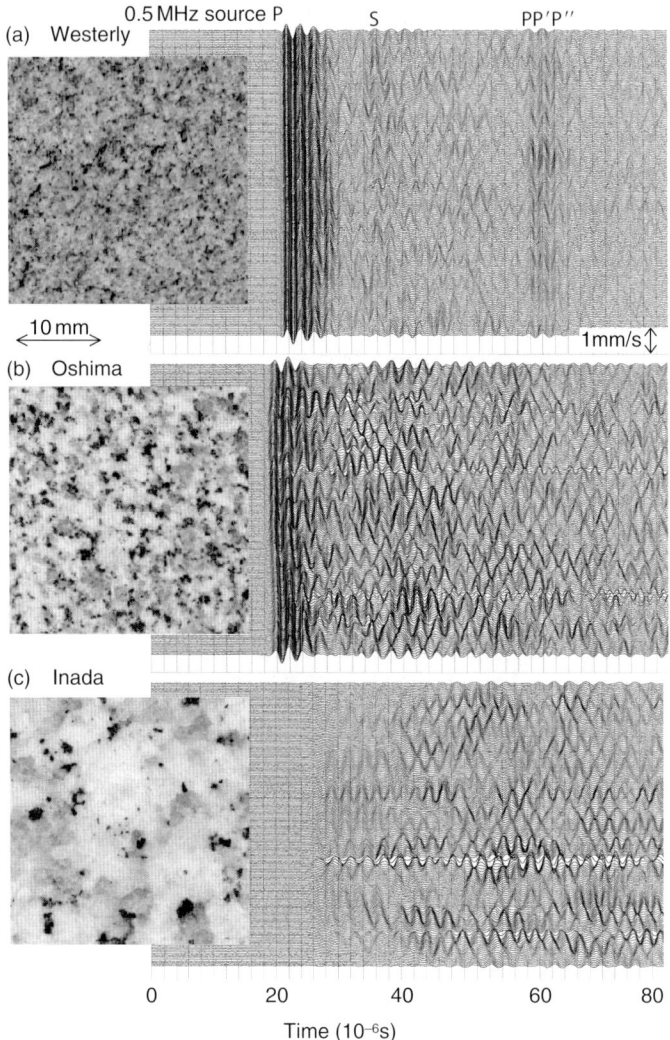

FIG. 9. Wave gathers along a circular array for the granite samples: (a) Westerly, (b) Oshima, and (c) Inada. Surface images of the three granites are shown in left. Source and array configurations are same as those for the steel sample. Source signal is same as that used for the steel experiment (Sivaji et al., 2002).

as shown in Fig. 10. The classification was made by using the scale-invariant values kL and ka which are the wave propagation distance (L) and the correlation length of heterogeneity (a) multiplied by the wave number k. k is expressed as $2\pi/\lambda$ by using the wavelength λ, or as $2\pi f/c$ by using the velocity c and frequency f of seismic wave.

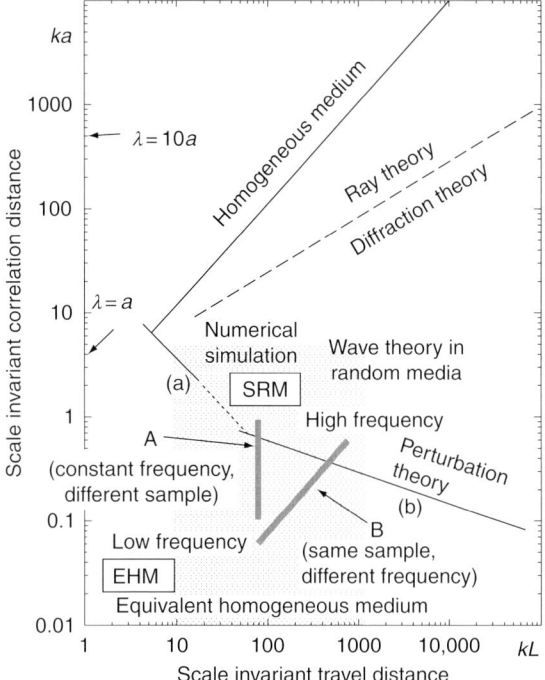

FIG. 10. An overview of approaches to wave propagation in random media. The classification of approaches is shown in terms of the scale-invariant parameters ka and kL (Aki and Richards, 1980). Laboratory scale-model experiments generally cover the shaded area. Line "A" which is parallel to the ka axis represents experiments using three-granite samples and 0.5 MHz source signal, whereas line "B" which has unit slope represents experiments using Westerly granite with different frequency source signal from 0.125 MHz to 2 MHz.

Subsurface heterogeneities could be modeled by considering different a values in horizontal and vertical directions. In such a case, wave fluctuations are affected by both a values in the vertical and horizontal directions. The diagram cannot be simply extended by changing ka values for each direction. For example, the wave propagating along the horizontal direction, which have a larger a value, strongly interact with the vertical heterogeneity having a smaller a value (Saito, 2006a). Interactions between wave and heterogeneity are therefore very complicated in a medium having different a values in different directions. This brings much more difficulties for classifying the wave propagations in a ka–kL relationship. Here, we restrict our discussion to isotropic cases where a is equal in all directions. We separate crack effects and consider that cracks are embedded in an isotropic random medium with preferred crack orientations. Cracks are considered to be discrete scatterers that become effective in high frequency.

The horizontal axis kL is associated with the number of chances of interaction between the propagating wave and scatterers. This is expressed as the number of cycles along the propagation path. The vertical axis corresponds to the sensitivity of interaction between the propagating wave and scatterers. Scattering intensity also affects the location of classification boundaries.

When ka is small, the bulk elastic properties of the medium are deemed to be equal to those of an equivalent homogeneous material (hereafter denoted as EHM). In this area, the static theory of elasticity is applicable for calculating bulk elastic constants. Most of the inclusion and crack models implicitly assume this condition (Eshelby, 1957; Budiansky and O'Connell, 1976; Nishizawa, 1982).

On the other hand, for large ka values, there is an area of geometrical optics where the ray theoretical approach is appropriate. In this region, the scale length of heterogeneity in the medium is much larger than the wavelength. Heterogeneities are regarded as a composite of homogeneous materials separated by sharp boundaries across which the velocity varies. Wave propagation is illustrated by ray paths on the basis of Snell's law. Curved ray paths are also used for continuous velocity variations.

Between the above two extreme regions, there is a region referred to as an area of scattering random medium (SRM). Theoretical approaches face difficulties for treating wave propagation of this intervening region. A number of approaches have been developed to extend from the upper (ray theory) or the lower (EHM) regions into this area. Modified ray theories are applicable for the cases close to the upper region (Müller et al., 1992; Roth et al., 1993). Some scattering theories are used for the cases close to the lower region (Sato and Fehler, 1998). In the SRM area around $kL = 10$–1000, numerical simulations (Frankel and Clayton, 1986) have to be used. Müller et al. (2002) studied in the area $ka \approx 1.5$–20 and $kL \approx 3.5$–120. Most of the numerical simulations are applicable to the large ka area. Attempts to study wave fluctuation from EHM to the middle of SRM are very limited.

In the region $ka = 0.1$–10, there are two negative-gradient boundaries between the areas EHM and SRM: -1 for $ka \gg 1$ and $-1/3$ for $ka \ll 1$, which are labeled as (a) and (b), respectively. The boundary between EHM and SRM moves upward or downward with decreasing or increasing values of ε^2, respectively. Aki and Richards (1980) assumed the intensity of the crustal heterogeneity, $\varepsilon^2 = 0.001$. The boundaries for the large and small ka values are from calculations on the basis of the plane-wave Born approximation by assuming scattering energy loss of 10% (Aki and Richards, 1980).

Laboratory scale-model experiments generally cover the shaded area in Fig. 10. For conducting different ka experiments, there are two simple alternatives: using the same wave frequency for different random media having different a but identical ε^2, or using the same random medium but with varying wave frequency. In the kL–ka plane, the former results in changes of ka for constant kL, while the latter results in changes of both kL and ka along a line with unit gradient. Both examples are presented in the following sections.

4. WAVEFORM ANALYSIS

For describing wave fluctuation in random media, we investigated the followings analysis: (1) travel-time fluctuations, (2) cross spectra of waves between different observation points, (3) three-component waveforms, and (4) wave envelope formation.

P-wave travel-time fluctuations for 180 points of the circular array in the three granites and steel are statistically investigated. Travel-time variations in random media are often referred to as the time shift (or the velocity shift) which is a decrease of travel time (an increase of velocity) with increasing frequency (Roth et al., 1993; Mukerji et al., 1995; Shapiro et al., 1996). Time shift is usually examined by numerical simulations for the ka values corresponding to the area around the boundary between SRM and ray theory in Fig. 10. The present experiments cover the area which crossing the boundary between EHM and SRM: $ka \sim 0.04$–1.0 and $kL \sim 60$–200.

Cross spectra was calculated between wave pairs observed in a circular array. The cross spectra indicate similarity between waves. Dissimilarity of wave is detected and characterized by statistically investigating the cross spectra of wave pairs observed by a circular array.

Perturbations of three-component particle velocities from the expected polarization of shear wave indicate scattering effects of random media on shear wave. Distortions in the S-wave polarization in the Oshima granite are closely related to a preferred orientation of cracks.

The total envelope formation was examined similarly to the approaches common in the analysis of field seismograms (Sato, 1984; Obara and Sato, 1995; Sato and Fehler, 1998). Shear-wave envelopes are examined in the Oshima granite and gabbro to compare the effects from characteristics of random heterogeneity and a preferred orientation of cracks.

4.1. Travel-Time Fluctuation

Sivaji *et al.* (2002) investigated P-wave travel times in steel and the three granites. They measured the P-wave onsets of 180 waveforms observed in a circular array (Figs. 8 and 9). The onset time determination is based on the AR model and the Akaike's information criterion (AIC), which determines the onset time as a separating point of different time-series AR models (Takanami and Kitagawa, 1988). Figure 11 (a)–(d) shows P-wave travel-time distributions for the granite and steel samples.

The width of the travel-time distribution in steel is interpreted as the uncertainties involved in waveform detection (beam-positioning errors and the electronic noise). The wider travel-time distributions of granites reflect the effects from random heterogeneities in rocks. Figure 12 shows the standard deviation of travel-time fluctuation (σ_T) in the three granites and steel. Increase of the standard deviation (σ_T) corresponds to the increase of correlation length a for the P-wave fractional velocity fluctuation, which are 0.22 (Westerly), 0.46 (Oshima), and 0.92 (Inada). σ_T gradually increases with increasing of ka. The slope of ka–σ_T relationship increases when $ka > 0.3$. This suggests that strong scattering appear above this value, which is expected from the strong irregularity in the wave field seen in Fig. 5. Spetzler *et al.* (2002) examined the difference between the measured travel times in random media and the travel times calculated from the average velocity values of random media. They compared the results of laboratory model experiments, numerical simulations, and the two theoretical predictions: scattering theory and ray theory. The experiments and numerical simulations suggest that the travel-time fluctuations in P wave through random media are in fairly good agreement with those predicted by scattering theory.

4.2. Cross Spectra Between Waves

Nishizawa and Kitagawa (2007) performed a statistical analysis of the cross spectra between waveform pairs of circular-array observations in Westerly granite. Source signals are the single-or double-cycle sine-wave pulses of 0.25, 0.5, 1, and 2 MHz. Dominant frequency of the observed waves changes depending on the source-signal frequencies. Time-series data are picked out from 180 waveform by using a 6-μs time

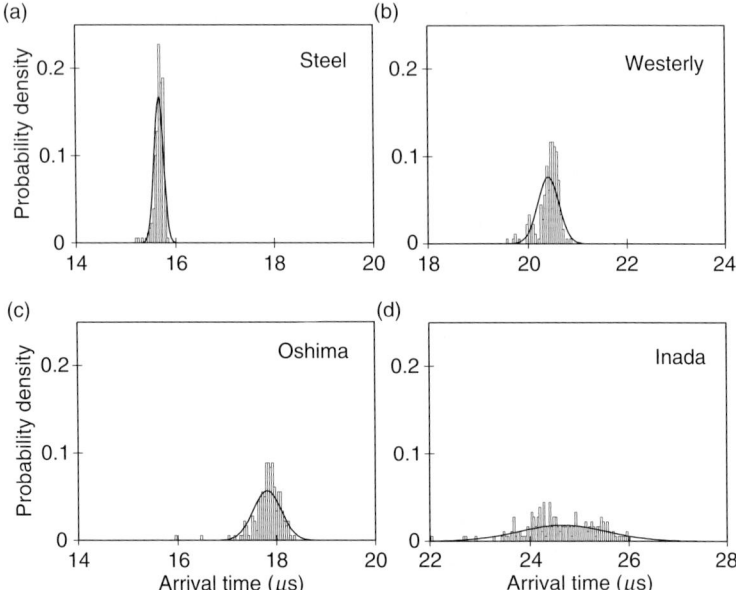

FIG. 11. Arrival time fluctuations observed along the circular array for the steel and the three-granite samples using 0.5-MHz source frequency (Sivaji et al., 2002).

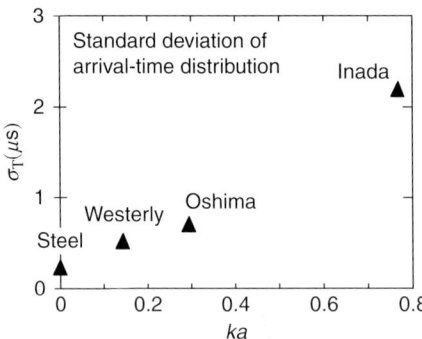

FIG. 12. Standard deviation of travel-time fluctuation for each sample plotted with respect to ka (Sivaji et al., 2002).

window at the same time point of waveform data. From the windowed wave data set, we prepare 180 data sets consisting of pairs of partial waves of which observation points are equally spaced on the circular array. Cross spectra are calculated for those wave pairs by employing the multivariate AR model (Kitagawa and Gersch, 1996). By moving the time window along the time axis of waveform, or by changing the interspacing between observation-point pairs, we obtain variations of cross spectra with respect to the time in waveform or the spatial distance in a random medium, respectively.

The cross spectrum of two waves is expressed by a matrix $p_{ij}(f)$, where f is the frequency and the suffixes $(i, j = 1, 2)$ indicate the first and second points that are equally spaced on the array, respectively. A non-diagonal element $p_{12}(f)$ or $p_{21}(f)$ represents the mutual relationship between the wave pair. Since the notation "first" and "second" is arbitrary, we use only $p_{12}(f)$ for describing cross spectrum between two waves. $p_{12}(f)$ is expressed by the amplitude $\alpha_{12}(f)$ and the phase $\phi_{12}(f)$ as

$$p_{12}(f) = \alpha_{12}(f)\exp[i\phi_{12}(f)], \qquad (3)$$

$$\alpha_{12}(f) = \sqrt{(\Re\{p_{12}(f)\})^2 + (\Im\{p_{12}(f)\})^2}, \qquad (4)$$

$$\phi_{12}(f) = \arctan\left(\frac{\Im\{p_{12}(f)\}}{\Re\{p_{12}(f)\}}\right), \qquad (5)$$

where $\Re\{\cdot\}$ and $\Im\{\cdot\}$ denote the real and imaginary parts of the term in braces, respectively.

Figure 13 (a), (b), and (c) shows the cross-spectral amplitude α_{12} (left) and the phase ϕ_{12} (right) for the wave pairs in $PP'P''$. One hundred eighty spectra are plotted to see the spectral distribution. Figure 13(a), 1–4 shows the spectra for different source-signal frequencies with a fixed interspacing denoted by the number 7, which is the number of array-point intervals between observation-point pairs. Figure 13(b) 1–3 shows the spectra for different interspacing numbers, 1, 3, and 5. Figure 13(c) shows the cross spectra of waves in steel with the interspacing number 7 and the source frequency 0.5 MHz.

Amplitude spectra have peaks near the frequency of source signal. The distributions of amplitude spectra show no characteristic changes with respect to the frequency and the interspacing. Actually phase difference can take any values, but we obtain phase values within $(-\pi, \pi)$ because of the cyclic character of phase. Phase spectra show symmetrical distributions around 0 rad. The distribution patterns of phase spectra are similar for the all source frequencies and they change with increasing the frequency.

Since the cross-spectral phase indicates phase differences between wave pairs in frequency domain, the distribution width of phase spectra is associated with probability distribution of the phase difference between waves. The narrow distribution of phase spectra in steel up to 1 MHz indicates high correlation between waveforms. This suggests that the waveforms are identical for all the wave traveling paths in steel. Spectra in steel show a small increase of the phase variation in high frequencies, which is associated with the errors of observation system. Beam positioning is sensitive to phase estimation in higher frequencies because observed phase in wave is sensitive to the location of observation point when wavelength becomes shorter. Differences in spectra between Westerly granite and steel represent fluctuation of waves in random heterogeneous media, particularly wider distributions of phase spectra in Fig. 13(a) and (b) compared to the distribution in Fig. 13(c).

Phase spectra of Westerly granite show a gradual broadenings of distribution width with increasing frequency up to around 0.5–1 MHz, and then the width increases suddenly in the frequency region 0.5–1 MHz for all source frequencies (Fig. 13 (b)-1 B, (b)-2 B, (b)-3 B, and (b)-4 B). The sudden increase of the distribution width is observed for all the phase spectra. Figure 13(b)-1 B, (b)-2 B, and (b)-3 B shows changes of phase spectra

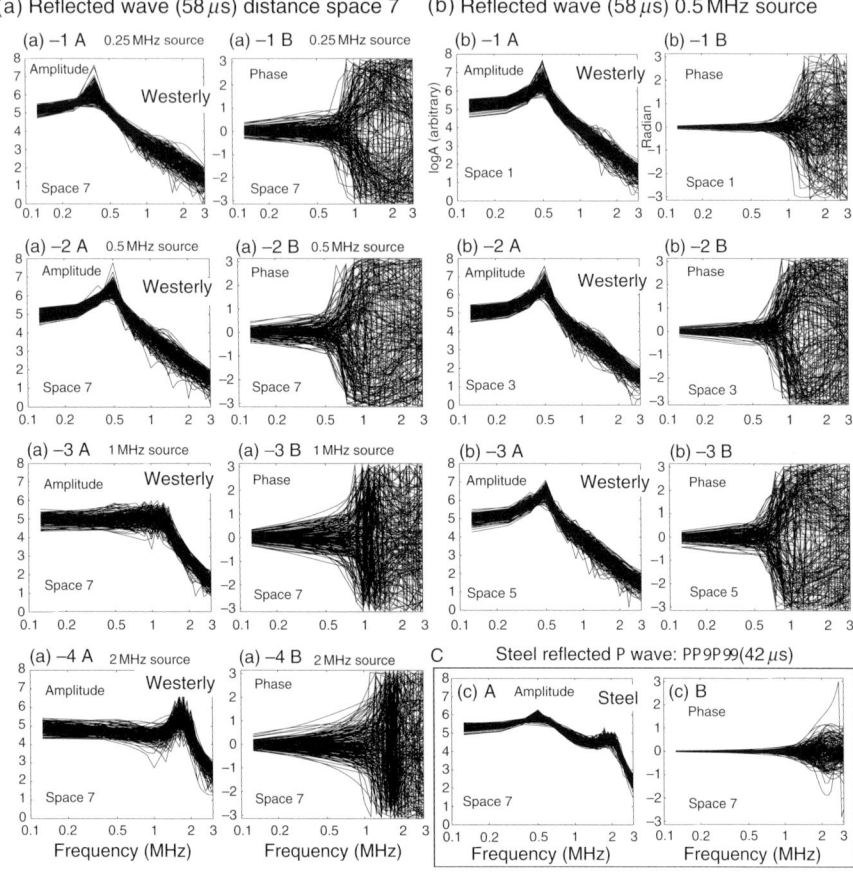

FIG. 13. Amplitude and phase spectra based on the multivariate AR model for 180 waveform pairs of the circular-array observations in Westerly granite (Nishizawa and Kitagawa, 2007).

with respect to the interspacing between observation points. In Fig. 13(b)-1 B, the distribution width shows narrow distribution in low frequencies below 0.5 MHz and increases at around 1 MHz. This suggests that waves become incoherent at high frequency even for the closest wave travel paths. The distribution width in low frequencies increase gradually with increasing interspacing. This suggests that phase coherency is gradually lost with increasing the observation distance. Loss of coherency with respect to the distance or frequency has been discussed in Nishizawa and Kitagawa (2007).

The distribution width is marked by the 67%-data area which spans from the median value of distribution to each 60th data point of positive and negative sides. The 67%-data area is shown in Fig. 14 for the phase spectra of 0.5 and 1 MHz. When the distribution is Gaussian, the 67% region roughly identical to the region inside twice the standard deviation (2σ), which includes about 68% of the total data. Since we calculated the

LABORATORY EXPERIMENTS

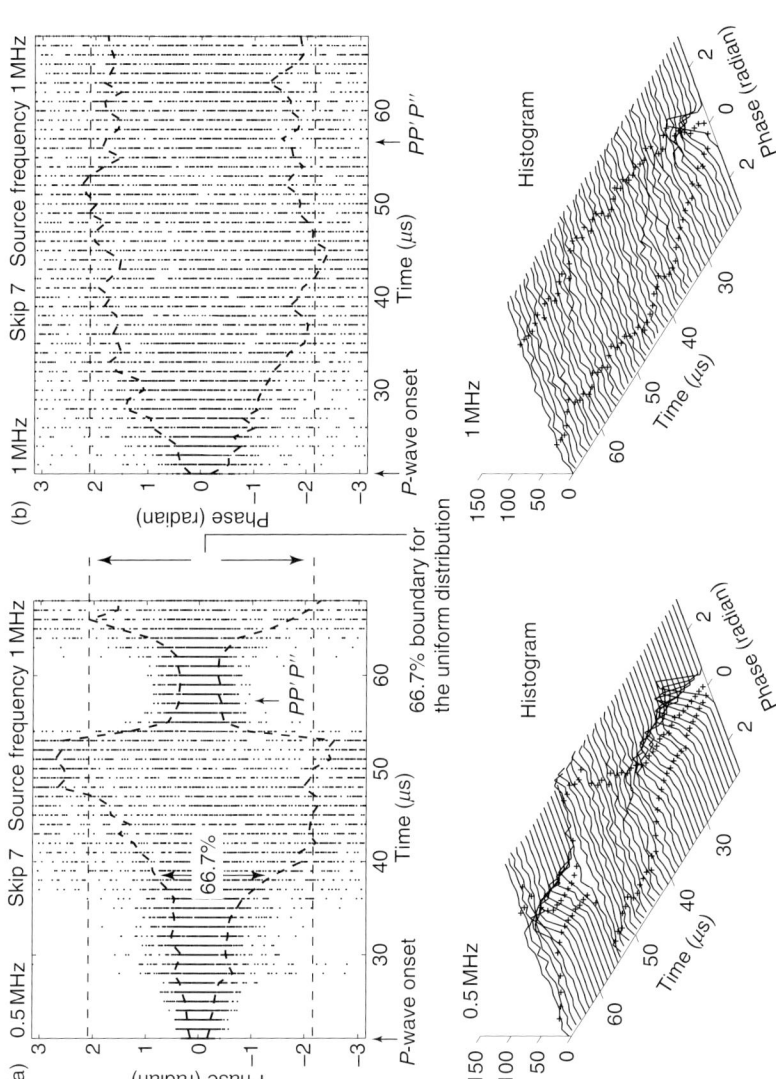

FIG. 14. Plots of phase spectral values with respect to lapse time for the frequency bands of (a) 0.5 MHz and (b) 1 MHz in Westerly granite. Source frequency is 1 MHz. The region inside the tick marks contains 67% data points (Nishizawa and Kitagawa, 2007).

phase value within $(-\pi, \pi)$, we consider a wrapped Gaussian distribution which is defined in the range $(-\pi, \pi)$ (Nishizawa and Kitagawa, 2007) by folding both tails of Gaussian distribution into $(-\pi, \pi)$. We then use the width of the 67%-data area as an index indicating loss of coherency between waves.

Figure 14(a) and (b) shows the changes of the width of the 67% region with respect to the waveform lapse time for the frequency bands 0.5 MHz and 1 MHz, which is calculated from the 1-MHz source-signal waveform data (Nishizawa and Kitagawa, 2007). The width of the 67% region increases after the P-wave onset. The increase of width just after the P-wave onset is due to the scattered waves induced during propagation. At the arrival of $PP'P''$ (54–62 μs), the 67% width for 0.5-MHz band becomes narrow, whereas this width for 1-MHz band is unchanged. The 67% width in 1-MHz band becomes close to $(-2\pi/3, 2\pi/3)$ at the lapse time around 35 μs. This shows that the phase differences between waves take random values between $(-\pi, \pi)$ (Nishizawa and Kitagawa, 2007) and waves are completely incoherent. In this case, the distribution becomes the uniform distribution between $(-\pi, \pi)$. The area inside the dashed lines in Fig. 14 is the 67% region for the uniform distribution between $(-\pi, \pi)$. If the tick marks are located near these lines, it suggests that waves are completely incoherent.

4.3. Shear-Wave Particle Velocities

As illustrated in Fig. 4, an AR reflection sheet reflects a laser beam along the incident path. The three-dimensional particle velocity can be obtained from waveforms in three independent directions. We measured vibrations in two mutually orthogonal plane perpendicular to the surface with an incident angle $\pm 45°$ off the surface normal. The measurement reproduces a three-component particle velocity with a vertical-component redundancy.

Figure 15 shows examples of S-wave particle velocities following the onset of the direct S-wave in steel, gabbro, and for the three orientations of Oshima granite. The source is a disk-shaped shear transducer that generates primarily a vibration parallel to x-axis with accompanying a weak radial vibration (Tang et al., 1994). Each observation point is located at the intersection of the disk axis and the sample surface opposite the source. The parameters of heterogeneity (a and ε for the S-wave fractional velocity fluctuation) in the Oshima granite and gabbro are given by (0.39 mm, 0.17) and (0.84 mm, 0.081), respectively. Gabbro has a large a but a small ε, whereas the Oshima granite has a small a but a large ε. The wave number k for the S wave is calculated from the velocity values listed in Table 1.

Since the Oshima granite contains oriented cracks inside quartz grains, it shows distinct velocity anisotropy (Sano et al., 1992). Most of crack planes are aligned parallel to a plane which is referred to as the rift plane. Three-component waveforms were measured for three combinations of the S-wave polarization direction and the rift-plane direction as shown in Fig. 15. Those are referred to as OS1 (polarization $\parallel$ rift) and (propagation $\parallel$ rift); OS2 (polarization $\perp$ rift) and (propagation $\parallel$ rift); and OS3 (polarization $\parallel$ rift) and (propagation $\perp$ rift), where $\parallel$ and $\perp$ denote parallel and perpendicular, respectively. The frequency shown in the figure represents the frequency of source signal.

In steel, the observed particle velocities retain the source polarization with a slight expansion along the y component at 1 MHz. As mentioned previously, the expansion is associated with inaccuracy in phase detection at high frequencies. In gabbro, particle velocities retain source polarization at 0.25 and 0.5 MHz but are appreciably distorted at

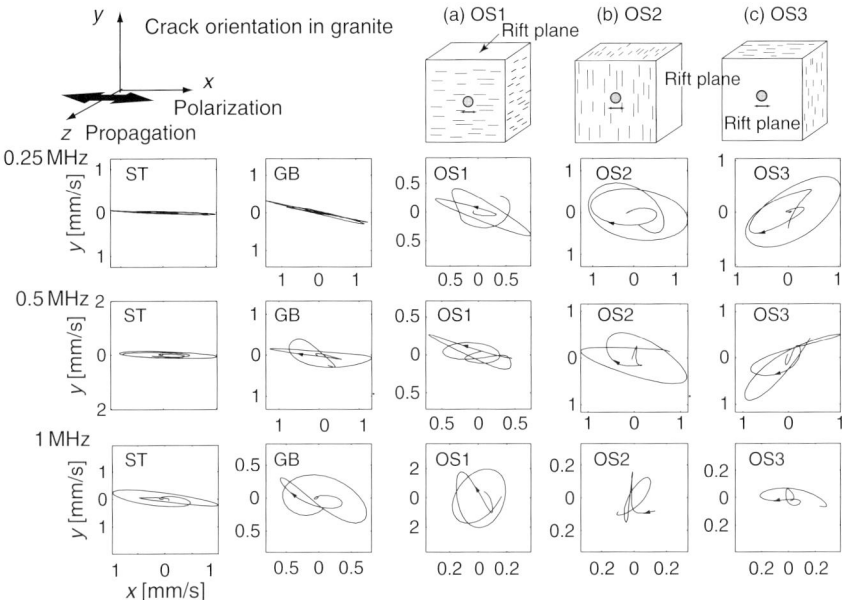

FIG. 15. Examples of S-wave particle velocities observed in steel, gabbro and Oshima granite at source frequencies 0.25, 0.5, and 1 MHz. Observations in the Oshima granite include three different cases in terms of the directions of propagation and S-wave polarization with respect to the rift (crack) plane. The S-wave source is a disk type shear vibration parallel to the surface (Fukushima et al., 2003).

TABLE 1. Velocities of P (c_P) and S (c_S) waves, and fluctuation parameters a and ε for the P and S waves, a (P), ε (P), and a (S), ε (S), respectively

Sample	c_P (km/s)	c_S (km/s)	a (P) (mm)	ε (P)	a (S) (mm)	ε (S)
Westerly granite	4.78	2.84	0.22	0.085	–	–
Oshima granite	4.81	2.56–3.0	0.46	0.093	0.39	0.170
Inada granite	3.78	2.53	0.92	0.079	–	–
Gabbro (Tamura)	4.78	2.84	–	–	0.84	0.081

1 MHz. Particle velocities in the Oshima granite show appreciable variations. Shear-wave polarization is evident in OS1 at 0.25 and 0.5 MHz with a little disturbance. Appreciable distortion appears at OS2 and OS3 at 0.25 Hz and 0.5 MHz. Polarizations at 1 MHz are generally missing or misleading for OS1, OS2, and OS3.

Three-component full waveforms observed in steel and the Oshima granite are shown in Fig. 16 (Fukushima et al., 2003). The polarization and propagation directions are x and z, respectively. In steel, signals are very weak in the y and z directions. Only the signals of x component show shear wave at arrival times of S and reflected S waves for 0.25 and 0.5 MHz, with weak disturbance at 1 MHz. Weak signals in the z component at the time of P

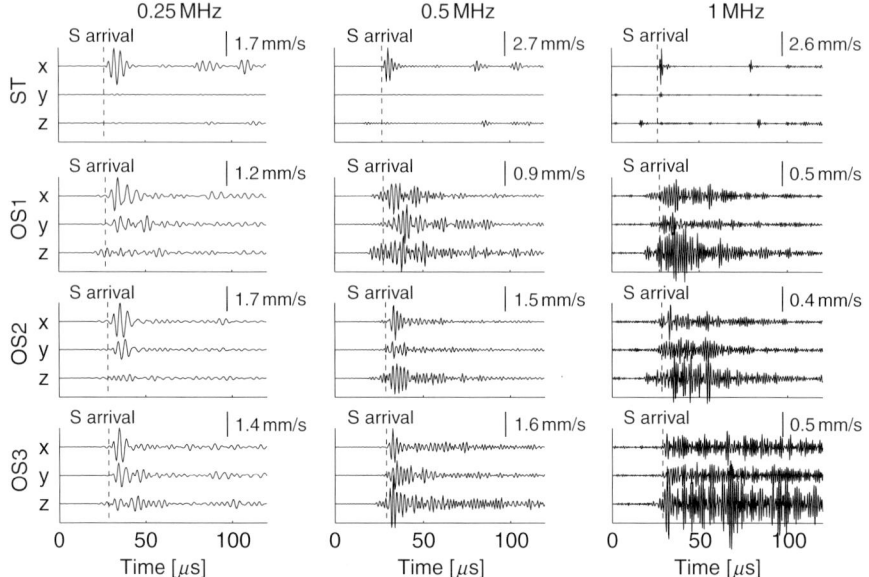

FIG. 16. Three-component waveforms from the S-wave source for three different orientations of the Oshima granite noted in Fig. 15 (Fukushima et al., 2003).

arrival and later arrivals are considered to be due to the emission of radial component from a shear source (Tang et al., 1994). Such weak vibrations can be negligible since wave fluctuations in random media are much lager than the emitted radial waves from the shear source.

In the Oshima granite, the x-component shear wave is strongly depolarized in all frequencies and orientations. Signal intensities are comparable to or even larger than the x component in both y and z directions at the time around the S arrival. The z-component signals prior to the S-wave arrival may be due to the S to P mode-converted scattered waves propagating in the forward direction. The depolarization intensity suggests strength of scattering, which increases with increasing of frequency. Energy partition into three components changes with increasing frequency. Particularly the energy partition in OS3 is characterized by larger amplitudes and long duration in the z component which is parallel to propagation direction. This suggests strong interaction between shear wave and cracks when crack planes are perpendicular to the S-wave propagation direction.

4.4. Waveform Envelope

Fukushima et al. (2003) studied envelope formation by using an eight-point circular array and a shear-wave source. The array spreading angle is 7° with respect to the axis of transducer. Actually, the radiation from the disk-shaped shear-mode transducer is not axially symmetric (Tang et al., 1994). However, the radiation energy is regarded as axially symmetric for this small spreading angle. Including the array center as the ninth observation point of the array, all the observation points are deemed equivalent.

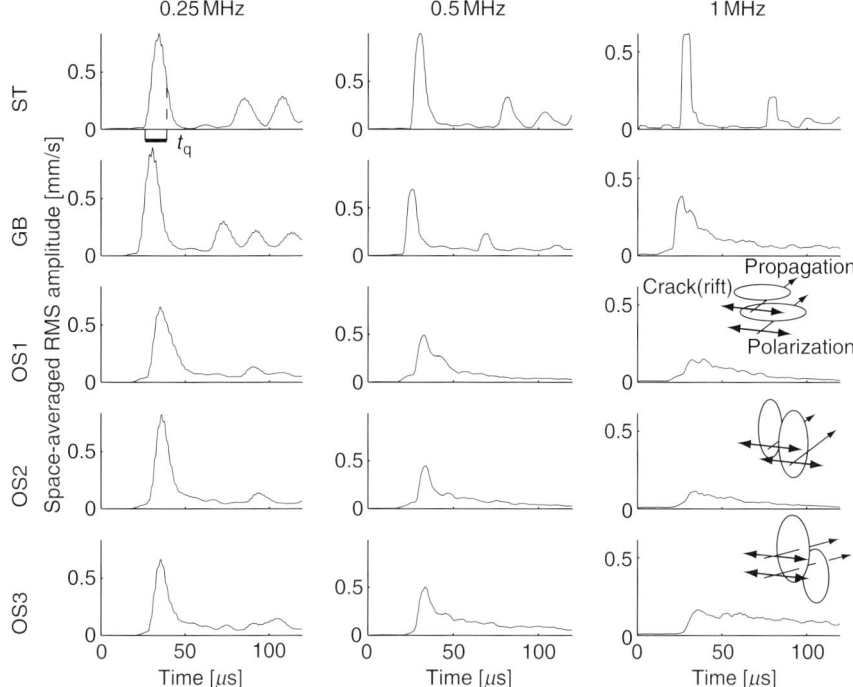

FIG. 17. RMS envelope of the shear wave (x component) observed in the steel sample and in three orientations of the Oshima granite sample (Fukushima et al., 2003).

Figure 17 shows RMS envelopes of the x component over the nine observation points for each source frequency. Samples are the same as those of the previous section: steel, gabbro, OS1, OS2, and OS3. Since Oshima granite shows velocity anisotropy generated from micro cracks (Sano et al., 1992), the values of k for the P and S waves in OS1, OS2, and OS3 show slight differences (Fukushima et al., 2003). However, the associated ka-value differences are not so large as to be considered for interpreting observed waveforms.

The half-amplitude duration from the onset of the S-wave envelope, t_q (Fig. 17), is often used as a measure of envelope broadening (Sato, 1989; Obara and Sato, 1995). To compare experimental wave envelopes, Fukushima et al. (2003) normalized the values of t_q by that observed in steel. The normalized value is denoted as $\bar{t}_q$. Figure 18 shows the change of $\bar{t}_q$ with respect to ka. The change in $\bar{t}_q$ includes the effects from scale-invariant parameters ka and kL. When frequency increases, the value of $\bar{t}_q$ increases, especially in OS3 where the S wave propagates perpendicular to the most of crack planes. The large value of $\bar{t}_q$ in the Oshima granite is primarily due to the large fluctuation intensity of S-wave velocity in the Oshima granite (ε in Table 1) and secondary due to the cracks.

The effects of oriented cracks on envelope are much pronounced in higher frequency, especially for the waves traveling perpendicular to the crack plane. This suggests strong scattering intensity for waves propagating perpendicular to the crack plane. Thus, a very

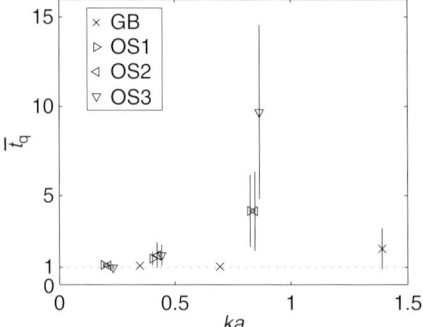

FIG. 18. Half-amplitude duration time of RMS envelope, $\bar{t}_q$, plotted with respect to ka for the gabbro sample and the three orientations in the Oshima granite sample. The vertical bars indicate the standard deviation (Fukushima et al., 2003).

small-scale heterogeneity produced by cracks has a strong effect on scattering at higher frequencies. Envelope formation is explained in connection with the diffraction from cracks and crack clusters in the next section, where intra-granular cracks have an important role.

5. Key Features of Wave Fluctuation in Random Media

5.1. Masking Signal Waves by Small-Scale Heterogeneities

Figure 19 summarizes the ka–kL relationship found in the present experimental studies. ka is obtained from the velocity values c_P and c_S in Table 1. kL is calculated from the distances between major surfaces of the sample block $d \sim 90$–100 mm. The sample thickness d corresponds to the characteristic length of the sample-scale heterogeneity which consists of a strong acoustic impedance contrast between rock and air at the sample surfaces. During propagation, the wave reflects at the distances about d and $2d$, which correspond to the sample-scale heterogeneity. $PP'P''$ is regarded as the wave produced by the sample-scale heterogeneity that is characterized by the scale length about d. The traveling distance of $PP'P''$ is a little larger than $3d$. The ka–kL relationship of $PP'P''$ for the sample-scale heterogeneity is shown by a wavy hatched zone located in the area of ray theory. The zone represents the frequency range from 0.125 MHz to 2 MHz.

The behavior of reflected wave, $PP'P''$, can be basically described by geometrical optics. However, waves are fluctuated by the small-scale heterogeneity in a rock sample. We can easily detect the reflected wave $PP'P''$ when the sample is homogeneous or regarded as EHM. However, $PP'P''$ is masked by scattered waves when scattered waves from the small-scale heterogeneity are very strong, especially, for the case of SRM. It is hard to find $PP'P''$ wave in the Oshima and Inada granites (Fig. 9). Thus, the present experiments represent a typical case of how the wave signals from the sample-scale (large-scale) heterogeneity are fluctuated by the grain-scale (small-scale) random heterogeneity during propagation.

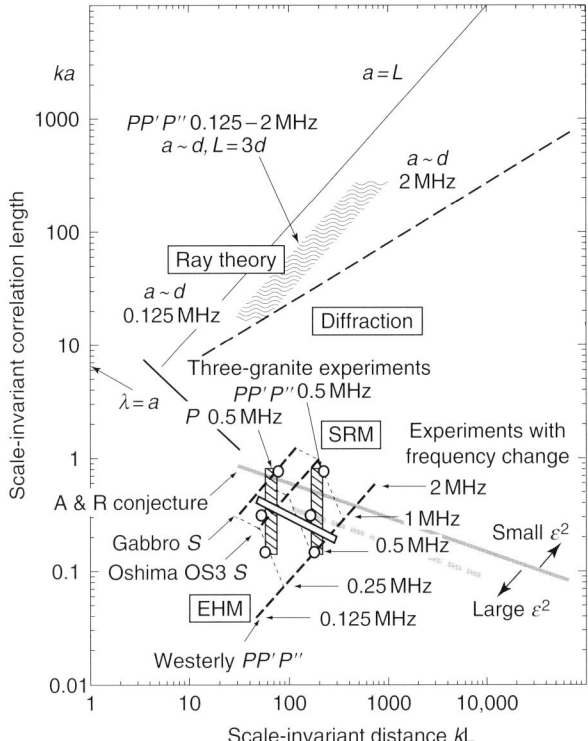

FIG. 19. Interpretation of experimental results on the ka–kL plane. The reflected waves from the sample surfaces can be treated by ray theory as shown in the wavy shaded area. Those waves are masked by scattered waves when scattering from the small-scale random heterogeneity is strong. Intensity of the scattered wave can be estimated from the ka–kL relationship for the small ka region. Present experimental results reveal the boundary between SRM and EHM regions.

Masking of signal waves may occur occasionally in field seismic observations and reduce the reliability of seismic signals produced from the large-scale heterogeneity. It is important to determine the location of the boundary between EHM and SRM in seismic data processing because the boundary illustrates how the masking effect on signal waves appears in heterogeneous media. In the following, we discuss the location of this boundary.

5.2. Boundary Between EHM and SRM

Since the sample blocks have almost the same thickness between the major surfaces (90–100 mm), each geometrical path length of P and $PP'P''$ is similar for all sample blocks. In addition to this, velocities of the three granites are almost same. If frequencies of propagating waves are equal, those yield almost the same kL value for each P and $PP'P''$ ray path. ka and kL values in the three-granite experiments are plotted as open circles on the narrow vertical hatched bands at the values of $kL \approx 65$ and 195, labeled as "three-granite experiments" in Fig. 19. The two vertical lines represent the waves P and $PP'P''$ and are labeled as "P-0.5 MHz" and "$PP'P''$-0.5 MHz," whose travel distances are about d and $3d$, respectively. ka value increases from the Westerly (bottom), then the Oshima, and the Inada (top) samples. The onset of P wave (wave front) is clear in the Westerly and Oshima granites (Fig. 9). However, for the Inada granite, the onset of P wave is much less distinct and the distribution width of the estimated P-wave arrival times increases (Figs. 11 and 12). Clear signals of the reflected $PP'P''$ wave are observed in the Westerly granite, but they are indiscernible in the Oshima and Inada granites, masked by the random wave trains. We consider that those differences represent a transition from EHM to SRM, and locate the EHM/SRM boundary as an outlined white band with a negative slope in Fig. 19.

Experiments by using the same sample but variable frequencies (0.125 MHz–2 MHz) yield a line with unit slope as shown by black dashed lines in Fig. 19. The line labeled as "Westerly $PP'P''$" represents $PP'P''$ wave in the Westerly granite. This line extends from $(kL, ka) \approx (40, 0.04)$ to $(kL, ka) \approx (700, 0.6)$, crossing the EHM/SRM boundary of the Aki and Richards' conjecture (labeled as "A & R conjecture"). Wave frequencies are shown on the line. The cross phase spectra in Fig. 13 show the sudden increases of phase distribution width in the frequency range 0.5–1 MHz. This may be regarded as the EHM/SRM transition. Since the transition appears between 0.5 MHz and 1 MHz, the boundary is located between these frequencies as shown by a negative-slope shaded line. The slope of the line is assumed to be same as that of Aki and Richards' conjecture.

Other two black dashed lines labeled as "gabbro S" and "Oshima OS3 S" represent the shear-wave experiments in gabbro and the Oshima granite of OS3 orientation. The ka and kL values for different frequencies are indicated by thin dashed lines that connect each frequency value on the line "Westerly $PP'P''$." Shear-wave depolarization, three-component waveforms, and envelope broadening in the Oshima granite suggest the location of the EHM/SRM boundary at around 0.5 MHz, which is almost close to the boundary obtained from the constant-frequency experiments (white band). The S-wave envelope broadening in gabbro is small compared to those in the Oshima granite. This may suggest larger ka values for the boundary. The variation of the boundary location between gabbro and the Oshima granite is resulted from the smaller ε^2 of gabbro than that of the Oshima granite (0.0064 and 0.029, respectively; Table 1). However, strong fluctuations in the Oshima granite are not solely explained by the difference of ε^2. In addition to this difference, effects from cracks may result in strong fluctuation in the Oshima granite as will be shown in the next section. We must note that the boundary determined by S wave in the Oshima granite may reflect crack effects on scattering as well as the effects from the heterogeneity associated with mineral distribution in rock.

5.3. Diffraction of Scattered Waves

Diffraction approaches are mostly applied for plane waves (Sato and Fehler, 1998) and extended to point-source waves by Spetzler and Snieder (2001). Diffraction is effective when ka values are in the range 1–10, which are larger than those obtained from the rock microstructures (Figs. 1 and 2). The observed heterogeneity in the granite rock samples are in the range $ka = 0.04$–1, which lies below the area of diffraction approach (Fig. 10). However, the apparent envelope broadening indicated by $\bar{t}_q$ in Fig. 17 may suggest diffraction from the heterogeneity lager than that given by the rock microstructure. RMS envelopes observed for all three orientations of the Oshima granite show increases of $\bar{t}_q$ with ka approaching 1. $\bar{t}_q$ values are strongly affected by crack orientations. These may suggest that diffraction is associated with cracks, which somehow generate a large-scale heterogeneity.

In our model, cracks behave as discrete scatterers which are embedded in an isotropic heterogeneous medium. The intensity of heterogeneity due to discrete scatterers is controlled by the effective scattering cross sections of scatterers and clusters of scatterer. The distributions of crack and crack-cluster control the scale length of heterogeneity. In the Oshima granite, intra-granular cracks are dominant (Sano et al., 1992), and they exist inside quartz grains. Each crack size is comparable to each grain size. This implies that the crack distribution in the Oshima granite is controlled by quartz grain distribution. The heterogeneity due to the fractional velocity fluctuation in the Oshima granite is primarily controlled by the size and distribution of biotite because biotite produces strong velocity contrast because of its anisotropy. Meanwhile, the heterogeneity caused by cracks is strongly related to the size and spatial distribution of quartz grain.

As shown in Fig. 1, grain size of quartz is mostly larger than that of biotite. This suggests that the crack distribution has a larger scale length compared to the microstructure in the Oshima granite. The cracks inside quartz grain will make up clusters. If crack clusters further make large-scale clusters, crack clusters will result in a hierarchical structure known as a fractal (Mandelbrot, 1982). Then crack clusters produce large-scale heterogeneity of which scale length is larger than that given by the microstructure in the Oshima granite.

The effective scattering cross section given by cracks or crack clusters depends on the wavelength and the vibration mode of propagating wave. If waves are strongly scattered by cracks, the scattered waves from cracks and crack clusters will be dominant in the transmitted wave. Transmitted waves will show diffraction associated with the crack clusters if the distributions of cracks and crack clusters produce large-scale heterogeneity. The envelope broadening of S-wave in the Oshima granite may be explained by diffraction caused by crack clusters. The effect of cracks is one possibility for explaining the large $\bar{t}_q$ values in the Oshima granite at 1 MHz. Further investigations will be needed for better understanding of crack effects on scattering.

6. Conclusions

6.1. Validity of Equivalent Homogeneous Medium Assumption

Subsurface structures are often modeled as combinations of regularly/irregularly shaped homogeneous materials, or gradual and continuous velocity variations. These models implicitly assume that each small-scale heterogeneity in each material is much smaller than the characteristic scale length of subsurface structure (target heterogeneity). Each small-scale heterogeneity is treated by the EHM approach. The whole waveform is explained by the target heterogeneity on the basis of ray theory. This approach faces a problem when some of the materials contain heterogeneities of which sizes are in the SRM area. In such a case, considerable fluctuations appear in waveforms and degrade quality of the seismic signals produced from the target heterogeneity.

If we assume that the subsurface ϵ values are almost same, EHM/SRM boundary is controlled by ka and kL. The resolution of the target heterogeneity mainly depends on the seismic wavelength (frequency). To reduce the effects from scattered waves and then to apply the EHM approach, long-wavelength seismic waves are required. There is generally a limitation in the use of high-frequency seismic waves for subsurface explorations.

6.2. Random Media Effect on Seismic Data Processing

We have shown that wave fluctuation in random media is primarily characterized by frequency dependence of phase fluctuation between waves. Phase fluctuation is more pronounced when a medium behaves as SRM by increases of ka, kL, and ε^2. The shift from EHM to SRM occurs when phase fluctuations of seismic waves appear randomly between $-\pi$ and π. Waveform stacking for signal enhancement assumes that noise components are incoherent. Incoherent character of "noise" is often mixed up with wave fluctuation, but not the same (Scales and Snieder, 1998). We must distinguish the two cases: stacking of waves along the same traveling path for repeated measurements and stacking of waves along different traveling paths to enhance signals from the target heterogeneity. For the first case, elimination of the time-series random noise is expected, but for the second case, the noise to be eliminated is the fluctuation due to the random heterogeneity around the traveling paths.

The second case is of our present concern. As shown in the previous sections, wave fluctuations are primarily characterized by the frequency-dependent character of phase fluctuations. When waves fluctuate due to small-scale heterogeneities, enhancement of signal waves by heterogeneities such as geologic discontinuities is expected only for low frequency components because high-frequency signals show almost random phase fluctuation, which balances out the expected signals by stacking. Thus, the reconstructed images reveal only the presence of larger heterogeneity even for denser observation networks.

6.3. Role of Laboratory Experiments for Studying Seismic Wave Propagation

We apply state of the art technology to experimentally investigate seismic wave propagation in random media. Numerical simulation is more popular than laboratory scale experiments for studying seismic wave propagation in random media. However, at present, numerical simulations for three-dimensional small-scale random heterogeneous media require huge computer resources and are not readily available. Laboratory studies have an advantage over numerical simulations that they can easily explore the three-dimensional behavior of wave propagation. Most of the numerical simulations are the approaches from the large ka values, larger than 1, whereas laboratory experiments cover the ka values less than 1. Laboratory experiments and numerical simulations are the two different approaches that are complementary to each other in terms of heterogeneity scale (ka value). Use of LDV as a waveform detector is a key of experimental studies since it allows us to accurately observe waveforms in a very small area which enables to detect wave fluctuations by scattering from the small-scale heterogeneity.

ACKNOWLEDGMENTS

We thank T. Wilson, T. Müller, K. van Wijk, H. Sato for their valuable comments and suggestions for improving the manuscript.

REFERENCES

Aki, K., Richards, P.G. (1980). Quantitative Seismology : W.H. Freeman and Co. San Francisco.
Aleksandrov, K.S., Ryzhova, T.V. (1961). The elastic properties of rock-forming minerals, II: Layered silicates. *Izv. Acad. Sci. USSR, Geophys. Ser.* **12**, 186–189.
Budiansky, B., O'Connell, R.J. (1976). Elastic moduli of a cracked solid. *Int. J. Solid Struct.* **12**, 81–97.
Eshelby, J. (1957). The determination of the elastic field of an ellipsoidal inclusion, and related problems. *Proc. R. Soc.* **A241**, 376–396.
Frankel, A., Clayton, R.W. (1986). Finite difference simulations of seismic scattering: Implications for the propagation of short-period seismic waves in the crust and models of crustal heterogeneity. *J. Geophys. Res.* **91**, 6465–6489.
Fukushima, Y., Nishizawa, O., Sato, H., Ohtake, M. (2003). Laboratory study on scattering characteristics of shear waves in rock samples. *Bull. Seism. Soc. Am.* **93**, 253–263.
Goff, J.A., Holliger, K. (1999). Nature and origin of upper crustal seismic velocity fluctuations and associated scaling properties: Combined stochastic analyses of KTB velocity and lithology logs. *J. Geophys. Res.* **104**, 13169–13182.
Goff, J.A., Holliger, K. (2003). Heterogeneity in the crust and upper mantle, nature, scaling, and seismic properties. In: Goff, J.A., Holliger, K., (Eds.), Kluwer Academic/Plenum Publishers, New York.
Kitagawa, G., Gersch, W. (1996). Smoothness Priors Analysis of Time Series Lecture Notes in Statistics, No. 116, New York: Springer-Verlag, New York.
Kravtsov, Y.A., Müller, T.M., Shapiro, S.A., Buske, S. (2003). Statistical properties of reflection travel-times in 3D randomly inhomogeneous and anisomeric media. *Geophys. J. Int.* **154**, 841–851.
Mandelbrot, B. (1982). The Fractal Geometry of Nature. W. H. Freeman and Co., New York.
Mukerji, T., Mavko, G., Mujica, D., Lucet, N. (1995). Scale-dependent seismic velocity in heterogeneous media. *Geophysics* **60**, 1222–1233.
Müller, G., Roth, M., Korn, M. (1992). Seismic-wave travel times in random media. *Geophys. J. Int.* **110**, 29–41.
Müller, T.M., Shapiro, S.A., Sick, C.M.A. (2002). Most probable ballistic waves in random media: A weak-fluctuation approximation and numerical results. *Waves in Random and Complex Media* **12**, 223–245.
Nishizawa, O. (1982). Seismic velocity anisotropy in a medium containing oriented cracks—Transversely isotropic case. *J. Phys. Earth* **30**, 331–347.
Nishizawa, O., Kitagawa, G. (2007). An experimental study of phase angle fluctuation in seismic waves in random heterogeneous media: Time-series analysis based on multivariate AR model. *Geophys. J. Int.* **169**, 149–160.
Nishizawa, O., Satoh, T., Lei, X., Kuwahara, Y. (1997). Laboratory studies of seismic wave propagation in inhomogeneous media using a laser Doppler vibrometer. *Bull. Seis. Soc. Am.* **87**, 809–823.
Nishizawa, O., Satoh, T., Lei, X. (1998). Detection of shear wave in ultrasonic range by using a laser Doppler vibrometer. *Rev. Sci. Instrum.* **69**, 2572–2573.
Obara, K., Sato, H. (1995). Regional differences of random inhomogeneities around the volcanic front in the Kanto-Tokai area, Japan, revealed from the broadening of S wave seismogram envelopes. *J. Geophys. Res.* **100**, 2103–2121.
Roth, M., Müller, G., Snieder, R. (1993). Velocity shift in random media. *Geophys. J. Int.* **115**, 552–563.
Saito, T. (2006a). Synthesis of scalar-wave envelopes in two-dimensional weakly anisotropic random media by using the Markov approximation. *Geophys. J. Int.* **165**, 501–515.
Saito, T. (2006b). Velocity shift in two-dimensional anisotropic random media using the Rytov method. *Geophys. J. Int.* **166**, 293–308.

Sano, O., Kudo, Y., Mizuta, Y. (1992). Experimental determination of elastic constants of Oshima granite, Barre granite, and Chelmsford granite. *J. Geophys. Res.* **97**, 3367–3379.
Sato, H. (1984). Attenuation and envelope formation of three-component seismograms of small local earthquake in randomly inhomogeneous lithosphere. *J. Geophys. Res.* **89**, 1221–1241.
Sato, H. (1989). Broadening of seismogram envelopes in the randomly inhomogeneous lithosphere based on the parabolic approximation: Southeastern Honshu, Japan. *J. Geophys. Res.* **94**, 17735–17747.
Sato, H., Fehler, M. (1998). Seismic Wave Propagation and Scattering in the Heterogeneous Earth. Spring Verlag, New York.
Scales, J.A., Malcolm, A.E. (2003). Laser characterization of ultrasonic wave propagation in random media. *Phys. Rev.* **E67**, 046618(7).
Scales, J.A., Snieder, R. (1998). What is noise. *Geophysics* **63**, 1122–1124.
Scales, J.A., van Wijk, K. (1999). Multiple scattering attenuation and anisotropy of ultrasonic surface waves. *Appl. Phys. Lett.* **74**, 3899–3901.
Shapiro, S.A., Schwarz, R., Gold, N. (1996). The effect of random isotropic inhomogeneities on the phase velocity of seismic waves. *Geophys. J. Int.* **127**, 783–794.
Shiomi, K., Sato, H., Ohtake, M. (1997). Broad-band power-law spectra of well-log data in Japan. *Geophys. J. Int.* **130**, 57–64.
Simmons, G., Todd, T., Baldridge, W.S. (1975). Toward a quantitative relationship between elastic properties and cracks in low porosity rocks. *Am. J. Sci.* **275**, 318–345.
Sivaji, C., Nishizawa, O., Kitagawa, G., Fukushima, Y. (2002). A physical-model study of the statistics of seismic waveform fluctuations in random heterogeneous media. *Geophys. J. Int.* **148**, 575–595.
Spetzler, J., Snieder, R. (2001). The effect of small-scale heterogeneity on the arrival time of waves. *Geophys. J. Int.* **145**, 786–796.
Spetzler, J., Sivaji, C., Nishizawa, O., Fukushima, Y. (2002). A test of ray theory and scattering theory based on a laboratory experiment using ultrasonic waves and numerical simulations by finite-difference method. *Geophys. J. Int.* **148**, 165–178.
Takanami, T., Kitagawa, G. (1988). A new efficient procedure for the estimation of onset times of seismic waves. *J. Phys. Earth* **36**, 267–290.
Tang, X.M., Zhu, Z., Toksöz, M.N. (1994). Radiation patterns of compressional and shear transducers at the surface of an elastic half-space. *J. Acoust. Soc. Am.* **95**, 71–76.
Wong, T.F., Fredrich, J.T., Gwanmesia, G.D. (1989). Crack aperture statistics and pore space fractal geometry of westerly granite and rutland quartzite: Implications for an elastic contact model of rock compressibility. *J. Geophys. Res.* **94**, 10267–10278.
Wu, R.S., Xu, Z., Li, X.P. (1994). Heterogeneity spectrum and scale-anisotropy in the upper crust revealed by the German Continental Deep-Drilling (KTB) Holes. *Geophys. Res. Lett.* **21**, 911–914.
Yamamoto, H., Iwasaki, S., Nagayama, H. (1992). Tape displacement in operating VCR measured with laser Doppler vibrometer. *Nikkei Electronics Acta* October, 62–66.

MEASUREMENTS OF THE EARTH AT THE SCALE OF LOGS, CROSSWELLS, AND VSPs

Arthur C.H. Cheng

Abstract

This chapter will review the measurement physics and geometry of three different types of downhole seismic measurements: acoustic logging, crosswell seismic, and vertical seismic profiling. These measurements cover a frequency range from about 15 kHz down to about 10 Hz, and can investigate heterogeneity in the earth from a scale of tens of centimeters to hundreds of meters. Moreover, because of the different geometries involved, the volume of earth being investigated and the information obtained are quite different for each method. This chapter will review the latest technologies in downhole measurements and what we can obtain from the data, and how we can potentially integrate these measurements into a consistent earth model.

Key Words: borehole, acoustic logging, crosswell seismic survey, vertical seismic profiling, VSP. © 2008 Elsevier Inc.

1. Introduction

Borehole measurements provide us with direct information of the properties of the earth's interior. As such, they complement the measurements we make on the earth's surface, and provide the "ground truth" at specific locations. However, borehole seismic methods measure the earth at a different scale and with a different wave frequency than earthquake seismology and active surface seismic reflection methods. In order to integrate these measurements with earthquake and surface reflection measurements, it is important to understand the size and volume of the earth that these methods investigate. In this paper, we will discuss the vertical and horizontal resolution of each of these measurements.

In this paper, we will discuss three commonly used borehole seismic and acoustic methods and the volume of the earth they sample. These are all active source methods. The main distinguishing feature between the three is the location of the source. The methods are acoustic logging, crosswell seismic survey, and vertical seismic profiling (VSP). Acoustic logging employs a tool that has both the sources and receivers on it, separated by around 3–4 m. It investigates the property of the earth immediately surrounding the borehole. Crosswell seismic survey uses an active seismic source in one borehole and a string of receivers in another borehole usually separated by less than 500 m. It measures the earth formation between the two boreholes. VSP uses a surface seismic source and a downhole receiver string. The source can be located directly on top of the borehole, or at a distance as far as several kilometers away. In all these applications, the depth of the borehole can range up to 5 km or more. Schematic figures of the three measurement techniques are shown in Fig. 1a–c.

There is also a passive borehole seismic measurement. It involves deploying one or more receiver strings down one or more boreholes and monitoring micro-seismic events

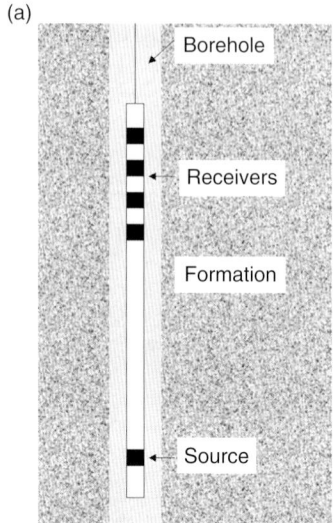
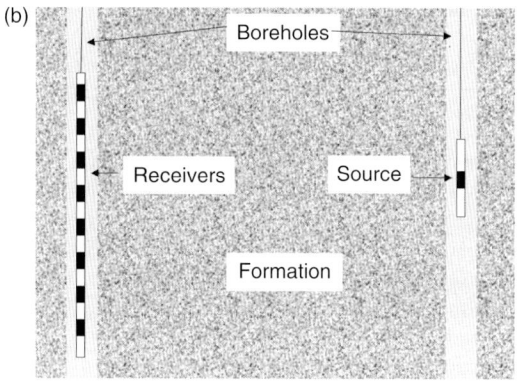
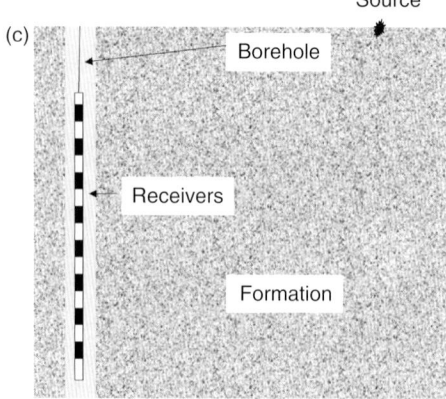

Fig. 1. (a) Schematic figure for acoustic logging. (b) Schematic figure for crosswell seismic survey. (c) Schematic figure for vertical seismic profiling.

induced from the production of the reservoir (usually but not limited to petroleum extraction) or from hydraulically induced fracturing of the formation. The scale of such a measurement depends on the size of the event. This technique is similar to earthquake location. This paper will only deal with measurement methods using active sources and will not discuss passive borehole monitoring.

2. ACOUSTIC LOGGING

Acoustic logs measure the properties of the earth immediately surrounding a borehole. An acoustic logging tool has sources and receivers embedded in a metal housing. The measurement is made by lowering the tool in a fluid-filled borehole and setting off an acoustic source (usually a pressure pulse) and measuring the refracted compressional head wave arrival at two receivers set at different distances away from the source. The formation compressional wave slowness (or the inverse of velocity) is simply given by the difference in arrival times divided by the separation between the two receivers. The resolution of such a measurement is simply given by the receiver spacing, which is typically 2 ft, or 0.6 m. (Fig. 2).

Modern acoustic logging is much more complex than the simple straightforward measurement described above. Modern acoustic logging tools generate different wave modes in the borehole, and use advanced signal processing algorithms to obtain information about the formation. A typical modern acoustic logging tool will have one or more monopole (or axi-symmetric) sources and a pair of orthogonally aligned dipole sources. The monopole sources generate a simple axi-symmetric pressure pulse in the borehole. The dipole source generates a directional displacement pulse in the borehole fluid. The tools have an array of receivers, typically eight levels (different distances from the sources), with each level having 4 receivers distributed around the circumference of the tool, positioned at about 3–4 m away from the source. The exact distance varies with different logging contractor. An example of a modern acoustic logging tool is shown in Fig. 3.

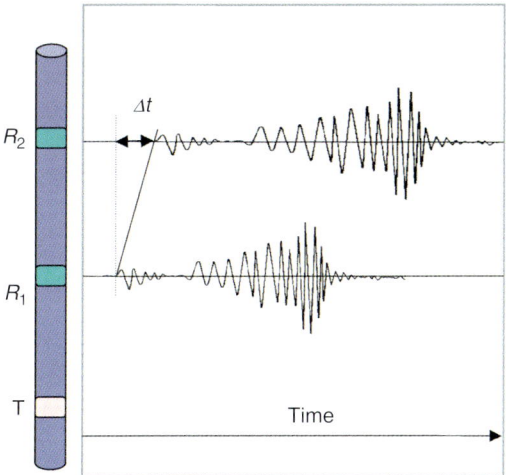

FIG. 2. Traditional measure of formation velocity from acoustic log: $v = (R_2-R_1)/\Delta t$. From Tang and Cheng (2004).

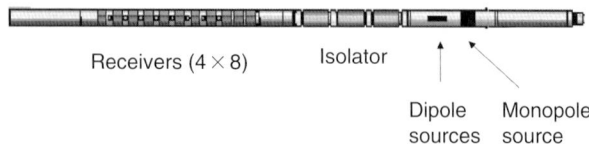

FIG. 3. Example of a modern acoustic logging tool, the WaveSonic™ from Halliburton (courtesy Halliburton Energy Services).

In order to understand the volume of investigation of acoustic logging measurements, we need to understand the physics of logging. We will start with monopole logging. Monopole logging is a direct extension of the traditional acoustic logging, with a pressure source generated in the borehole, and the resulting wavefield picked up by pressure sensitive receivers.

There are two types of waves generated in the borehole and detected by the receivers. One type is the compressional and shear head waves refracted along the surface of the borehole. The other type is the guided waves. There are two guided waves, the pseudo-Rayleigh wave and the Stoneley wave. The pseudo-Rayleigh is a properly guided wave that exists because of the borehole, while the Stoneley is an interface wave that exists because of the fluid-solid borehole interface. The head waves and the guided waves sample the formation in different manners. A monopole source generates compressional and shear head waves for compressional and shear wave slownesses. Certain applications, such as permeability estimation and fracture detection, utilize the Stoneley wave.

2.1. Dipole Logging

In monopole logging, we rely on the refracted head waves to determine the compressional and shear wave velocity of the formation around the borehole. However, in "soft" or "slow" formations, where the shear wave velocity is lower than the compressional wave velocity of the borehole fluid, there is no refracted shear head wave. Zemanek et al. (1984) introduced shear wave logging in a "slow" formation by means of the dipole logging tool. The dipole source generates a directional displacement in the borehole fluid. This excites a flexural wave mode in the borehole/formation system. The flexural wave is dispersive, but it travels at the formation shear wave velocity at the low frequency limit (usually around 2–4 kHz for nominal borehole radius and formation properties). The flexural wave is related to the guided wave modes in the borehole, and is not a head wave. This flexural wave is detected with receivers sensitive to the directional displacement in the fluid. In modern logging tools, this is achieved by the subtraction of the waveforms received by pressure transducers located on opposite side of the tool body. Since the flexural wave is a guided wave, we should bear that in mind when considering the depth of investigation of the shear wave velocity measurement from a dipole tool. It will be different from a refracted shear head wave from a monopole tool.

2.2. Modern Array Processing

How well and at what scale we can sample the formation is related intimately with the methodology we use to process the received waveforms. In the past, the formation slowness is determined simply by dividing the arrival time differences at two receivers by the distance separating them. With the advance of modern tools, we use array processing to determine the formation slownesses, including those for both head and interface waves.

The basic methodology in modern acoustic logging processing is the semblance cross-correlation, in the slowness-arrival time domain (Kimball and Marzetta, 1986). The basic algorithm involves calculating the coherence of the waveform in a window with a specific moveout across the receiver array. An example is given in Fig. 4. This process is repeated for a range of slownesses and arrival times, and plotted in the slowness-arrival time space. Peaks in the semblance coherence values are identified as arrivals.

For the semblance cross-correlation method, we are estimating the velocity of the formation over the length of the receiver array. In most modern tools, that array is made up of 8 receivers separated 15 cm (6 in.) from each other, giving an effective array length of 1.05 m (or 3.5 ft). However, we can improve on this resolution by using a particular property of the modern acoustic logging practice, namely, the data is acquired at constant depth intervals. In particular, the acoustic data is acquired at intervals of 0.3 m (6 in), corresponding to the distance between receivers.

Using successive shots and different combinations of the receivers of the array, we can estimate the velocity of the formation over a smaller interval than that of the full array. An example of such combinations is given in Fig. 5, (from Tang and Cheng, 2004). The resulting sub-arrays can be combined to obtain a high resolution estimate of the formation velocities (Hsu and Chang, 1987; Zhang et al., 2000). An example of the improvement in vertical resolution by the use of sub-arrays is shown in Fig. 6 (from Tang and Cheng, 2004).

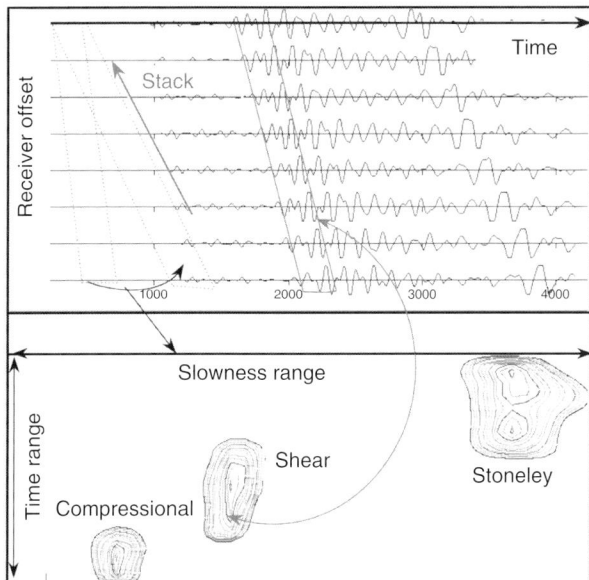

FIG. 4. Illustration of semblance coherence processing to determine formation slowness (inverse of velocity) from acoustic logs (from Tang and Cheng, 2004).

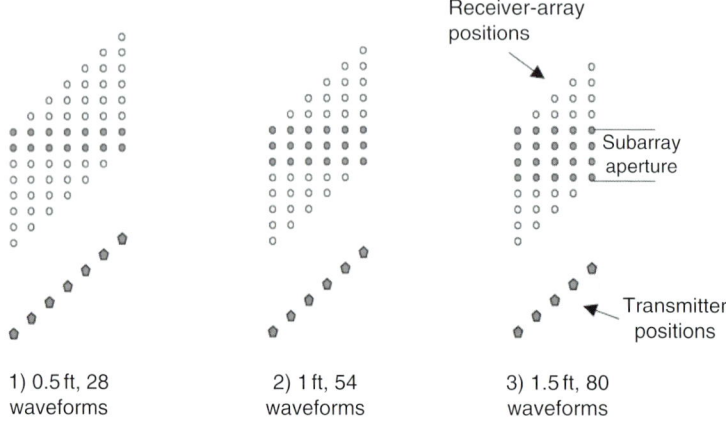

Fig. 5. Illustration of the use of sub-arrays in resolution refinement for acoustic logs (from Tang and Cheng, 2004).

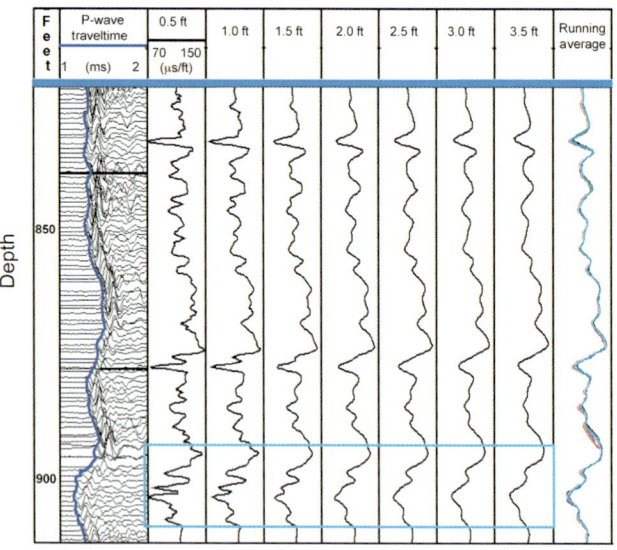

Fig. 6. An example of the enhancement of resolution using different aperture sub-arrays for acoustic logs (from Tang and Cheng, 2004). Panels for different length scales represent results from using differing sub-arrays as shown in Fig. 5.

2.3. Depth of Investigation

As discussed earlier, the depth of investigation of the acoustic logging tool is a complicated issue. There are two distinct approaches to answer this question. One is to consider the head waves, and the other the guided waves. These two approaches give different answers to the depth of investigation of modern acoustic logs.

Let's first consider the head waves. In a homogeneous formation, the depth of investigation is infinite. In a heterogeneous formation, the depth of investigation depends on the velocity variation away from the borehole. If the velocity is increasing away from the borehole, then the depth of investigation can be estimated using classical ray bending theory (Hornby, 1993; Zeroug et al., 2006). For a formation with a velocity decreasing away from the borehole, the situation is more complex. Since the fastest velocity is next to the borehole, the head wave energy is confined to a cylinder around the borehole. An estimate of the depth of investigation in such cases is that it will be approximately equal to the half wavelength of the head wave. For a P head wave at 8 kHz traveling in a formation with a velocity of 4.5 km/s (typical sandstone), the corresponding wavelength is about 0.5 m.

However, things are a bit more complicated than that. Since we are dealing with wave propagation with a wavelength on the order of the scale of the heterogeneities, ray theoretical approaches are not applicable. One way to test the sensitivity of the head waves to velocity variations around the borehole is to generate synthetic waveforms (Cheng and Toksöz, 1981; Tubman et al., 1984) and then analyze them using the standard semblance cross-correlation approach (Kimball and Marzetta, 1986). Figure 7 shows the schematic of such a synthetic formation. From the bottom to the top, the formation consists of progressively deeper changes in velocity away from the borehole into the formation. Figures 8a and b shows the results from semblance analysis for a "damaged" formation, where the near borehole velocities are lower than the unchanged formation, and those from an "invaded" formation, where the near borehole velocities are higher. The center

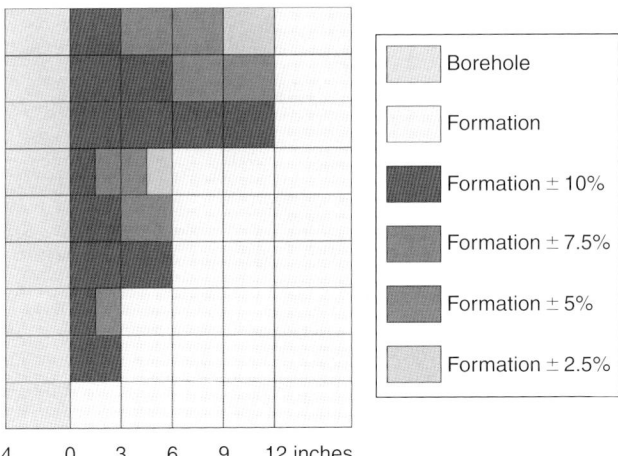

FIG. 7. Synthetic model used to study the depth of investigation of acoustic logging tools. The velocities increase or decrease away from the borehole as a percentage of the velocity of the undisturbed formation. The borehole radius is 4 in.

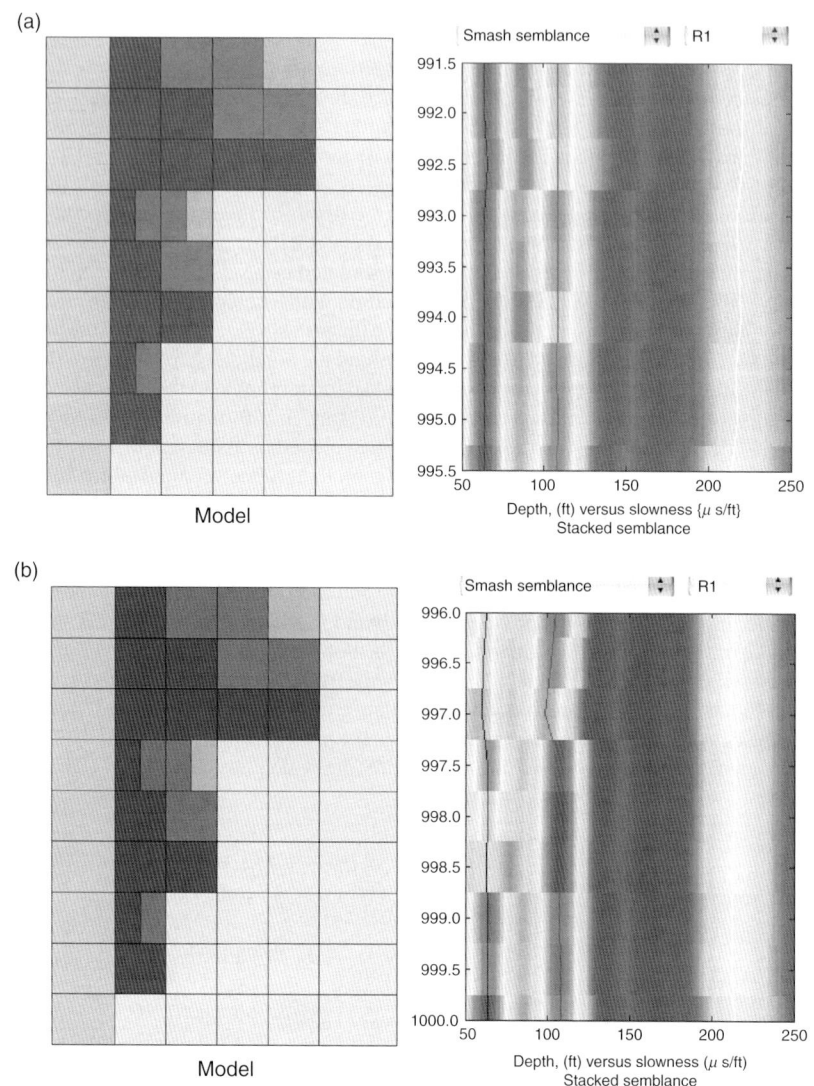

FIG. 8. (a) Results of semblance estimation of P, S, and Stoneley wave velocities (thick black, thin black, and white lines on the right panel) from the synthetic "damaged zone" model on the left. The near borehole velocities are lower than those in the unaltered formation (lowest depth). (b) Results of semblance estimation of P, S, and Stoneley wave velocities (thick black, thin black, and white lines on the right panel) from the synthetic "invaded zone" model on the left. The near borehole velocities are higher than those in the unaltered formation (lowest depth).

frequency is 8 kHz. The panels in Fig. 8 show the peaks in semblance values corresponding to the P, S, and Stoneley wave arrivals.

From Fig. 8a, we see that using semblance we can barely detect the decrease in near borehole P wave velocity even at a "damaged" zone depth of 12 in. (0.3 m) into the formation. The changes are not detectable in the S wave velocity. From Fig. 8b, we can see that, we cannot detect any changes in the measured velocities until the "invaded" zone is greater than 6 in. (0.15 m, the top three depths). This is somewhat consistent with our estimates presented above.

The depths of investigation for the guided waves (flexural and Stoneley) are still more complex. Because of the geometry of the borehole, the depth of investigation is not a simple analysis of the wavelength of the guided wave, as it is in the case of earthquake surface waves. An example in Tubman *et al.* (1984) clearly shows that the Stoneley wave "sees" mostly the casing in a cased well and not changes in the formation. Our analysis above shows similar results when the near borehole velocity is higher than the formation. For the case with slower "damaged" zone, the Stoneley wave appears to feel the effect even when the zone is only 3 in (0.125 m) deep.

The changes in near borehole velocities can be estimated using high resolution velocity dispersion analysis of the guided wave (Araya *et al.*, 2003) followed by model based inversion. Because of the complex interaction between the borehole and the propagation waves, simple analysis based on ray theoretical approaches may not always give the right answer. For the purposes of this review, it is sufficient to say that the depth of investigation of the modern acoustic logging tool usually does not go more than one wavelength of the propagating wave, and is usually of the order of 0.5 m.

3. CROSSWELL SEISMIC SURVEY

For crosswell seismic surveys, the situation is much different from the acoustic logging case. Here the source is placed in one well, with the receiver string in another well. There is no "standard" crosswell configuration, as there is in acoustic logging. Thus the resolution of a crosswell seismic survey depends somewhat on the exact geometry used.

There are two main "products" of a crosswell survey: a seismic velocity tomogram, and a reflection image. Both primarily use the body wave (usually the P wave, but one can also use the S wave) arrivals from the source in one well to the receivers in the other well. The scale of the measurement is typically in the hundreds of meters to less than a kilometer between the wells. The vertical extent is usually about the same dimension as the horizontal well separation, but these scales are changeable depending on the objective of the experiment.

There are two commercially available sources for crosswell surveys: the piezoelectric source from Z-Seis (www.z-seis.com), and the orbital vibrator (Cole, 1997) from Oyo Geospace (www.oyogeospace.com). The piezoelectric source is a swept-frequency type source, going from about 100 Hz up to 4 kHz. The orbital vibrator is also a sweeping type source, but the upper frequency is limited to around 400 Hz. Z-Seis uses a hydrophone string as a receiver string. The receivers are located nominally at 3 m (10 ft) intervals. Oyo Geospace uses a wall-locking 3-component receiver string, with variable spacing. A typical receiver separation for the Oyo Geospace string is 3 m (10 ft). This corresponds to the lower frequency content of the orbital vibration signal.

3.1. Resolution of a Crosswell Seismic Survey

We shall discuss the resolution of a crosswell seismic survey in two parts: the first part relates to the velocity tomogram, and the second part relates to the reflection imaging.

For the velocity tomogram, the geometry of the survey plays an important part in determining the resolution of the resulting image. The tomogram is reconstructed from the first arrival travel times from the source to the receiver array. The travel (or ray) paths are primarily horizontal or sub-horizontal, limiting the resolution in the horizontal direction. The reconstruction of such a tomogram is in general an ill-posed inverse problem, and various regularization techniques are used to obtain a stable velocity image. One of the more common techniques is the use of Tikhonov regularization, with smoothing parameters much larger in the horizontal versus the vertical direction (as much as 10 or 100 to 1, Matarese and Rodi, 1991; Matarese et al., 1992). It follows that the ratio of the horizontal to vertical resolution of the tomogram will be constrained by the ratio of the regularization parameters.

In the vertical direction, we can generally take the resolution to be the separation between the receivers, assuming that we have a band-limited signal and adequate digitization. For the Z-Seis equipment, the receiver separation and thus the vertical resolution is 3 m (10 ft), for the velocity tomograms. Z-Seis uses a different approach to address the ill-posed nature of the reconstruction. They use a cubic equation to describe the horizontal velocity variation (Washbourne and Rector, 1998; Washbourne

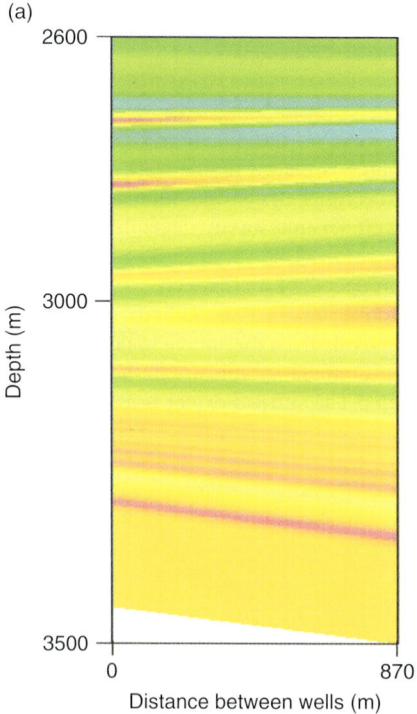

FIG. 9. continued.

MEASUREMENT OF THE EARTH USING BOREHOLE METHODS 257

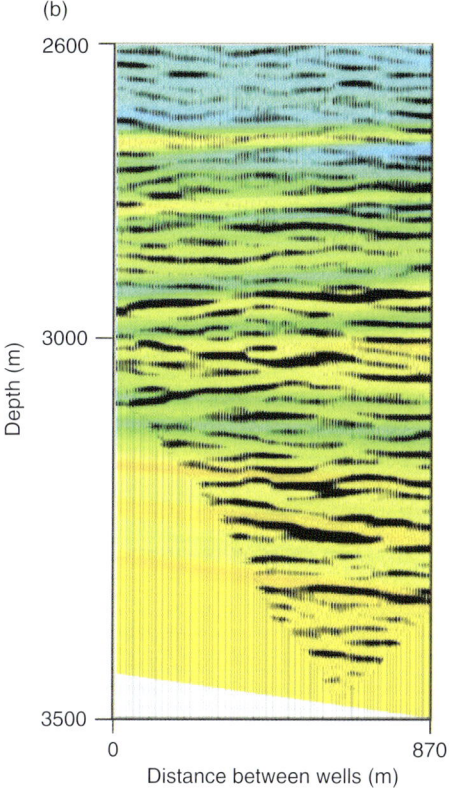

FIG. 9. (a) Tomographic velocity image from crosswell seismic survey (Courtesy of Gang Yu and Bruce Marion, Z-Seis Corp.). The horizontal distance between the wells is 870 m. The vertical distance in the image is about 900 m on the left and 850 m on the right. (b) Crosswell seismic reflection imaging corresponding to the tomographic velocity image shown in Fig. 8a (Courtesy of Gang Yu and Bruce Marion, Z-Seis Corp.) Notice that the reflection image covers a larger area (shallower and deeper) than the velocity image. This is because both up- and down-going reflections are used.

et al., 2002). As a result, the horizontal resolution of their tomogram is usually around one-quarter of the horizontal distance between the wells, of the order of 50–100 m. An example of the vertical and horizontal tomographic velocity image resolution is shown in Fig. 9a (see figure caption for details). It is clear that in such an image the vertical velocity variations are much better resolved than the horizontal velocity variations.

For the reflection image, the considerations are a bit different. The most commonly used algorithm is the CDP (also known as VSP-CDP) transform, which transforms the seismic reflection signals in time into waveforms in space by the use of a specific velocity model (Lazaratos et al., 1995). Other commonly used method for reflection imaging is the prestack Kirchhoff depth migration (Qin and Schuster, 1993; Byun et al., 2002). Either way, the methods used are adopted from standard seismic reflection techniques, and the vertical

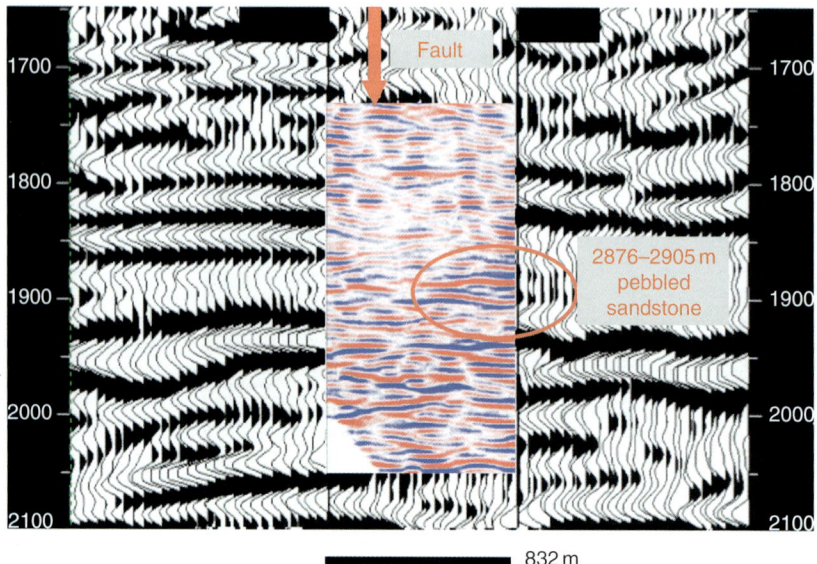

FIG. 10. A comparison of the resolution of the reflection images from a surface seismic survey (background) and a crosswell seismic survey (inserted). The surface seismic survey is recorded in two-way reflection time (in seconds). The crosswell survey shows a vertical resolution of about 3–5 m. The circled section shows a formation of pebbled sandstone that is not evident in the surface seismic data (Courtesy of Gang Yu and Bruce Marion, Z-Seis Corp.).

resolution is generally taken to be about one quarter of the wavelength. In the case of the crosswell seismic survey, assuming a high center frequency of 500 Hz, and an average formation velocity of 4 km/s, the best vertical resolution we can get is about 2 m, on the order of the receiver separation. The reflection image corresponding to the tomographic image shown in Fig. 9a is shown in Fig. 9b. A comparison of the resolutions of a surface seismic survey and a crosswell reflection image is shown in Fig. 10.

For the horizontal resolution of a reflection image, the issue is a little more complex. In general, we can potentially get a better horizontal resolution than the velocity tomogram because of the many source-receiver combinations and the number of reflection points in the image. The resolution is the best in the middle between the two wells, and degrades rapidly toward either side. The actual resolution depends on the source-receiver pattern as well as the formation geology. A way to estimate the horizontal resolution is to do ray tracing from the source to the reflection horizon and then to the receiver. Separation between adjacent reflection points provides an estimate to the horizontal resolution in each particular case. In addition the Fresnel zone of the wave at the reflecting horizon sometimes needs to be considered as well.

It is important to point out, when discussing horizontal resolution of crosswell images, that very seldom an individual change in a reflector position will lead to a distinct interpretation of the underlying geology. Usually coherent groups of reflections across a number of traces make up interpretable events. Thus the horizontal resolutions are more related to the proper positioning of such events, rather than a small perturbation in a noisy background.

There are other methods for imaging crosswell seismic data. The two best known are diffraction tomography (Wu and Toksöz, 1987; Lo *et al.*, 1988; Huang *et al.*, 1992), and full waveform inversion (Song and Williamson, 1995; Song *et al.*, 1995; Pratt, 1999; Pratt and Shipp, 1999). However, because of the complexity of the techniques involved, these are not currently applied routinely in commercial applications. I refer the interested reader to the above references for more information.

4. Vertical Seismic Profiling

For VSP, we have the lowest resolution of the three methods discussed here. VSPs use seismic sources on the surface of the earth. As a result, the frequency content of the signals is much lower than acoustic logs and crosswell seismic surveys. The frequency content is usually up to a factor of 2 higher than standard seismic reflection signal because of the fact that the receivers are located in a borehole and thus we are looking at one way rather than two-way propagation through the earth. Typical VSP signals are of the order of 10–100 Hz, about an order of magnitude less than a typical crosswell seismic signal.

The receivers used in VSPs are usually 3-component clamped geophones. The distance between the receivers is variable, but is usually set at 7.5 or 15 m (25 or 50 ft). A standard VSP product is the velocity profile along the borehole. The vertical resolution of the measurement is given by the receiver separation. This measurement is usually acquired using a seismic source next to the borehole, commonly known as a zero-offset VSP.

The main product from a zero-offset VSP is a velocity log (Fig. 11). This is commonly generated by taking the travel time difference between the first arrival picks of the direct arriving P waves at two adjacent receivers. S wave picks are also possible (even in a zero-offset VSP there are some S waves since the source is never truly zero offset) by properly rotating the three components of the receiver string. These velocities can be somewhat refined by the use of formation top locations picked from well logs, and by formal inversion algorithms (Stewart, 1984; Wyatt and Wyatt, 1981; Salo and Schuster, 1989), but the general vertical resolution is limited by the receiver separations.

In many cases, a simpler measurement is taken. Its aim is to measure the travel time to specific target depths rather than the more detailed velocity structure along the borehole. It will use a much large receiver spacing (as large as 100 m or more), and it is known as a check shot survey. It is important to know whether the survey is a check shot survey or a zero-offset VSP.

Another product from a zero-offset VSP is the corridor stack. It is generated by isolating the reflected wavefield observed at each receiver, correct it to equivalent two way travel time from the surface to the receiver and back to the surface, and then stacked. It is then equivalent to a zero offset surface seismic trace, and used to directly compare with the surface seismic section to identify major reflection horizons. Hardage (2000) gives a detailed description of the process. Since it is a reflection trace, the resolution of the corridor stack is similar to a seismic trace, and is limited to about one quarter of the dominant wavelength of the trace. In most geology, that is usually larger than the receiver separation of 15 m.

The horizontal depth of investigation of zero-offset VSP is usually taken to be about one Fresnel zone or wavelength away from the borehole. Namely, the velocity determined from the VSP is taken to be the average of the velocities of the heterogeneities of the order of one wavelength surrounding the borehole. Reflections in the corridor stack are also assumed to come from this zone.

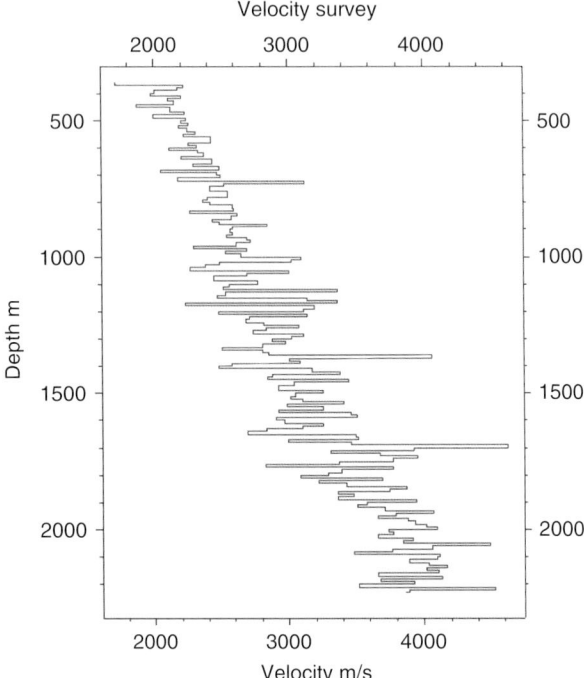

FIG. 11. An example of the velocity log generated from the P wave first arrival in a zero-offset VSP. The separation between the geophones is 15 m.

There are also offset VSPs, where the sources are placed a distance away from the well head, and more complex variations such as a walk-away VSP, where the sources are sequentially placed further and further away in a line; and 3D VSP, where the sources are placed around the borehole in a 2D grid. In all cases, because of the restriction of the placement of source and receivers, it is not routine to obtain an estimate of the velocity structure away from the borehole. In a walkaway VSP, the volume of the earth sampled by the first arrival travel time measurements is an area bounded by the borehole, the deepest receiver, and the farthest source. In general it is a triangular area. In a 3D VSP, the volume is a cone shape similarly defined. In both cases there are very few areas within the triangle/cone that is sampled by more than a very few ray paths (see Fig. 1c). Furthermore, these ray paths are all concentrated in a very narrow range of incidence angles. Under such conditions, we have a very ill-posed problem for tomographic inversion for the velocity structure away from the borehole, worse than the crosswell case. The general result from such a velocity analysis is a smooth velocity structure away from the borehole, in the direction of the source. Thus we have very poor velocity resolution away from the borehole, even with walkaway and 3D VSPs.

A routine product of the offset, walkaway, and 3D VSP is the reflection image. These are obtained using either a CDP transform or a pre-stack migration algorithm (Dillon and Thomson, 1984; Dillon, 1988), similar to imaging of crosswell seismic surveys. An example of a VSP-CDP image is shown in Fig. 12. As discussed before, these are

MEASUREMENT OF THE EARTH USING BOREHOLE METHODS 261

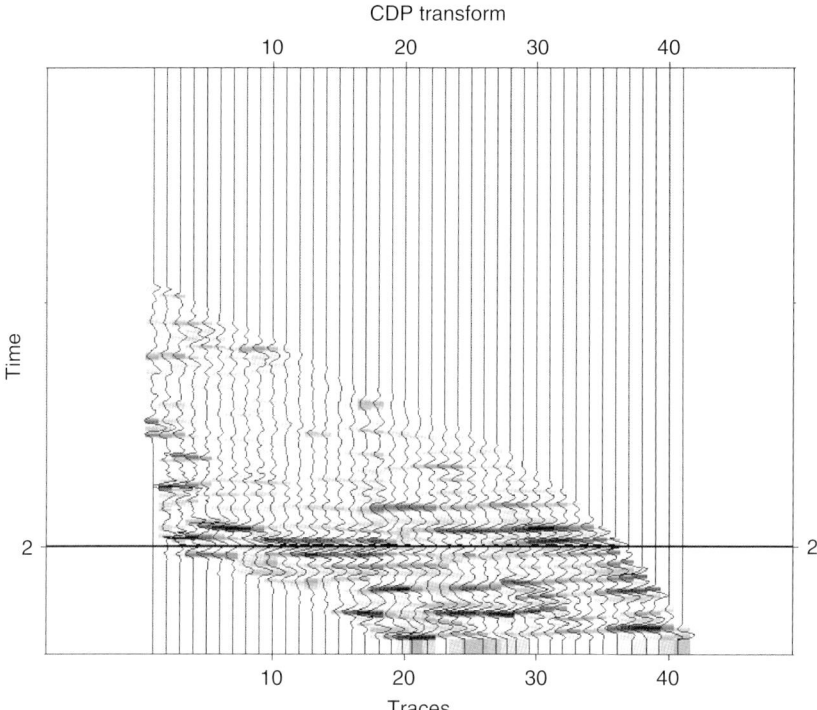

FIG. 12. An example of VSP imaging using the VSP-CDP transform. The image can be displaced in either time or depth domain on the vertical axis (time shown here). Because of the non-linear nature of the VSP-CDP transform, it is common to resort and stack the traces into constant width (horizontal distance) bins for display. In this example the horizontal traces are 25 m apart. Time is in seconds.

standard seismic techniques, and the vertical resolution is about one quarter of the wavelength of the signal, which in this case will be around 20–30 m, depending on the geology and source signature. The horizontal resolution will be dependent on the source locations and the depth of the reflector relative to the receiver position. This horizontal resolution relates only to the resulting image, and not to the velocity variations. Discussions in the previous section regarding the horizontal resolution of the crosswell reflection image apply in a similar way to VSP images. For a more detailed discussion of VSP acquisition and processing, the interested reader is referred to Hardage (2000); Hinds et al. (1996); and Hinds and Kuzmiski (2001).

5. DISCUSSIONS AND SUMMARY

The three different borehole measurement techniques described in this paper measure the earth at very different scales, ranging from the 0.15-m scale of the acoustic logs to the hundred-meter scale of VSPs. In addition, the measurements have the highest resolution along the borehole, in the vertical direction (most of these measurements are made in borehole that are vertical or near vertical). Because of the measurement geometries, the horizontal resolutions are in general much poorer than the vertical resolution.

Another question that is still not totally resolved is that how to relate these measurements to one another. It is well known that integrated travel times from acoustic logs and VSPs do not in general agree with each other (Stewart et al., 1984). The same statement can be generalized to crosswell measurements also. The differences have been attributed to a number of factors, including interbed multiples and intrinsic attenuation and frequency dispersion. The simple matter is that because of the differences in the acquisition geometry and frequency content, these measurements sample different volumes of the sub-surface. Unless the sub-surface is truly homogeneous and isotropic, the results from the three types of measurements will be different, and can be used to build a more realistic image of the earth.

Acknowledgements

The author would like to thank Dr. Xiaoming Tang, Baker Atlas and Halliburton Energy Services for the use of some of the figures. Dr. Gang Yu, Bruce Marion, and Z-Seis provided the latest crosswell images. Dr. Mark Willis of MIT contributed to the depth of investigation study of acoustic logs. Dr. Janusz Peron generated the VSP velocity and CDP images. Chris Payton provided many stimulating discussions on crosswell seismic imaging over the years.

References

Araya, K., Blanch, J., Cheng, A., Varsamis, G. (2003). Evaluation of dispersion estimation methods for borehole acoustic data. 73rd Annual ternational Meeting, SEG, Expanded Abstracts, pp. 305–308.

Byun, J., Rector, J.W., III, Nemeth, T. (2002). Postmap migration of crosswell reflection seismic data. *Geophysics* **67**, 135–146.

Cheng, C.H., Toksöz, M.N. (1981). Elastic wave propagation in a fluid-filled borehole and synthetic acoustic logs. *Geophysics* **46**, 1042–1053.

Cole, J.H. (1997). The orbital vibrator, a new tool for characterizing interwell reservoir space. *The Leading Edge* **16**, 281–283.

Dillon, P.B. (1988). Vertical seismic profile migration using the Kirchhoff integral. *Geophysics* **53**, 786–799.

Dillon, P.B., Thomson, R.C. (1984). Offset-source vertical-seismic-profile (VSP) surveys and their image reconstruction. *Geophys. Prospect.* **32**, 790–811.

Hardage, B.A. (2000). Vertical seismic profiling. *In* Handbook of Geophysical Exploration, (K. Helbig, S. Treitel, Eds.), vol. 14, p. 352. Pergamon.

Hinds, R.C., Kuzmiski, R.D. (2001). VSP for the interpreter/processor for 2001 and beyond. Part 1. *Recorder* **26**, 84–95.

Hinds, R.C., Anderson, N.L., Kuzmiski, R.D. (1996). VSP interpretive processing. Theory and practice. Open File Publications No. 3, *Society of Exploration Geophysicists* .

Hornby, B.E. (1993). Tomographic reconstruction of near-borehole slowness using refracted borehole sonic arrivals. *Geophysics* **58**, 1726–1783.

Hsu, K., Chang, S.K. (1987). Multiple-shot processing of array sonic waveforms. *Geophysics* **52**, 1376–1390.

Huang, L.J., Wu, R.S., Araujo, F.V. (1992). Multifrequency backscattering tomography: Extension to the case of vertically varying background. 62nd Annual International Meeting, SEG, Expanded Abstracts, pp. 766–769.
Kimball, C.V., Marzetta, T.L. (1986). Semblance processing of borehole acoustic array data. *Geophysics* **49**, 274–281.
Lazaratos, S.K., Harris, J.M., Rector, J.W., van Schaack, M. (1995). High-resolution crosswell imaging of a West Texas carbonate reservoir: Part 4: Reflection imaging. *Geophysics* **60**, 702–711.
Lo, T.W., Toksöz, M.N., Xu, S.H., Wu, R.S. (1988). Ultrasonic laboratory tests of geophysical tomographic reconstruction. *Geophysics* **53**, 947–956.
Matarese, J.R., Rodi, W.L. (1991). Nonlinear traveltime inversion of cross-well seismics: A minimum structure approach. 61st Annual ternational Meeting, SEG, Expanded Abstracts, pp. 917–921.
Matarese, J.R., Rodi, W.L., Toksöz, M.N. (1992). Nonuniqueness in nonlinear traveltime tomography. 54th Meeting, EAGE, Explanded Abstracts, pp. 110–111.
Pratt, R.G. (1999). Seismic waveform inversion in the frequency domain, Part 1: Theory and verification in a physical scale model. *Geophysics* **64**, 888–901.
Pratt, R.G., Shipp, R.M. (1999). Seismic waveform inversion in the frequency domain, Part 2: Fault delineation in sediments using crosshole data. *Geophysics* **64**, 902–914.
Qin, F., Schuster, G.T. (1993). Constrained Kirchhoff migration of crosswell seismic data. 63rd Annual International Meeting, SEG, Expanded Abstracts, pp. 99–102.
Salo, E.L., Schuster, G.T. (1989). Traveltime inversion of both direct and reflected arrivals in vertical seismic profile data. *Geophysics* **54**, 49–56.
Song, Z.M., Williamson, P.R. (1995). Frequency-domain acoustic-wave modeling and inversion of crosshole data: Part I: 2.5-D modeling method. *Geophysics* **60**, 784–795.
Song, Z.M., Williamson, P.R., Pratt, R.G. (1995). Frequency-domain acoustic-wave modeling and inversion of crosshole data: Part II: Inversion method, synthetic experiments and real-data results. *Geophysics* **60**, 796–809.
Stewart, R.R. (1984). Vertical-seismic-profile (VSP) interval velocities from traveltime inversion. *Geophys. Prospect.* **32**, 608–628.
Stewart, R.R., Huddleston, P.D., Kan, T.K. (1984). Seismic versus sonic velocities—A vertical-seismic-profiling study. *Geophysics* **49**, 1153–1168.
Tang, X.M., Cheng, A.C.H. (2004). Quantitative Acoustic Logging Methods : *Elsevier*.
Tubman, K.M., Cheng, C.H., Toksöz, M.N. (1984). Synthetic full waveform acoustic logs in cased boreholes. *Geophysics* **49**, 1051–1059.
Washbourne, J.K., Rector, J.W. (1998). Crosswell seismic imaging in three dimensions. 68th Annual International Meeting, SEG, Expanded Abstracts, pp. 334–337.
Washbourne, J., Rector, J.W., Bube, K. (2002). Crosswell traveltime tomography in three dimensions. *Geophysics* **67**, 853–871.
Wu, R.S., Toksöz, M.N. (1987). Diffraction tomography and multisource holography applied to seismic imaging. *Geophysics* **52**, 11–25.
Wyatt, K.D., Wyatt, S.B. (1981). Determination of subsurface structural information using the vertical seismic profile. 51st Annual ternational Meeting, SEG, Expanded Abstracts, p. S5.2.
Zemanek, J., Angona, F.A., Williams, D.M., Caldwell, R.L. (1984). Continuous shear wave logging. 25th Annual Logging Symposium Transactions. Society of Professional Well Log Analysts, Paper U.
Zeroug, S., Valero, H.P., Bose, S., Yamamoto, H. (2006). Monopole radial profiling of compressional slowness. 76th Annual ternational Meeting, SEG, Expanded Abstracts, pp. 354–357.
Zhang, T., Tang, X.M., Patterson, D.L. (2000). Evaluating laminated thin beds in formations using high-resolution slowness logs. 41st Annual Logging Symposium Transactions. Society of Professional Well Log Analysts, Paper XX.

CODA ENERGY DISTRIBUTION AND ATTENUATION

KAZUO YOSHIMOTO AND ANSHU JIN

ABSTRACT

Coda waves in local and/or regional seismograms consist of scattered waves caused by randomly distributed small-scale heterogeneities in the lithosphere. Observations and analysis of coda waves have grown up into a well-developed branch of seismology since the pioneer observations by Aki in late 1960s. Under the assumption of uniform distribution of isotropic scatterers, coda energy should be uniformly distributed in space at large lapse times. This has been confirmed, observationally, in areas where the lateral variation of seismic structure is small and smooth. This characteristic of coda waves has been utilized to investigate earthquake source, attenuation of direct P- and S-waves, and receiver site amplification by means of the coda normalization method. The coda Q^{-1} (Q_C^{-1}) is a parameter that characterizes the temporal decay rate of observed coda energy. Numerous studies, worldwide, on Q_C^{-1} revealed the following observational facts: (1) Q_C^{-1} decreases with frequency for > 1 Hz, (2) Q_C^{-1}, in general, lies between direct S-wave Q_S^{-1} and intrinsic absorption Q_{int}^{-1}, and (3) Q_C^{-1} is correlated with tectonic activity spatially and temporally. With the development of the seismic observation system, recent studies using data from dense seismic networks revealed that coda energy is not uniformly distributed in tectonically active regions, and the non-uniformity increases with increasing frequency. The spatial correlation between Quaternary volcanoes and high heat flux suggests that the thermal structure (or intrinsic absorption structure) of the crust and/ or the uppermost mantle characterizes the regional variation of coda energy in high frequencies.

KEY WORDS: Coda Q^{-1} (Q_C^{-1}), Coda Waves, High-frequency Seismic Waves, Heterogeneity, Intrinsic Absorption, Scattering Attenuation.   © 2008 Elsevier Inc.

1. INTRODUCTION

One of the most striking features of short period seismograms of local earthquakes is the regular manner in which the amplitude in the tail of a seismogram decays long after the direct P and S, and other direct phases. Aki (1969) was the first to highlight such continuous wavetrains in local seismograms and named this phenomenon "(seismic) coda waves," a term that has been used in seismology ever since to describe the tail portion of local and regional seismograms. He observed that the coda waves have a common decay curve, independent of the location of the source and recording station, and the event's magnitude within a seismic region, although it varies from region to region.

Aki (1969) proposed a phenomenological model of coda excitation, which is called a "single back-scattering model." This model explains coda waves as back scattered seismic waves by the small-scale heterogeneities randomly and uniformly distributed in the propagation medium. Coda energy density, $E_{ij}(f|t)$, from a local earthquake j and recorded at the station i can be best described by

$$E_{ij}(f|t) = W_j(f)R_i(f)C(f|t) \quad t > 2t_S, \tag{1}$$

where f is the frequency and t is the lapse time measured from the origin time of the source. t_S is the travel time of S-waves from the source to the station. $W_j(f)$ is the total energy at frequency f radiated from the source, and $R_i(f)$ is the amplification factor at the recording site. The term $C(f|t)$ is common to all sources and recording sites in a given seismic region. Equation (1) was first recognized by Aki (1969) by investigating the aftershocks of the Parkfield Earthquake, 1966 and the empirical condition, whereby Eq. (1) holds for $t > 2t_S$, emerged following extensive study of coda waves in central Asia by Rautian and Khalturin (1978).

The first attempt to predict the explicit form of $E(f|t)$ for a mathematical model of earthquake source and earth medium was made by Aki and Chouet (1975), with a model based on the following assumptions:

- Both primary and scattered waves are S-waves.
- The scattering is isotropic and multiple scatterings are neglected. (Since the scattering is considered to be a weak process, the Born approximation is applicable.)
- Scatterers are distributed randomly with uniform density.
- The background elastic medium is uniform and unbounded.

The first assumption has been validated, based on the theoretical prediction that the S to P conversion scattering due to small-scale heterogeneities is an order of magnitude smaller than the P to S scattering (Knopoff and Hudson, 1964, 1967; Aki, 1992). Zeng (1993) showed that the above difference in conversion scattering between P to S and S to P leads to the dominance of S-waves in the coda. This assumption is also supported by various observations, such as common site amplification (Tsujiura, 1978) and common attenuation (Aki, 1980b; Sarker and Abers, 1998) between S-waves and coda waves.

Under the simplification by co-locating source and receiver, and adopting the form for the propagating term

$$C(f|t) \propto \exp(-2\pi f t Q_C^{-1})/t^2. \tag{2}$$

Aki and Chouet (1975) and Aki (1981) adopted the following expression:

$$E(f|t) = \frac{W g(\pi) H(t)}{2\pi V_S^2 t^2} \exp(-2\pi f t Q_C^{-1}), \tag{3}$$

where $H(t)$ is a step function, and V_S is the average S-wave velocity. $g(\theta)$ is the directional scattering coefficient that is defined as 4π times the fractional loss of energy by scattering per unit travel distance of primary waves in per unit solid angle in radiation direction θ measured from the direction of primary wave propagation. It is called the "back-scattering" when $\theta = \pi$. Parameter Q_C^{-1} is called coda Q^{-1}, which characterizes the temporal decay rate of observed coda energy density. By means of a 2D finite difference method, Jannaud et al. (1991) confirmed that the coda power spectrum is proportional to the back-scattering coefficient, $g(\pi)$, as predicted by Eq. (3) for random media with small fluctuations of wave velocity.

Based on the single isotropic scattering approximation, Sato (1977) derived the equation in the case of the co-locating assumption being invalid. The mean energy density of the scattered waves, recorded at a distance r from the source, can then be expressed as

$$E(r,f|t) = \frac{Wg_0}{4\pi r^2} K\left(\frac{t}{t_S}\right) \exp(-2\pi f t Q_C^{-1}), \tag{4}$$

where

$$K\left(\frac{t}{t_S}\right) \equiv \frac{t_S}{t} \ln \frac{t+t_S}{t-t_S} \quad \text{for } t > t_S.$$

The total scattering coefficient g_0 that characterizes a scattering power per unit volume of the heterogeneous medium is defined as the average of the directional scattering coefficient over a solid angle Ω:

$$g_0 \equiv \frac{1}{4\pi} \oint g(\theta) d\Omega = l^{-1} = Q_{\text{scat}}^{-1} k, \tag{5}$$

where l is the mean free path (l/V_S is the mean free time), defined as a reciprocal of the total scattering coefficient and it is another important parameter describing the characteristics of a randomly heterogeneous medium. Q_{scat}^{-1} is the reciprocal of the quality factor for the scattering attenuation of incident waves with wavenumber k.

Peng (1989) calculated the spatial auto-correlation function of Q_C^{-1} in southern California and found that "the coherence distance," defined as the distance at which the auto-correlation first comes to zero, is 135 km for the lapse time window 30–60 s, 90 and 45 km for lapse time window 20–45 and 15–30 s, respectively. The above coherence distances are close to the S-wave travel distance for the middle lapse time corresponding to each of the above three time windows. This observation, indicating that the later time window gives the slower decay in the autocorrelation, offers strong support for the assumption that coda waves in these lapse time windows are primarily composed of S-to-S back scattering waves (see Aki, 1995). In other words, the Q_C^{-1} measured from a time window represents the seismic attenuation property of the lithosphere, averaged over an ellipsoidal volume, with the source and receiver as the foci and the radius as $V_S t/2$.

Approaches have been developed for modeling coda excitation, including multiple scattering, for instance, numerical experiments by Frankel and Clayton (1986), Frankel and Wennerberg (1987), and Hoshiba (1991), and theoretical studies by Wu (1985), Shang and Gao (1988), and Gao and Aki (1996). Zeng et al. (1991) derived an integral equation that expresses the energy density of seismic waves at a given location $\mathbf{r}$ and lapse time t due to an impulsive point source located at the origin:

$$E(\mathbf{r},f|t) = \frac{We^{-(g_0 V_S + 2\pi f Q_{\text{int}}^{-1})t}}{4\pi V_S |\mathbf{r}|^2} \delta\left(t - \frac{|\mathbf{r}|}{V_S}\right) + \int_{-\infty}^{\infty}\int_{-\infty}^{\infty}\int_{-\infty}^{\infty}\int_{-\infty}^{\infty} g_0 E(\mathbf{r}',f|t') \\ \times \frac{e^{-(g_0 V_S + 2\pi f Q_{\text{int}}^{-1})(t-t')}}{4\pi |\mathbf{r}-\mathbf{r}'|^2} \delta\left(t - t' - \frac{|\mathbf{r}-\mathbf{r}'|}{V_S}\right) dt' \, d\mathbf{r}', \tag{6}$$

where $\delta(t)$ is the Dirac delta function and Q_{int}^{-1} is the reciprocal of the quality factor of the intrinsic absorption. The first term on the right side represents the propagation of direct wave energy from the source to the receiver, while the second term is the sum of scattered energy from all possible scatterers located at **r'**. This integral equation can be solved in the integral transform domain (see Zeng et al., 1991). In Eq. (6), the total scattering attenuation coefficient and intrinsic absorption coefficient are explicitly specified, and all multiple scatterings are included, although, the background medium remains uniform and unbounded and scattering is assumed to be isotropic. Equation (6) is a function of distance and time, unlike Eq. (3) in which coda energy depends on time only.

Due to the extraordinary separability, as shown in Eq. (1), of the source, path and site effects in coda waves, coda analyses have been applied for studies on earthquake sources (e.g., Aki and Chouet, 1975; Su et al., 1991; Mayeda and Walter, 1996; Morasca et al., 2005); recording site (e.g., Phillips and Aki, 1986; Su and Aki, 1995; Takahashi et al., 2005). However, the most extensive application of this equation is in the study of the heterogeneity and/or seismic attenuation characteristics of the lithosphere.

To date, most of our knowledge concerning the small-scale heterogeneity of the lithosphere comes from studies of coda waves. Without recognizing the coda waves, the enormous difficulty in deciphering numerous factors affecting primary waves has prevented us from even recognizing the existence of small-scale heterogeneity related to the structure of the wave propagation medium. For example, the single scattering model has been used in the analysis to estimate the total scattering coefficient of the lithosphere. Figure 1 shows that the estimations of g_0 vary roughly between 10^{-3} and 10^{-1} km^{-1}, taking an average of about 10^{-2} km^{-1}. In addition, due to the simplicity of Eqs. (3) and (4), the measurements of Q_C^{-1}, its geographical variation over a large region as well as its temporal variation over a long time period can be studied relatively easily (e.g., Herraiz and Espinosa, 1987; Sato and Fehler, 1998). On the other hand, inversion schemes have been developed to generate images of a three-dimensional distribution of scatterers by comparing the coda envelopes between those observed to that predicted by a given model (e.g., Revenaugh, 1995; Nishigami, 2000; Asano and Hasegawa, 2004).

This paper intends to emphasize a review of the observed characteristics of the spatial distribution and temporal decay of coda energy.

2. Coda Energy Distribution and Measurement on $Q_{P,S}^{-1}$ using Local Seismograms

2.1. Uniformity of Coda Energy Distribution

As discussed above, the scattered waves recorded at large lapse times traversed a wide volume of the lithosphere, where the scattering process averages the elastic properties over the sampled volume. Thus, the coda amplitudes observed in a local area are independent from the source-receiver locations.

From the analysis of S-coda waves recorded by a local seismic array, Tsujiura (1978) found good agreement in the amplification factors of S-coda waves and direct S-waves. This result, based on the observed data analysis, suggested that the energy of the S-coda waves is distributed uniformly in the lithosphere beneath the seismic array deployed. Further evidence supporting this interpretation is given by the stable determination of earthquake magnitude from the S-coda duration, irrespective of the source-receiver location (e.g., Tsumura, 1967).

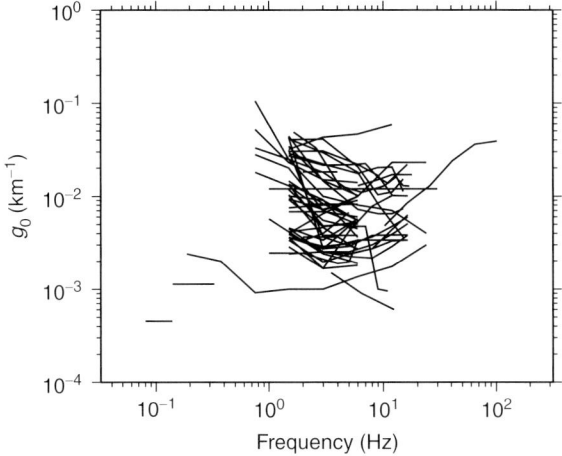

FIG. 1. Total scattering coefficient g_0 of the lithosphere in the world: Sato (1978) in Kanto, Japan; Felher et al. (1992) in Kanto-Tokai region, Japan; Mayeda et al. (1992) in central California, Hawaii, and Long Valley, California; Hoshiba (1993) in Japan; Jin et al. (1994) in southern California; Leary and Abercrombie (1994) in southern California; Hatzidimitriou (1994) in northern Greece; Akinci et al. (1995) in southern Spain; Gusev and Abubakirov (1997) in Kamchatka; Adams and Abercrombie (1998) in southern California; Ugalde et al. (1998) in northeastern Venezuela; Gusev and Abubakirov (1999) in Kamchatka; Akinci and Eyidoğan (2000) in eastern Turkey; Bianco et al. (2002) in southern Apennines, Italy; Lacombe et al. (2003) in central France; Lee et al. (2003) in central Asia; Dutta et al. (2004) in south-central Alaska; Goutbeek et al. (2004) in southern Netherlands; Vargas et al. (2004) in northwestern Colombia; Bianco et al. (2005) in northeastern Italy; Giampiccolo et al. (2006) in southeastern Sicily, Italy; Sens-Schönfelder and Wegler (2006) in Germany.

The multiple scattering models have been developed to interpret the characteristics of observed coda energy distribution (e.g., Zeng et al., 1991). It is mathematically verified that the solution of the multiple isotropic scattering models converge to diffusion model at large lapse times. Unlike the field observations, however, these models predict a concentration of energy around the source. Such contradiction between the observations and the theoretical prediction is mainly due to the assumption of isotropic scattering. Using radiative transfer theory, Sato (1995) developed a multiple nonisotropic scattering model, and demonstrated that the uniform energy distribution around the source at large lapse times can be explained when the scattering is much stronger in the forward direction. All these scattering models assume the source radiation to be isotropic, however, the nonspherical radiation of actual earthquake source may cause the nonuniform spatial distribution of scattered energy. Based on the radiative transfer theory, Sato et al. (1997) demonstrated that the predicted nonuniform distribution of coda energy would be smeared out by the multiple scattering processes with increasing lapse time. Their result shows that the coda energy distribution from a point shear-dislocation source converges to that for a spherical radiation source, asymptotically, when the lapse time exceeds twice the travel time of the direct S-waves. Their results, again, support the validity of the Eq. (1).

One of the difficulties in data analyses in seismology is to separate the effects of the source, path, and site. The uniform distribution of coda energy within a region for lapse times after sufficient time offers a reliable tool to overcome such difficulty. Eq. (1) has evolved into a technique called "the coda normalization method," which has been applied to study earthquake sources, propagation path property and site amplification as summarized by Sato and Fehler (1998). The coda normalization method for measuring the direct S-wave attenuation with travel distance was initially proposed by Aki (1980a), and later extended by Yoshimoto et al. (1993) to measure the attenuation of the direct P-waves with travel distance.

This method was designed to normalize source spectral amplitude, using coda spectra at some fixed lapse time to measure the reciprocal of the quality factors of P-wave and S-wave (Q_P^{-1} and Q_S^{-1}, respectively), using single station data. The coda normalization method is free from the site effect correction and does not require any assumption concerning source spectral shape (except for the source spectral ratio of P-wave and S-wave in Q_P^{-1} estimation). The following equations are applied to a data set that consists of a number of earthquakes with variable hypocentral distribution and a variety of fault plane solutions:

$$\left\langle \ln \frac{A_P(f,r)r}{A_C(f,t_C)} \right\rangle_{r \pm \Delta r} = -\frac{\pi f}{V_P} Q_P^{-1}(f) r + \text{const}(f) \tag{7}$$

$$\left\langle \ln \frac{A_S(f,r)r}{A_C(f,t_C)} \right\rangle_{r \pm \Delta r} = -\frac{\pi f}{V_S} Q_S^{-1}(f) r + \text{const}(f), \tag{8}$$

where V_P is the velocity of P wave. $A_P(f,r)$ and $A_S(f,r)$ is the spectral amplitude of the direct P- and S-waves at hypocentral distance r, respectively. $A_C(f,t_C)$ is the spectral amplitude of S-coda waves at a lapse time t_C, where $t_C > 2t_S$. The operator $\langle \rangle_{r \pm \Delta r}$ denotes the average within a small hypocentral distance range of $r \pm \Delta r$. $Q_P^{-1}(f)$ and/or $Q_S^{-1}(f)$ can be estimated from a linear regression of $\langle \ln [A_{P,S}(f,r)/A_C(f,t_C)] \rangle_{r \pm \Delta r}$ versus r, using the least-squares method. As summarized by Sato and Fehler (1998), the coda normalization method has been frequently used to estimate Q_P^{-1} and Q_S^{-1} values of various regions in the world (e.g., Chung and Sato, 2001).

Figure 2 shows the results of Q_S^{-1} measurements from various regions of the world. Despite significant regional variations, it is clear that Q_S^{-1} decreases with increasing frequency at frequencies 1–10 Hz and the Q_S^{-1} curve of several high frequency resolution studies exhibits a bend around 5 Hz, implying a significant change in the decreasing rate. The decreasing rate is weak at high frequencies. This characteristic is distinct in the result of Kinoshita and Ohike (2002) (No. 9 in this figure), based on strong motion data analysis carried out in Kanto, Japan. Interestingly, Yoshimoto et al. (1998) applied the coda normalization method to the borehole recordings of local earthquakes in central Japan and found that the Q_S^{-1} tends to become nearly constant for frequencies above 25 Hz (No. 8). Using local earthquake seismograms recorded at a deep borehole (> 2 km in depth) in California, Adams and Abercrombie (1998) also found that Q_S^{-1} decreases with frequency very weakly between 10 and 100 Hz (No. 6b).

On the other hand, we know that Q_S^{-1} is very low for long period shear waves from the global measurements on surface waves (see Mitchell and Romanowicz, 1998 for a review on Q^{-1} of the Earth). Thus, Aki (1980a) suggested that there should be a peak

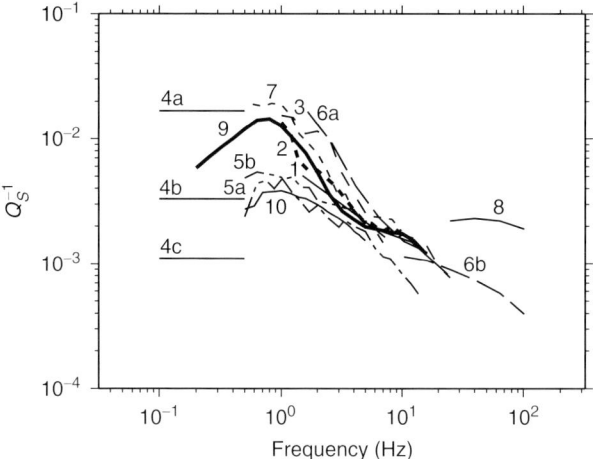

FIG. 2. Reported values of Q_S^{-1} for the lithosphere. 1: Aki (1980a) in Kanto, Japan; 2: Takemura et al. (1991) in southern Tohoku region, Japan; 3: Castro and Munguía (1993) in the Oaxaca, Mexico; 4a–c: Mitchell (1995) for the upper crust of tectonically active regions, the upper crust of tectonically stable regions, and the lower continental crust, respectively; 5a and b: Benz et al. (1997) in southern and south-central California, and in the Basin and Range Province, respectively; 6a and b: Adams and Abercrombie (1998) in southern California (surface and borehole data, respectively); 7: Kato et al. (1998) in southern Tohoku region, Japan; 8: Yoshimoto et al. (1998) for the shallow crust at western Nagano, Japan; 9: Kinoshita and Ohike (2002) in Kanto, Japan; 10, Ojeda and Ottemöller (2002) in Colombia.

of Q_S^{-1} at a frequency range of around 0.5–1 Hz. Clear evidence for such a peak at a frequency of around 0.8 Hz was first reported by Kinoshita and Ohike (2002) (originally Kinoshita, 1994), via the spectral inversion analysis of strong motion borehole seismograms. The Q^{-1} of Lg waves in Columbia determined by Ojeda and Ottemöller (2002) (No. 10 in Fig. 2) also shows a peak at around 1 Hz.

For a medium in which both intrinsic absorption and scattering attenuation occur, the total attenuation Q_S^{-1} can be written as $Q_S^{-1} = Q_{int}^{-1} + Q_{scat}^{-1}$ (e.g., Dainty, 1981). In order to clarify the contribution of the intrinsic and scattering losses, studies have been carried out in various regions of the world by using "the Multiple Lapse Time Window Analysis (MLTWA)" (Felher et al., 1992; Mayeda et al., 1992; Hoshiba, 1993; Hoshiba et al., 1991; Jin et al., 1994). Figure 3 compiles the frequency dependence of S-wave attenuation, the contribution from intrinsic absorption and scattering attenuation measured in various seismic regions by different investigations. As shown in Fig. 3b, the seismic albedo B_0, defined as the ratio of the scattering attenuation to the total attenuation, takes 1/2 at about 3 Hz, indicating that the contributions of the intrinsic absorption and scattering attenuation are almost comparable at this frequency. Moreover, this ratio decreases with increasing frequency showing that the intrinsic absorption dominates scattering attenuation in frequencies above 3 Hz, approximately. This change in the dominant attenuation mechanism is caused by the difference in frequency dependence of Q_{int}^{-1} and Q_{scat}^{-1}: both parameters decrease with increasing frequency, however, the decreasing rate is weaker for the former. These results are consistent with the bend of Q_S^{-1} curve at around 3 Hz observed by Kinoshita and Ohike (2002).

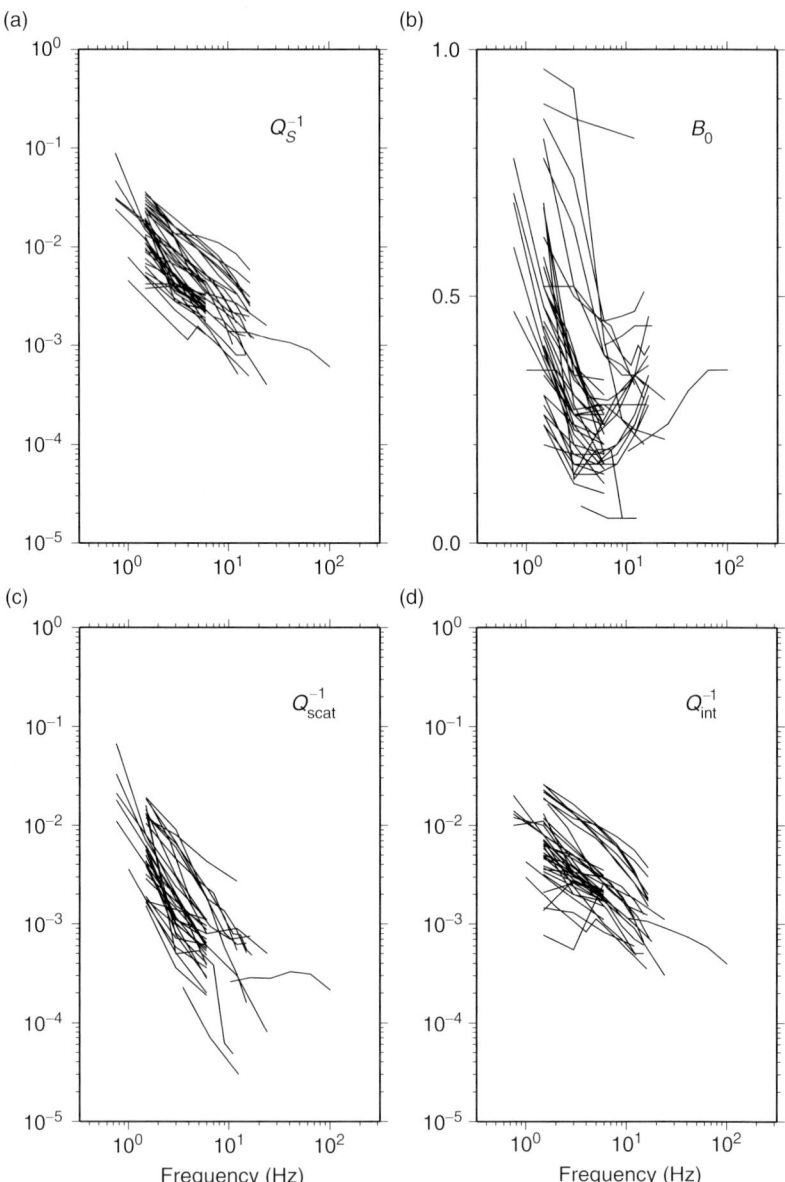

FIG. 3. Summary of the application of the multiple lapse time window analysis to various regions in the world: (a) Q_S^{-1} for apparent attenuation of S waves; (b) the seismic albedo B_0 of S waves; (c) Q_{scat}^{-1} for the scattering attenuation of S waves; (d) Q_{int}^{-1} for the intrinsic absorption of S waves. Results are from: Felher et al. (1992) in Kanto-Tokai region, Japan; Mayeda et al. (1992) in central California, Hawaii, and Long Valley, California; Hoshiba (1993) in Japan; Jin et al. (1994) in southern California; Hatzidimitriou (1994) in northern Greece; Akinci et al. (1995) in southern Spain; Adams and Abercrombie (1998) in southern California; Ugalde et al. (1998) in northeastern Venezuela; Akinci and Eyidoğan (2000) in eastern Turkey; Bianco et al. (2002) in southern Apennines, Italy; Dutta et al. (2004) in south-central Alaska; Goutbeek et al. (2004) in southern Netherlands; Vargas et al. (2004) in northwestern Colombia; Bianco et al. (2005) in northeastern Italy; Giampiccolo et al. (2006) in southeastern Sicily, Italy.

2.2. Nonuniform Coda Energy Distribution in Tectonically Active Regions

It is important to note that the coda normalization method is based on empirical evidence observed in local seismic networks. The agreement between the results obtained from this method and the regional array analyses supports the uniform coda energy distribution, at least within local areas.

Traditionally, coda wave analysis was firstly performed at a single station, and then its spatial variation was studied, combining the single-station results obtained at different stations (e.g., Mayeda *et al.*, 1991). However, this procedure does not take advantage of the spatial smoothness inherent to the coda amplitude as a diffusive process. By using coda waves of the volcano-tectonic events to trace the magma movement in the volcano Piton de la Fournaise in La Reunion, Aki and Ferrazzini (2000) revised the procedure by first studying the spatial distribution of coda wave characteristics for a single event, recorded at different stations of a dense array, and then synthesizing results from many events. They used the coda energy in a fixed later lapse time window to normalize the coda energy for a fixed earlier lapse time window for each station. Aki and Ferrazzini (2000) found that the coda energy at a frequency band of 1–4 Hz peaked at the summit area, which they attributed to slow waves trapped in the fluid–solid 2-phase system of the magma body. Such coda energy concentration phenomena have also been observed in the Hida region, central Japan for frequency band 1–2 Hz, related to active volcanoes, by Jin and Ando (1998) and at Merapi volcano, Indonesia by Friedrich and Wegler (2005). However, the physical explanation remains open.

Yoshimoto *et al.* (2006) analyzed the seismograms of a local event of magnitude 6.4, recorded via the high density seismic network (Hi-net) in the Tohoku region, Japan, where the volcanic front is running from north to south. They found a significant lateral variation of coda energy, for frequencies above 1 Hz, across the volcanic front from east (forearc) to west (backarc) (Fig. 4). As shown in Fig. 4b, the coda energy at frequencies 16–32 Hz is uniformly distributed in the forearc, whereas an exponential decrease with horizontal offset from the volcanic front is found in the backarc. They interpreted this observation as indicating that the intrinsic absorption in the backarc is significantly greater compared to those in the forearc.

By introducing a diffusion–absorption model that consists of two welded half-spaces with the same diffusivity but different absorption strength (Fig. 5), they evaluated the spatial variation of energy density from the following equations:

$$\frac{\partial E(x,t)}{\partial t} = D\frac{\partial^2 E(x,t)}{\partial x^2} - qE(x,t) \tag{9}$$

$$\frac{\partial E'(x,t)}{\partial t} = D\frac{\partial^2 E'(x,t)}{\partial x^2} - q'E'(x,t), \tag{10}$$

where E and E' are the energy density in the left and right half-spaces, respectively. Parameters q and q' characterize the intrinsic absorption in the left and right half-spaces, respectively. Parameter D is the diffusivity, and x is the distance from the boundary of the two half-spaces. Applying initial and boundary conditions to Eqs. (9) and (10), they demonstrated that the energy density in the strong absorptive half-space at large lapse times can be characterized as

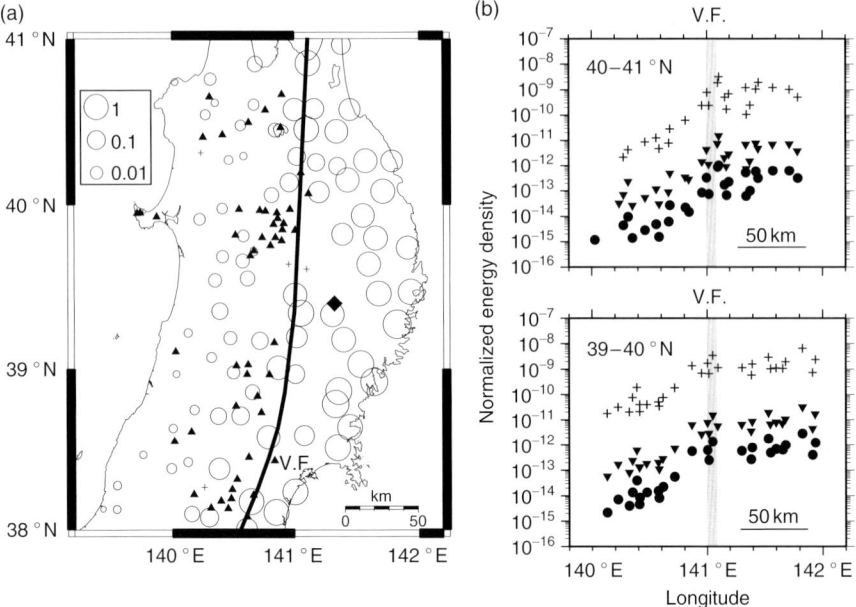

FIG. 4. (a) Map showing the spatial distribution of coda energy at a lapse time of 180 s in northern Honshu, Japan. The epicenter of the earthquake analyzed is denoted by a solid diamond. The open circles are normalized and scaled in proportion to the coda energy calculated at each borehole station. The solid triangles show locations of the Quaternary volcanoes. The thick solid line indicates the volcanic front. (b) The east–west variation of coda energy at frequency of 16–32 Hz for two regions (39–40 °N and 40–41 °N). Coda energies at lapse times of 60, 120, and 180 s are denoted by crosses, solid triangles, and solid circles, respectively. The shaded line shows the location of the volcanic front. (Reproduced from Yoshimoto et al., 2006.)

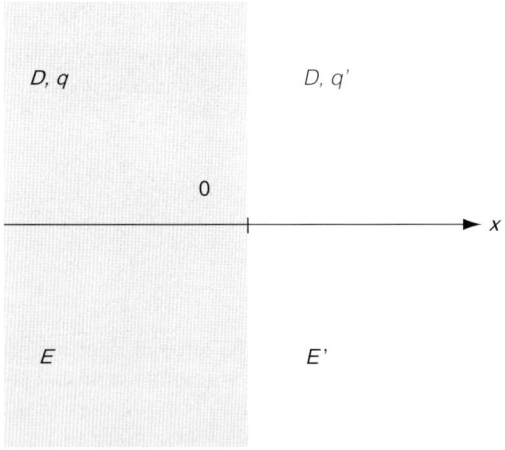

FIG. 5. A diffusion–absorption model for calculating the spatial variation of coda energy. This model consists of two half-spaces with the same diffusivity (D) but different absorption strengths ($q > q' \geq 0$). (Reproduced from Yoshimoto et al., 2006.)

$$E(x) \propto \exp\left(-\sqrt{\frac{2\pi f Q_{\text{int}}^{-1}}{D}}|x|\right),\tag{11}$$

where Q_{int}^{-1} is the intrinsic absorption parameter of the strong absorptive half-space. Equation (11) indicates that the energy density in the strong absorptive medium decreases, exponentially, to the weak absorptive medium with increasing distance from the boundary. This corresponds to the observation shown in Fig. 4b. Assuming $D = 133$ km^2/s ($g_0 = 0.01$ km^{-1} and $V_S = 4.0$ km/s), Yoshimoto *et al.* (2006) estimated the intrinsic absorption parameter of the lithosphere in the backarc of the Tohoku region, Japan and found $Q_{\text{int}}^{-1} \approx 2 \times 10^{-3}$ for $f = 10$ Hz. This value is about twice of those reported from previous studies for the forearc (e.g., Hoshiba, 1993), indicating that the S-wave attenuation in the backarc is significantly stronger than that in the forearc.

To investigate the characteristics of coda energy distribution over a greater region, in this article, we newly analyzed the data from 18 local events (Fig. 6) recorded by the Hi-net, which consists of about 800 borehole stations with average station spacing of 20 km distributed almost uniformly over Japan. To avoid possible contamination from

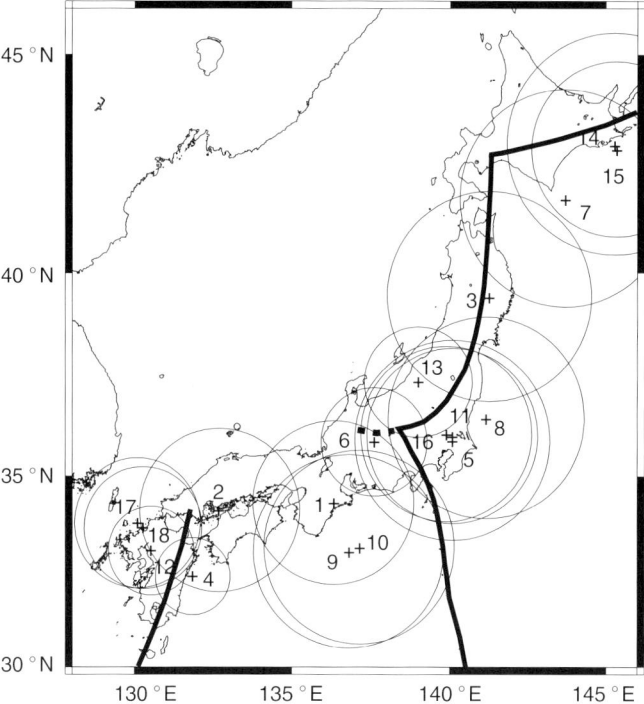

FIG. 6. Earthquakes occurred in Japan used in this study. The crosses show the epicenters of 18 local events. Stations included in the circles were used in the analysis. The bold line represents the volcanic front.

aftershocks, we selected earthquakes with very few aftershocks, of which the hypocentral parameters are listed in Table. 1. Nonuniform hypocentral distribution should not result in serious problems because any assumptions on coda wave distribution are not adopted and coda energy measurement is carried out at large lapse times. The velocity seismograms from about 700 borehole seismometers were used to measure the squared amplitudes of S-coda waves at lapse times at least 1.5 times after the S-wave travel time. In the analysis, a coda energy factor ($\propto R_i(f)C_i(f|t)$) was introduced as an observational parameter to characterize the lateral variation of coda energy. We assumed that $C_i(f|t)$ in Eq. (1) depends on the local absorption strength of S waves around station i. The coda energy factor can be estimated by an ordinary inversion algorithm employed in the analyses of site amplification (e.g., Phillips and Aki, 1986). We also considered that the contribution of the site amplification factor $R_i(f)$ on the lateral variation of the coda energy factor is relatively small and random, because all the seismograms are recorded at borehole stations.

Figure 7a shows a regional variation of the coda energy factor for a frequency band of 16–32 Hz, revealing that the coda energy is not uniformly distributed throughout Japan at this frequency band. In general, in the northeastern part of the Honshu island (Tohoku and Kanto regions), the coda energy level is higher and uniform in the forearc, and lower and nonuniform in the backarc, respectively. It is worth noting that the area in which the coda normalization method had been successfully applied for Q_P^{-1} and/or Q_S^{-1} estimation (Aki, 1980a,b; Yoshimoto et al., 1993) is restricted in the forearc side, where coda energy is distributed uniformly in space. This is probably why Takahashi et al. (2005) restricted their Q_S^{-1} estimation in the forearc of the Tohoku and Kanto regions by applying the coda normalization method.

TABLE. 1 Event source parameters for earthquakes used for the study of coda energy distribution (after the Japan Meteorological Agency)

No.	Time	Latitude (°N)	Longitude (°E)	Depth (km)	Magnitude
1	2000/10/31, 01:42:52.98	34.2987	136.3215	38.7	5.7
2	2001/03/26, 05:40:53.47	34.1172	132.7092	45.9	5.2
3	2001/12/02, 22:01:55.25	39.3983	141.2632	121.5	6.4
4	2002/11/04, 13:36:00.02	32.4127	131.8695	35.2	5.9
5	2003/05/12, 00:57:06.08	35.8688	140.0857	46.9	5.3
6	2003/05/18, 03:23:25.10	35.8672	137.5958	7.2	4.7
7	2003/09/26, 06:08:01.84	41.7098	143.6915	21.4	7.1
8	2003/11/15, 03:43:51.64	36.4325	141.1652	48.4	5.8
9	2004/09/05, 19:07:07.50	33.0332	136.7977	37.6	7.1
10	2004/09/05, 23:57:16.81	33.1375	137.1413	43.5	7.4
11	2004/10/06, 23:40:40.16	35.9888	140.0898	66.0	5.7
12	2004/11/04, 03:13:21.19	33.0775	130.5438	14.2	4.2
13	2004/11/09, 04:15:59.73	37.3540	138.9993	0.0	5.0
14	2004/11/29, 03:32:14.53	42.9460	145.2755	48.2	7.1
15	2004/12/06, 23:15:11.81	42.8477	145.3428	45.8	6.9
16	2005/02/16, 04:46:36.13	36.0385	139.8888	46.2	5.3
17	2005/03/21, 23:59:21.95	33.7853	130.1008	12.0	4.8
18	2005/04/10, 20:34:37.87	33.6685	130.2822	4.7	5.0

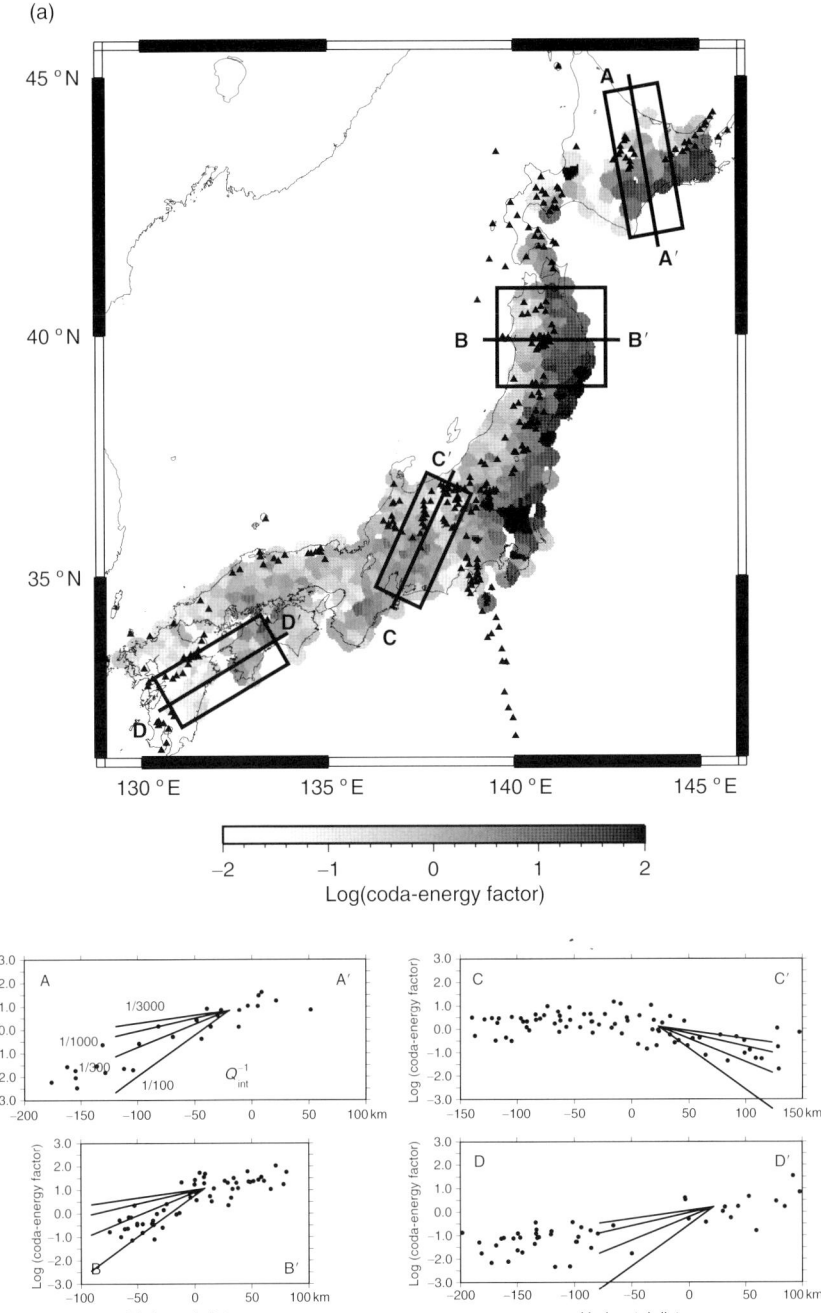

FIG. 7. (a) Regional variation of coda energy in the frequency band of 16–32 Hz in Japan. It should be noted that, as shown in Fig. 6, there are some areas where data is lacked: northern and eastern Hokkaido, southern Kyushu, etc. The solid triangles show locations of the Quaternary volcanoes. (b) Estimation of the intrinsic absorption parameter by using a diffusion–absorption model (the diffusion coefficient is assumed to be 133 km^2/s).

Figure 7a indicates that the observed coda energy level is low, systematically, in areas where the Quaternary volcanoes exist, such as central Hokkaido, western Tohoku, Hokuriku, in and around the Izu peninsula, and southwestern Kyushu. The spatial variation in the coda energy factor increases with increasing frequency, with the variation in magnitude reaching up to 40 dB across the volcanic front at a frequency band of 16–32 Hz. As shown in Fig. 7b, the estimated Q_{int}^{-1} using a diffusion–absorption model (Yoshimoto et al., 2006) is in the order of 10^{-3} for regions where the coda energy level is systematically low. The spatial correlation between the Quaternary volcanoes and the high heat-flux suggests that the thermal structure (or intrinsic absorption structure) of the crust and the uppermost mantle characterizes the regional variation of coda energy in high frequencies.

3. Temporal Decay Rate of Coda Energy: Q_C^{-1}

While Eq. (3) or (4) is valid for a single frequency f, Q_C^{-1} is estimated, typically, from octave-width bandpass-filtered seismograms from plots of the logarithm of $E(f|t)$ versus lapse time t within a given lapse time window. The long-standing puzzle among those who studied Q_C^{-1} is the significant variation and strong erratic behavior (e.g., Aster et al., 1996) from event to event at a single station within a short time period. Ouyang and Aki (1994) conducted an intensive investigation, focusing on both earthquakes and quarry blasts having detonated in the eastern Mojave California. They found that the variation in Q_C^{-1} measurements does not appear to decrease by the use of the blast sources compared with that measured from earthquakes. The inherent large fluctuation of every single measurement can only be stabilized by averaging over many events. Many investigations (e.g., Jin and Aki, 1988, 1993; Peng, 1989; Su and Aki, 1990) have found that the average value of Q_C^{-1} stabilizes when the number of single measurements reaches 10–20. Using clustered earthquakes with similar waveforms recorded by the U.S. Geological Survey Parkfield Dense Seismograph Array, Hellweg et al. (1995) investigated the stability of Q_C^{-1}, and suggested that, to obtain a stable Q_C^{-1} value, the observations should include array averaged measurements from a single lapse time, regardless of the source location.

In this section, we shall summarize the characteristics of Q_C^{-1} observed by numerous studies worldwide.

3.1. Lapse-Time Dependence

Observationally, Q_C^{-1} decreases with increasing lapse time, and takes almost a constant value after a certain lapse time (Hatzidimitriou, 1993; Hellweg et al., 1995; Kanao and Ito, 1990; Kosuga, 1992; Su et al., 1991). However, based on the multiple scattering models in unbounded uniform scattering media, the theoretical investigation predicts an opposite temporal variation: namely, at early lapse times Q_C^{-1} should increase with increasing lapse time (Hoshiba, 1991; Wennerberg, 1993). Wennerberg (1993) suggested that the observed decreases in Q_C^{-1} is due to the decrease of the intrinsic absorption with depth in the lithosphere.

Using the lapse time dependence of Q_C^{-1} at earlier lapse times, Gagnepain-Beyneix (1987) estimated the depth variation of Q_S^{-1} of the lithosphere beneath the western Pyrenean range, using local earthquake analyses. To interpret the apparent coda-wave attenuation estimated from local Kamchatka earthquakes, Gusev (1995a) proposed a model in which the scattering coefficient decayed with depth by an inverse-power-law. However, apparently, the early S-coda may be contaminated by the P-coda waves and also affected by the source radiation pattern (Sato, 1984).

As an example, Fig. 8 shows the results of Q_C^{-1} by Kosuga (1992) based on the analysis of aftershocks of the 1984 Western Nagano Prefecture Earthquakes of magnitude 6.8 in Japan. He used the data recorded at hypocentral distances of less than 23 km. Lapse time windows from 1.5 t_S to a series of t_{max} were used in his analysis over a wide frequency range of 1–64 Hz. As shown in the figure, the estimated value varies with both lapse time and frequency. With increasing lapse time Q_C^{-1} decreases for lapse times of less than 10 s, then approaches a constant value at a lapse time of about 24 s, at least. There seems to be a tendency for higher frequencies to require an earlier lapse time for Q_C^{-1} to reach a stable value, implying a fast approach to the diffusion regime where multiple scattering is dominant. In other words, the characteristic lapse time for Q_C^{-1} to become a constant value is both frequency and location dependent.

On the other hand, Q_C^{-1} at large lapse times are measured globally as a local structural parameter to characterize the spatial average of the seismic attenuation of the lithosphere.

3.2. Frequency Dependence

For a frequency range of 1–10 Hz, Q_C^{-1} can be expressed as $Q_C^{-1} = Q_0^{-1}(f/f_0)^{-\gamma}$, where Q_0^{-1} is the value of Q_C^{-1} at a reference frequency f_0, usually taken as 1 Hz. The power γ ranges from 0.5 to 1 among different regions, as summarized by Sato and Fehler (1998). It has been reported that tectonically active regions are generally characterized by high values of Q_0^{-1} and strong frequency dependence (e.g., Singh and Herrmann, 1983; Mitchell, 1995). Mitchell (1995) compiled the results from Lg coda attenuation measurements in the world and reported values of $Q_0^{-1} = 1/400$–$1/150$ and $\gamma = 0.3$–1 for tectonically active regions, such as western North America, western South America, and

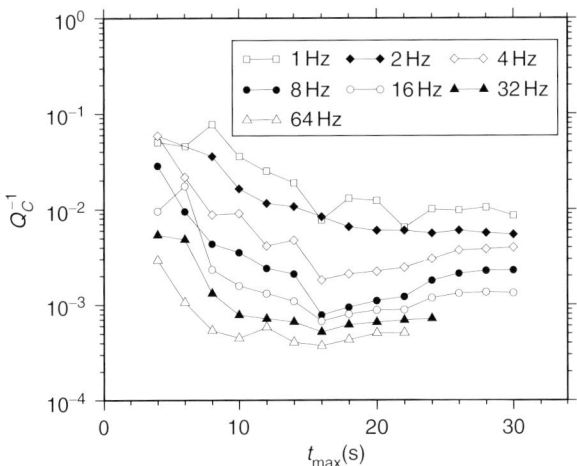

FIG. 8. Q_C^{-1} versus the end lapse time of the data window (t_{max}) measured from the earthquake origin time. (Reproduced from Kosuga, 1992.)

the Himalaya. In addition to the regional variations, local variations in Q_0^{-1} and γ in terms of the site geology were reported from the coda analysis of local network data (e.g., Steck et al., 1989). Meanwhile, a recent coda study in northwestern Colombia by Vargas et al. (2004) reported a clear correlation between Q_0^{-1} and γ, indicating a strong frequency dependence of Q_C^{-1} in the region where the value of Q_0^{-1} is large.

There are many observations that report the decrease in observed γ with increasing lapse time, or the length of time window used for Q_C^{-1} estimation (e.g., Gupta et al., 2006). Thus, it is important to minimize the effect of lapse time variation on Q_C^{-1} (not only γ but also Q_0^{-1}) when we compare the results from different observations. Figure 9 shows the frequency dependencies of Q_C^{-1} that have been reported by the analyses of coda waves of local earthquakes (epicentral distances less than 100 km) at large lapse times (> 60 s, except for the measurements for earthquakes with very small hypocentral distances such as Kosuga, 1992 and Giampiccolo et al., 2004) in the world. Selection of the similar Q_C^{-1} studies in conditions of data and measurements reduces the scatter of plot points, for example, as compared to Fig. 3.13 of Sato and Fehler (1998). Figure 9 indicates that, for the observational condition specified above, in average, Q_C^{-1} is about 10^{-2} at 1 Hz and decreases to about 10^{-3} at 10 Hz.

Interestingly, the frequency dependence tends to diminish for frequencies above 10 Hz or so, compared to that at lower frequencies. This tendency is similar to that observed for

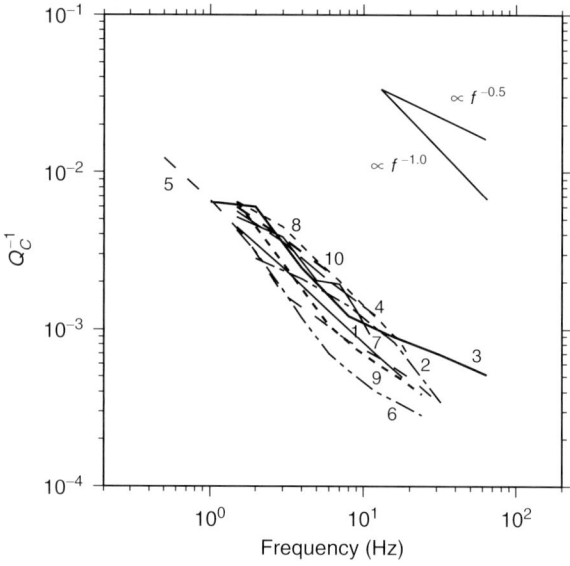

FIG. 9. Reported values of Q_C^{-1} for the lithosphere in the world. 1: Ibáñez et al. (1990) in the Granada Basin, southern Spain; 2: Kanao and Ito (1990) in central and northern Kinki region, Japan; 3: Kosuga (1992) in western Nagano region, central Japan; 4: Hatzidimitriou (1993) in northern Greece; 5: Woodgold (1994) in the Charlevoix, Quebec; 6: Gupta et al. (1998) in the Koyna region, India; 7: Bianco et al. (2002) in the Southern Apennine zone, Italy; 8: Giampiccolo et al. (2004) in southeastern Sicily, Italy; 9: Gupta et al. (2006) in the Kachchh Basin, India; 10: Tuvè et al. (2006) in the Straits of Messina area, Italy.

Q_S^{-1}, as summarized in the previous section. Based on seismic-frequency laboratory measurements using crustal and upper mantle rocks, Gribb and Cooper (1998) and Lu and Jackson (1998) had reported that the strain energy dissipation Q^{-1} is generally low and shows very weak frequency dependence at temperatures below a certain threshold: e.g., about 400 °C for Cape Sorell quartzite. These laboratory measurements may support the interpretation that Q_C^{-1} at frequencies exceeding 10 Hz is characterized mainly by intrinsic absorption in the lithosphere.

The physical mechanism of Q_C^{-1} has long been debated. Within the context of the single scattering theory, Q_C^{-1} is an attenuation parameter for average S-wave attenuation including both intrinsic absorption and scattering loss. On the other hand, numerical experiments by Frankel and Clayton (1986), laboratory experiments by Matsunami (1991), and theoretical studies including multiple scattering effects (e.g., Shang and Gao, 1988) concluded that Q_C^{-1} measured from the time window later than the mean free time should correspond only to intrinsic absorption, and should not include the effect of scattering loss. The debate concerning this issue was summarized by Aki (1991). Later, Jin et al. (1994) compiled the results from several areas/regions around the earth using MLTWA and found that in general, Q_C^{-1} lies between Q_S^{-1} and Q_{int}^{-1}. However, the domination varies with region and frequency.

Several studies (e.g., Gao and Aki, 1996; Yomogida et al., 1997) have suggested that the leakage of seismic energy from the scattering layer into the non-scattering deeper layers might cause a bias of the Q_C^{-1} observed. More recently, by applying the radiative transfer theory to a lithosphere model that consists of a scattering crust over a homogeneous mantle, Margerin et al., (1999) suggested that the effect of the leakage of the scattered S-wave energy from the scattering crust on Q_C^{-1} is not strong at frequencies exceeding 10 Hz. This result implies that Q_C^{-1} at frequencies exceeding 10 Hz is dominated by intrinsic absorption. Matsunami and Nakamura (2004) reported consistent observation that the intrinsic absorption becomes predominat above 4 Hz on the basis of an envelope analysis of shallow crustal earthquakes in Wakayama, southwestern Japan. In spite of these recent studies, for a more complete understanding of Q_C^{-1} for the whole seismic frequency band, we need more realistic models, including three-dimensional variation of scattering strength and intrinsic absorption, and worldwide field studies.

3.3. Geographic Variation

Results from numerous studies on Q_C^{-1} over the last two decades (e.g., Singh and Herrmann, 1983; Jin and Aki, 1988; Hoshiba, 1993; Mitchell et al., 1997; Baqer and Mitchell, 1998) show that Q_C^{-1} varies systematically with the tectonic activity. The differences on Q_C^{-1} between tectonic active and stable regions can be as large as more than an order of magnitude as summarized by Sato and Fehler (1998) and by Mitchell and Cong (1998). For example, in high frequencies ($f \gg 1$ Hz), large Q_C^{-1} have been observed around active volcanoes, implying strong intrinsic absorption due to high temperature volcanic medium (e.g., Matsumoto and Hasegawa, 1989; O'Doherty et al., 1997; Wu et al., 2006). In this section, as examples for lower frequencies, we shall review several studies concerning the spatial distribution of Q_C^{-1} at 1 Hz in the world.

Singh and Herrmann (1983) are the first to have constructed a Q_0^{-1} map over the continental United States, although the spatial resolution of their map was rather poor, since they had to use distant earthquakes for some stations. Figure 10 shows the 2D image of Lg coda Q_0^{-1} over the continental United States constructed by Baqer and

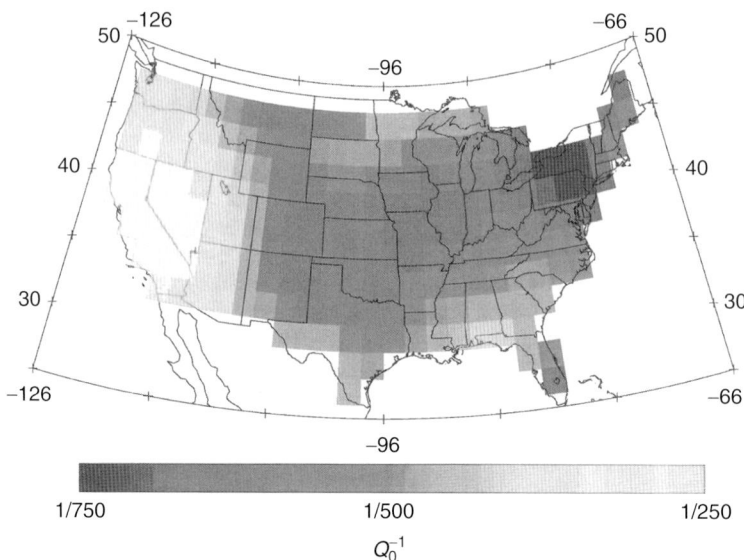

FIG. 10. 2D image of Lg coda Q_0^{-1} over the continental United States (note that the original measurements is not Q_0^{-1} but Q_0). (Reproduced from Baqer and Mitchell, 1998.)

Mitchell (1998). Clearly, California has the highest Q_0^{-1} and the northeast US has the lowest, indicating that Q_0^{-1} reflects the current seismicity better than other geophysical parameters, such as low P_n and S_n velocity and/or a thin lithosphere lid. For example, the area with the lowest S_n velocity and the thinnest lithosphere is probably the Basin and Range, but this is not the most seismically active area of California. However, Q_0^{-1} is higher in California than in the Basin and Range. The relatively high Q_0^{-1} region is also found in the southeast area along the Gulf coast, where there is no current seismicity but a so-called "growth fault" exists, generated by sedimentation in the Gulf of Mexico (Martin, 1978; Davis, 1984). Studies on coda waves in Alaska (Biswas and Aki, 1984; Steensma and Biswas, 1988; Dutta et al., 2004) revealed that Q_0^{-1} and the frequency dependence of Q_C^{-1} are similar to those in California, while both regions have comparable high seismicity. Such observation suggests that Q_C^{-1} at low frequencies, of say 1–5 Hz, may be indicating the degree of fracture in the lithosphere rather than the tectonic activity originating in the asthenosphere (Jin and Aki, 1989).

Jin and Aki (1988) constructed a map of Q_0^{-1} for mainland China with a spatial resolution of 150–200 km by using earthquakes at short distances from each station (Fig. 11). The variation in Q_0^{-1} at individual stations, as estimated from the time window for $2t_S$ to 100 s, is smooth enough to draw contours of equal Q_0^{-1}. This contour map and the epicenters of major earthquakes with $M > 7$ show a strong spatial correlation. For example, seismically active regions, such as Tibet, western Yunnan, and North China, correspond to high Q_0^{-1} regions, while stable regions, such as the Ordos plateau, middle-eastern China, and the desert in Southern Xinjiang have very low Q_0^{-1}. The difference between the peak and lowest Q_0^{-1} values amounts to a factor of more than a factor 20. In this figure, two different symbols are used to distinguish earthquakes occurring before

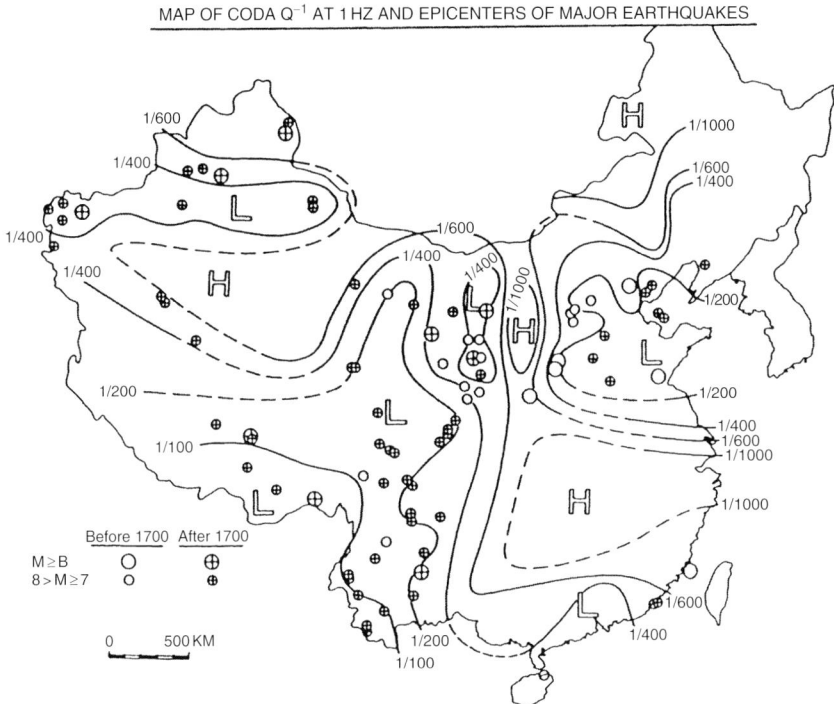

FIG. 11. Map of Q_C^{-1} at 1 Hz and epicenters of major earthquakes with $M \geq 7$ in China (note that the original measurements is not Q_C^{-1} but Q_C). Different symbols are used for $M \geq 8$ and $M < 8$, and before and after 1700. (Reproduced from Jin and Aki, 1988.)

1700 from those having occurred after 1700. There has been a migration of epicenters from west to east over the past 300 years in North China, and the Q_0^{-1} value for the region active before 1700 is about twice as high as that for the region currently active. Jin and Aki (1988) suggested that the high Q_0^{-1} region might also have migrated, together with the high seismicity, citing the high Q_0^{-1} values estimated by Chen and Nuttli (1984), based on the intensity maps for previous major earthquakes in the region. Since the other geophysical conditions of the lithosphere cannot migrate several hundred kilometers during a few hundred years, Jin and Aki (1988) again find that Q_0^{-1} is most likely presenting a fracture condition in the lithosphere related to seismicity.

The high local seismicity, together with the high quality seismograms recorded by the Hi-net, offers an opportunity to study the spatial distribution of Q_C^{-1} in Japan at unprecedented high resolution. Jin and Aki (2005) measured Q_C^{-1} at different frequency bands by using seismograms recorded at 582 Hi-net stations for earthquakes located within 30 km from each station. Figure 12 shows their Q_C^{-1} map for frequency band 1–2 Hz (referred to hereafter as Q_0^{-1}). Figure 12 shows a significant spatial variation of Q_0^{-1} within Japan, up to a factor of 3. The high Q_0^{-1} regions are correlated with the currently high seismic activity. However, the most conspicuous high Q_0^{-1} zone is a narrow belt from Niigata running in a south-westerly direction towards Biwa lake along the Japan Sea coast, where

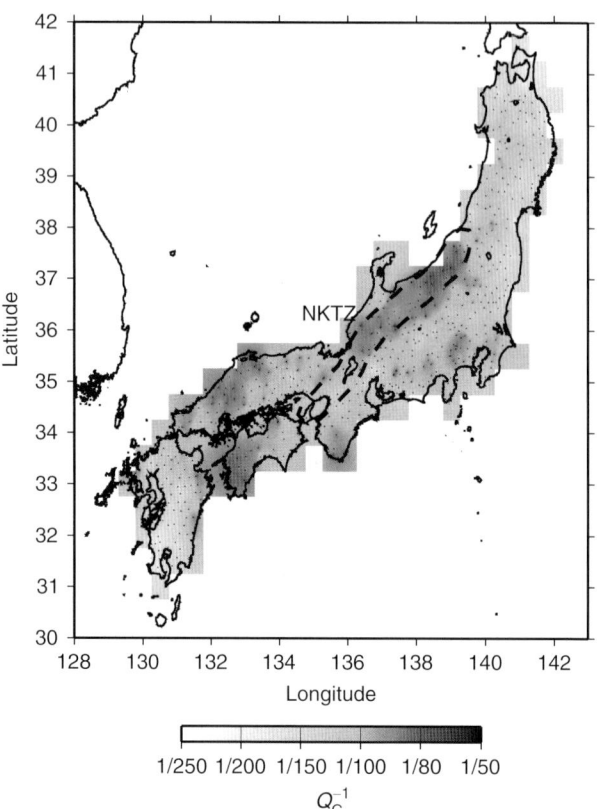

FIG. 12. Q_C^{-1} map in Japan for frequency band 1–2 Hz, where dots are Hi-net stations (note that the original measurements is not Q_C^{-1} but Q_C). Broken line indicates location of the Niigata-Kobe Tectonic Zone (NKTZ). (Reproduced from Jin and Aki, 2005.)

the GPS data reveals a high crustal deformation rate (Sagiya et al., 2000). The high deformation rate zone has been named as the Niigata-Kobe Tectonic Zone (NKTZ) by Sagiya et al. (2000). The observed difference in deformation rates between the NKTZ and its surrounding is almost one order of magnitude. To produce such a significant regional difference in deformation rate an anomalous stress concentration is required within the NKTZ, unless there is strong lateral variation in the stiffness/viscosity in the region. Iio et al. (2002, 2004) proposed a model that attributed the concentrated deformation to the low viscosity in weak zones existing in the lower crust of the lithosphere. The spatial coincidence observed by Jin and Aki (2005) may imply that Q_0^{-1} strongly reflects the plate-driving loading from the ductile part of the lithosphere.

3.4. Temporal Variation

Numerous studies revealed that the temporal correlation between Q_C^{-1} and seismicity is not as simple as the spatial correlation described above (see Sato, 1988, for a critical review of early works on temporal changes in Q_C^{-1}). In a number of cases (Gusev and Lemzikov, 1984; Novelo-Casanova et al., 1985; Jin and Aki, 1986; Sato, 1986; Faulkner, 1988; Su and Aki, 1990), Q_C^{-1} showed a peak during a period of 1–3 years before the occurrence of a major earthquake. For example, Tsukuda (1988) found in the epicentral area of the 1983 Misasa earthquake that a period of high Q_C^{-1} from 1977 to 1980 corresponds to a low rate of seismicity (quiescence). Meanwhile, the period of the high Q_C^{-1} before the Tangshan earthquake from 1973 to 1976, as observed by Jin and Aki (1986), also coincides with the period of quiescence in the epicentral area of the mainshock.

Gusev (1995b) investigated temporal changes in the relative amplitude levels of backscattered shear waves from local earthquakes in Kamchatka (ΔK_C in Fig. 13). He analyzed 24 years data from 1967 to 1990 and found two significant anomalies in the coda magnitude residuals (the deviation of coda magnitude at a station from network average magnitude): (1) proceeding to two $M8$ class earthquakes, a negative deviation of 3 years duration was observed at a station (KBG) within 100 km of the epicenters and (2) an anomaly with the same sign but 1.5 years duration was detected at a station (APH) in advance of a major fissure volcanic eruption located 70 km from the station. To investigate the temporal change in the scattering properties of the volcano, Grêt et al. (2005) applied a coda wave interferometry to reproducible seismic events recorded at Mount Erebus Volcano, Antarctica, and found a temporal variation of coda wave coherency, suggesting changes in the near-summit magma/conduit system.

On the other hand, such a characteristic pattern preceding some earthquakes is lacking before others, and sometimes a similar pattern was not followed by a major earthquake. For example, in order to minimize the possible effects from hypocentral parameters on Q_C^{-1} measurements, Beroza et al. (1995) analyzed earthquake doublets that span the preseismic, coseismic, and postseismic intervals to search temporal changes of coda attenuation in the vicinity of the Loma Prieta earthquake and found no significant difference of Q_C^{-1} among those three time periods.

The controversial observations on Q_C^{-1} behavior related to the occurrence of major earthquakes lead to the conclusion that Q_C^{-1} cannot be recognized as a reliable earthquake precursor (Wyss, 1991). However, this judgment concerning the reliability of the Q_C^{-1} precursor was based on the preconception that the physical system governing the precursor phenomena should be stationary in time.

Several convincing cases have been made for the temporal correlation between Q_C^{-1} and b-value. The result was initially puzzling because the correlation was negative in some cases (e.g., Aki, 1985; Jin and Aki, 1986; Robinson, 1987) and positive in others (e.g., Tsukuda, 1988; Jin and Aki, 1989).

Jin and Aki (1989, 1993) studied the temporal variation of Q_C^{-1}, using more than 50 years of records in southern and central California. Again, they found no certain pattern between Q_C^{-1} variation and the occurrence of major earthquakes. To characterize seismic activity in a seismic region, instead of using the b-value, Jin and Aki (1989) defined a new index as the partial number of earthquakes with magnitude of $M_i \leq M \leq M_i + 0.5$. In their study, the numbers were counted for $M_i = 3.0$, 3.5, 4.0, and 4.5, individually, using the ANSS catalog for central California and the SCEC catalog for southern California respectively. The windows of 100 consecutive earthquakes are overlapped by 25 events with the neighbors. Each time series of $N(M_i)$ is then used to calculate the cross-correlation with that of Q_C^{-1}. They found that the cross-correlation coefficient peaked (> 0.85) for $M_i = 3.0$ in southern California and $M_i = 4.0$ in central California

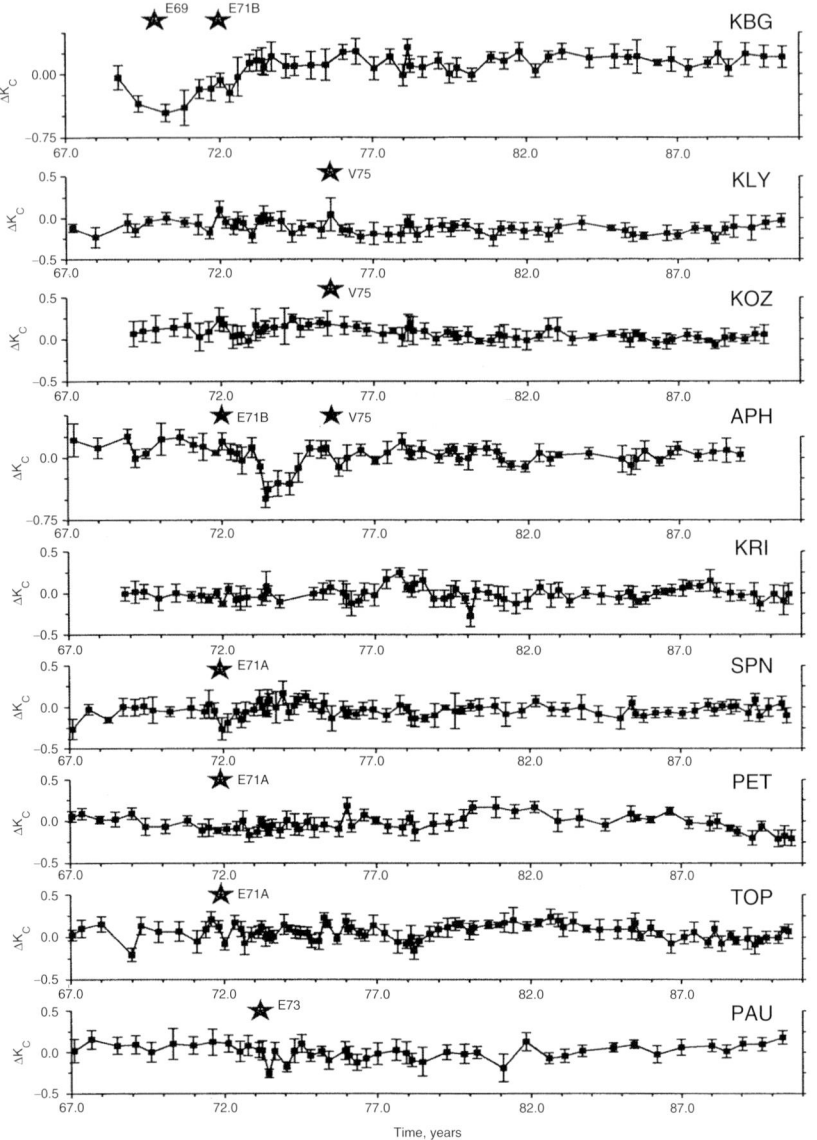

FIG. 13. Coda magnitude residual (ΔK_C) for nine Kamchatka stations, with error bars denoting $\pm 1\sigma$. Stars denote major geophysical events such as large earthquakes and volcanic eruptions in the vicinity of each station. (Reproduced from Gusev, 1995b.)

with zero-time shift (Fig. 14a, b); and that the correlation coefficients are significantly lower, sometimes even negative, for the other choices of M_i. Such a specific magnitude that characterizes the temporal correlation between Q_C^{-1} and $N(M_i)$ for a seismic region is called M_C.

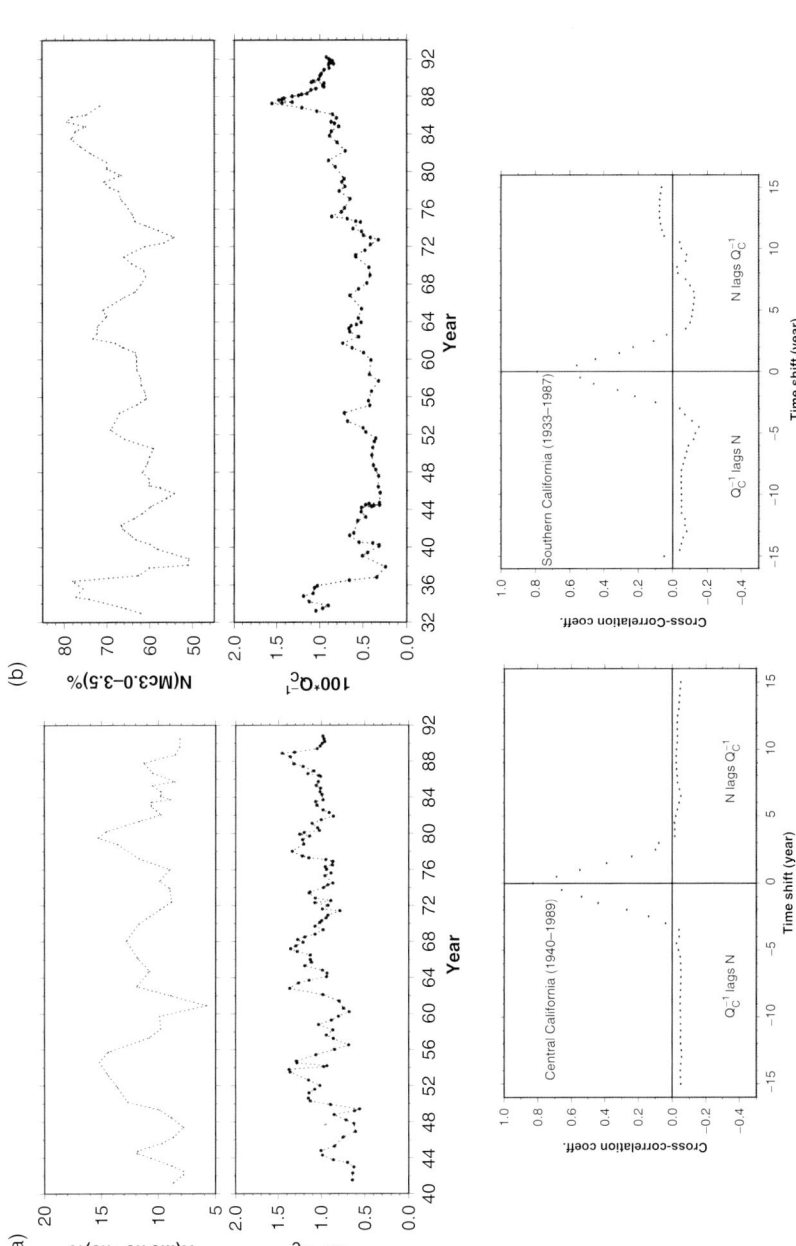

FIG. 14. Temporal change of Q_C^{-1} and $N(M_C)\%$ and the corresponding cross-correlation function: (a) central California and (b) southern California. (Reproduced from Jin and Aki, 1989, 1993.)

Using a high quality data set, produced by DRPI, Kyoto University, Hiramatsu *et al.* (2000) observed the temporal variation of Q_C^{-1} for the two frequency bands centered at 3 and 4 Hz, of which the temporal change associated with the occurrence of the Hyogo-ken Nanbu (Kobe) earthquake of 1995 is the most significant (Fig. 15). A Q_C^{-1} change of about 20% was observed for the 2-year period before and after the Hyogo-ken Nanbu earthquake, with a confidence level of 99% based on the Student's t test. They calculated the change in shear stress at a depth of 10 km, close to the depth of the brittle–ductile

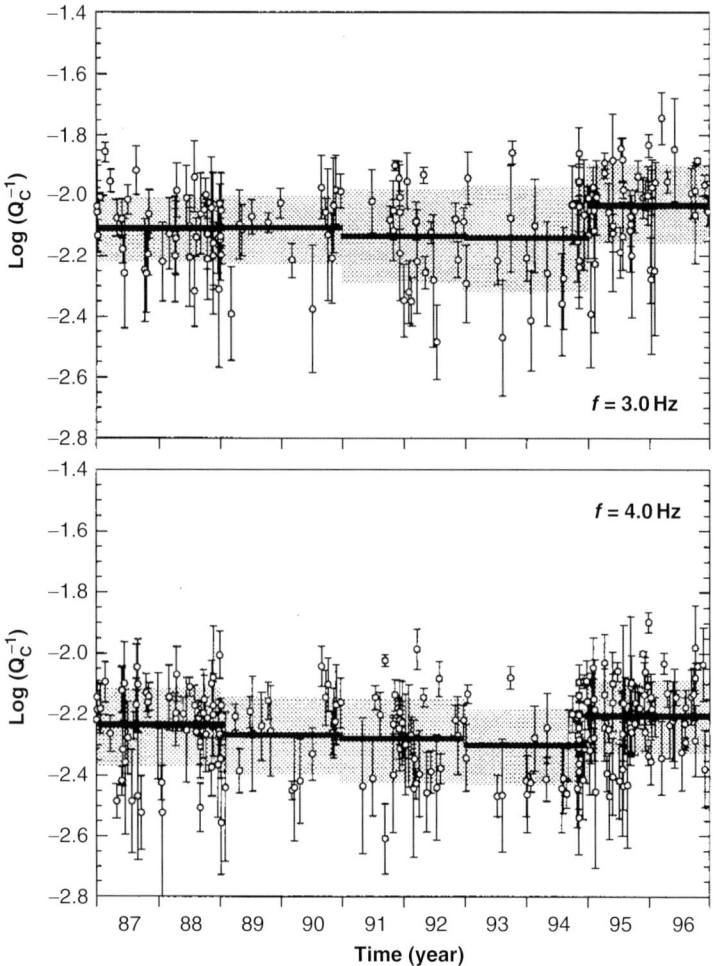

FIG. 15. Temporal variation of Q_C^{-1} at the frequency bands of 3.0 and 4.0 Hz in the Tamba region, southwestern Japan. Bold lines show the average, bars are the individual errors, and hatched zones are the standard deviation of data of each time window. Significant increase in Q_C^{-1} for 1995 to 1997 corresponds to the occurrence of the Kobe earthquake of 1995. (Reproduced from Hiramatsu *et al.*, 2000.)

transition zone, and found that the average increase in the shear stress over the study area was estimated to be about 0.02 MPa, implying that the stress-sensitivity of Q_C^{-1} is about 10% change per 0.01 MPa (0.1 bar). The frequency dependence of the change in Q_C^{-1} found by Hiramatsu et al. (2000) may be attributed to scattering by fractures. According to Benits (1990), for example, the scattering of seismic waves by a fracture is most effective when the wavelength is comparable to twice the characteristic length of the fracture. The wavelength of S waves for which the change in Q_C^{-1} was most significant is around 1 km corresponding to the characteristic size of the fracture around 500 m. Similarly, the corresponding M_C in this case is around 3.0.

3.5. Models to Explain the Spatio-Temporal Correlation Between Q_C^{-1} and Seismicity

To explain the observed temporal correlation between Q_C^{-1} and $N(M_C)$, Jin and Aki (1989, 1993) proposed a creep model, in which the creep fractures in the ductile part of the lithosphere are assumed to have a characteristic size within a given seismic region. The increased creep activity would then cause the seismic attenuation to increase and simultaneously generate stress concentration in the adjacent brittle part, likely to result in earthquakes of magnitude M_C corresponding to the characteristic size of the creep fracture. Thus, if M_C is at the lower end of the magnitude range from which the b-value is evaluated, the b-value would show a positive correlation with Q_C^{-1} and in the case of M_C being at the upper end, the correlation would be negative. This model explains the above mentioned puzzle correlation between the Q_C^{-1} and b-value, while a positive correlation is also expected between Q_C^{-1} and $N(M_C)$. The characteristic magnitude M_C corresponds to a characteristic scale length of heterogeneity in the brittle–ductile transition zone. The existence of a characteristic scale length within the brittle–ductile transition zone has been supported by various observations (see Aki, 1995, 2003, for the summary).

After a thorough survey of the tectonic stress in the Earth's lithosphere, based on observations concerning its brittle part, Zoback and Zoback (2002) concluded that the tectonically stable region is stable because of the low rate of deformation in its ductile part, and the active region is active because of the high rate of deformation in its ductile part. They stated that the force applied to the lower lithosphere will result in steady-state creep in the lower crust and upper mantle and hence the stress will be accumulated in the upper brittle crust due to the drag of the layers below.

The above view suggests a simultaneous occurrence of the higher (lower) rate of stress increase in the brittle part and the higher (lower) rate of deformation in the ductile part due to plate-driving forces. This coincides, exactly, with the creep model proposed by Jin and Aki (1989, 1993) and indicates that the Q_C^{-1} variation may reflect the dynamic loading process in the ductile layer and that the $N(M_C)$ may represent the response from the brittle layer.

Based on this fresh perspective, Aki (2003) re-examined the temporal correlation between Q_C^{-1} and $N(M_C)$ observed in California (Jin and Aki, 1989, 1993) and found a clear delay of Q_C^{-1} change relative to the $N(M_C)$; several years before both the M 7.1 Loma Prieta earthquake of 1989 and the M 7.3 Kern County earthquake of 1952. Figure 16a represents the time series and its correlation function 10-years before the Loma Prieta earthquake. Clearly, the correlation between the two time series is no longer simultaneous, but the Q_C^{-1} change is delayed by about 1 year relative to that of the $N(M_C)$ before both earthquakes.

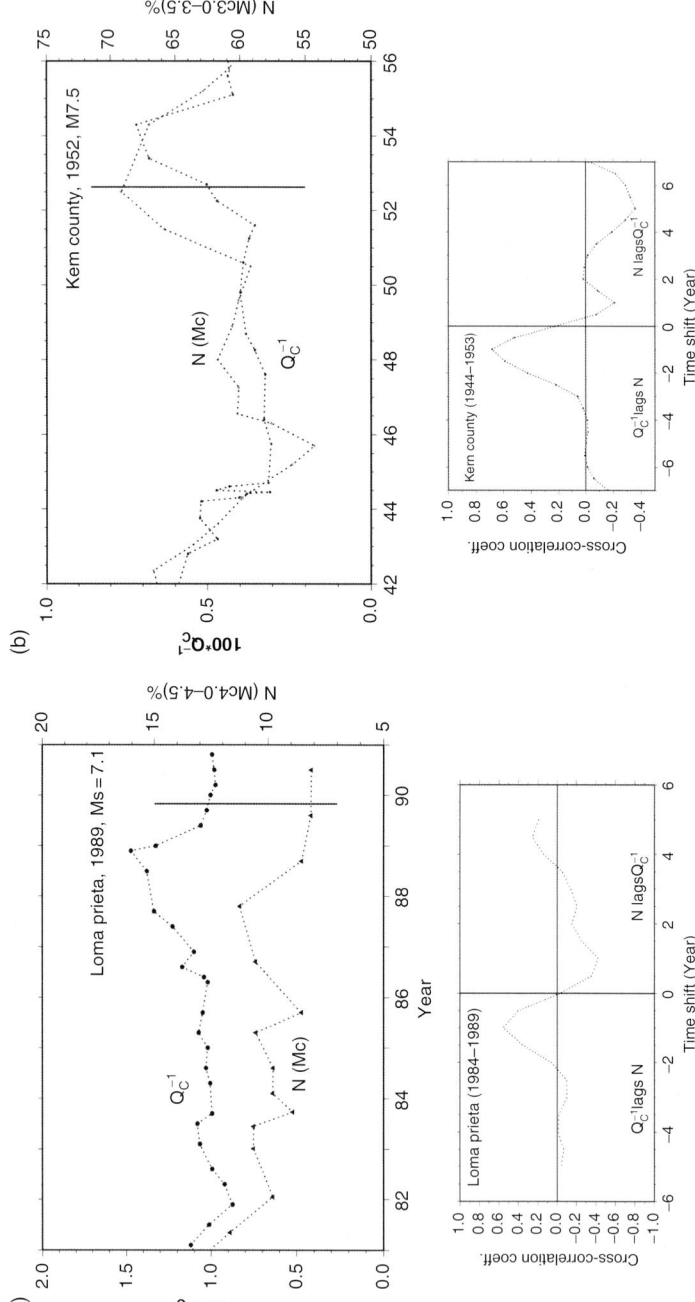

FIG. 16. (a) Time series of Q_C^{-1} and $N(M_C)$ 10 years before the occurrence of the Loma Prieta Earthquake, 1989, central California (upper). Cross-correlation function (lower) is calculated for 5-year time period before the occurrence of the M 7.1 event. (b) The same plot but for 10 years before the occurrence of the Kern County earthquake of 1952, southern California (upper). Cross-correlation function (lower) is calculated for 8-year time period before the occurrence of the M 7.5 event. (Reproduce from Aki, 2003.)

Jin et al. (2004) extended the analyses of the temporal variation of Q_C^{-1} and the seismicity for central and southern California to the year 2003. The cross-correlation function between Q_C^{-1} and $N(M_C)$ is calculated using a 10-year moving time window. The correlation coefficient for the entire period of about 60 years is peaked at the zero-time shift, with the value close to 0.8 for both regions. They found, however, that the simultaneous correlation was disturbed before every single major earthquakes, say $M > 7.0$, in both regions. As an example, Fig. 16b shows the disturbance before the Kern County Earthquake, 1952. The disturbances are, consistently, a delay in the change of Q_C^{-1} relative to that of $N(M_C)$ before the occurrence of a major earthquake. They attribute the temporal change in Q_C^{-1} to fractures in the ductile part of the lithosphere and that in $N(M_C)$ to the response of the brittle part to the ductile fracture. The M_C characteristic to a seismic region originates from the characteristic size of fractures in the ductile zone of the lithosphere. The observed delay of Q_C^{-1} change relative to $N(M_C)$ before a major earthquake can be explained simply by the strain energy stored in the brittle part of lithosphere reaching its saturation limit and starting to flow back to the ductile part. Aki (2003) pointed out that "When the stress in the brittle part builds up over time to the point of failure preparing for a major earthquake, we may expect a change in its mechanical property as a whole, as suggested in various laboratory experiments on rock samples. There is, however, an important difference between the laboratory and nature. During controlled laboratory experiments, the loading is a fixed condition given externally by the experimental device, whereas in nature, the loading is an internal process that is likely to be influenced by the change in property of the material being loaded. One may subsequently expect a change in the mode of loading as the brittle part undergoes such a change in preparation for failure. Thus, it might be responsible for a breakdown in the positive simultaneous correlation between Q_C^{-1} and $N(M_C)$."

Aki (2003) proposed a possible physical model called the "Brittle-Ductile Interaction Hypothesis (BDIH)." This model suggests that the observed temporal change in Q_C^{-1} may be unrelated to fractures in the brittle part, where the earthquakes are occurring, but primarily related to those in the ductile part of the lithosphere or in the transition zone from the brittle part to the ductile part.

Within the BDIH, the positive correlation between Q_C^{-1} and $N(M_C)$ might be a tool to monitor the status of the reaction of the ductile part of the lithosphere to the plate-driving force loading, whereupon the brittle part responds to the change of the ductile part within a seismic region. Jin et al. (2004) summarize the results from the studies on correlations between the temporal change of Q_C^{-1} and seismicity in Table. 2. In this table, the duration is defined as the time length during which the simultaneous correlation between the two time series is disturbed. The model has 4 parameters: (1) duration of the abnormal period, (2) delay time of the change of Q_C^{-1} and that of the $N(M_C)\%$, (3) the characteristic magnitude M_C, and (4) the dominant frequency, f_P, at which the peak Q_C^{-1} change occurs. It is interesting that M_C varies inversely with f_P in harmony with the brittle–ductile interaction model of earthquake loading as described in Aki (2003). According to this model, f_P corresponds to the fracture size in the ductile part of the lithosphere, which must be comparable to the size of an earthquake of magnitude M_C. The data listed in Table. 2 indicate that such requirement is at least qualitatively met.

TABLE. 2 The parameters of the brittle–ductile interaction model (after Jin et al., 2004)

Target earthquake	Reference	Duration (year)	Delay time (year)	M_C	f_P (Hz)
Stone Canyon ($M5$)	Chouet (1979)	(Normal period)		1–2	24
Misasa, Japan ($M6.2$)	Tsukuda (1988)	>8	2–3	2–3	5–10
Loma Prieta ($M7.1$)	Jin and Aki (1993)	7	1	4–4.5	1–3
Kobe, Japan ($M7.2$)	Hiramatsu et al. (2000)	6	2	2.6–3.6	1.5–4
Kern County ($M7.5$)	Jin and Aki (1989)	8	1	3–3.5	1–3
Landers ($M7.3$)	Jin et al. (2004)	10	4	3–3.5	1–3
Hector Mine ($M7.1$)	Jin et al. (2004)	6	3.5	3–3.5	1–3
Tangshan ($M7.8$)	Jin and Aki (1986); Li and Chen (1981)	?	3	4.5–5	1–2

4. Closing Remarks

1. The common decay rate of the coda envelope after a certain lapse time observed in a given seismic region supported, empirically, a simple separation of the effects of source, propagating path, and the recording site on coda spectra. This simplicity leads to a powerful technique for seismological studies called the "coda normalization method." The application of this method on a single station records enables us to measure attenuation factor of direct P- and S-waves, Q_P^{-1} and Q_S^{-1}, free from the source and/or site correction.
2. In the context of the single scattering model coda Q^{-1} (Q_C^{-1}) estimated from a given lapse time window around t represents the S-wave attenuation of the lithosphere averaged over an ellipsoidal volume, with the source and receiver as the foci and a radius of $V_S t/2$. On the other hand, the multiple scattering model implies that Q_C^{-1} measured from the time window later than the mean free time should correspond only to intrinsic absorption, and should not include the effect of scattering loss.
3. Numerous observations demonstrated the following characteristics of Q_C^{-1} :
(i) Observed Q_C^{-1} exhibits frequency dependence: Q_C^{-1} estimated from coda waves of local earthquakes at large lapse times is about 10^{-2} at 1 Hz and decreases to about 10^{-3} at 10 Hz. In general, Q_C^{-1} lies between direct S-wave Q_S^{-1} and intrinsic absorption Q_{int}^{-1}.
(ii) Observed Q_C^{-1} shows systematical correlation with tectonic activity: high Q_C^{-1} in tectonically active regions such as California, Japan, and low Q_C^{-1} in tectonically stable regions, such as the northeastern US and Norway.
(iii) Attempts have been made to correlate the temporal variations of Q_C^{-1} with the occurrence of major earthquakes, although the observations are controversial. The observed positive temporal correlation between Q_C^{-1} and $N(M_C)$ may offer a possible clue to monitor the stress status of the brittle–ductile transition zone under the plate-driving loading processes.

4. Analyses of high density seismic network data have revealed non-uniform distribution of coda energy. The non-uniformity increases with increasing frequency. The spatial correlation between the Quaternary volcanoes and the high heat flux suggests that the thermal structure (or intrinsic absorption structure) of the crust and the uppermost mantle characterizes the regional variation of coda energy at high frequencies.

ACKNOWLEDGMENTS

The authors are grateful to H. Sato of Tohoku University for his valuable discussions and suggestions. Insightful reviews by E. Del Pezzo and N. Biswas greatly improved the content of this article. The authors are also indebted to the National Research Institute for Earth Science and Disaster Prevention for providing the Hi-net data used in this study. This work was partially (for K.Y.) supported by the Grant-in-Aid for Scientific Research No. 17740291 of the Ministry of Education, Culture, Sports, Science and Technology (MEXT) from 2005 to 2007. The GMT software (Wessel and Smith, 1998) was used to create some figures.

REFERENCES

Adams, D.A., Abercrombie, R.E. (1998). Seismic attenuation above 10 Hz in southern California from coda waves recorded in the Cajon Pass borehole. *J. Geophys. Res.* **103**, 24257–24270.
Aki, K. (1969). Analysis of seismic coda of local earthquakes as scattered waves. *J. Geophys. Res.* **74**, 615–631.
Aki, K. (1980a). Attenuation of shear-waves in the lithosphere for frequencies from 0.05 to 25 Hz. *Phys. Earth Planet. Inter.* **21**, 50–60.
Aki, K. (1980b). Scattering and attenuation of shear waves in the lithosphere. *J. Geophys. Res.* **85**, 6496–6504.
Aki, K. (1981). Attenuation and scattering of short-period seismic waves in the lithosphere. In: Husebye, E.S., Mykkeltveit, S. (Eds.), *Identification of Seismic Sources-Earthquake or Underground Explosion*, D. Reidel, Dordrecht, Holland, pp. 515–541.
Aki, K. (1985). Theory of earthquake prediction with special reference to monitoring of the quality factor of lithosphere by the coda method. *Earthquake Predict. Res.* **3**, 219–230.
Aki, K. (1991). Summary of discussion on coda waves at the Istanbul IASPEI meeting. *Phys. Earth Planet. Inter.* **67**, 1–3.
Aki, K. (1992). Scattering conversions P to S versus S to P. *Bull. Seismol. Soc. Am.* **82**, 1969–1972.
Aki, K. (1995). Interrelation between fault zone structures and earthquake processes. *Pure Appl. Geophys.* **145**, 647–676.
Aki, K. (2003). Seismology of earthquake and volcano prediction, lecture notes for the Seventh International Workshop on Non-linear dynamics and earthquake prediction. Workshop on Volcanic Disaster Mitigation, Tokyo, Japan.
Aki, K., Chouet, B. (1975). Origin of coda waves: Source, attenuation, and scattering effects. *J. Geophys. Res.* **80**, 3322–3342.
Aki, K., Ferrazzini, V. (2000). Seismic monitoring and modeling of an active volcano for prediction. *J. Geophys. Res.* **105**, 16617–16640.
Akinci, A., Del Pezzo, A., Ibáñez, J.M. (1995). Separation of scattering and intrinsic attenuation in southern Spain and western Anatolia (Turkey). *Geophys. J. Int.* **121**, 337–353.
Akinci, A., Eyidoğan, H. (2000). Scattering and anelastic attenuation of seismic energy in the vicinity of north Anatolian fault zone, eastern Turkey. *Phys. Earth Planet. Inter.* **122**, 229–239.

Asano, Y., Hasegawa, A. (2004). Imaging the fault zones of the 2000 western Tottori earthquake by a new inversion method to estimate three-dimensional distribution of the scattering coefficient. *J. Geophys. Res.* **109**, doi: 10.1029/2003JB002761.

Aster, R.C., Slad, G., Henton, J., Antolik, M. (1996). Differential analysis of coda Q using similar microearthquakes in seismic gaps. Part 1: Techniques and application to seismograms recorded in the Anza Seismic Gap. *Bull. Seismol. Soc. Am.* **86**, 868–889.

Baqer, S., Mitchell, B.J. (1998). Regional variation of Lg coda Q in the continental United States and its relation to crustal structure and evolution. *Pure Appl. Geophys.* **153**, 613–636.

Benits, R. (1990). Seismological Applications of Boundary Integral and Gaussian Beam Methods. PhD thesis, MIT, Massachusetts.

Benz, H.M., Frankel, A., Boore, D.M. (1997). Regional Lg attenuation for the continental United States. *Bull. Seismol. Soc. Am.* **87**, 606–619.

Beroza, G.C., Cole, A.T., Ellsworth, W.L. (1995). Stability of coda wave attenuation during the Loma Prieta, California, earthquake sequence. *J. Geophys. Res.* **100**, 3977–3987.

Bianco, F., Del Pezzo, E., Castellano, M., Ibanez, J., Di Luccio, F. (2002). Separation of intrinsic and scattering seismic attenuation in the Southern Apennine zone, Italy. *Geophys. J. Int.* **150**, 10–22.

Bianco, F., Del Pezzo, E., Malagnini, L., Di Luccio, F., Akinci, A. (2005). Separation of depth-dependent intrinsic and scattering seismic attenuation in the northeastern sector of the Italian Peninsula. *Geophys. J. Int.* **161**, doi: 10.1111/j.1365–246X.2005.02555.x.

Biswas, N.N., Aki, K. (1984). Characteristics of coda waves: Central and southcentral Alaska. *Bull. Seismol. Soc. Am.* **74**, 493–507.

Castro, R.R., Munguía, L. (1993). Attenuation of P and S waves in the Oaxaca, Mexico, subduction zone. *Phys. Earth Planet. Inter.* **76**, 179–187.

Chen, P., Nuttli, O.W. (1984). Estimates of magnitudes and short-period wave attenuation of Chinese earthquakes from modified Mercalli Intensity data. *Bull. Seismol. Soc. Am.* **74**, 957–968.

Chouet, B. (1979). Temporal variation in the attenuation of earthquake coda near Stone Canyon, California. *Geophys. Res. Lett.* **6**, 143–146.

Chung, T.-W., Sato, H. (2001). Attenuation of high-frequency P- and S-waves in the crust of the southeastern Korea. *Bull. Seismol. Soc. Am.* **91**, 1867–1874.

Dainty, A.M. (1981). A scattering model to explain seismic Q, observations in the lithosphere between 1 and 30 Hz. *Geophys. Res. Lett.* **8**, 1126–1128.

Davis, G.H. (1984). Structure Geology of Rocks and Regions. Wiley, New York.

Dutta, U., Biswas, N.N., Adams, D.A., Papageorgiou, A. (2004). Analysis of S-wave attenuation in south-central Alaska. *Bull. Seismol. Soc. Am.* **94**, 16–28.

Faulkner, J. (1988). Temporal variation of coda Q MS Thesis, University of Southern California. Los Angeles.

Felher, M., Hoshiba, M., Sato, H., Obara, K. (1992). Separation of scattering and intrinsic attenuation for the Kanto-Tokai region, Japan, using measurements of S-wave energy versus hypocentral distance. *Geophys. J. Int.* **108**, 787–800.

Frankel, A., Clayton, R.W. (1986). Finite difference simulations of seismic scattering: Implications for the propagation of short-period seismic waves in the crust and models of crust heterogeneity. *J. Geophys. Res.* **91**, 6465–6489.

Frankel, A., Wennerberg, L. (1987). Energy-flux model of seismic coda: Separation of scattering and intrinsic attenuation. *Bull. Seismol. Soc. Am.* **77**, 1223–1251.

Friedrich, C., Wegler, U. (2005). Localization of seismic coda at Merapi volcano (Indonesia). *Geophys. Res. Lett.* **32**, L14312, doi: 10.1029/2005GL023111.

Gagnepain-Beyneix, U. (1987). Evidence of spatial variations of attenuation in the western Pyrenean range. *Geophys. J. R. Astron. Soc.* **89**, 681–704.

Gao, L.S., Aki, K. (1996). Effect of finite thickness of scattering layer on coda Q of local earthquakes. *J. Geodyn.* **21**, 191–213.

Giampiccolo, S., Gresta, S., Rasconà, F. (2004). Intrinsic and scattering attenuation from observed seismic codas in Southeastern Sicily (Italy). *Phys. Earth Planet. Inter.* **145**, 55–66.
Giampiccolo, E., Tuvè, E., Gresta, S., Patanè, D. (2006). S-waves attenuation and separation of scattering and intrinsic absorption of seismic energy in southeastern Sicily (Italy). *Geophys. J. Int.* **165**, doi: 10.1111/j.1365–246X.2006.02881.x.
Goutbeek, F.H., Dost, B., van Eck, T. (2004). Intrinsic absorption and scattering attenuation in the southern part of The Netherlands. *J. Seismol.* **8**, 11–23.
Grêt, A., Snieder, R., Aster, R.C., Kyle, P.R. (2005). Monitoring rapid temporal change in a volcano with coda wave interferometry. *Geophys. Res. Lett.* **32**, L06304, doi: 10.1029/2004GL021143.
Gribb, T.T., Cooper, R.F. (1998). Low-frequency shear attenuation in polycrystalline olivine: Grain boundary diffusion and the physical significance of the Andrade model for viscoelastic rheology. *J. Geophys. Res.* **103**, 27267–27279.
Gupta, S.C., Teotia, S.S., Rai, S.S., Gautam, N. (1998). Coda Q estimates in the Koyna region, India. *Pure Appl. Geophys.* **153**, 713–731.
Gupta, S.C., Kumar, A., Shukla, A.K., Suresh, G., Baidya, P.R. (2006). Coda Q in the Kachchh Basin, western India using aftershocks of the Bhuj earthquake of January 26, 2001. *Pure Appl. Geophys.* **163**, 1583–1595, doi: 10.1007/s00024–006–0086–2.
Gusev, A.A. (1995a). Vertical profile of turbidity and coda Q. *Geophys. J. Int.* **123**, 665–672.
Gusev, A.A. (1995b). Baylike and continuous variations of the relative level of the late coda during 24 years of observation on Kamchatka. *J. Geophys. Res.* **100**, 20311–20319.
Gusev, A.A., Abubakirov, I.R. (1997). Study of the vertical scattering properties of the lithosphere based on the inversion of P- and S-wave pulse broadening data. *Volc. Seis.* **18**, 453–464.
Gusev, A.A., Abubakirov, I.R. (1999). Vertical profile of effective turbidity reconstructed from broadening of incoherent body-wave pulses—II. Application to Kamchatka data. *Geophys. J. Int.* **136**, 309–323.
Gusev, A.A., Lemzikov, V.K. (1984). The anomalies of small earthquake coda wave characteristics before the three large earthquakes in the Kuril–Kamchatka zone (in Russian). *Vulk. Seism.* **4**, 76–90.
Hatzidimitriou, P.M. (1993). Attenuation of coda waves in northern Greece. *Pure Appl. Geophys.* **140**, 63–78.
Hatzidimitriou, P.M. (1994). Scattering and anelastic attenuation of seismic energy in northern Greece. *Pure Appl. Geophys.* **143**, 587–601.
Hellweg, M., Spudich, P., Fletcher, J.B., Baker, L.M. (1995). Stability of coda Q in the region of Parkfield, California: View from the U.S. Geological Survey Parkfield Dense Seismograph array. *J. Geophys. Res.* **100**, 2089–2102.
Herraiz, M., Espinosa, A.F. (1987). Coda waves: A review. *Pure Appl. Geophys.* **125**, 499–577.
Hiramatsu, Y., Hayashi, N., Furumoto, M., Katao, H. (2000). Temporal changes in coda Q^{-1} and b value due to the static stress change associated with the 1995 Hyogo-ken Nanbu earthquake. *J. Geophys. Res.* **105**, 6141–6151.
Hoshiba, M. (1991). Simulation of multiple-scattered coda wave excitation based on the energy conservation law. *Phys. Earth Planet. Inter.* **67**, 123–136.
Hoshiba, M. (1993). Separation of scattering attenuation and intrinsic absorption in Japan using the multiple lapse time window analysis form full seismogram envelope. *J. Geophys. Res.* **98**, 15809–15824.
Hoshiba, M., Sato, H., Fehler, M. (1991). Numerical basis of the separation of scattering and intrinsic absorption from full seismogram envelope—A Monte-Carlo simulation of multiple isotropic scattering. *Pa. Meteorol. Geophys. Meteorol. Res. Inst.* **42**, 65–91.
Ibáñez, J.M., Del Pezzo, E., De Miguel, F., Herraiz, M., Alguacil, G., Morales, J. (1990). Depth-dependent seismic attenuation in the Granada zone (Southern Spain). *Bull. Seismol. Soc. Am.* **80**, 1232–1244.
Iio, Y., Sagiya, T., Kobayashi, Y., Shiozaki, I. (2002). Water-weakened lower crust and its role in the concentrated deformation in the Japanese Islands. *Earth Planet. Sci. Lett.* **203**, 245–253.

Iio, Y., Sagiya, T., Kobayashi, Y. (2004). Origin of the concentrated deformation zone in the Japanese Islands and stress accumulation process of intraplate earthquakes. *Earth Planets Space* **56**, 831–842.

Jannaud, L.R., Adler, P.M., Jacqin, C.G. (1991). Spectral analysis and inversion of codas. *J. Geophys. Res.* **96**, 18215–18231.

Jin, A., Aki, K. (1986). Temporal change in coda Q before the Tangshan earthquake of 1976 and the Haicheng earthquake of 1975. *J. Geophys. Res.* **91**, 665–673.

Jin, A., Aki, K. (1988). Spatial and temporal correlation between coda Q and seismicity in China. *Bull. Seismol. Soc. Am.* **78**, 741–769.

Jin, A., Aki, K. (1989). Spatial and temporal correlation between coda Q^{-1} and seismicity and its physical mechanism. *J. Geophys. Res.* **94**, 14041–14059.

Jin, A., Aki, K. (1993). Temporal correlation between coda Q^{-1} and seismicity—Evidence for a structure unit in the brittl–ductile transition zone. *J. Geodyn.* **17**, 95–120.

Jin, A., Aki, K. (2005). High-resolution maps of coda Q in Japan and their interpretation by the brittle–ductile interaction hypothesis. *Earth Planets Space* **57**, 403–409.

Jin, A., Aki, K., Liu, Z., Keilis-Borok, V.I. (2004). Seismological evidence for the brittle–ductile interaction hypothesis on earthquake loading. *Earth Planets Space* **56**, 823–830.

Jin, A., Ando, M. (1998). Coda energy distribution in the Hida region, central Japan Proceeding of the Seismological Society of Japan Annual Meeting, 1998, (Abstract).

Jin, A., Mayada, K., Adams, D., Aki, K. (1994). Separation of intrinsic and scattering attenuation in southern California using TERRAscope data. *J. Geophys. Res.* **99**, 17835–17848.

Kanao, M., Ito, K. (1990). Attenuation property of coda waves in the middle and northern parts of Kinki district (in Japanese with English abstract). *Zisin 2* **43**, 311–320.

Kato, K., Takemura, M., Yashiro, K. (1998). Regional variation of source spectra in high-frequency range determined from strong motion records (in Japanese with English abstract). *Zisin 2* **51**, 123–138.

Kinoshita, S. (1994). Frequency-dependent attenuation of shear waves in the crust of the Southern Kanto area, Japan. *Bull. Seismol. Soc. Am.* **84**, 1387–1396.

Kinoshita, S., Ohike, M. (2002). Scaling relations of earthquakes that occurred in the upper part of the Philippine Sea Plate beneath the Kanto region, Japan, estimated by means of borehole recordings. *Bull. Seismol. Soc. Am.* **92**, 611–624.

Knopoff, L., Hudson, J.A. (1964). Scattering of elastic waves by small inhomogeneities. *J. Acoust. Soc. Am.* **36**, 338–343.

Knopoff, L., Hudson, J.A. (1967). Frequency dependence of amplitude of scattered elastic waves. *J. Acoust. Soc. Am.* **42**, 18–20.

Kosuga, M. (1992). Dependence of coda Q on frequency and lapse time in the western Nagano region, central Japan. *J. Phys. Earth* **40**, 421–445.

Lacombe, C., Campillo, M., Paul, A., Margerin, L. (2003). Separation of intrinsic absorption and scattering attenuation from Lg coda decay in central France using acoustic radiative transfer theory. *Geophys. J. Int.* **154**, 417–425.

Leary, P., Abercrombie, R. (1994). Frequency dependent crustal scattering and absorption at 5–160 Hz from coda decay observed at 2.5 km depth. *Geophys. Res. Lett.* **21**, 971–974.

Lee, W.S., Sato, H., Lee, K. (2003). Estimation of S-wave scattering coefficient in the mantle from envelope characteristics before and after the ScS arrival. *Geophys. Res. Lett.* **30**, doi: 10.1029/2003GL018413.

Li, Q., Chen, B. (1981). Spatial and temporal change of b-value before the Tangshan earthquake of 1976 (in Chinese with English abstract). *J. Geophys.* **21**, 89–96.

Lu, C., Jackson, I. (1998). Seismic-frequency laboratory measurements of shear mode viscoelasticity in crustal rocks II: Thermally stressed quartzite and granite. *Pure Appl. Geophys.* **153**, 441–473.

Margerin, L., Campillo, M., Shapiro, N.M., van Tiggelen, B. (1999). Residence time of diffuse waves in the crust as a physical interpretation of coda Q: Application to seismograms recorded in Mexico. *Geophys. J. Int.* **138**, 343–352.
Martin, R.G. (1978). Northern and eastern Gulf of Mexico continental margin: Stratigraphic and structural framework. In: Bouma, A.H., Moore, G.T., Coleman, J.M. (Eds.), *Framework, Facies, and Oil-Trapping Characteristics of the Upper Continental Margin. AAPG Stud. Geol.*, Vol. 7, pp. 21–42.
Matsumoto, S., Hasegawa, A. (1989). Two-dimensional coda Q structure beneath Tohoku, NE Japan. *Geophys. J. Int.* **99**, 101–108.
Matsunami, K. (1991). Laboratory tests of excitation and attenuation of coda waves using 2D models of scattering media. *Phys. Earth Planet. Inter.* **67**, 104–114.
Matsunami, K., Nakamura, M. (2004). Seismic attenuation in a nonvolcanic swarm region beneath Wakayama, southwest Japan. *J. Geophys. Res.* **109**, doi: 10.1029/2003JB002758.
Mayeda, K., Koyanagi, S., Aki, K. (1991). Site amplification from S-wave coda in the Long Valley caldera region, California. *Bull. Seismol. Soc. Am.* **81**, 2194–2213.
Mayeda, K., Koyanagi, S., Hoshiba, M., Aki, K., Zeng, Y. (1992). A comparative study of scattering, intrinsic and coda Q^{-1} for Hawaii, Long Valley, and central California between 1.5 and 15.0 Hz. *J. Geophys. Res.* **97**, 6643–6659.
Mayeda, K., Walter, W.R. (1996). Moment, energy, stress drop, and source spectra of western United States earthquakes from regional coda envelopes. *J. Geophys. Res.* **101**, 11195–11208.
Mitchell, B.J. (1995). Anelastic structure and evolution of the continental crust and upper mantle from seismic surface wave attenuation. *Rev. Geophys.* **33**, 441–462.
Mitchell, B.J., Cong, L. (1998). Lg coda Q and its relation to the structure and evolution of continents: A global prospective. *Pure Appl. Geophys.* **153**, 655–663.
Mitchell, B.J., Pan, Y., Xie, J., Cong, L. (1997). Lg coda Q variation across Eurasia and its relation to crustal evolution. *J. Geophys. Res.* **102**, 22767–22780.
Mitchell, B.J., Romanowicz, B. (Eds) (1998). Q of the Earth. *Pure Appl. Geophys.* **153**, 235–713.
Morasca, P., Mayeda, K., Malagnini, L., Walter, W.R. (2005). Coda derived source spectra, moment magnitudes, and energy-moment scaling in the western Alps. *Geophys. J. Int.* **160**, 263–275, doi: 10.1111/j.1365–246X.2005.02491.x.
Nishigami, K. (2000). Deep crustal heterogeneity along and around the San Andreas fault system in central California and its relation to the segmentation. *J. Geophys. Res.* **105**, 7983–7998.
Novelo-Casanova, D.A., Berg, E., Hsu, Y., Helsley, C.E. (1985). Time-space variation seismic S-wave coda attenuation (Q^{-1}) and magnitude distribution (b-values) for the Petatlan earthquake. *Geophys. Res. Lett.* **12**, 789–792.
O'Doherty, K.B., Bean, C.J., Closkey, J.M. (1997). Coda wave imaging of the Long Valley caldera using a spatial stacking technique. *Geophys. Res. Lett.* **24**, 1547–1550.
Ojeda, A., Ottemöller, L. (2002). Q_{Lg} tomography in Colombia. *Phys. Earth Planet. Inter.* **130**, 253–270.
Ouyang, H., Aki, K. (1994). *EOS* **75**, 186. (Abstract).
Peng, J.Y. (1989). Spatical and Temporal Variation of Coda Q^{-1} in California, Ph.D Thesis, University of Southern California, Los Angeles.
Phillips, W.S., Aki, K. (1986). Site amplification of coda waves from local earthquakes in central California. *Bull. Seismol. Soc. Am.* **76**, 627–648.
Rautian, T.G., Khalturin, V.I. (1978). The use of coda for determination of the earthquake source spectrum. *Bull. Seismol. Soc. Am.* **68**, 923–948.
Revenaugh, V.I. (1995). A scattered-wave image of subduction beneath the transverse ranges. *Science* **268**, 1888–1892.
Robinson, R. (1987). Temporal variations in coda duration of local earthquakes in the Wellington region, New Zealand. *Pure Appl. Geophys.* **125**, 579–596.

Sagiya, T., Miyazaki, S., Tada, T. (2000). Continuous GPS array and present-day crustal deformation of Japan. *Pure Appl. Geophys.* **157**, 2003–2322.
Sarker, G., Abers, G.A. (1998). Comparison of seismic body wave and coda wave measure of Q. *Pure Appl. Geophys.* **153**, 665–684.
Sato, H. (1977). Energy propagation including scattering effect: Single isotropic scattering approximation. *J. Phys. Earth* **25**, 27–41.
Sato, H. (1978). Mean free path of S-waves under the Kanto district of Japan. *J. Phys. Earth* **26**, 185–198.
Sato, H. (1984). Attenuation and envelope formation of three-component seismograms of small local earthquakes in randomly inhomogeneous lithosphere. *J. Geophys. Res.* **89**, 1221–1241.
Sato, H. (1986). Temporal change in attenuation intensity before and after Eastern Yamanashi earthquake of 1983, in central Japan. *J. Geophys. Res.* **91**, 2049–2061.
Sato, H. (1988). Temporal change in scattering and attenuation associated with the earthquake occurrence: A review of recent studies on coda waves. *Pure Appl. Geophys.* **126**, 465–497.
Sato, H. (1995). Formulation of the multiple non-isotropic scattering process in 3D space on the basis of the energy transport theory. *Geophys. J. Int.* **121**, 523–531.
Sato, H., Fehler, M.C. (1998). Seismic Wave Propagation and Scattering in the Heterogeneous Earth. Springer-Verlag, New York.
Sato, H., Nakahara, H., Ohtake, M. (1997). Synthesis of scattered energy density for non-spherical radiation from a point shear-dislocation source based on the radiative transfer theory. *Phys. Earth Planet. Inter.* **104**, 1–13.
Sens-Schönfelder, C., Wegler, U. (2006). Radiative transfer theory for estimation of the seismic moment. *Geophys. J. Int.* **167**, doi: 10.1111/j.1365–246X.2006.03139.x.
Shang, T., Gao, L.S. (1988). Transportation theory of multiple scattering and its application to seismic coda waves of impulse source. *Sci. Sin. Ser. V* **31**, 1503–1514.
Singh, S., Herrmann, R.B. (1983). Regionalization of crustal coda Q in the continental United States. *J. Geophys. Res.* **88**, 527–538.
Steck, L.K., Prothero, W.A., Scheimer, J. (1989). Site-dependent Coda Q at Mono Craters, California. *Bull. Seismol. Soc. Am.* **79**, 1559–1574.
Steensma, G.J., Biswas, N.N. (1988). Frequency dependent characteristics of coda wave quality factor in central and southcentral Alaska. *Pure Appl. Geophys.* **128**, 295–307.
Su, F., Aki, K. (1990). Spatial and temporal variation in coda Q^{-1} associated with the North Palm Springs earthquake of 1986. *Pure Appl. Geophys.* **133**, 23–52.
Su, F., Aki, K. (1995). Site amplification factors in central and Southern California determined from coda waves. *Bull. Seismol. Soc. Am.* **85**, 452–466.
Su, F., Aki, K., Biswas, N.N. (1991). Discriminating quarry blasts from earthquakes using coda waves. *Bull. Seismol. Soc. Am.* **81**, 162–178.
Takahashi, T., Sato, H., Ohtake, M., Obara, K. (2005). Scale dependence of apparent stress for earthquakes along the subducting pacific plate in Northeastern Honshu, Japan. *Bull. Seismol. Soc. Am.* **95**, doi: 10.1785/0120040075.
Takemura, M., Kato, K., Ikeura, T., Shima, E. (1991). Site amplification of S-waves from strong motion records in special relation to surface geology. *J. Phys. Earth.* **39**, 537–552.
Tsujiura, M. (1978). Spectral analysis of the coda waves from local earthquakes. *Bull. Earthq. Inst. Univ. Tokyo* **53**, 1–48.
Tsumura, K. (1967). Determination of earthquake magnitude from duration of oscillation (in Japanese with English abstract). *Zisin 2* **20**, 30–40.
Tsukuda, T. (1988). Coda Q before and after the 1983 Misasa earthquake of M6.2, Tottori Pref., Japan, 1988. *Pure Appl. Geophys.* **128**, 261–280.
Tuvè, T., Bianco, F., Ibáñez, J., Patanè, D., Del Pezzo, E., Bottari, A. (2006). Attenuation study in the Straits of Messina area (southern Italy). *Tectonophysics* **421**, 173–185.
Ugalde, A., Pujades, L.G., Canas, J.A., Villaseñor, A. (1998). Estimation of the intrinsic absorption and scattering attenuation in northeastern Venezuela (Southeastern Caribbean) using coda waves. *Pure Appl. Geophys.* **153**, 685–702.

Vargas, C.A., Ugalde, A., Pujades, L.G., Canas, J.A. (2004). Spatial variation of coda wave attenuation in northwestern Colombia. *Geophys. J. Int.* **158**, doi: 10.1111/j.1365–246X.2004.02307.x.

Wennerberg, L. (1993). Multiple-scattering interpretations of coda-Q measurements. *Bull. Seismol. Soc. Am.* **83**, 279–290.

Wessel, P., Smith, W.H.F. (1998). New, improved version of the Generic Mapping Tools released. *EOS Trans. AGU* **79**, 5–9.

Woodgold, C.R.D. (1994). Coda Q in the Charlevoix, Quebec, region: Lapse-time dependence and spatial and temporal comparisons. *Bull. Seismol. Soc. Am.* **84**, 1123–1131.

Wu, J., Jiao, W., Ming, Y., Su, W. (2006). Attenuation of coda waves at the Changbaishan Tianchi volcanic area in Northeast China. *Pure Appl. Geophys.* **163**, 1351–1368.

Wu, R.-S. (1985). Multiple scattering and energy transfer of seismic waves—Separation of scattering effect from intrinsic attenuation—1. Theoretical modeling. *Geophys. J. R. Astron. Soc.* **82**, 57–80.

Wyss, M. (Ed.) (1991). *Evaluation of proposed earthquake precursors*. AGU, Washington, DC.

Yomogida, K., Aki, K., Benites, R. (1997). Coda Q in two-layer random media. *Geophys. J. Int.* **128**, 425–433.

Yoshimoto, K., Sato, H., Iio, Y., Ito, H., Ohminato, T., Ohtake, M. (1998). Frequency-dependent attenuation of high-frequency P- and S-waves in the upper crust in western Nagano, Japan. *Pure Appl. Geophys.* **153**, 489–502.

Yoshimoto, K., Sato, H., Ohtake, M. (1993). Frequency-dependent attenuation of P and S waves in the Kanto area, Japan, based on the coda-normalization method. *Geophys. J. Int.* **114**, 165–174.

Yoshimoto, K., Wegler, U., Korn, M. (2006). A volcanic front as a boundary of seismic-attenuation structures in northeastern Honshu, Japan. *Bull. Seismol. Soc. Am.* **96**, doi: 10.1785/0120050085.

Zeng, Y. (1993). Theory of scattered P and S waves energy in a random isotropic scattering medium. *Bull. Seismol. Soc. Am.* **83**, 1264–1277.

Zeng, Y., Su, F., Aki, K. (1991). Scattering wave energy propagation in a random isotropic scattering medium: 1. Theory. *J. Geophys. Res.* **96**, 607–619.

Zoback, M.D., Zoback, M.L. (2002). State of Stress in the Earth's Lithosphere, International Handbook of Earthquake and Engineering Seismology. Academic Press, Amsterdam, pp. 559–568.

IMAGING INHOMOGENEOUS STRUCTURES IN THE EARTH BY CODA ENVELOPE INVERSION AND SEISMIC ARRAY OBSERVATION

Kin'ya Nishigami and Satoshi Matsumoto

Abstract

In this chapter, we introduce two kinds of deterministic analyses of coda waves, that is, inversion analyses of coda envelopes and seismic array observations, and we show several studies that effectively estimate the inhomogeneous structures in the crust and uppermost mantle. The first one analyzes wave data obtained by local or regional seismographic networks. Nishigami (1991) presented an inversion analysis of coda waves from local earthquakes, to estimate 3-D distribution of relative scattering coefficients. The deviation of coda envelopes from average decay curves is measured as the observational data, assuming a single isotropic scattering model. This method was applied to central California and the inhomogeneous structure around the San Andreas fault system was revealed (Nishigami, 2000). Asano and Hasegawa (2004) revised this method to estimate the absolute scattering coefficients. Revenaugh (1995a) proposed another analysis method, called Kirchhoff coda migration, in which the forward-scattered energy in teleseismic P coda observed by a regional seismographic network is stacked. The second approach is seismic array observation with station spacing shorter than the wavelength of seismic waves. We first summarize several analysis methods of seismic waves propagating through the array. For example, scattered waves with weak energy can be detected by beam-forming techniques. Coda waves are also decomposed into wave trains with various ray directions using analyses such as multiple signal classification or semblance coefficients. The energy of scattered waves in the coda can be evaluated by processing the slant-stacked waveforms under the assumption of a single-scattering model. For example, Matsumoto et al. (1998) applied this method to the source area of the 1995 Kobe earthquake (M7.3), and revealed the existence of strong scatterers just beneath the hypocenter of the mainshock. These studies analyzing the seismic network or array observation data seem to be effective to estimate the Earth's inhomogeneous structures.

KEY WORDS: Coda wave, seismic scattering, coda envelope, inversion, seismic array, crustal inhomogeneity. © 2008 Elsevier Inc.

1. Introduction

Coda waves from local earthquakes are considered to be waves scattered from inhomogeneities in the crust and uppermost mantle (e.g., Aki, 1969). The Earth's medium contains various scale lengths of inhomogeneities as summarized by Wu and Aki (1988). Inhomogeneous structures with scale length greater than the seismic wavelength cause a fluctuation of travel times and amplitudes of the direct waves. These observations can be inverted to estimate a 3-D distribution of seismic wave velocities and attenuation properties (e.g., Aki and Lee, 1976). Detailed inhomogeneous images of the lithosphere have been extensively studied and their relationships to the earthquake generating properties have been discussed (e.g., Zhao et al., 2000). On the other hand, inhomogeneous structures with scale length comparable to or smaller than the seismic wavelength cause strong scattering of the seismic waves, such that we observe an excitation of coda waves after direct P and S waves. Analysis of coda waves can allow us to extract small-scale inhomogeneities in much wider area than analysis of the direct waves. Analysis methods of coda waves to estimate the statistical or deterministic medium properties have been extensively developed, as described in Sato and Fehler (1998).

In this chapter, we show two kinds of deterministic analyses to image the inhomogeneous structures in the Earth, in terms of the inhomogeneous distribution of seismic scattering properties. The first one is an inversion analysis of coda envelopes observed by local or regional seismographic networks, and the other is the seismic array observation with station spacing shorter than the wavelength of seismic waves. The former analysis has an advantage to estimate 3-D distribution of scattering coefficients in a relatively wider area surrounding a seismic network, while the latter can obtain more detailed images of scattering properties below the array.

2. Analysis of Seismic Network Data

2.1. Inversion of Coda Envelope

We analyze the S coda waves from local earthquakes based on a single, S–S, isotropic scattering model. If we assume a spherical source radiation, the energy density $E(t)$ of single-scattered waves at lapse time t, measured from the event origin time, is expressed as follows, referring to Sato (1977):

$$E(t) = \iiint \frac{W_0}{\beta 4\pi r_1^2} g(\mathbf{x}) \frac{1}{4\pi r_2^2} \exp\left(-\frac{\omega t}{Q}\right) \delta\left(t - \frac{r_1 + r_2}{\beta}\right) d\mathbf{x}, \quad (1)$$

where W_0, the energy radiated from the source; $g(\mathbf{x})$, scattering coefficient at a point of the coordinate vector $\mathbf{x}$; r_1 and r_2, distances between hypocenters and scatterers, and scatterers and stations, respectively; and β, S-wave velocity. The exponential term in Eq. (1) expresses the attenuation effect along the propagation path with its quality factor Q and the angular frequency ω. The volume integral is taken over the "scattering shell" where the travel times of single-scattered waves equal to the lapse times of coda. When the spatial distribution of scattering coefficient is random and uniform with the averaged value g_0, Eq. (1) is approximated as follows for lapse times greater than about twice the S wave travel times,

$$E(t) \approx \frac{W_0 g_0}{4\pi \beta^2} \frac{1}{t^2} \exp\left(-\frac{\omega t}{Q}\right). \quad (2)$$

This equation has been extensively used for estimating coda Q values in many regions in the world (e.g., Jin et al., 1985). The actual observations of coda envelopes, however, show a fluctuation due to nonuniform or localized distribution of scatterers. Nishigami (1991) used Eq. (2) as a master curve for coda envelopes and defined the deviation of observed coda envelope from this master curve as "coda energy residuals." Figure 1

IMAGING INHOMOGENEOUS STRUCTURES IN THE EARTH

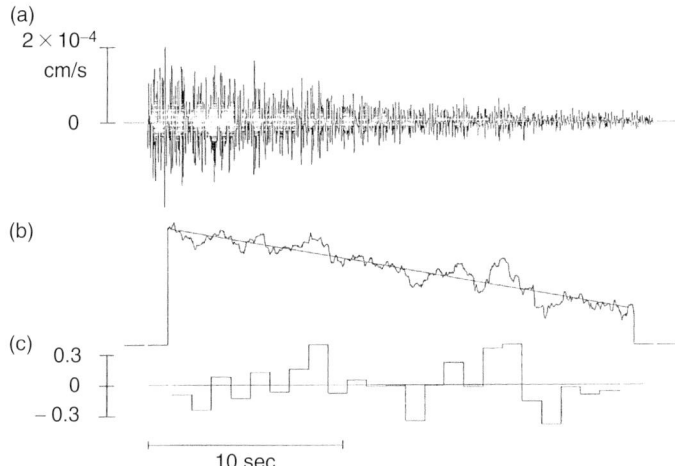

FIG. 1. An example of waveform processing to obtain the coda energy residuals. (a) Band-pass-filtered velocity seismogram of the coda part after direct S wave. (b) Running-mean-squared coda trace after correcting for the geometrical spreading effect, with the best fit straight line, plotted in arbitrary log scale. (c) Logarithm of the coda energy residuals $a(t_j)$ averaged with a time window of 1 s [modified after Nishigami (1991)].

shows an example of the procedure to obtain the coda energy residuals. The coda envelope after the direct S wave is calculated by band-pass filtering (6.7–15 Hz, in case of Fig. 1), correcting the geometrical spreading effect, and taking a running-mean-square of velocity seismograms. Taking the ratio of the envelope to its average decay, which is based on Eq. (2) and shown by a straight line in log scale in Fig. 1(b), and averaging it with an appropriate time window (1 s, in case of Fig. 1), coda energy residuals are obtained, as shown in Fig. 1(c).

In Eq.(1), spatial variation of the scattering coefficient $g(\mathbf{x})$ has much greater effect on the temporal variation of $E(t)$ than that of Q values do, as explained in Nishigami (1991). Therefore, the coda energy residuals are considered to reflect mostly the nonuniform distribution of scattering coefficients, and from the coda energy residuals observed by a seismographic networks, we can estimate a spatial distribution of scattering coefficients. We divide the analysis area into small blocks with one side of 5–10 km, mostly depending on distribution of stations and events, and then the volume integral in Eq. (1) is changed to a summation of all the blocks concerning the scattering shell. Finally we obtain the following observational equation, which describes the relationships between the relative scattering coefficient α_i in the ith block ($i = 1, M$) and the coda energy residuals $a(t_j)$ at lapse time t_j ($j = 1, N$) for all of the waveforms analyzed (Nishigami, 1991):

$$\frac{1}{\sum_i \frac{1}{(r_{1,i} \cdot r_{2,i})^2}} \sum_i \frac{\alpha_i}{(r_{1,i} \cdot r_{2,i})^2} = a(t_j), \quad (3)$$

where $r_{1,i}$ and $r_{2,i}$ represent the distances from the center of the ith block to the hypocenters and stations, respectively, and α_i is a scattering coefficient (g_i) in the ith block normalized by the background or averaged scattering coefficient g_0 in the analysis area. We can solve the linear Eq. (3) for a large number of events and stations. Nishigami (1997) solved these equations by a recursive stochastic inversion (e.g., Zeng, 1991).

Nishigami (2000) applied this method to the San Andreas fault system in central California. He analyzed 3801 waveforms from 157 local earthquakes recorded at 140 stations of the Northern California Seismic Network. The block size was taken as 10 km in horizontal and 5 km in depth. Figure 2 shows the estimated distribution of relative scattering coefficient. The inversion was made four times by shifting the horizontal block assignment by half an interval, and the results were superposed for this figure. Diagonal

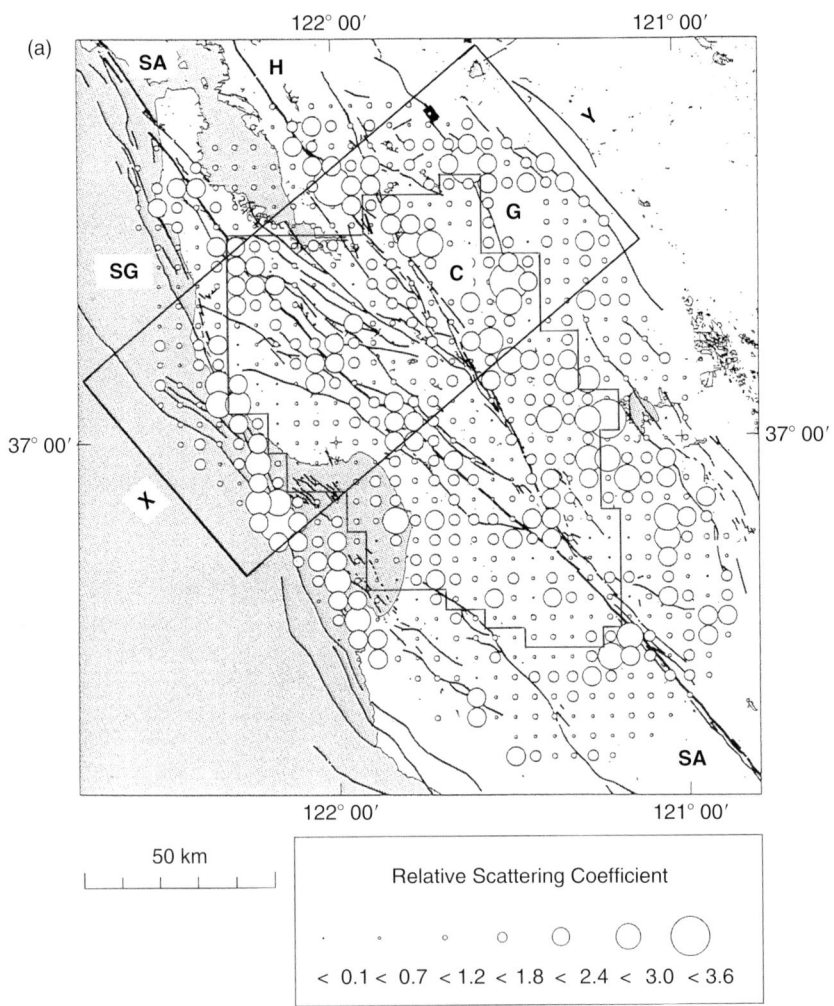

FIG. 2. Continued.

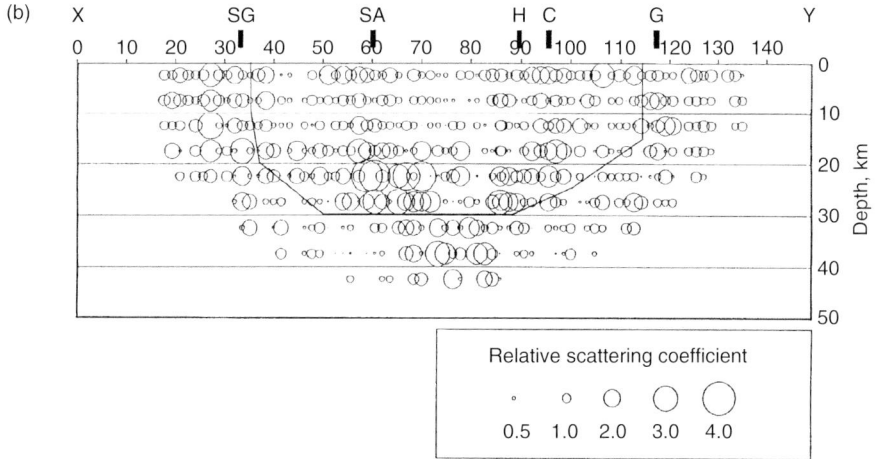

FIG. 2. (a) Distribution of relative scattering coefficient estimated at a depth of 0–5 km in central California. The solutions with resolution >0.2 are plotted and the area enclosed by solid line segments indicates the resolution >0.5. Solid lines represent the active faults, SG, San Gregorio; SA, San Andreas; H, Hayward; C, Calaveras; and G, Greenville faults. (b) Vertical cross section of relative scattering coefficient within the rectangular area shown in (a). Surface location of active faults is approximately indicated by thick bars. The area above the solid line segments indicates the resolution >0.5 [modified after Nishigami (2000)].

elements of the resolution matrix were almost all greater than 0.5 except for the peripheral of the analysis area. Figure 2(a) shows a good correlation between subparallel active faults and relatively stronger scattering zones in the crust. The result also suggested a segmentation of the San Andreas fault, where segment boundaries were characterized by relatively stronger scattering. This inversion analysis was also applied to several regions in Japan and crustal heterogeneities in active fault zones and active volcanoes were discussed (e.g., Nishigami, 1997, 2006). Nishigami (1997) revealed two zones with stronger scattering at a depth of 7–17 km just below Mt. Ontake, central Japan, and showed that they agree well with the two reflectors (or reflection planes) of S waves previously estimated by the normal moveout correction analysis (Inamori et al., 1992). Nishigami (2007) also showed a good correlation between the distribution of strong scattering zones and S wave reflectors estimated at a depth from 20 to 30 km in the Kinki district, southwest Japan. These results show that the scattering inversion analysis is effective to estimate inhomogeneous structures in the crust.

Asano and Hasegawa (2004) revised the inversion method stated above. They took into account the source radiation pattern based on the focal mechanism and also the intrinsic and scattering attenuation effects in the observational equation, and inverted the absolute coda energy to estimate the absolute value of scattering coefficient. Asano and Hasegawa (2004) applied their inversion analysis to the aftershock seismograms, band-passed for 4.5–9.0 Hz, of the 2000 Western Tottori earthquake (M 7.3), and showed larger scattering coefficients along the mainshock fault plane and also relatively smaller scattering coefficients possibly related to the asperity with large coseismic slip.

We assumed a single-scattering model in the inversion analyses shown above. The coda energy due to multiple scattering, however, becomes dominant over that of single scattering at large lapse times, as shown in Sato (1988). The lapse time of coda waves that is available for a single-scattering model is considered to be approximately within the mean free time of the inhomogeneous medium. The mean free path of S waves in the crust and uppermost mantle was estimated as $\sim$100 km at 10 Hz (e.g., Sato, 1978), and therefore the corresponding mean free time is $\sim$30 s. The inversion analyses stated above (e.g., Nishigami, 1997, 2000; Asano and Hasegawa, 2004) obeyed this limitation of lapse times. Also they took the beginning of coda analysis at $\sim$1.5–2 times the S wave travel times, which is a little earlier than the limitation of lapse times that satisfies Eq. (2). The early coda part contains effective data to estimate detailed scattering properties in the crust. It will be relevant to use the "K function," which expresses the decay curve of coda energy and its asymptote gives Eq. (2) for lapse times greater than about two times the S wave travel time (Sato, 1977), in order to include the early coda part in the inversion analyses.

2.2. Kirchhoff Coda Migration

In the previous section, we described the analysis method of back-scattered (or side-scattered) waves from local earthquakes. Forward-scattered waves in the seismograms from teleseismic events have also been analyzed to estimate scattering properties in the crust and uppermost mantle beneath the local or regional seismographic networks. In the Kirchhoff coda migration, we stack the absolute amplitude or the nth root of teleseismic coda at the arrival times of forward-scattered waves, for each grid point of supposed scatterers assigned in the analysis area (e.g., Revenaugh, 1995b). Nth-root stacking is considered to suppress incoherent energy in coda waves (Revenaugh, 1995b). P–P, P–S, or P–R_g scattering is assumed in this stacking. For example, Revenaugh (1995b) analyzed teleseismic P coda recorded by a short-period seismometer network, that is, 5606 seismograms from 120 teleseismic events recorded at 232 stations of the Southern California Seismic Network (SCSN). Horizontal grid spacing was taken as 0.1° for several depth layers from 50 to 400 km. He detected a zone with strong P–P scattering at depths from 50 to 200 km, and interpreted it as a slab subducting beneath the Transverse Ranges. Revenaugh (1995a) also analyzed teleseismic P coda recorded at the SCSN and pointed out a correlation of P–R_g scattering strength with topographic roughness. Revenaugh (1995c) analyzed the area surrounding the 1992 M7.3 Landers earthquake sequence, using the SCSN data, and suggested a correlation between P–S scattering strength and the aftershock distribution. In these studies, seismograms low-pass-filtered below 1–2 Hz were analyzed.

From a viewpoint of imaging heterogeneous structure of vertical faults in the crust, inversion analyses of back-scattered waves from local earthquakes seem to be more effective than the Kirchhoff coda migration of forward-scattered waves from teleseismic events. As described above, however, the inversion analysis has a limitation of maximum lapse times analyzed, Kirchhoff coda migration may be more appropriate to estimate inhomogeneous structures in the uppermost mantle.

IMAGING INHOMOGENEOUS STRUCTURES IN THE EARTH

3. Analysis of Seismic Array Data

3.1. Detection of Seismic Signals by Array Observations

A seismic array is composed of many seismometers deployed with intervals shorter than a target wavelength of seismic waves, and is used for investigating the characteristics of coda waves. Aki (1969) first showed that coda waves consist of wave trains arriving to the array from various directions. Following this pioneering work, many studies have investigated the characteristics of coda waves. Array analyses have been used in the studies of the rupture process of earthquakes and the origin of volcanic tremors, and the analysis techniques, for example, detecting waves with various ray directions have been remarkably improved in these studies.

In the reflection survey for seismic exploration, many sophisticated techniques have been developed to determine subsurface structures just beneath a seismic array profile. These are the procedures that the wave field observed at the surface is back-propagated downward, that is, downward continuation, Kirchoff migration, and so on. However, the reflection survey can determine the fine structure only beneath the profile. Such surveys usually focus on finding a deterministic structure beneath the array.

Detectability of a seismic array depends on the configuration of the array, that is, the aperture, shape, and number and interval of stations. The array usually consists of individual stations, which sample the wave field at discrete spatial locations, so that array response function W is defined as a function of wave number vector $\mathbf{k}$ of steered direction:

$$W(\mathbf{k}) = \sum_{n=1}^{N} w_n \exp(i\mathbf{k} \cdot \mathbf{x}_n) = \sum_{n=1}^{N} \frac{1}{N} \exp(i\mathbf{k} \cdot \mathbf{x}_n), \tag{4}$$

where $\mathbf{x}_n$ is a positioning vector of the nth station, and N denotes the number of stations. A weighting factor is defined as w_n, which equals to $1/N$ when we adopt equal weight for all stations. For convenience, we introduce a matrix expression for the above formula and a complex expression for the wave field. The complex waves consist of waves in the real part and Hilbert-transformed ones in the imaginary part. This expression is often used in signal processing. In an ordinary beam forming, the vector of weighting factor $\mathbf{W}$ is a function of wave number $\mathbf{k}$ in the direction where the beam is steered:

$$\mathbf{W}(\mathbf{k}) = [\exp(-i\mathbf{k} \cdot \mathbf{x}_1), \exp(-i\mathbf{k} \cdot \mathbf{x}_2), \ldots, \exp(-i\mathbf{k} \cdot \mathbf{x}_N)]^T. \tag{5}$$

$\mathbf{k}$ is also called the steering vector. The array response function W is related with $\mathbf{W}$ by the formula:

$$W(\mathbf{k}) \equiv \sum_{i=1}^{N} \mathbf{W}_i(\mathbf{k}). \tag{6}$$

An ideal array aperture function is a delta function, which has no directional dependency.

A standard technique of beam forming is a slant stacking. This is also called the "delay-sum beam former," which shifts the phases and stacks the waveforms among the stations. The phase shift is calculated from an assumed ray direction (i.e., defined by slowness vector $\mathbf{s} = \mathbf{k}/\omega$, $\mathbf{k}$: wave-number vector). The array aperture function for an incident wave with $\mathbf{k}_0$ and ω is defined as a function of stacking slowness W ($\omega\mathbf{s}$-$\mathbf{k}_0$). The aperture function is also called a beam pattern or a steered response. The array response depends on the difference between the direction vector of incident wave and that of stacking. Johnson and Dudgeon (1993) described the beam-forming and the array response in detail.

The power of slant-stacked waveforms can be expressed as:

$$P(\mathbf{k},t) = \mathbf{W}(\mathbf{k})^H \mathbf{d}\mathbf{d}^H \mathbf{W}(\mathbf{k}), \tag{7}$$

where $\mathbf{d}$ is a complex input wave vector in a narrow frequency band and $\mathbf{W}$ is a complex weighting vector given in Eq. (5). In this case, each column of vector $\mathbf{W}$ is a function of $\mathbf{k}$ defined as a wave number vector of the stacking direction. The superscript H denotes the Hermitian transposition. $\mathbf{d}$ is a column vector composed of the complex input waves at stations in the array; $\mathbf{d} = [d_1(t), d_2(t), \ldots, d_N(t)]^T$, where N is the number of stations and $d_i(t)$ is the waveform observed at ith station.

$\mathbf{W}^H \mathbf{d}$ implies a slant-stacked waveform since $\mathbf{W}$ is a delay operator for an assumed stacking direction with the wave-number vector $\mathbf{k}$. The phase delay at each station can be calculated by the scalar product between $\mathbf{k}$ and $\mathbf{x}$. Multiplying $\mathbf{W}^H$ and $\mathbf{d}$, we obtain a delay and stacked waveform, namely the slant-stacked output.

There are many methods to determine ray direction with high resolution. For example, the multiple signal classification (MUSIC) spectrum (Schmidt, 1986) has a high resolution in the array signal detection. The ordinary beam-forming is a beam-sensing method by using a lobe of the array response function, as described above. In contrast, the MUSIC method uses "null sensing" of the array response function based on eigen vector decomposition. This is a reason why the MUSIC has much higher resolution than the classical method. Goldstein and Archuleta (1991) studied the MUSIC spectrum in detail, providing effective applications to the adoptive alignment of seismograms and also spatial averaging of seismograms in the sub-array.

A semblance coefficient is one of the simple methods to estimate ray directions (Neidel and Taner, 1971; Nikolaev and Troitskiy, 1987). This processing is often used for a velocity analysis of reflected waves in the seismic reflection survey (Yilmaz, 1987). The ray direction is determined by searching a maximum semblance value among possible $\mathbf{k}$ vectors. The semblance coefficient for an assumed slowness vector $\mathbf{s}$ is calculated from the waveform vector $\mathbf{d}$ observed by an array (Neidel and Taner, 1971). Using a matrix expression, semblance is defined as:

$$S(\mathbf{k},t) = \frac{1}{N} \frac{\mathbf{W}^H \mathbf{d}\mathbf{d}^H \mathbf{W}}{\mathbf{d}^H \mathbf{d}}, \tag{8}$$

where N is number of traces stacked. The semblance coefficient has a characteristic similar to a coherence function between waves within the time window. The semblance value ranges from 0 (no correlation) to 1 (identical waveform). Random noise provides a semblance value of $1/N$ since the stacked power is equal to its total power. It should be noted that the semblance analysis detects the waves with high correlation among stations, not always with large amplitudes. In addition, a scattering strength can not be directly obtained from the semblance coefficient due to its independence of the absolute power of the signal. However, the analysis has an advantage that it is a simple algorithm to

IMAGING INHOMOGENEOUS STRUCTURES IN THE EARTH   309

implement and it requires little computation. The zero-lag cross-correlation coefficient has been developed as a method similar to the semblance coefficient. This method is also useful in estimating the crustal inhomogeneity (e.g., Frankel *et al.*, 1991; Del Pezzo *et al.*, 1997).

3.2. Single-Scattering Model for Seismic Array

As mentioned above, there are several techniques to determine the ray directions of wave trains by means of the array processing. The high resolution method such as MUSIC and the semblance can not always evaluate the absolute energy approaching to the array. In this section, we describe the energy of slant-stacked waveforms under the assumption of a single-scattering model. We simply assume that a source radiates energy in a spherical symmetry and the scattering coefficient g is homogeneous in a medium. Energy density $E(r_0,t)$ at a hypocentral distance r_0 and a lapse time t from the origin time can be expressed as follows in a coordinate system shown in Fig. 3:

$$E(r_0,t) = \int\int\int \frac{W_0 g(\psi) L(\theta, \varphi, \theta_0, \varphi_0)}{\beta 4\pi r_1^2 4\pi r_2^2} \delta\left(t - \frac{r_1 + r_2}{\beta}\right) d\mathbf{x}, \qquad (9)$$

where W_0 is the radiated energy from the source, β is a seismic velocity of the medium, $g(\psi)$ is the scattering coefficient having anisotropic property on scattering angles ψ, and $\mathbf{x}$ is a location vector of scatterers. The intrinsic attenuation of scattered waves during propagation in the medium is neglected in Eq. (9) for simplicity. This is a general form of Eq. (1), taking an effect of array beam forming into account. It means that the coda energy is, as described in Sato (1977), obtained by summing up the scattered wave

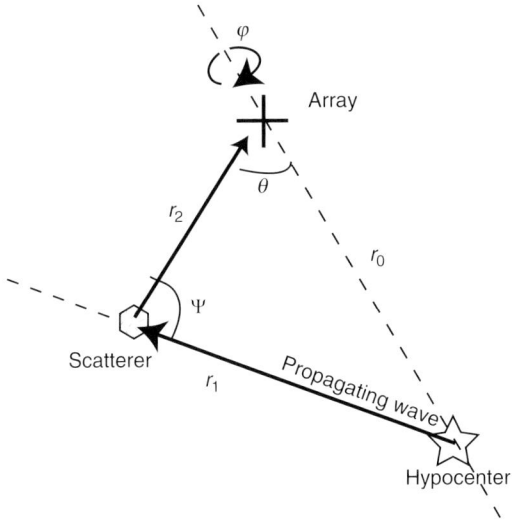

FIG. 3. Coordinate system and geometrical relationship among hypocenter, array, and scatterer.

energy from scatterers distributed in the space. For the envelopes of slant-stacked waveforms by the seismic array, the sensitivity of the array varies as a function of the ray direction coming to the array as well as the direction of slant stacking. Therefore, we have to add a directional weighting function $L(\theta, \varphi, \theta_0, \varphi_0)$ in the kernel of the integration. This reflects the array response for the slant-stack directions θ_0 and φ_0, which are obtained from **k** in Eq. (5).

We transform Eq. (9) into prolate-spheroidal coordinates (see Sato and Fehler, 1998) and obtain

$$E(r_0, t) = H(v-1)\frac{W_0}{4\pi r_0^2 2\pi}\int_0^{2\pi} d\varphi \int_{-1}^{1} dw \frac{g(\psi)L(\theta, \varphi, \theta_0, \varphi_0)}{v^2 - w^2}, \qquad (10)$$

$$\cos\theta = \frac{1+vw}{v+w}$$

$$\cos\psi = \frac{2-v^2-w^2}{v^2-w^2},$$

where v is a travel distance normalized by a hypocentral distance ($v = \beta t/r_0$). For an envelope of a single station record, the energy is obtained in the case of $L = 1$. According to Sato (1977), Eq.(10) becomes under an isotropic scattering hypothesis,

$$E(r_0, t) = H(v-1)\frac{g_0 W_0}{4\pi r_0^2}\frac{1}{v}\ln\left(\frac{v+1}{v-1}\right)$$

$$v = \beta t/r_0, \quad K = 1,$$

$$g_0 = \frac{1}{4\pi}\oint g(\Psi)d\Omega, \qquad (11)$$

where $H(x)$ is Heaviside step function. For the slant-stacked trace, (10) can be rewritten using the angle from the array θ,

$$E(r_0, t) = H(v-1)\frac{W_0}{4\pi r_0^2 2\pi}\int_0^{2\pi} d\varphi \int_0^{\pi} d\theta \frac{g_0 \sin\theta}{v^2 - 2v\cos\theta + 1}L(\theta, \varphi, \theta_0, \varphi_0). \qquad (12)$$

The above equation means that the slant-stacked coda wave energy is expressed by a summation of a product of energy sequence propagating from the direction θ, φ and the array energy response. If we write in the matrix expression used in the previous section, the energy would become $P(\mathbf{k},t) = (\mathbf{W}^H \mathbf{W})(\mathbf{d}^H\mathbf{d})$. However, the exact expression described before is $P(\mathbf{k},t) = \mathbf{W}^H\mathbf{dd}^H\mathbf{W}$. The right hand sides in both equations agree with each other, when non-diagonal components of $\mathbf{dd}^H$ become zero. It should be noted that this situation is realized only when either the scattered waves come from various directions randomly or the array response function is close to the delta function.

IMAGING INHOMOGENEOUS STRUCTURES IN THE EARTH   311

On the basis of the above formula, Matsumoto *et al.* (2001a) analyzed the seismic array records and found that the coda energy levels for slant-stacked records from explosion sources are smaller than that for single station records as expected from the single-scattering model, as shown in Fig. 4. This implies that the slant-stacking can eliminate energy coming from directions other than the direction steered by the array because of the envelope in Fig. 4 normalized by direct P wave energy. Moreover, the energy level of the slant-stacked waveform depends on the stacking direction. The differences in the energy level of the envelopes stacked in the directions of the hypocenter and the opposite one attribute to the area size of the scattering shell, which is related to the geometries of the source and array locations, and also to the lapse times. Their result showed that the single-scattering model for the slant-stacked waveform was applied to the observations effectively and that the single-scattering model is applicable to the observed records.

3.3. Characteristics of Coda Waves Based on Array Observations

The array observation can decompose the ray directions of wave trains in the coda part, and this can reveal the characteristics of coda waves and inhomogeneous structures in the Earth. We introduce some of the recent studies with remarkable results in this section.

3.3.1. Wave Composition of Seismogram

Wave types (P or S waves) and propagating directions of wave trains in the coda part of seismograms have been studied (e.g., Scherbaum *et al.*, 1997; Kuwahara *et al.*, 1997). Especially, detailed studies based on the array observations with three-component seismometers were performed by Wagner (1998) and Taira (2004). Wagner (1998) analyzed both P and S coda waves and showed that P coda waves from local earthquakes are composed of P waves with the same ray directions with the direct P waves.

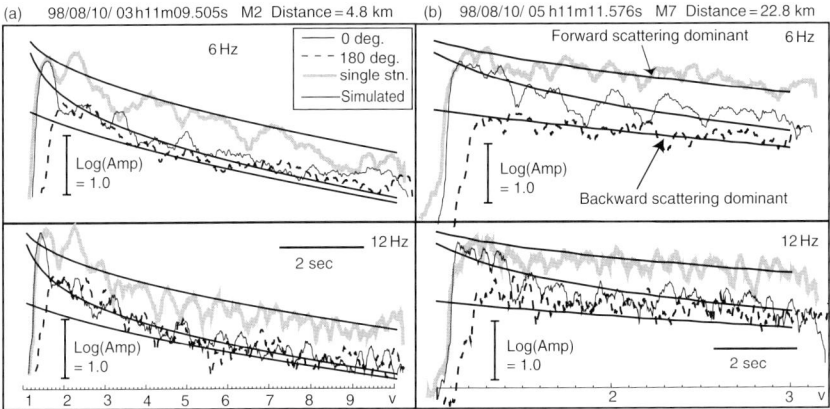

FIG. 4. Envelopes observed and theoretically expected for 6 and 12 Hz ranges. (a) and (b) are those for epicentral distance of 4.8 and 22.8 km, respectively. Gray lines are envelopes of the single station record. Solid and dashed lines are slant-stacked envelopes in the direction of 0 and 180 degree. Thick solid lines are theoretical curves of the single isotropic scattering model [modified after Matsumoto *et al.* (2001a)].

He interpreted this result as a propagation in some waveguide. On the other hand, he showed that S coda waves consist of scattered S waves with random arrival-directions to the array. Taira (2004) analyzed seismograms recorded by a dense three-component seismic array from explosion sources, and determined both slowness and polarization vectors of coda waves. He showed that P–S scattering, but not P–P scattering, was dominant near the surface trace of the fault. This means that composition of the coda waves is not always simple but affected by strong inhomogeneities such as active faults.

3.3.2. Number Density of Scatterers and Strength of Scattering

In the basic expression of coda waves, seismic energy in the coda part can be expressed as a summation of energy propagating to the station from randomly distributed scatterers. Based on the single-scattering model, as reviewed by Sato and Fehler (1998), we briefly describe the relationships between a scatterer density and a scattering strength. We consider a medium with average velocity β and point-like scatterers distributed randomly and uniformly with a number density n. The energy density in the coda part is expressed by a summation of energy contribution from each scatterer distributed in the medium as follows:

$$E_c(r_0, t) = \sum_{\text{scatterer}} \frac{W_0}{\beta 4\pi r_1^2 4\pi r_2^2} \sigma_0 \delta\left(t - \frac{r_1 + r_2}{\beta}\right) \exp(-g_0 \beta t), \tag{13}$$

where r_0 is the hypocentral distance, W_0 is the energy radiated from the source, and r_1 and r_2 are distances from a hypocenter to a scatterer and from a scatterer to a station, respectively. σ_0 is called as the total scattering cross section, which is in integral form of the differential scattering cross section over a solid angle. σ_0 is a product of n-value and the scattering coefficient g_0. Both the n-value and the scattering coefficient g_0 are important parameters in modeling the crustal inhomogeneity. Matsumoto (2005) showed that the n-value can be estimated from the lapse time dependency of the semblance coefficient in coda part. Figure 5 shows the relationships between the semblance values and the number of waves incident to the linear-aligned array. Matsumoto (2005) analyzed the data observed in the aftershock area of the 2000 Western Tottori earthquake, and evaluated the n-value and g_0 as 0.03 km^{-3} and 0.001 km^{-1}, respectively, by comparing the simulated and observed sequences of the semblance and power. The estimated n-value (0.03 km^{-3}) means that, for example, scatterers are distributed with average 3-D spacing of 3.2 km. These scatterers are most effective in scattering of seismic waves at 20 Hz. This is the first estimation of n-value in the crust, so that studies about regional differences in n-value are needed.

3.4. Scatterer/Inhomogeneity Distribution Inferred from Seismic Array Data

The location of scatterers can be estimated from the ray directions and travel times of wave trains, which were extracted from the coda using the array analyses described in the previous section. The wave type of coda waves, that is, P–P, P–S, S–P, or S–S scattering, is usually assumed in determining the location of scatterers. However, in case of analyzing the three component seismograms, the wave type can be estimated from the polarization of coda waves.

IMAGING INHOMOGENEOUS STRUCTURES IN THE EARTH 313

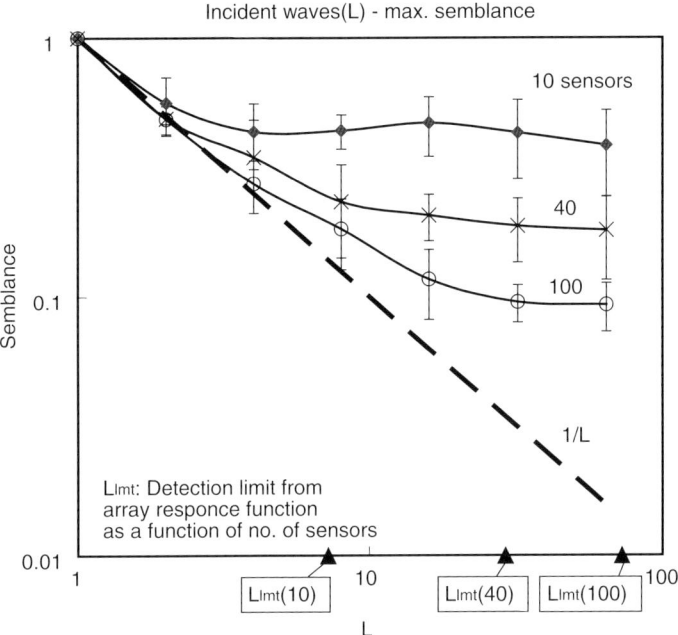

FIG. 5. Semblance variation for number of wave incidence L, for a linear array with 10, 40, and 100 sensors, respectively. Sensor interval is 0.1 km and wavelength is 0.25 km. Triangle on the lower horizontal axis indicates detection limit for each array defined by array aperture function. L increases with lapse time because of extending scattering shell, so that semblance value decreases with lapse time [modified after Matsumoto (2005)].

The analysis method of scatterer distribution has been studied (e.g., Nikolaev and Troitskiy, 1987). Spudich and Bostwick (1987) developed an analysis method that used an earthquake cluster well localized as a seismic source array. Krüger et al. (1993) proposed an array analysis, called the "Double Beam Imaging," which uses both the source and the station arrays. Scherbaum et al. (1997) estimated a detailed scatterer distribution in the lower mantle by applying the Double Beam Imaging method as well as the travel time analyses. Similar analyses were also made, for example, by Thomas et al. (1999).

The slant stack is a transformation of array records from a space-time domain to a slowness-time domain. This has an advantage that many processings for a time series can also be applied to the slant-stacked waveforms. Matsumoto et al. (1998) obtained an image of scatterers in the source area of the 1995 Kobe earthquake (M7.3). They applied the processing used in seismic exploration such as filtering, gain recovery, and depth conversion, to the slant-stacked waveforms, and revealed the existence of strong scatterers just beneath the hypocenter of the mainshock. Furthermore, Matsumoto et al. (1999, 2006) imaged scatterer distributions beneath a high seismicity region in northeastern Japan and also in the aftershock area of the 2005 West off Fukuoka Pref. earthquake (M7.0) by a semblance-weighted slant stacking. The semblance-weighted stacking provides a higher-resolution image than the usual slant stacking, although the stacked waveforms in this analysis reflect no longer the energy of the scattered waves correctly, due to use of semblance weighting. In northeastern Japan, Matsumoto *et al.* (1999) obtained the scatterer distribution from the records of seismic explosions and found relatively strong inhomogeneities just beneath the high micro-seismicity regions, as shown in Fig. 6. In the aftershock area of the 2005 West off Fukuoka Pref. earthquake,

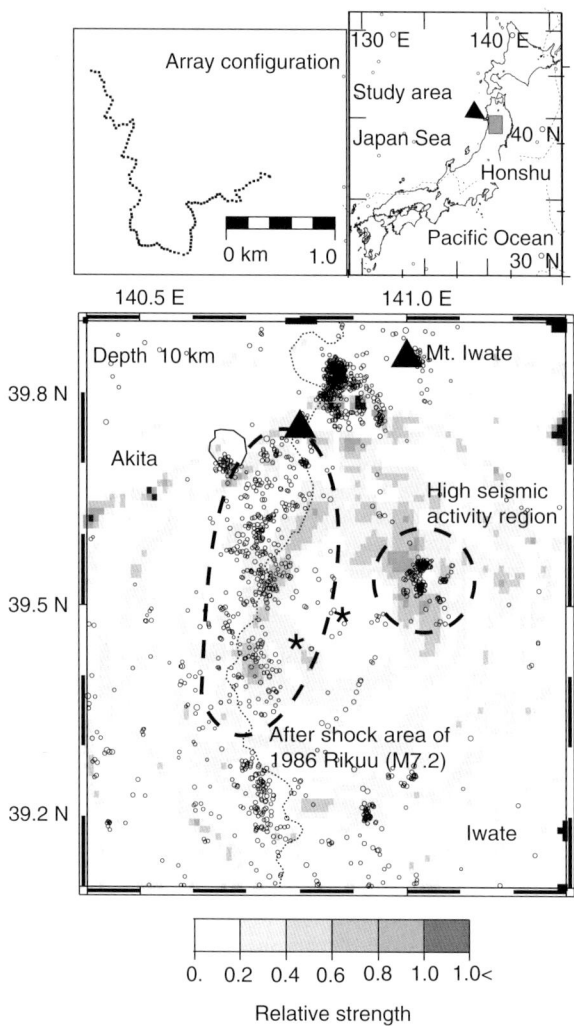

FIG. 6. Map showing a distribution of relative P–P scattering strength at a depth of 10 km in northeastern Japan estimated from seismic explosion data. Dark areas correspond to those with high strengths of scattering. Solid triangles denote locations of active volcano. Asterisks are locations of the array. Open circles are hypocenters determined by Tohoku University. Upper figures show configuration of the array and location of the target region [modified after Matsumoto *et al.* (1999)].

they showed that the scatterer distribution was partly correlated with the low velocity zone at the edge of the mainshock fault plane. The scatterer distribution inferred from array observations revealed a possible relationship between strong inhomogeneities and the generation of earthquakes. These results are similar to those estimated from the analyses of seismic network data, as described in Section 2.

In addition, there are a number of important findings obtained from the array analyses. These include a temporal variation of scatterer distribution (Matsumoto et al., 2001b), detecting an anisotropic structure (Bear et al., 1999), and a spatial distribution of P–P and P–S scatterers (Taira, 2004). Referring to the findings mentioned above, the array processing can focus on the areas of interest and, therefore, is considered to be a powerful tool for detecting inhomogeneous structures in the Earth.

4. Summary

In this chapter, we introduced two deterministic analyses of coda waves, that is, inversion analyses of coda envelopes and seismic array observations, and we showed several studies that effectively estimated the inhomogeneous structures in the crust and uppermost mantle.

The first one analyzes the wave data obtained by local or regional seismographic networks. Nishigami (1991) presented an inversion analysis of coda envelopes from local earthquakes. In this method, the deviation of coda envelopes from average decay curves is measured as the observational data, assuming a single isotropic scattering model, and then 3-D distribution of relative scattering coefficient is estimated by solving the observational equations. This method was applied to central California and the deep structure of the San Andreas fault system was revealed as a structure with strong scattering (Nishigami, 2000). Asano and Hasegawa (2004) revised this inversion analysis to estimate a distribution of absolute scattering coefficients. Revenaugh (1995a) proposed another method, called Kirchhoff coda migration, which stacks the forward-scattered energy within the coda of teleseismic P waves observed by a regional seismographic network. This method was applied to southern California and revealed, for example, strong P–P scattering from the slab subducting beneath the Transverse Ranges at depths from 50 to 200 km. These methods, analyzing the seismic network data, seem to be effective to estimate the inhomogeneous structures in the crust and uppermost mantle.

As to the second approach, that is, seismic array observations, many studies have developed the analysis techniques and revealed the images of inhomogeneity from the lower mantle to the surface. Scattered waves with weak energy can be detected by beam-forming techniques. Coda waves can be decomposed into the wave trains with various ray directions using the array analyses such as MUSIC or semblance coefficients. Analyses of three-component seismograms are especially effective. For example, Wagner (1998) showed that P coda waves from local earthquakes are composed of P wave trains coming from the direction of hypocenters, while S coda waves consist of scattered S waves coming from random directions. The scattered wave energy in the coda can be evaluated by processing the slant-stacked waveforms under the assumption of a single-scattering model. For example, Matsumoto et al. (1998) applied this method to the source area of the 1995 Kobe earthquake (M7.3), and revealed the existence of strong scatterers just beneath the hypocenter of the mainshock.

Many studies shown here have been performed under the ill-conditions, such as sparse distribution of hypocenters and stations, and large hypocentral distances. Consequently, these studies have well exemplified the effectiveness of the seismic network or array observations, providing lots of information of the Earth's inhomogeneity. For further study, newer and more reliable techniques for analyzing natural earthquakes are necessary in order to improve the images of inhomogeneities and understand their physical properties.

ACKNOWLEDGMENTS

The authors are grateful to Prof. H. Sato and Dr. M. Fehler for inviting us to this publication. Comments by an anonymous reviewer and Dr. M. Fehler were helpful in improving the chapter.

REFERENCES

Aki, K. (1969). Analysis of the seismic coda of local earthquakes as scattered waves. *J. Geophys. Res.* **74**, 615–631.
Aki, K., Lee, W.H.K. (1976). Determination of three-dimensional velocity anomalies under a seismic array using first P-arrival times from local earthquakes, 1, A homogeneous initial model. *J. Geophys. Res.* **81**, 4381–4399.
Asano, Y., Hasegawa, A. (2004). Imaging the fault zones of the 2000 western Tottori earthquake by a new inversion method to estimate three-dimensional distribution of the scattering coefficient. *J. Geophys. Res.* **109**, B06306, doi:10.1029/2003JB002761.
Bear, L.K., Pavlis, G.L., Bokelmann, G.H.R. (1999). Multi-wavelet analysis of three-component seismic arrays: Application to measure effective anisotropy at Pinon Flats, California. *Bull. Seism. Soc. Am.* **89**, 693–705.
Del Pezzo, E., Rocca, M.L., Ibanez, J. (1997). Observations of high-frequency scattered waves using dense arrays at Teide Volcano. *Bull. Seismol. Soc. Am.* **87**, 1637–1648.
Frankel, A., Hough, S., Friberg, P., Busby, R. (1991). Observations of Loma Prieta aftershocks from a dense array in Sunnyvale, California. *Bull. Seismol. Soc. Am.* **80**, 1900–1922.
Goldstein, P., Archuleta, R.J. (1991). Deterministic frequency-wavenumber method and direct measurements of rupture propagation during earthquakes using a dense array: Theory and method. *J. Geophys. Res.* **96**, 6173–6185.
Inamori, T., Horiuchi, S., Hasegawa, A. (1992). Location of mid-crustal reflectors by a reflection method using aftershock waveform data in the focal area of the 1984 Western Nagano Prefecture earthquake. *J. Phys. Earth* **40**, 379–393.
Jin, A., Cao, T., Aki, K. (1985). Regional change of coda Q in the oceanic lithosphere. *J. Geophys. Res.* **90**, 8651–8659.
Johnson, D.H., Dudgeon, D.E. (1993). Array signal processing—Concepts and techniques. Oppenheim, A.V. (Ed.), *Signal Processing Series*. Prentice Hall, Tokyo.
Krüger, F., Weber, M., Scherbaum, F., Schlittenardt, J. (1993). Double beam analysis of anomalies in the core-mantle boundary region. *Geophys. Res. Lett.* **20**, 1475–1478.
Kuwahara, Y., Ito, H., Kawakatsu, H., Ohminato, T., Kiguchi, T. (1997). Crustal heterogeneity as inferred from seismic coda wave decomposition by small-aperture array observation. *Phys. Earth Planet. Inter.* **104**, 247–256.
Matsumoto, S. (2005). Scatterer density estimation in the crust by seismic array processing. *Geophys. J. Int.* **163**, 622–628.

Matsumoto, S., Obara, K., Hasegawa, A. (1998). Imaging P-wave scatterer distribution in the focal area of the 1995 M7.2 Hyogo-ken Nanbu (Kobe) Earthquake. *Geophys. Res. Lett.* **25**, 1439–1442.
Matsumoto, S., Obara, K., Yoshimoto, K., Saito, T., Hasegawa, A., Ito, A. (1999). Imaging of crustal inhomogeneous structure of the crust beneath Ou Backbone Range, northeastern Japan, based on small aperture seismic array observations. *J. Seism. Soc. Jpn.* **2**(52), 293–297, (in Japanese with English abstract).
Matsumoto, S., Obara, K., Hasegawa, A. (2001a). Characteristics of coda envelope for slant-stacked seismogram. *Geophys. Res. Lett.* **28**(6), 1111–1114.
Matsumoto, S., Obara, K., Yoshimoto, K., Saito, T., Ito, A., Hasegawa, A. (2001b). Temporal change in P-wave scatterer distribution associated with M6.1 earthquake near Iwate volcano, NE Japan. *Geophys. J. Int.* **145**, 48–58.
Matsumoto, S., Watanabe, A., Matsushima, T., Miyamachi, H., Hirano, S. (2006). Imaging S-wave scatterer distribution in south-east part of the focal area of the 2005 West Off Fukuoka Prefecture Earthquake (M_{JMA}7.0) by dense seismic array. *Earth Plant. Space* **58**, 1627–1632.
Neidel, N.S., Taner, M.T. (1971). Semblance and other coherency measures for multichannel data. *Geophysics* **36**, 482–497.
Nikolaev, A.V., Troitskiy, P.A. (1987). Lithospheric studies based on array analysis of P- coda and microseisms. *Tectonophysics* **140**, 103–113.
Nishigami, K. (1991). A new inversion method of coda waveforms to determine spatial distribution of coda scatterers in the crust and uppermost mantle. *Geophys. Res. Lett.* **18**, 2225–2228.
Nishigami, K. (1997). Spatial distribution of coda scatterers in the crust around two active volcanoes and one active fault system in central Japan: Inversion analysis of coda envelope. *Phys. Earth Planet. Inter.* **104**, 75–89.
Nishigami, K. (2000). Deep crustal heterogeneity along and around the San Andreas fault system in central California and its relation to the segmentation. *J. Geophys. Res.* **105**, 7983–7998.
Nishigami, K. (2006). Crustal heterogeneity in the source region of the 2004 Mid Niigata Prefecture earthquake: Inversion analysis of coda envelopes. *Pure Appl. Geophys.* **163**, 601–616.
Nishigami, K. (2007). Crustal heterogeneity and its relation to seismic activity in the Kinki district, southwest Japan. *EOS Trans. AGU, 88, Fall Meet. Suppl., Abstract* S31A–0213.
Revenaugh, J. (1995a). The contribution of topographic scattering to teleseismic coda in southern California. *Geophys. Res. Lett.* **22**, 543–546.
Revenaugh, J. (1995b). A scattered-wave image of subduction beneath the Transverse Ranges. *Science* **268**, 1888–1892.
Revenaugh, J. (1995c). Relation of the 1992 Landers, California, earthquake sequence to seismic scattering. *Science* **270**, 1344–1347.
Sato, H. (1977). Energy propagation including scattering effects: Single isotropic scattering approximation. *J. Phys. Earth* **25**, 27–41.
Sato, H. (1978). Mean free path of S-waves under the Kanto district of Japan. *J. Phys. Earth* **26**, 185–198.
Sato, H. (1988). Is the single scattering model invalid for the coda excitation at long lapse time. *Pure Appl. Geophys.* **128**, 43–47.
Sato, H., Fehler, M.C. (1998). Seismic Wave Propagation and Scattering in the Heterogeneous Earth. Springer-Verlag, New York.
Scherbaum, F., Krüger, F., Weber, M. (1997). Double beam imaging: Mapping lower mantle heterogeneities using combinations of source and receiver arrays. *J. Geophys. Res.* **102**, 507–522.
Schmidt, R.O. (1986). Multiple emitter location and signal parameter estimation. *IEEE Trans. Antennas Propag. AP* **34**, 276–280.
Spudich, P., Bostwick, T. (1987). Studies of the seismic coda using an earthquake cluster as a deeply buried seismograph array. *J. Geophys. Res.* **92**, 10526–10546.

Taira, T. (2004). Quantitative imaging of small-scale heterogeneities around active fault system as seismic wave scatterers. Ph. D. Thesis, Hokkaido University, Sapporo, Japan.

Thomas, C., Weber, M., Wicks, C.W., Scherbaum, F. (1999). Small scatterers in the lower mantle observed at German broadband arrays. *J. Geophys. Res.* **104**, 15073–15088.

Wagner, G.S. (1998). Local wave propagation near the San Jacinto fault zone, southern California: Observations from a three-component seismic array. *J. Geophys. Res.* **103**, 7231–7246.

Wu, R.S., Aki, K. (1988). Introduction: Seismic wave scattering in three-dimensionally heterogeneous Earth. *Pure Appl. Geophys.* **128**, 1–6.

Yilmaz, Ö. (1987). Seismic Data Processing, SEG Investigations in Geophysics No. 2, Society of Exploration Geophysicists, Tulsa.

Zeng, Y. (1991). Deterministic and stochastic modeling of the high frequency seismic wave generation and propagation in the lithosphere. Ph.D. Thesis, University of Southern California, Los Angeles, USA.

Zhao, D., Ochi, F., Hasegawa, A., Yamamoto, A. (2000). Evidence for the location and cause of large crustal earthquakes in Japan. *J. Geophys. Res.* **105**, 13579–13594.

SOURCE EFFECTS FROM BROAD AREA NETWORK CALIBRATION OF REGIONAL DISTANCE CODA WAVES

William Scott Phillips, Richard Jerome Stead, George Edward Randall, Hans Edward Hartse, and Kevin Mitsuo Mayeda

Abstract

We have applied *regional coda* techniques to a network of 64 stations across central and east Asia to isolate source effects for bands from 0.03 to 8 Hz. The heterogeneity of the study region required us to determine two-dimensional (2-D) path and transfer function corrections. The importance of the 2-D path corrections increased with frequency and distance, and for continental paths, became critical beyond 500 km for 1 Hz data. We propose a new *spreading* model for coda amplitudes, termed the extended Street-Herrmann (ESH) model, to which *attenuation* can be added, facilitating the use of tomographic techniques for path correction. The 2-D transfer function varied between continents and oceans, as well as within continents in areas of poor Lg propagation, reflecting differing excitation of Lg and Sn coda. We also demonstrate the use of empirically determined coda shapes, or *type curves*, to measure coda amplitudes, adding precision and flexibility for source regions of special interest. We applied these techniques to 112,000 records from 35,000 events, magnitudes 2–7, depths 0–50 km, between latitudes 0° and 60° and longitudes 60° and 150°. The resulting coda source spectra were used to derive moments and, for the better recorded events, corner frequencies, allowing computation of *apparent stress* for just under 6700 earthquakes. Preliminary apparent stress results ranged from 10^{-2} to 1 MPa and showed some regional variation. For example, stress increased from south to north across the Tian Shan, perhaps reflecting deformation in varying crustal rheology or effects of prior slip history. Low stress observed in Tibet could be an artifact of under correction for high attenuation; however, the correlation also could be physical. Low stress observed in oceanic regions is inconsistent with local studies and indicates that upgrades to the coda methodology to more explicitly account for mixed Lg and Sn coda will be needed. The regional network coda results should be further tested by comparing to ground-truth spectra obtained by applying coda techniques to data from local scale networks within the study region.

Key Words: Coda, source parameters, apparent stress, calibration, tomography, Asia.
© 2008 Elsevier Inc.

1. Introduction

Interest in the origin and analysis of the seismic coda was sparked by the work of Dr. Keiiti Aki, who discussed the topic in the literature as early as 1956 (Aki, 1956). Since that time, seismologists have applied significant effort to model coda wave behavior, as well as to analyze coda wave data for purposes of understanding the stochastic nature of earth materials and to isolate source, propagation, and site effects on seismic waves.

The scattering origin of seismic coda waves leads to their observed stability. Stability refers to the lack of dependence of coda amplitudes and coda shapes on source-receiver distance, and other path details, and on source mechanism. While direct waves would be expected to differ for a given event recorded at a number of stations, the coda waves are observed to be similar, outside of slight differences in overall level due to local site response. This allows the coda to be of great utility in studies of the seismic source, and motivated much of the early basic research on coda waves (Aki, 1969; Aki and Chouet, 1975; Tsujiura, 1978).

In the predigital era, seismologists deployed so-called spectral analyzing seismometers that produced a series of band-passed traces over a broad range of frequencies. These traces could be measured manually, and the coda provided redundant measures of amplitude that helped to improve precision in spectral studies of the earthquake source. Chouet et al. (1978) developed a coda source isolation method that applied an empirical Green's function technique to amplitude spectra, followed by a shift to match independently determined moments, allowing the direct comparison of earthquake scaling behavior and source parameters at sites around the world where such instruments had been deployed (California, Japan, and Hawaii). Similar instrumentation (ChISS) and analysis methods had been developed independently, arguably prior to the western work, and applied to data from the Garm region of Tajikistan by Rautian and Khalturin (1978).

Coda techniques were developed for use with local network data (source-receiver distance less than 100 km) and relied on measurements in the late coda (beginning at twice the S-wave travel time). This allowed time for the scattered wavefield to homogenize and reduce dependence on path and source radiation effects. At regional distances (up to 2000 km), however, the late coda is measurable only for the largest events, and we must adjust by measuring earlier sections of the coda. This introduces distance-dependent (1-D) effects. Mayeda (1993), Mayeda and Walter (1996), and Mayeda et al. (2003) developed a new technique that made it possible to analyze regional distance coda by empirically calibrating distance-dependent effects on coda start time, coda shape, and coda amplitude. Their results showed that early coda methods retain the precision and the source radiation averaging of the late coda methods. This technique has been applied to study source scaling issues, finding nonconstant stress scaling in a number of regions (Eken et al., 2004; Malagnini et al., 2006; Mayeda et al., 2005a; Morasca et al., 2005a,b), and to monitoring studies, as a basis for event identification as well as magnitude and yield estimation. The technique appears to work well over broad areas for local to near regional distances (up to 500 km or so) and to study events from the same source region, traveling similar paths, for far regional distances.

In applying these methods to studies in central and east Asia, however, we have seen that 2-D effects on coda amplitudes can be strong. By spatially interpolating distance-corrected coda amplitudes, we observed patterns that were similar to known lateral variations in regional phase propagation characteristics (e.g., Phillips et al., 1998; Phillips, 1999). These observations encouraged us to extend the regional coda methodology to 2-D to be more effective over broad areas and far regional distances, and allow the simultaneous calibration of an extensive network. Here, we further allow the source-to-coda transfer function to vary regionally, under the presumption that coda generation efficiency may be affected by structure in the vicinity of the source.

In the following text, we describe our work with coda wave data from a network of stations covering central and east Asia, including oceanic regions, for frequencies from 0.03 to 8 Hz. We emphasize methodology that consists of multiple quality control and calibration steps. The extensive calibration process may surprise the reader, given the simplicity of the original, local distance methods developed by Dr. Aki and coworkers.

We point out that high precision, as well as the lack of source radiation effects, relative to direct wave approaches, motivate our effort. We hope that our experience will benefit others who seek to improve coda analysis further. We also point out challenges that we hope the community will help to solve. Finally, we present source spectral results and preliminary regional variations in apparent stress that we hope will help to understand rupture physics across this region, as well as aid in predicting source spectra for use in monitoring studies.

2. Data Analysis

We analyze coda wave data from a network of 64 stations across central and east Asia (Fig. 1). While station density varies significantly, with high densities in Kyrgyzstan, Taiwan, and Japan, we find interstation distances of 1000 km or greater over much of the region. Waveform data and instrument response information are obtained from many sources, chief of which is the IRIS DMC, housing data from GSN, FDSN, IDA, and Geoscope global networks. Unique data sets such as the Borovoye Archive (Kim and Ekstrom, 1996) are also included.

We segmented waveforms for processing based on a merged set of global, regional, and special event bulletins. Event merging was carried out by comparing epicenters and origin times of events defined by different bulletins. When more than one origin was available for a given event, the origin with the highest ranking was selected. The ranking

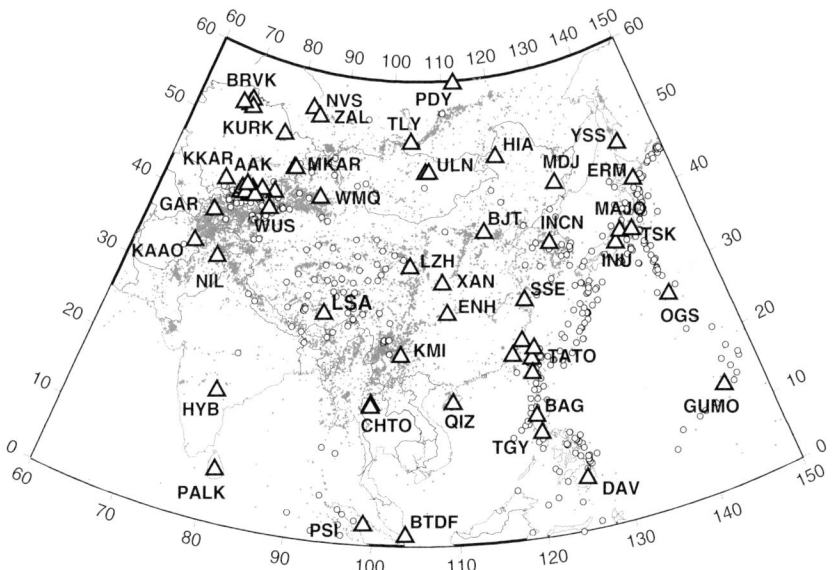

Fig. 1. Stations (triangles), events (gray crosses), and ground-truth moment events (open circles) used in the regional coda study. Stations named in the text are labeled; however, not all stations used (Table 2) are labeled, in particular those in or near the Kyrgyz (AAK) and Taiwan (TATO) networks.

was based on estimates of location quality, with explosions of known location and origin time ranked highest, followed by special studies based on dense local or temporary network data, then regional and global or teleseismic sources. The majority of our origins were taken from, in rank order, EHB (Engdahl et al., 1998), ISC, EDR, and REB global event catalogs, with additional origins from regional Chinese (ABCE) networks (Lee et al., 2002).

We collected broadband and short-period, three-component data for this study. Our primary focus was on the broadband data; however, short-period data were included when coverage could be increased, due to higher short-period triggering rates or gaps in the broadband archives, or when data quality was higher such as for the short-period borehole data from Lanzhou.

Events were selected, for a given station, based on the following magnitude–distance criteria: we accepted all events within 500 km of the station, and any event with magnitude greater than 2.5 plus distance/1250 between 500 and 2000 km. Depths were restricted to 50 km and under. For this study, we processed 112,000 records from 35,000 events between latitudes 0° and 60° and longitudes 60° and 150°. Magnitudes ranged from 2 to 7. The data set included records from 154 underground nuclear tests from Semipalatinsk and Lop Nor. These events helped constrain calibrations in less seismically active regions such as the Kazakh platform, and were used in all calibration steps except for the transfer function. Explosions have been shown to produce codas with shapes indistinguishable from those of earthquakes in southern Nevada (Hartse et al., 1995). Events yielding at least one coda measurement, nearly 19,000 in number, are plotted in Fig. 1.

Data were processed by stacking smoothed, narrow band, Hilbert transform envelopes. Processing was performed using Seismic Analysis Code software. We followed the suboctave band recipes of Mayeda et al. (2003; Table 1). After deconvolving the instrument, we decimated the trace and then applied a Butterworth, four-pole, two-pass band-pass filter. Decimation included an FIR antialias filter and was performed to maintain the bandwidth to Nyquist frequency ratio greater than 0.05, ensuring stability of the band-pass filter. Decimation also stabilized the subsequent Hilbert transform envelope processing, which Seismic Analysis Code software implements in the time domain. After taking a base-ten logarithm, the resulting envelopes were smoothed using a boxcar shape function with smoother width dependent on the frequency band (Table 1, column 2). Horizontal component envelopes were averaged together. Sample envelopes are shown in Fig. 2.

We controlled data quality using a combination of manual and automatic methods. Certain stations were scanned manually for saturation, dropouts, and glitches (e.g., Borovoye archive data) prior to processing, and poor data were repaired or eliminated from consideration. Following processing, all events for selected stations, as well as all records for ground-truth moment events, were manually reviewed to determine coda start times and measurement window limits (Fig. 2). The start time picks were made at the envelope peak, or break in slope, that was most closely associated with the final coda. The latter picks are not easily made using automatic methods (e.g., Fig. 2, NIL, 3–4 Hz). The coda start time picks associate the coda with different phases, most often Lg, but also P, Sn, and surface wave phases. We found that P and Sn coda are more common at higher frequencies and greater distances, whereas surface waves become important below periods of 5 s. An optional measurement window starting point can be chosen to avoid compromised data. Finally, we picked the end of the measurement window prior to the

TABLE 1. Band-dependent processing and calibration parameters

Band (Hz)	Smoother width (s)	Window start (s)	Window end (s)	Measurement time (s)	Initial spreading	Final spreading	Transition distance (km)	Transition factor	Lambda
0.03–0.05	50	60.0	460	325	0.0	1.0	4900	2.0	16.0
0.05–0.1	20	40.0	405	285	0.0	1.0	2470	2.0	16.0
0.1–0.2	20	40.0	360	255	0.1	1.0	1250	2.0	8.0
0.2–0.3	20	30.0	320	225	0.12	1.0	760	2.0	6.0
0.3–0.5	14	20.0	285	195	0.22	1.0	760	2.0	3.0
0.5–0.7	14	10.0	255	175	0.24	1.0	730	2.0	4.0
0.7–1.0	14	5.0	230	155	0.24	1.0	550	2.0	4.0
1.0–1.5	8	2.0	202	135	0.24	1.0	340	2.0	4.0
1.5–2.0	8	2.0	177	120	0.24	1.0	330	2.0	3.0
2.0–3.0	4	1.0	156	105	0.12	1.0	320	2.0	3.0
3.0–4.0	4	1.0	131	90	0.08	1.0	290	2.0	3.0
4.0–6.0	4	1.0	111	75	0.07	1.0	250	2.0	3.0
6.0–8.0	4	1.0	81	55	0.06	1.0	280	2.0	2.0

Window start, end, and coda measurement time are all relative to the calibrated coda start time. Initial and final spreading refer to short-distance and long-distance spreading parameters, respectively. Lambda is the regularization (smoothing constraint) applied to attenuation tomography.

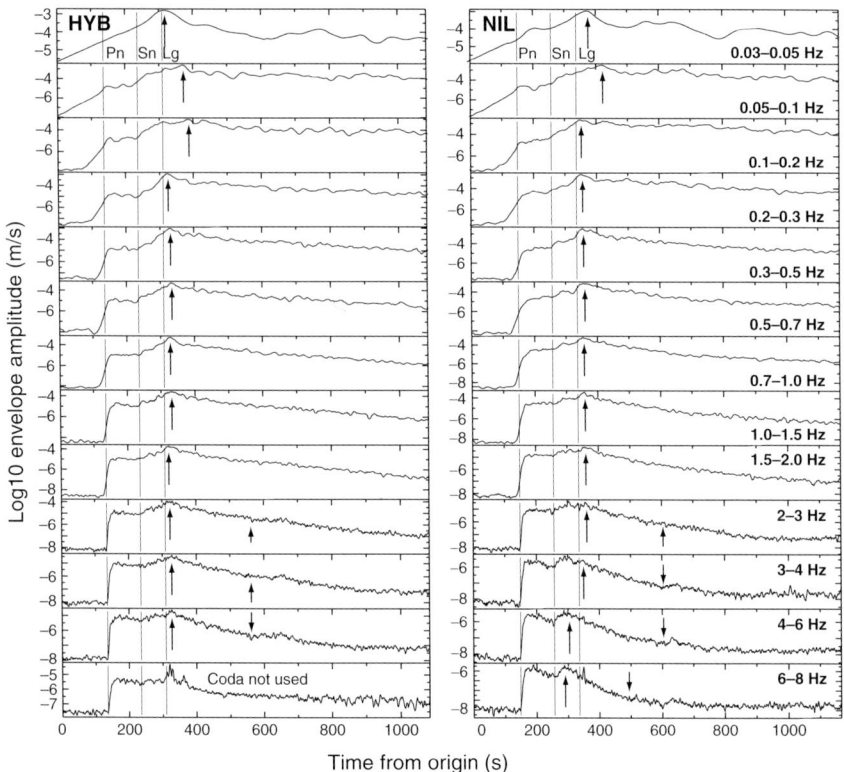

FIG. 2. Sample mean horizontal channel envelopes from the Hyderabad (HYB) and Nilore (NIL) stations for the Bhuj mainshock, showing manually set coda window picks. Frequency bands increase from top to bottom as noted. Coda start and window endpoint picks (arrows) are shown if within the plot window (limited by group velocity 1 km/s). A secondary event influenced endpoint picks for high-band envelopes. Predicted Pn, Sn, and Lg arrivals are marked (8.0, 4.6, and 3.5 km/s, respectively). Note the shift from Lg to Sn coda for high bands along the path to NIL.

point at which the coda decays into noise, avoiding secondary events and other transient sources of noise. This pick indicated good quality coda to later processing procedures. Conversely, if an envelope had been reviewed, but no window endpoint chosen, we did not use that envelope further. In addition to controlling processing, the window end picks were used to tune automatic methods.

We also tested for undocumented changes in instrument response by plotting 1-Hz envelope noise levels against time. This method allowed us to observe shifts on the order of a factor of two or greater, for well-sampled stations. In such cases, we chose one or more intervals to eliminate. This is a crude method, but it allowed us to identify and eliminate one or more time segments for 29 of our 64 stations. Clearly, minor response issues may remain undetected with this method.

Further, arrivals from secondary events must be avoided in coda analysis. Using the event catalog information described earlier, we estimated arrival times of any direct

SOURCE EFFECTS FROM CODA WAVES 325

phases that could contaminate the coda. Automatically determined measurement windows were not allowed to include a predicted secondary arrival, whereas manually chosen endpoints were left alone. We found this especially useful with aftershock sequences. Of course, the available catalogs are not complete enough to predict all secondary arrivals, but we do want to avoid those that we can.

To aid the reader in following the numerous quality control and subsequent calibration steps, we present a flowchart in Fig. 3.

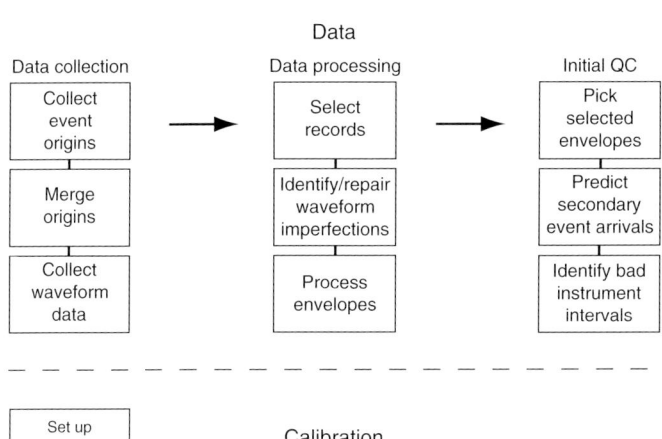

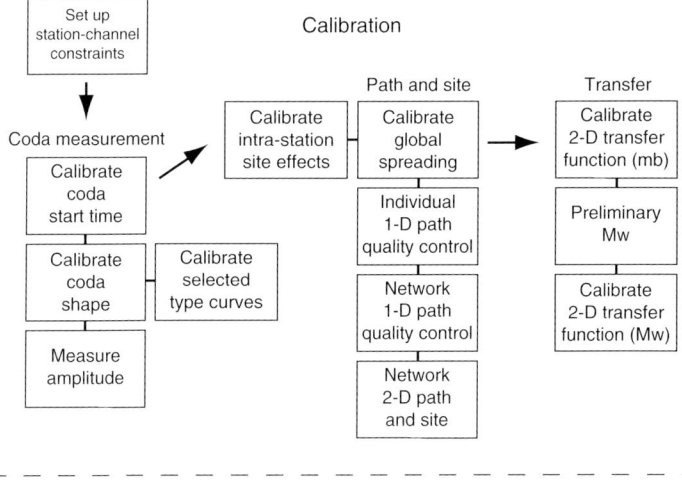

FIG. 3. Flowchart representing processing, quality control, and calibration steps taken to obtain source spectra from coda envelope data.

3. CODA CALIBRATION METHODOLOGY

Calibration consists of a series of steps that convert high precision, coda amplitude measurements to absolute source spectral estimates. We followed Mayeda et al. (2003) with noted modifications, to extend the coda method to far regional distances. Briefly, the calibration steps estimated: (1) group velocity of the coda start time, (2) coda shape functions, (3) relative site effects for different recording channels of the same station, (4) path and relative site effects between stations, and (5) absolute source-to-coda transfer function (Fig. 3). Extensions to the Mayeda et al. (2003) method included the integration of purely empirical (type curve) decay functions for source regions of special interest, a coda spreading function extending that of Street et al. (1975), and the two-dimensional (2-D) attenuation and transfer functions that are needed to describe coda behavior over the broad study area.

To explicitly summarize the calibrated effects, we write the measured coda amplitude,

$$A_{ij}(f) = S_i(f)T(\phi_i, \theta_i, f)P'(\phi_i, \theta_i, \phi_j, \theta_j, f)R'_j(f)D'_j[x, f, t_c(x,f) + t_m(f)], \quad (1)$$

where f is the frequency, i and j are the source and site indices, respectively; ϕ is the latitude, θ is the longitude, x is the source-receiver distance, and t_m is the coda measurement time, relative to the coda start time,

$$t_c(x,f) = t_0 + \frac{x}{v_g(x,f)}, \quad (2)$$

where t_0 is the event origin time and v_g is the calibrated, range- and frequency-dependent coda start time group velocity. In Eq. (1), S represents the source spectrum that will be recovered by calibrating the remaining terms on the right-hand side, specifically T, a 2-D, frequency-dependent, source-to-coda transfer term; P', a 2-D path term; R', a site amplification term; and D', a coda decay function. The primes indicate relative or dimensionless terms. Individual terms will be further defined in appropriate sections to follow.

We employed a station network of uneven density, with multiple recording channels for each station, and variable operation periods and numbers of records for each channel. Thus, for practical reasons, we combined certain stations and channels together during calibration and such groups could change with each calibration step. For example, if two or more stations are closely located, or located in similar geological terrain, we might constrain them to have the same coda shape functions. Stations with limited recordings can be grouped with better-populated stations. We set up station and channel groups to constrain coda start time group velocity, coda shape functions, coda decay type curves, intrastation site effects, 1-D path effects, and source spectral fitting. Groups used in this study are indicated by a master station for each calibration step (Table 2), such that all stations and channels with the same master will be constrained the same. In constructing such a table, care must be taken that at any calibration stage, each group will have been constrained the same for all previous stages upon which the current calibration step relies. A group constrained to have the same coda decay function, for example, must all rely on the same coda-origin group velocities. In this study, master station constraint groups were the same for all frequency bands.

TABLE 2. Master station groups

Station and channel	Master station	Coda origin master	Intrastation site master	Shape master	Type curve master	1-D path master
AAK_BHH	AAK	WMQ_BHH	AAK_BHH	AAK_BHH	AAK_BHH	AAK_BHH
AAK_BHH00	AAK	WMQ_BHH	AAK_BHH	AAK_BHH	AAK_BHH	AAK_BHH
AAK_BHHKN	AAK	WMQ_BHH	AAK_BHH	AAK_BHH	AAK_BHH	AAK_BHH
AML_BHH	AML	WMQ_BHH	AML_BHH	AAK_BHH	AML_BHH	AAK_BHH
BAG_BHH	BAG	WMQ_BHH	BAG_BHH	BAG_BHH	BAG_BHH	BAG_BHH
BJI_BHH	BJI	WMQ_BHH	BJI_BHH	BJT_BHH	BJT_BHH	BJT_BHH
BJI_SHH	BJI	WMQ_BHH	BJI_SHH	BJT_BHH	BJT_BHH	BJT_BHH
BJT_BHH	BJT	WMQ_BHH	BJT_BHH	BJT_BHH	BJT_BHH	BJT_BHH
BJT_BHH00	BJT	WMQ_BHH	BJT_BHH00	BJT_BHH	BJT_BHH	BJT_BHH
BRVK_BHH	BRVK	WMQ_BHH	BRVK_BHH	BRVK_BHH	BRVK_BHH	BRVK_BHH
BRVK_BHH00	BRVK	WMQ_BHH	BRVK_BHH	BRVK_BHH	BRVK_BHH	BRVK_BHH
BRVK_BHZ	BRVK	WMQ_BHH	BRVK_BHZ	BRVK_BHH	BRVK_BHH	BRVK_BHH
BRVK_BHZ00	BRVK	WMQ_BHH	BRVK_BHZ	BRVK_BHH	BRVK_BHH	BRVK_BHH
BRVK_SHH03	BRVK	WMQ_BHH	BRVK_SHH03	BRVK_BHH	BRVK_BHH	BRVK_BHH
BRVK_SHH07	BRVK	WMQ_BHH	BRVK_SHH07	BRVK_BHH	BRVK_BHH	BRVK_BHH
BRVK_SHZ01	BRVK	WMQ_BHH	BRVK_SHZ01	BRVK_BHH	BRVK_BHH	BRVK_BHH
BRVK_SHZ03	BRVK	WMQ_BHH	BRVK_SHZ03	BRVK_BHH	BRVK_BHH	BRVK_BHH
BRVK_SHZ07	BRVK	WMQ_BHH	BRVK_SHZ07	BRVK_BHH	BRVK_BHH	BRVK_BHH
BRVK_SHZV6	BRVK	WMQ_BHH	BRVK_SHZV6	BRVK_BHH	BRVK_BHH	BRVK_BHH
BVA0_BHH	BVAR	WMQ_BHH	BVA0_BHH	BRVK_BHH	BVA0_BHH	BRVK_BHH
CHKZ_BHH	CHKZ	WMQ_BHH	CHKZ_BHH	BRVK_BHH	CHKZ_BHH	BRVK_BHH
CHMS_BHH	CHMS	WMQ_BHH	CHMS_BHH	AAK_BHH	CHMS_BHH	AAK_BHH
CHTO_BHH	CHTO	WMQ_BHH	CHTO_BHH	CHTO_BHH	CHTO_BHH	CHTO_BHH
CHTO_BHH00	CHTO	WMQ_BHH	CHTO_BHH	CHTO_BHH	CHTO_BHH	CHTO_BHH
CMAR_BHH	CMAR	WMQ_BHH	CMAR_BHH	CHTO_BHH	CMAR_BHH	CHTO_BHH
DAV_BHH00	DAV	WMQ_BHH	DAV_BHH00	BAG_BHH	DAV_BHH00	BAG_BHH
EKS2_BHH	EKS2	WMQ_BHH	EKS2_BHH	AAK_BHH	EKS2_BHH	AAK_BHH
ENH_BHH	ENH	WMQ_BHH	ENH_BHH	ENH_BHH	ENH_BHH	ENH_BHH

(continued on next page)

TABLE 2. Continued.

Station and channel	Master station	Coda origin master	Intrastation site master	Shape master	Type curve master	1-D path master
ENH_BHH00	ENH	WMQ_BHH	ENH_BHH	ENH_BHH	ENH_BHH	ENH_BHH
ERM_BHH00	ERM	WMQ_BHH	ERM_BHH00	MAJO_BHH00	ERM_BHH00	MAJO_BHH00
GAR_BHH00	GAR	WMQ_BHH	GAR_BHH00	KKAR_BHH	GAR_BHH00	KKAR_BHH
GUMO_BHH00	GUMO	WMQ_BHH	GUMO_BHH00	GUMO_BHH00	GUMO_BHH00	GUMO_BHH00
HIA_BHH	HIA	WMQ_BHH	HIA_BHH	HIA_BHH	HIA_BHH	HIA_BHH
HIA_BHH00	HIA	WMQ_BHH	HIA_BHH	HIA_BHH	HIA_BHH	HIA_BHH
HIA_SHH	HIA	WMQ_BHH	HIA_SHH	HIA_BHH	HIA_BHH	HIA_BHH
HIA_SHH10	HIA	WMQ_BHH	HIA_SHH	HIA_BHH	HIA_BHH	HIA_BHH
HYB_BHH	HYB	WMQ_BHH	HYB_BHH	HYB_BHH	HYB_BHH	HYB_BHH
INCN_BHH	INCN	WMQ_BHH	INCN_BHH00	KSRS_BHH	INCN_BHH00	KSRS_BHH
INCN_BHH00	INCN	WMQ_BHH	INCN_BHH00	KSRS_BHH	INCN_BHH00	KSRS_BHH
INCN_BHH10	INCN	WMQ_BHH	INCN_BHH10	KSRS_BHH	INCN_BHH00	KSRS_BHH
INU_BHH	INU	WMQ_BHH	INU_BHH	MAJO_BHH00	INU_BHH	MAJO_BHH00
KAAO_SHZ	KAAO	WMQ_BHH	KAAO_SHZ	NIL_BHH	KAAO_SHZ	NIL_BHH
KBK_BHH	KBK	WMQ_BHH	KBK_BHH	AAK_BHH	KBK_BHH	AAK_BHH
KKAR_BHH	KKAR	WMQ_BHH	KKAR_BHH	KKAR_BHH	KKAR_BHH	KKAR_BHH
KMI_BHH	KMI	WMQ_BHH	KMI_BHH	KMI_BHH	KMI_BHH	KMI_BHH
KMI_BHH00	KMI	WMQ_BHH	KMI_BHH	KMI_BHH	KMI_BHH	KMI_BHH
KMI_SHH	KMI	WMQ_BHH	KMI_SHH	KMI_BHH	KMI_BHH	KMI_BHH
KMNB_BHH	KMNB	WMQ_BHH	KMNB_BHH	TATO_BHH	KMNB_BHH	TATO_BHH
KSRS_BHH	KSRS	WMQ_BHH	KSRS_BHH	KSRS_BHH	KSRS_BHH	KSRS_BHH
KURK_BHH	KURK	WMQ_BHH	KURK_BHH	KURK_BHH	KURK_BHH	KURK_BHH
KURK_BHH00	KURK	WMQ_BHH	KURK_BHH	KURK_BHH	KURK_BHH	KURK_BHH
KURK_BHHKZ	KURK	WMQ_BHH	KURK_BHH	KURK_BHH	KURK_BHH	KURK_BHH
KZA_BHH	KZA	WMQ_BHH	KZA_BHH	AAK_BHH	KZA_BHH	AAK_BHH
LSA_BHH	LSA	WMQ_BHH	LSA_BHH	LSA_BHH	LSA_BHH	LSA_BHH
LSA_BHH00	LSA	WMQ_BHH	LSA_BHH	LSA_BHH	LSA_BHH	LSA_BHH
LZH_BHH	LZH	WMQ_BHH	LZH_BHH	LZH_BHH	LZH_BHH	LZH_BHH

SOURCE EFFECTS FROM CODA WAVES 329

LZH_SHH	LZH	WMQ_BHH	LZH_SHH	LZH_BHH	LZH_BHH
MAJO_BHH00	MAJO	WMQ_BHH	MAJO_BHH00	MAJO_BHH00	MAJO_BHH00
MAKZ_BHH	MAKZ	WMQ_BHH	MAKZ_BHH	MKAR_BHH	MKAR_BHH
MAKZ_BHH00	MAKZ	WMQ_BHH	MAKZ_BHH	MKAR_BHH	MKAR_BHH
MAKZ_BHHKZ	MAKZ	WMQ_BHH	MAKZ_BHH	MKAR_BHH	MKAR_BHH
MAKZ_SHH	MAKZ	WMQ_BHH	MAKZ_BHH	MKAR_BHH	MKAR_BHH
MATB_BHH	MATB	WMQ_BHH	MATB_BHH	TATO_BHH	TATO_BHH
MDJ_BHH	MDJ	WMQ_BHH	MDJ_BHH	MDJ_BHH	MDJ_BHH
MDJ_BHH00	MDJ	WMQ_BHH	MDJ_BHH	MDJ_BHH	MDJ_BHH
MDJ_BHH10	MDJ	WMQ_BHH	MDJ_BHH10	MDJ_BHH	MDJ_BHH
MDJ_SHH	MDJ	WMQ_BHH	MDJ_SHH	MDJ_BHH	MDJ_BHH
MKAR_BHH	MKAR	WMQ_BHH	MKAR_BHH	MKAR_BHH	MKAR_BHH
NIL_BHH	NIL	WMQ_BHH	NIL_BHH	NIL_BHH	NIL_BHH
NIL_BHH00	NIL	WMQ_BHH	NIL_BHH	NIL_BHH	NIL_BHH
NIL_BHH10	NIL	WMQ_BHH	NIL_BHH10	NIL_BHH	NIL_BHH
NIL_SHH	NIL	WMQ_BHH	NIL_BHH10	NIL_BHH	NIL_BHH
NVS_BHH	NVS	WMQ_BHH	NVS_BHH	NVS_BHH	NVS_BHH
NVS_BHH00	NVS	WMQ_BHH	NVS_BHH	NVS_BHH	NVS_BHH
OGS_BHH	OGS	WMQ_BHH	OGS_BHH	OGS_BHH	OGS_BHH
PALK_BHH10	PALK	WMQ_BHH	PALK_BHH10	PALK_BHH10	HYB_BHH
PALK_BHZ00	PALK	WMQ_BHH	PALK_BHZ00	PALK_BHH10	HYB_BHH
PALK_BHZ10	PALK	WMQ_BHH	PALK_BHZ10	PALK_BHH10	HYB_BHH
PDGK_BHH	PDGK	WMQ_BHH	PDGK_BHH	PDGK_BHH	PDGK_BHH
PDY_SHH	PDY	WMQ_BHH	PDY_SHH	PDY_SHH	YAK_BHH00
PDY_SHZ	PDY	WMQ_BHH	PDY_SHZ	PDY_SHH	YAK_BHH00
PSI_BHH	PSI	WMQ_BHH	PSI_BHH	PSI_BHH	BAG_BHH
QIZ_BHH	QIZ	WMQ_BHH	QIZ_BHH	QIZ_BHH	QIZ_BHH
QIZ_BHH00	QIZ	WMQ_BHH	QIZ_BHH	QIZ_BHH	QIZ_BHH
QIZ_SHH	QIZ	WMQ_BHH	QIZ_SHH	QIZ_BHH	QIZ_BHH
SSE_BHH	SSE	WMQ_BHH	SSE_BHH	SSE_BHH	SSE_BHH
SSE_BHH00	SSE	WMQ_BHH	SSE_BHH00	SSE_BHH	SSE_BHH
SSE_SHH	SSE	WMQ_BHH	SSE_SHH	SSE_BHH	SSE_BHH

(continued on next page)

TABLE 2. Continued.

Station and channel	Master station	Coda origin master	Intrastation site master	Shape master	Type curve master	1-D path master
SSLB_BHH	SSLB	WMQ_BHH	SSLB_BHH	TATO_BHH	SSLB_BHH	TATO_BHH
TATO_BHH	TATO	WMQ_BHH	TATO_BHH	TATO_BHH	TATO_BHH	TATO_BHH
TATO_BHH00	TATO	WMQ_BHH	TATO_BHH00	TATO_BHH	TATO_BHH	TATO_BHH
TATO_BHH10	TATO	WMQ_BHH	TATO_BHH10	TATO_BHH	TATO_BHH	TATO_BHH
TGY_BHH	TGY	WMQ_BHH	TGY_BHH	BAG_BHH	TGY_BHH	BAG_BHH
TKM2_BHH	TKM2	WMQ_BHH	TKM2_BHH	AAK_BHH	TKM2_BHH	AAK_BHH
TLG_BHH	TLG	WMQ_BHH	TLG_BHH	AAK_BHH	TLG_BHH	AAK_BHH
TLY_BHH	TLY	WMQ_BHH	TLY_BHH	TLY_BHH	TLY_BHH	TLY_BHH
TLY_BHH00	TLY	WMQ_BHH	TLY_BHH00	TLY_BHH	TLY_BHH	TLY_BHH
TSK_BHH	TSK	WMQ_BHH	TSK_BHH	MAJO_BHH00	TSK_BHH	MAJO_BHH00
TWK1_BHH	TWK1	WMQ_BHH	TWK1_BHH	TATO_BHH	TWK1_BHH	TATO_BHH
UCH_BHH	UCH	WMQ_BHH	UCH_BHH	AAK_BHH	UCH_BHH	AAK_BHH
ULHL_BHH	ULHL	WMQ_BHH	ULHL_BHH	AAK_BHH	ULHL_BHH	AAK_BHH
ULN_BHH	ULN	WMQ_BHH	ULN_BHH	ULN_BHH	ULN_BHH	ULN_BHH
ULN_BHH00	ULN	WMQ_BHH	ULN_BHH	ULN_BHH	ULN_BHH	ULN_BHH
USP_BHH	USP	WMQ_BHH	USP_BHH	AAK_BHH	USP_BHH	AAK_BHH
VOSK_BHH	VOSK	WMQ_BHH	VOSK_BHH	BRVK_BHH	VOSK_BHH	BRVK_BHH
WMQ_BHH	WMQ	WMQ_BHH	WMQ_BHH	WMQ_BHH	WMQ_BHH	WMQ_BHH
WMQ_BHH00	WMQ	WMQ_BHH	WMQ_BHH	WMQ_BHH	WMQ_BHH	WMQ_BHH
WMQ_SHH	WMQ	WMQ_BHH	WMQ_BHH	WMQ_BHH	WMQ_BHH	WMQ_BHH
WUS_BHH	WUS	WMQ_BHH	WUS_BHH	WUS_BHH	WUS_BHH	WUS_BHH
XAN_BHH	XAN	WMQ_BHH	XAN_BHH	XAN_BHH	XAN_BHH	XAN_BHH
XAN_BHH00	XAN	WMQ_BHH	XAN_BHH	XAN_BHH	XAN_BHH	XAN_BHH
YSS_BHH00	YSS	WMQ_BHH	YSS_BHH00	MAJO_BHH00	YSS_BHH00	MAJO_BHH00
ZAL_SHH	ZAL	WMQ_BHH	ZAL_SHH	NVS_BHH	ZAL_SHH	NVS_BHH
ZAL_SHZ	ZAL	WMQ_BHH	ZAL_SHZ	NVS_BHH	ZAL_SHH	NVS_BHH
ZRNK_BHH	ZRNK	WMQ_BHH	ZRNK_BHH	BRVK_BHH	ZRNK_BHH	BRVK_BHH

Standard station names are listed, while channel names are constructed from IRIS channel names and location codes. Exceptions are the Borovoye (BRVK) archive channel names, for which, the original Russian channel code is appended to a standard channel name. An "H" as the third character of a channel name indicates an N plus E channel stack.

In the following sections, we describe calibration steps in detail, with emphasis on techniques we have developed to extend the Mayeda *et al.* (2003) method to a broad area network of stations and to far regional distances.

3.1. Coda Start Time Calibration

The first calibration step determined the starting point of the coda [Eq. (2)], and thus defined the coda type (P, Sn, Lg, or surface wave). Mayeda *et al.* (2003) suggested a three-parameter hyperbolic function to describe the distance dependence of the group velocity,

$$v_g(x,f) = v_0(f) - \frac{v_1(f)}{v_2(f) + x}. \tag{3}$$

Note that the hyperbola coefficients all have different units, in spite of our naming convention. This function was fit to manually determined starting times for each frequency band. For this study, coda start times were obtained via manual review of stations WMQ, MDJ, BJI, SSE, MKAR/MAKZ, BRVK, NIL, QIZ, ULN, TATO, and KKAR, in order of number of reviewed events, with over 6000 events reviewed at WMQ and over 300 at KKAR. We combined these data together and applied results to all stations, as indicated by the master station grouping (Table 2, column 3). Figure 4 shows group velocities for selected bands and includes the calibrated curves. Data were restricted by an upper bound on group velocity that varied slightly with band in order to isolate the Lg or surface wave branch from mantle shear phases. We found that group velocities were relatively independent of distance for the lowest three bands, as expected for surface waves. Group velocity increased with distance, reflecting transition from shallow crustal shear to Lg phases for higher bands. We calibrated the Lg and surface wave branches in this study; however, Sn and P branches were also apparent. Sn and P were more common at greater distance and at higher frequency, likely due to stripping of the Lg energy by attenuating crustal materials.

3.2. Coda Shape Calibration and Amplitude Measurement

We obtained coda amplitudes by fitting envelopes to a calibrated coda shape function. This shape function varied with frequency and distance via a simple model for broad area application. We also determined empirical coda shapes or type curves to measure amplitudes in certain source regions. Both techniques and a method to integrate their results will be described next.

As proposed by Mayeda *et al.* (2003), coda shape functions follow a standard, spreading, and attenuation model,

$$D'_j(x,f,t) = [t - t_c(x,f)]^{-\alpha(x)} \exp[-\beta(x)\{t - t_c(x,f)\}], \tag{4}$$

where t is greater than t_c, and the distance dependence of both $\alpha(x)$ and $\beta(x)$ is identical to that of Eq. (3). This standard form is known to fit lengthy coda poorly (e.g., Roecker *et al.*, 1982), and we restricted measurement window lengths to avoid this effect. Measurement window limits used in this study are given in Table 1.

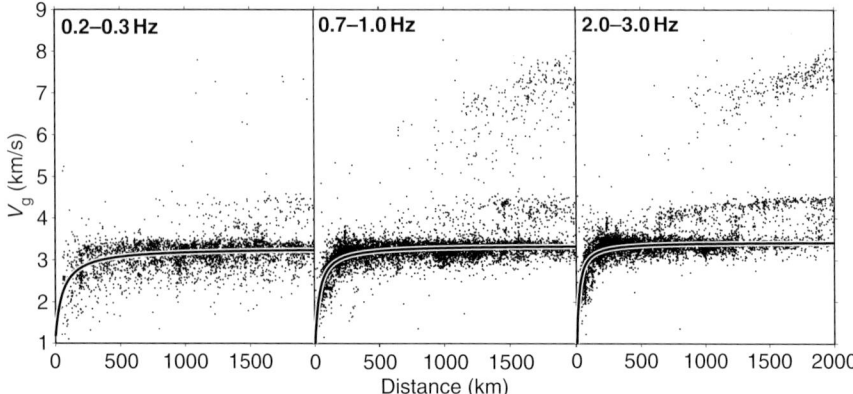

FIG. 4. Group velocities of manually determined coda start times (dots) versus distance, and calibrated curves (solid lines). Data are shown from all reviewed envelopes, irrespective of station. Bands 0.2–0.3 Hz, 0.7–1 Hz, and 2–3 Hz are shown, from left to right. Note the increasing importance of Sn and P branches as frequency increases, especially at distance. Fits were made to the main, Lg, branch.

We determined shape functions for a station or group of stations (Table 2, column 5). In this way, we crudely accounted for regional scale, 2-D effects on a station-by-station basis. Our calibration procedure fitted parameters of six cubic polynomials, each describing the variation of one shape hyperbola coefficient with log-base-ten frequency, using the highest quality coda decay data binned by radius and azimuth. Data quality was based on manual review, coda length, and event magnitude, in that order. This selection procedure reduced the amount of data used, and thus speeded up computation, while retaining a broad geographical selection. The polynomial parameterization reduced the number of free parameters from 78 to 24, adding stability between results for adjacent bands.

While calibrating shape functions, we automatically determined coda endpoints for records that had not been manually reviewed. To do this, we stopped the coda at the first point falling below four times the noise level. Noise was estimated by taking the minimum envelope after applying a second smoothing operation, with smoother widths five times the original widths listed in Table 1 (column 2) or 30 s, whichever is greater. This noise technique is flexible, but can be fooled by large dropouts, which we assume are rare.

Once calibrated, we used the shape functions to measure amplitudes at a consistent point in the coda [t_m, Eq. (1); Table 1, column 5]. We chose t_m near the end of the maximum prescribed coda window. This produced flat local distance spreading that was similar for amplitudes measured at different stations, even in cases where coda shapes might vary between stations. If the coda fell below the noise level prior to t_m, the fitted coda shape was extrapolated to t_m. This is acceptable as long as the shape functions are well calibrated over the full, prescribed window. If we had no manual endpoint pick, we took advantage of the calibrated decay shapes to determine endpoints automatically. This method was more versatile than applying the signal-to-noise cutoff described earlier, as coda can be analyzed without a good background noise measurement, a common situation with triggered data. Furthermore, this method was more sensitive to secondary arrivals and other sources of noise in the coda. To apply, we fitted a short section of the earliest coda and extended to a point where the absolute difference exceeded 0.2. The new length of coda was fitted in the same way, and the process iterated until the coda endpoint failed to increase further. The end of the measurement window was set to the earliest of the automatically determined endpoint, the maximum allowed duration from Table 1, and any predicted secondary event arrival.

We also employed a second decay shape method: the type curve (Aki,1969; Hartse et al., 1995). A type curve is a purely empirical decay curve, with no model assumptions, that is obtained by combining coda decay data from a source region of limited extent. The type curve defines the coda shape for a particular path and has arbitrary level. Amplitudes were measured by directly comparing data to the type curve and taking the median difference. To incorporate type curve amplitude measurements into our broad area data set, we fitted the type curve with the appropriate shape function, over the window limits, as if the type curve were actual coda data, and corrected using that offset. Advantages of the type curve approach include increased precision, as the coda shape is not limited by a model. Even more important for monitoring work, the type curve can be defined outside the window limits applied to standard shape function calibration [Eq. (4)], and can be used with older data sets in which instrument saturation is common. For example, the Borovoye Archive data set contains saturated records, and saturation often extends beyond the window restrictions we have set; yet relative amplitudes can still be measured using the type curve method and integrated into the study.

We constructed type curves by minimizing the difference, less the mean difference, between the type curve model and each envelope coda segment, in a least squares sense. To ensure that we work with the main body of envelopes, we determined the most heavily represented point in time, given a group of envelopes of various offsets and lengths, then traced overlapping envelope continuity forwards and backwards to determine time limits for the type curve model. Disconnected segments were left out. As the model level is unconstrained, we added an equation to the inversion that damps the sum of model parameters to zero. Type curve results are shown for Semipalatinsk explosions at the Kyrgyz network station AAK (Fig. 5). Also shown are shape function fits, as described above. We note that the shape functions are determined using earthquake data primarily, and fit the explosion codas well.

Only stations that are very close together can be combined in the type curve calculations, and this was controlled using master station groups (Table 2, column 6). Type curve groups can be finer than for later calibration stages, seemingly violating the grouping rules discussed earlier. Recall, however, that we readjust the type curve results using shape functions, which is the calibration step in the main progression. The type curve can be considered a detour in this sense, as indicated in the flowchart (Fig. 3).

3.3. Intrastation Site Calibration

Seismic stations typically record many channels of data. Channels can be physically separate, on a pier, or in a shallow borehole, and the arrangement can change over time, leading to different site amplification effects that must be accounted for if we want the most comprehensive coverage. We accounted for intrastation site effects using a combination of three methods: (1) site terms were constrained equal given knowledge of the recording environment, (2) relative site effects were measured by direct comparison of

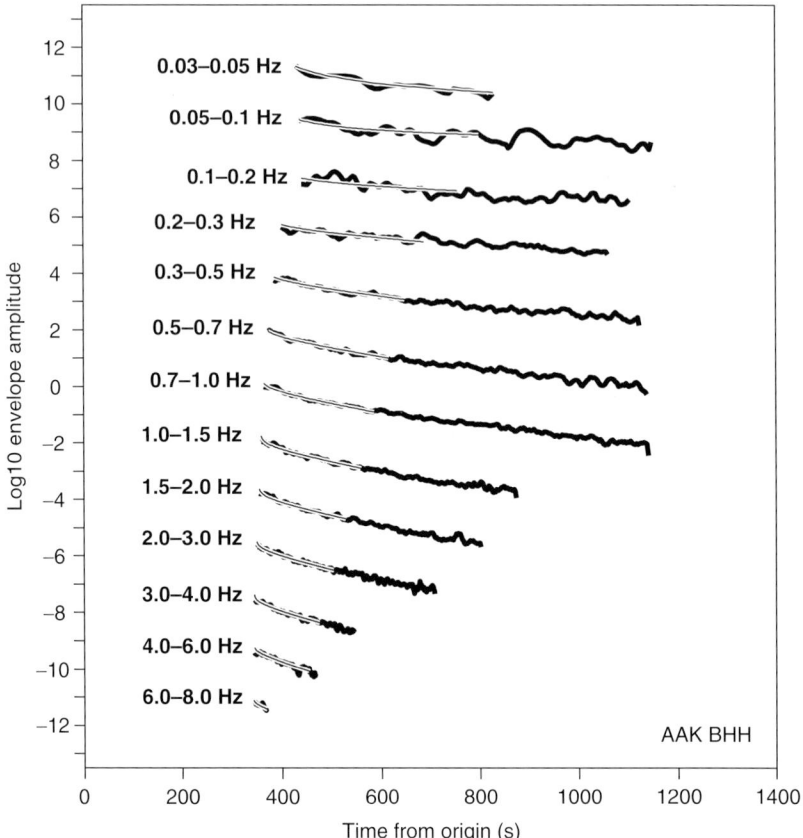

FIG. 5. Type curves (thick lines) and fitted shape functions (thin white lines). This example shows results for Semipalatinsk underground explosions recorded at Kyrgyz network station AAK. Log-base-ten amplitudes are offset arbitrary amounts for display purposes. Type curves are given for a range of frequencies, increasing from the top down, as noted. Time is relative to event origin time for the cluster centroid.

coda envelopes, or (3) relative site effects were estimated along with path effects in a later calibration stage. Site constraints (method 1) were implemented using master station and channel groups (Table 2, column 4) such that channels with the same master were constrained to have the same site terms.

For channels that run concurrently, we estimated relative site terms by direct comparison of coda envelopes for common events using

$$\delta a_{ij} = s_i - s_j, \qquad (5)$$

where s is the intrastation relative site term (i and j are site indices) and δa is the median amplitude difference over the coda window common to both channels, for one band. We solved for s using singular value decomposition, which allowed the separation of unconstrained subgroups by analysis of the eigenvector null space. If channels have no common events, often the case when more than one channel is changed at one time, we were left with more than one site effect group for that station. This procedure set the mean relative site term to zero in each band.

Examples of intrastation site effect calculations are shown for stations NIL and HIA in Fig. 6. At NIL, channels BHH and BHH00 are borehole seismometers, whereas channels SHH and BHH10 are sited on a nearby pier. We constrained BHH and BHH00 to have the same site term, likewise for SHH and BHH10 (Table 2, column 4). The relative site terms between groups peak between 3 and 4 Hz, perhaps the result of a resonance in the amplified surface channels. Relative site terms for HIA also peak at about 1 Hz, but differences are of lower magnitude. Relevant deployment details of the HIA channels are not known to the authors. The site terms show symmetric behavior because the mean is constrained to zero by our procedure, as mentioned above.

3.4. 2-D Path and Interstation Site Calibration

We performed path calibration in four steps, updating starting models and data quality control as we proceeded: (1) basic spreading parameters were chosen for each band, using a grid search in which 2-D tomography was repeatedly applied to amplitudes that were corrected for source effects using event magnitude, (2) 1-D attenuation and relative site terms were determined for each station and band individually, using the basic spreading parameters and source corrected amplitudes, (3) 1-D attenuation and relative site terms were updated by considering amplitude differences for all network stations simultaneously, for each band, and (4) 2-D attenuation and relative site terms were obtained from an amplitude difference, network inversion, for each band. Steps 2 and 3 used robust minimization techniques, whereas steps 1 and 4 used least squares inversion. Steps 1 and 2 used amplitudes corrected for source based on magnitude, whereas steps 3 and 4 relied on relative amplitudes from multiply recorded events. Given our sparse network, we rarely obtained more than one local distance record per event, and had to

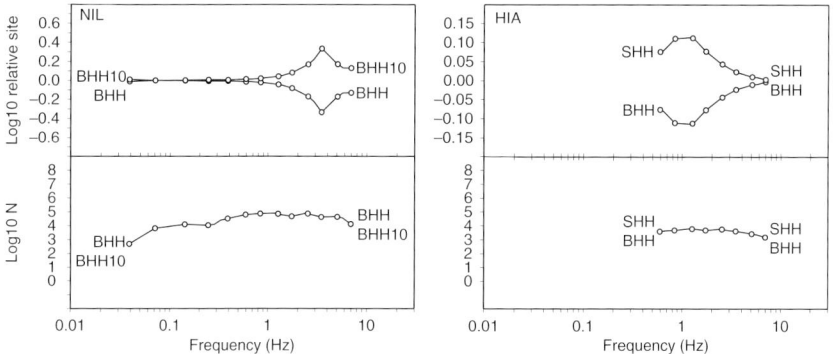

FIG. 6. Relative, intrastation site effects (log base ten) for stations NIL (*left*) and HIA (*right*) versus frequency. Site terms (*top*) and number of amplitude difference measurements (*bottom*) are shown. Broadband results are available at NIL whereas only short period results are available at HIA. Note that the NIL site term scale is broader than the HIA scale.

perform the more crude magnitude-based source correction to determine the spreading parameters that control local distance behavior. The relative amplitude techniques are more accurate, but can only be used to constrain path parameters that control long distance behavior, such as attenuation. We anticipate that focusing on data from local scale networks within our study region will allow us to use amplitude differences to more accurately determine short distance spreading parameters in future work. All inversions were based on a new spreading model for coda, to be described next. We constrained stations to have the same attenuation for 1-D path steps via master station grouping as for earlier calibration steps (Table 2, column 7). In this study, path effect constraint groups were set the same as for coda shape. In the following, we discuss the new spreading model and then provide detail concerning application and results of each of the four path calibration steps outlined above.

We based our coda spreading model on that of Street et al. (1975), commonly referred to as the Street-Herrmann model, that uses two branches to define the transition from 3-D to 2-D spreading with distance:

$$a(x) = x^{-\alpha_1}, \quad x \leq X_0$$
$$a(x) = X_0^{-\alpha_1}(x/X_0)^{-\alpha_2}, \quad x \geq X_0, \quad (6)$$

where amplitude, a, is a function of three parameters: short distance and long distance spreading coefficients, α_1 and α_2, respectively, and a transition distance, X_0. Measuring coda amplitudes at a fixed time from the beginning of the coda, as was done here, requires spreading to be close to zero at short distances (e.g., Aki, 1969), and to approach the spreading of the direct wave at long distances. Equation (6) could be used to describe this behavior; however, the transition at X_0 is very sharp. Mayeda et al. (2003) showed that the transition between short and long distances is gradual for coda amplitudes. Therefore, we add one more parameter, F, the transition factor, to smooth over the sharp corner. The new spreading function, which we refer to as the ESH model, is as follows:

$$a(x) = x^{-\alpha_1}, \quad x \leq X_1$$
$$a(x) = X_1^{-\alpha_1}(x/X_1)^{-\alpha'(x)}, \quad X_1 < x < X_2$$
$$a(x) = X_1^{-\alpha_1}(X_2/X_1)^{-\alpha'(X_2)}(x/X_2)^{-\alpha_2}, \quad x \geq X_2, \quad (7)$$

where

$$X_1 = X_0/F$$
$$X_2 = X_0 F \quad (8)$$

and

$$\alpha'(x) = \alpha_1 + 1/2(\alpha_2 - \alpha_1)\log(x/X_1)/\log(X_2/X_1). \quad (9)$$

Equations (7)–(9) define a four-parameter model specified by α_1, α_2, X_0, and F. The X_1 and X_2 mark the edges of a smooth transition range. The F parameter must be greater than one, and the ESH model approaches the original Street-Herrmann model as F approaches one. For the same α_1, α_2, and X_0, the ESH and Street-Herrmann models are equal for all distances outside the transition range. The α' is an effective spreading parameter that is applied as if the spreading were uniform between X_1 and x.

A suite of ESH curves are plotted in Fig. 7 for short distance spreading 0, long distance spreading 1, transition distance 100 km, and transition factors from 1 (no transition) to 40. Also included is the path effect model used by Mayeda et al. (2003),

$$a(x) = \frac{1}{1 + (x/X_0)^p} \qquad (10)$$

for $X_0 = 100$ and $p = 1$. The two model types have slightly different character as the Mayeda et al. (2003) curve is sharper at transition, but tails away more slowly than comparable ESH curves. ESH parameters can be adjusted to fit the Mayeda et al. (2003) model nearly identically over practical distance ranges (distance greater than 1 km), although such fits will diverge at zero distance. We assume distances less than 1 km are meaningless in a regional study, given location error and source finiteness.

We added attenuation and site terms to the ESH model to fully characterize path and relative site effects in our calibration procedures. Nonzero spreading at local distances can be introduced by complex shape effects and measurement points near the beginning of the coda window. For this study, we used a measurement point deeper in the coda to lessen this effect, as discussed earlier.

We now return to describing details of our four path calibration steps. Our first step determined basic spreading parameters that were used for the entire study region. We started by assuming the long distance spreading to be 1.0 (Yang, 2002) as we expect an early coda measurement to eventually spread at the same rate as the direct phase. We set

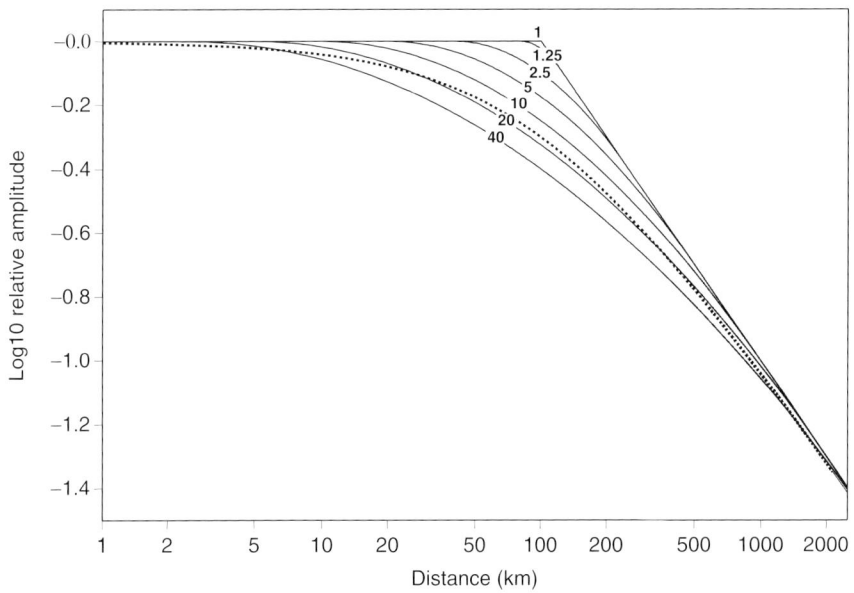

FIG. 7. Extended Street-Herrmann (ESH, solid lines) and Mayeda et al. (2003) path curves (dotted line) as described in the text. Numerals denote different transition factors, F, for the ESH curves.

the transition factor to 2.0, based on fits to data from shield regions, and general similarity to the Mayeda et al. (2003) path model, recognizing that this parameter will trade off with attenuation to some extent; thus, its exact value is less consequential. The initial spreading and transition distance were adjusted in a grid search wherein amplitude data, corrected for source size using event magnitude and a central Asia scaling model (Taylor and Hartse, 1998), were inverted for 2-D attenuation and site terms. Data were weighted inversely with distance to emphasize behavior at short distances. The spreading parameters we obtained are listed in Table 1. Initial spreading was close to zero, although slightly above zero for higher bands, whereas transition distances increased with decreasing frequency.

Once basic spreading parameters were set, we performed our second path calibration step: fitting the ESH model, with variable attenuation and site terms, to the source-corrected amplitudes for each station group and band. We used an L_1 fit and, again, weighted inversely to distance so that the close-in data are emphasized (Fig. 8). This facilitated visual evaluation and quality control of the amplitude data. We observed much scatter, which is due to error in magnitude as well as to regional path variation. We eliminated extreme outliers by applying a residual cutoff (log base ten) of 1.0 (above) and 1.5 (below) at this stage. The cutoff was set more tightly for high measurements to eliminate noise. We often observed a leveling of the source corrected amplitudes at long distances, especially for higher bands and for stations in attenuating regions (Fig. 8). This could be a noise effect; however, spot checks showed that coda-like envelopes well above noise levels are often present. These could be mantle coda that exhibit flatter path behavior. Although this behavior would be of interest to quantify, we have not investigated further and simply set a distance limit for each station and band so that long distance data that do not follow the spreading and Q models are not used. The shortest such limit is 400 km at station KMI for the 6–8 Hz band. Clearly, including such data will contaminate path effect inversions, while we lose data from large, distant events, which are likely recorded well elsewhere.

In our third path calibration step, we inverted for 1-D path and relative site terms using amplitude differences from multiply recorded events, as proposed by Mayeda et al. (2003). The inversion was performed as an L_1 minimization, and we solved for station (or station group)-dependent attenuation and relative site terms. Using amplitude differences eliminated our dependence on magnitude-based source corrections, and mean absolute residuals ranged from 0.9 to 1.1 log-ten units. Using L_1 techniques allowed easy identification of outliers, which were removed using tighter thresholds at this stage (0.5 log base ten), in preparation for the use of least squares in our 2-D inversions to follow.

In our fourth and final path calibration step, we applied amplitude tomography to multiply recorded event data that pass all previous, quality control steps, to recover 2-D attenuation ($1/Q$) and relative site terms. We assumed great circle ray paths, presuming that the early coda samples the earth within an elongated ellipsoid that stretches further as path length increases. The inversion was regularized by applying a first-difference smoothing constraint to the attenuation parameters and damping the sum of site terms to zero. The magnitude of the smoothing constraint was chosen to minimize variance without introducing geologically unreasonable, high wave number model fluctuations (lambda, Table 1). This approach was developed for application to Lg data in central and eastern Asia (Phillips et al., 2005), and was applied to coda wave amplitudes from northern California by Mayeda et al. (2005b). Mayeda et al. (2005b) found that the 2-D inversion improves residuals only slightly, but for distances that are relatively short

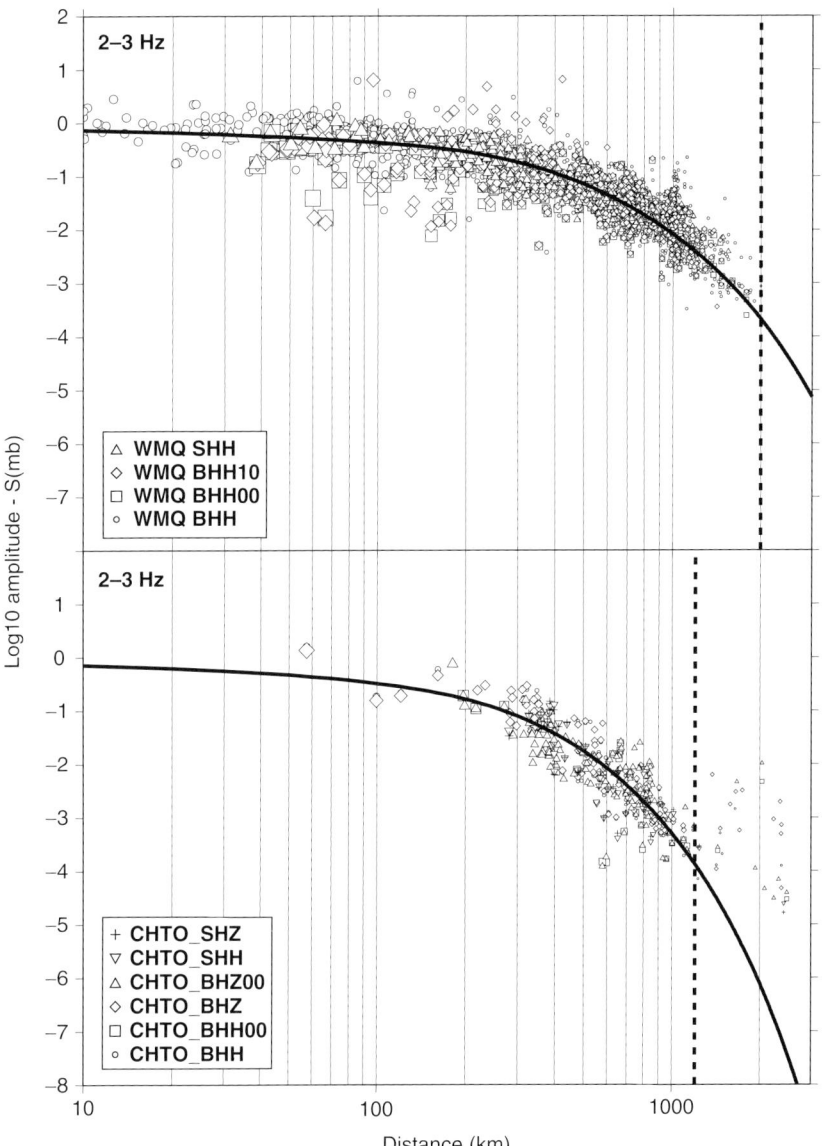

FIG. 8. Path model fits (solid lines) to source corrected, 2–3 Hz coda amplitudes (various symbols for different channels, as noted in legends), for stations WMQ (*top*) and CHTO (*bottom*). Source corrections are made using magnitudes and a scaling law as described in the text. Symbol size is related to distance-based weighting factors. Q is 740 for WMQ and 380 for CHTO. The vertical dashed line represents the distance limit set for these stations and this band, by visual inspection of these plots.

(under 500 km), compared to our Asia study. The tomography yielded RMS residuals of 0.07–0.11 log-ten units, with smallest residuals for bands around 1 Hz. Results for the 0.7–1.0 Hz band are shown in Fig. 9. Attenuation results followed regional geology, with high Q in stable regions such as India, the southeast China platform, and northern regions of Kazakhstan and Siberia, and for microcontinent regions such as the Tarim, Ordos, and Sichuan basins. Low Q was observed in plateau, mountain belt, and rifting areas, including Tibet, Qinghai, Yunnan and Burma, the Pamir, Tian Shan and Hindu Kush ranges, the east China basin, and oceanic regions. The low Q observed for oceans is an artifact of Lg blockage in otherwise high Q oceanic crust (Knopoff et al., 1979). Continental attenuation patterns were similar to those of Lg (Mitchell et al., 1997; Phillips et al., 2000, 2001, 2005). Figure 10 shows the resulting 2-D path corrections for station WMQ and band 0.7–1 Hz for events at all azimuths. Clearly, the 2-D corrections were small for distances less than 500 km, but became more important beyond.

3.5. Source to Coda Transfer Function

Our final calibration step estimated the source-to-coda transfer function [T, Eq. (1)], which allowed us to produce absolute source spectra. The transfer function was obtained by applying a multiple event, empirical Green's function technique to amplitude spectra, including constraints to match ground-truth moments for certain events. We estimated earthquake corner frequencies based on magnitudes, and obtained corrections that fit the flat portion of a Brune source model for frequencies well below the estimated corner. If this is performed using a number of earthquakes and a sufficient range of magnitudes,

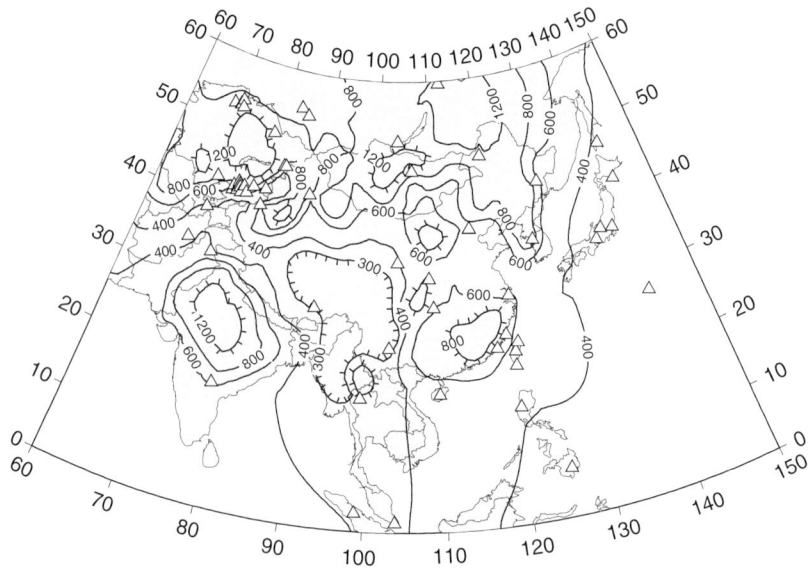

FIG. 9. Two-dimensional Q map obtained by applying tomography techniques to 0.7–1 Hz coda amplitudes. Stations used in the inversion are shown as triangles. Contour ticks point in downhill directions. Q is poorly resolved in oceanic regions.

SOURCE EFFECTS FROM CODA WAVES 341

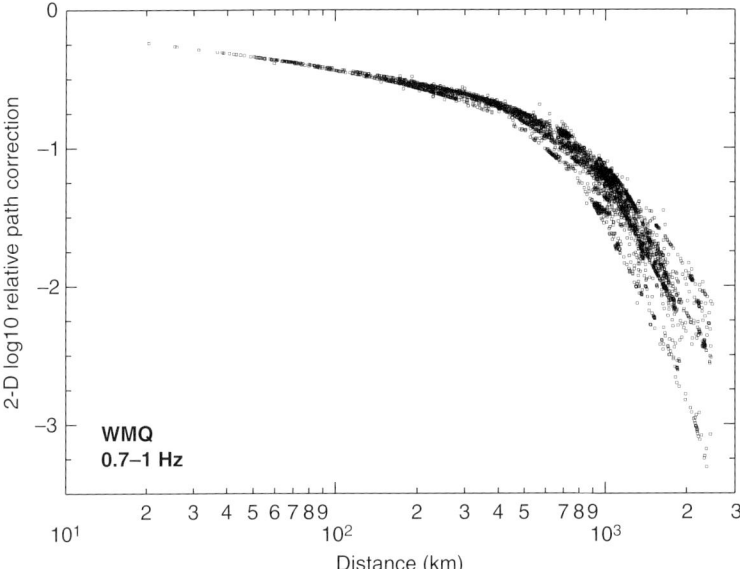

Fig. 10. Path (plus site) corrections applied to 0.7–1 Hz data for events recorded at station WMQ, based on the two-dimensional Q map shown in Fig. 9.

the full transfer spectrum can be recovered. We first estimated the transfer function using catalog magnitudes, which allowed us to determine initial moments for our data set, whereupon the transfer function was recalculated using coda Mw in place of catalog magnitude. The Mw-based calculation extended the transfer function to higher bands, as more small events could be included. As our study covered a broad area, we allowed the transfer function to vary in 2-D by source epicenter. The transfer function converted velocity amplitude (m/s) to moment rate (Nm), thus has units of Ns.

Only earthquake data can be used to compute the transfer function. We eliminated known explosions, required source depths greater than 5 km, and only allowed events under magnitude 3 that occurred at night to avoid any unidentified mine blasts that might be present. Absolute levels were constrained using 165 continental events with independently determined moments (Randall et al., 1995; Zhu et al., 1997, Ghose et al., 1998; Patton and Randall, 2002; Ammon et al., 2003; Saikia, 2006). To calibrate oceanic regions, we added 217 CMT moments from offshore events of depth less than 50 km. Locations of ground-truth moment events are shown in Fig. 1.

We recovered the 2-D transfer function using tomographic techniques, applying relative constraints to amplitude data between bands and adding absolute constraints for ground-truth moments. We set first difference regularization parameters that varied with band, in the same manner as for 2-D path tomography. 2-D grids for all bands were determined simultaneously. For low bands, the transfer function is most sensitive to differences between continental and oceanic crust. For high bands, the transfer function is quite flat across continental regions with dramatic exceptions in limited areas (0.7–1.0 Hz, Fig. 11), primarily reflecting differing excitation of Lg and Sn coda. In the 0.7–1.0 Hz example shown, we observed rapid changes between Tibet and the Assam

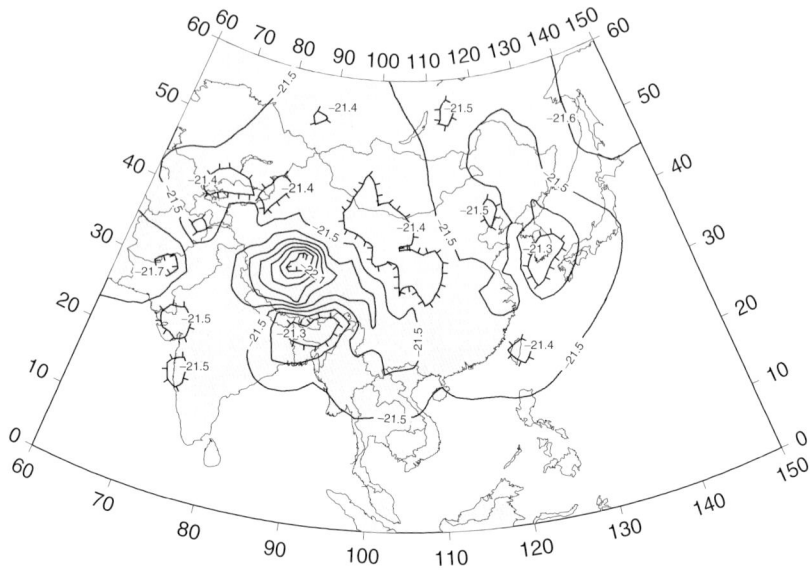

FIG. 11. Transfer function for the 0.7–1 Hz band. Contour intervals are 0.1 Ns, the −21.5 Ns contour and high and low points are annotated. Contour ticks point in downhill directions.

region of India. Remarkably, we discovered that low-frequency spectral slopes were not as flat as anticipated, biasing the transfer function. We assumed that events with more negative, low-frequency spectral slopes result from low stress, or low band noise contamination, and down weight those data, requiring iterative recalculation of the transfer function. Prior to undertaking the iterative weighting procedure, we observed slight spectral peaking (1 Hz) for many events and small coda moments that deviated slightly from one-to-one scaling with ground-truth moments. The iteratively reweighted transfer function corrected these problems, and Fig. 12 shows that the final coda moment magnitudes vary one-to-one with the ground-truth measures. Scatter was 0.13 magnitude units for the continental, regionally modeled events. Scatter increased to 0.24 for oceanic moment magnitudes, derived using teleseismic data and global models. The oceanic coda moments are larger and relied heavily on the lowest bands, which are less well calibrated. Broadband spectral results are compared between stations ENH and XAN, 421 km apart, in Fig. 13, showing nice correspondence over an order of magnitude in frequency.

4. Coda Spectral Results

The coda-derived, absolute source spectra allowed us to determine source parameters. We only analyzed bands for which the transfer function is resolved, for a given source region. To obtain Mw, we fix the scaling law, weight amplitudes by their proximity to the second lowest measured band, and fit an ω^2 source model using L_1 minimization.

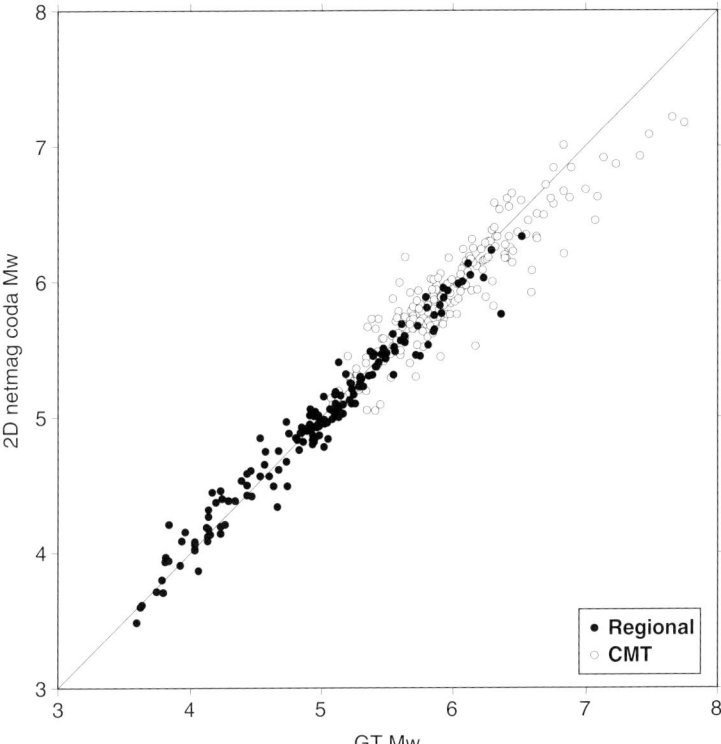

Fig. 12. Network coda versus ground-truth moment magnitudes. Continental events for which regional waveform modeling was performed are shown as filled circles, whereas oceanic CMT events are shown as open circles. A one-to-one line is included for reference.

This scheme is insensitive to departures from the assumed scaling law when signal is measured below the corner frequency, but will increase in error if signal is only available above the corner and we have mischaracterized scaling. To obtain Mw and corner frequency simultaneously, we used a two-parameter L_1 minimization with no band weighting. The quality of the result for calculation of source parameters, such as apparent stress, was judged by the number of spectral points or bandwidths on both sides of the corner, as well as a crude L_2 estimate of error at the L_1 solution point. Figure 14 gives spectra from selected, ground-truth moment events, showing variations in corner frequencies, thus stress, with respect to a constant scaling model fit (apparent stress 1 MPa).

Apparent stress (σ_a; Wyss, 1970) can be obtained from moment (M_0) versus shear wave corner frequency (f_0) measurements, following Walter and Taylor (2001):

$$f_0 = \frac{1}{2\pi} \left(\frac{K\sigma_a}{M_0} \right)^{1/3}, \quad K = \frac{16\pi}{\beta_s^2 \left(R_{0\varphi P}^2 \zeta^3 / \alpha_s^5 + R_{0\varphi S}^2 / \beta_s^5 \right)}, \tag{11}$$

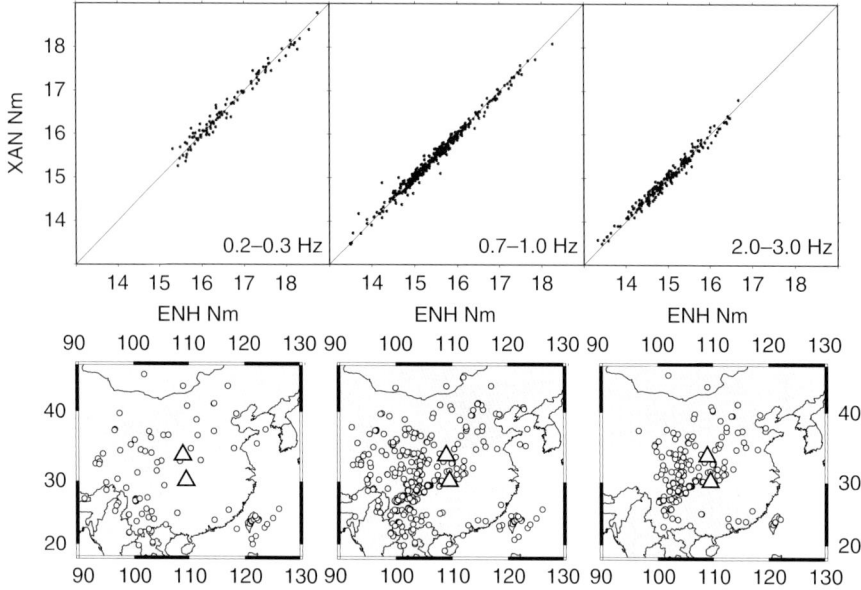

FIG. 13. Final moment-rate spectral levels (Nm) compared between stations ENH and XAN, for bands 0.2–0.3 Hz, 0.7–1 Hz, and 2–3 Hz, left to right. One-to-one lines are added for reference. Maps show the distribution of events (circles) for each band, as well as the two stations (triangles).

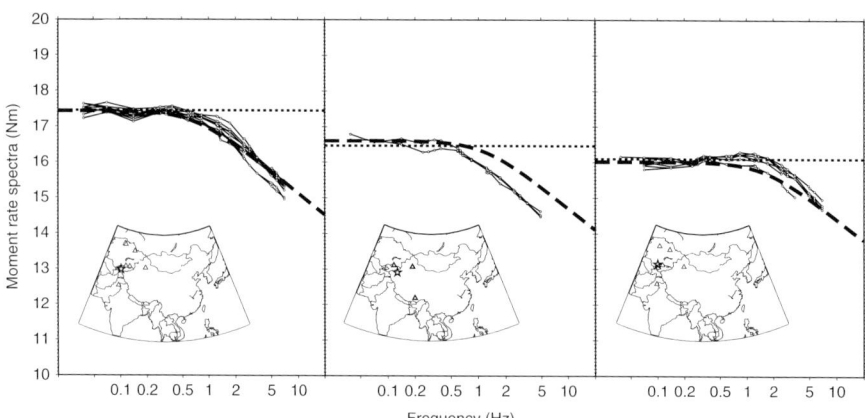

FIG. 14. Coda moment-rate spectra (Nm) examples for three events. Results are shown for multiple stations and channels, for each event. These three events provided ground-truth moments, which are indicated by the horizontal, dotted lines. Source model fits for an apparent stress of 1 MPa are shown by the dashed lines. Event locations (stars) and stations used (triangles) are shown on the inset maps.

where α_s and β_s are compressional and shear velocities at the source point (6.0 and 3.5 km/s), respectively; $R_{0,P}$ and $R_{0,S}$ are average P and S radiation patterns for earthquakes (0.44 and 0.6), respectively; and ζ relates P and S corner frequencies (1.0). Walter and Taylor (2001) assumed an ω^2 source model to derive Eq. (11).

We plot apparent stress in map view for network spectral fits in Fig. 15. We required bandwidths of a factor of two above and below the corner frequency, yielding nearly 6700 measurements. The majority fall between 10^{-2} and 1 MPa. Much scatter is apparent, and physically interesting variations may occur on scales too small to observe here (Shearer *et al.*, 2006). High stresses appear to be associated with the Tian Shan, especially the eastern and northwestern extensions, also the Altay range, edges of the Tarim, and Assam, Bhuj, and Koyna regions of India. Low stress appears in Tibet, central Pakistan, and certain trench areas.

5. Discussion

We are working to extend 1-D coda methods to be effective over broad areas and far regional distances. We begin by applying 2-D techniques to the path and transfer function portions of our coda analysis. However, our heterogeneous region of study

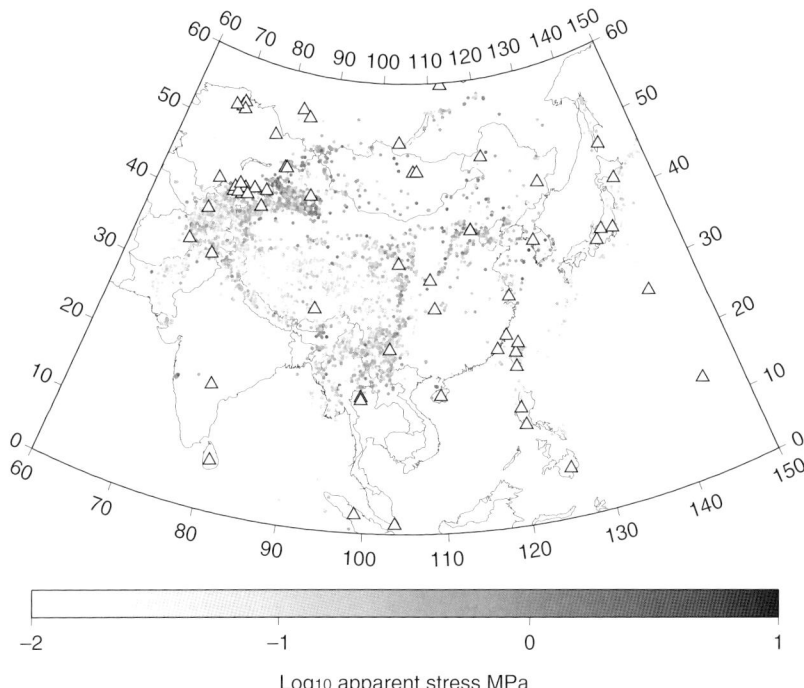

Fig. 15. Distribution of apparent stress (log-ten MPa) by event location. Stations are indicated by triangles.

has provided many opportunities to observe 2-D behavior that may affect other calibration stages. It will be important to quantify the effect of ignoring additional 2-D variations, and, if substantial, we must develop coda techniques further. The most important of these are variations in coda type and coda start time.

It is well known that surface waves are affected by regional geology and maps can be found in the literature that show the effects of near surface structure, including basins, on Rayleigh and Love group velocities (e.g., Pasyanos, 2005). Because we set coda origin at the peak arrival, if it exists, we must pay attention to lateral variations in group velocity, if only to avoid having a late peak fall in our coda window and contaminate the measurement. Our long-period (0.03–0.05 Hz) coda start time picks show variations consistent with the literature maps of group velocity. The effects are more dramatic for shorter periods (5–20 s), with extremely late envelope peaks common in basin regions such as the Tarim and the Junggar. If we are to investigate further, we should consider using radial and/or transverse components, rather than the stacked horizontals, so that the direct surface waves will be more predictable. These low-frequency bands are influential when setting absolute levels using ground-truth moments. We also observe regional variations in higher bands. At 1 Hz, we observe late group arrivals across basins such as the Tarim and Junggar, but the phenomenology is different as we see normal arrivals for more distant events along the same azimuths (Fig. 16; Phillips and Stead, 2006). This implies a modal effect for events occurring under the basins, rather than an integrated velocity effect as for the long periods. We can also expect degradation and delay of envelope peaks in higher bands (above 4 Hz) due to high path heterogeneity, as shown in Honshu by Saito *et al.* (2005).

Of equal or greater importance are coda types, which can vary, for high frequencies, between Lg, Sn, and P coda, or even Rg in special cases. The dominant coda type can be regionally dependent (Fig. 16). We would be wise to measure Sn coda in areas where Lg is absent, in order to start the coda at the earlier Sn time and recover more measurement length, and thus, include smaller events. When Sn and Lg are present, the Lg will contaminate the Sn coda, and what appears to be Sn to Lg conversion contaminates the Sn coda prior to arrival of the Lg, which makes the calibration problem difficult. Effective strategies have yet to be worked out, but we could start by calibrating Sn coda over limited areas where Lg is strictly absent, and similarly for P. Of further interest are models of where the various coda types sample the earth. We assume that the Lg coda are crustal scattered waves, whereas P and Sn coda are scattered in the upper mantle. Array studies could help understand the composition of the different, regional coda types.

Any attention to 2-D variations in coda type must also include a focus on lateral variations in coda shape. If we fit a shape function representative of Lg coda to an Sn coda, or vice versa, the measured amplitude will change with coda length, which means varying signal-to-noise levels will bias the measurements. This effect may be significant for only the smallest measurable events, but currently remains to be quantified.

Formal error propagation is another topic for future work with regional coda. We currently treat results with equal weight and assign errors based on final calibration fits when calculating Mw, corner frequency, or calibrating to magnitude. Residual variance from path and transfer function calculations is 0.1 log-ten units or so, but represents a lower bound on absolute error. The scatter in coda versus ground-truth moments are generally 0.2 log-ten Nm, but reflect errors in the moments as well as the coda. Mayeda *et al.* (2003) suggested setting the data error based on coda length, using interstation scatter measurements. Whether such scatter measurements give accurate

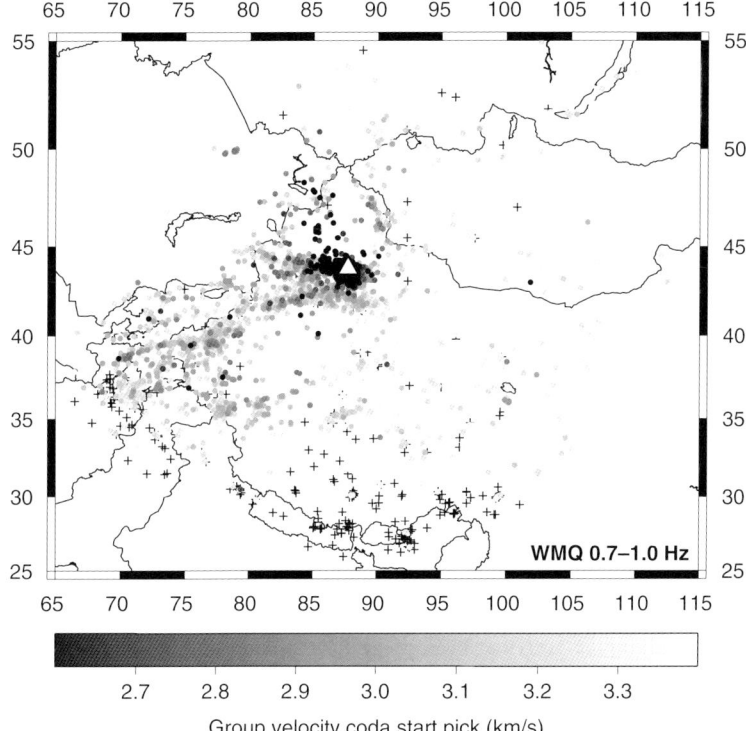

FIG. 16. Distribution of coda origin group velocity (km/s) by event location for station WMQ, band 0.7–1 Hz. The location of station WMQ is shown as a triangle. Lg arrivals are represented by the gray scale and Sn arrivals, which fall beyond the high end of the gray scale bar, by crosses.

errors can be argued, but, at minimum, effective relative errors can be established and carried through the calibration procedure.

Apparent stress computed from the coda spectra show much scatter and hints of physical variations in central and eastern Asia. We observe low stresses in central Pakistan, Tibet, and certain trench areas. These patterns have evolved as we have included 2-D effects on the coda, in particular, the 2-D transfer function (Phillips et al., 2003, 2004). The 2-D transfer function has affected ocean results the most, but stresses observed for Japan by this study (0.1 MPa) still fall at the low end of ranges determined by focused studies (0.1–10 MPa; Kinoshita and Ohike, 2002; Takahashi et al., 2005). Little manual review was performed for Japanese events, but even so, we are concerned that the current method does not yet adequately account for cases where Lg and Sn coda must be considered together. We believe that it remains a challenge to calibrate such regions of oceanic and continental crust. Interestingly, low stress for continental events is somewhat correlated with low Q crust (Phillips et al., 2005) such as for Pakistan and Tibet. One could argue that the lower stresses are artifacts of undercorrection for low Q path effects. It is well known that tomographic calculations smooth high and low extremes. Conversely, the low stress and low Q may be correlated for physical reasons. We speculate that earth materials may be less brittle or faults are more lubricated, have more recent slip history, or are less heterogeneous in such areas. Furthermore, higher stress events are observed in low-stress regions, which should not be the case if Q is poorly corrected for, unless extremely low Q materials are concentrated in certain, limited source regions and are, for this reason, not captured by the tomographic calculations. Increasing station density in remote regions of Asia, such as Tibet, and surrounding regions, by including temporary deployment data will help resolve the stress-attenuation trade-off issue.

We observe high stresses along edges of seismic regions, most dramatically at the eastern and northwestern extensions of the Tian Shan, but also in the Altay ranges to the north and around the western edges of the Tarim basin. Small clusters of high stress are also seen in the Indian plate, most notably in the Assam region. We are unsure of how to interpret this, except to note that these areas are less seismic than neighboring areas, are associated with higher Q crust, and deformation may occur in a more brittle manner. We note that our high-stress Assam cluster is in the vicinity of the high-stress events discussed by Tatham *et al*. (1976), who considered the concentration of stress between the Himalaya and the Andaman arc and speculated on the formation of a new fault.

Spectral ratio studies of far regional P waves from Tibet and northwest Tian Shan regions failed to find stress differences between source regions (J. Granville, personal communication, 2003) and this remains to be reconciled.

Fehler and Phillips (1991) mapped static stress drops during pressurization of a geothermal reservoir, finding that high-stress drops occurred along the edge of a cloud of seismicity, whereas low-stress drops occurred in the central regions, where previous injection experiments had already caused much slip. Thus, previous slip history perhaps related to mechanical smoothing of fault heterogeneity may affect stress drop. Fehler and Phillips (1991) also discussed the lowering of stress drop by increased heterogeneity related to injection (after Madariaga, 1979). Although stress patterns are remarkably similar to those shown here, other than scale, we have no reason to believe that similar mechanisms are at work.

Shearer *et al*. (2006) mapped static stress drops obtained from stacked P spectra for an extensive southern California data set, showing striking variations at small scales. A correlation between lower stress and slipped fault surfaces was noted in one case. Such stress variations would be on scales too small to observe, if similar effects are present in our Asia data set, given the larger and less well-located event set we have used.

We hope to further quantify regional variations in scaling behavior and high frequency roll off, in addition to stress. Monitoring work relies on comparisons with earthquake scaling models to identify anomalous events (e.g., Taylor and Hartse, 1998). If we can quantify regional scaling variations, our ability to identify non-earthquake sources, and, of course, anomalous natural events, will greatly improve.

6. Conclusions

We have worked to apply coda techniques to an extensive network, with the attendant issues of data quality control, multiple channels, and time periods of operation, automation, and the beginning of 2-D calibration. Much of our discussion covered practical details that we hope will help others that wish to improve network-based coda analysis further.

We have shown that 2-D path calibration becomes important at far regional distances. We also find that transfer functions are not generally transportable, with large differences between continents and oceans, and also for regions such as Tibet where Lg attenuation is high. Thus, much of the variation may be due to differences in how well earthquakes excite Sn versus Lg coda. Learning how to calibrate when both Lg and Sn are present, in a 2-D sense, is an important future step in extending coda methods to broad areas.

We observed regional variations in apparent stress that tend to correlate with poor crustal propagation. We believe that the correlation could be physical, although undercorrection of path effects remains a possibility. We propose further tests using local network coda studies to produce ground-truth source spectra for comparison with regional results. Networks such as HIMNT (de la Torre and Sheehan, 2005), which straddles Tibet and Indian plate regions, hold promise. We observed low stress for earthquakes in Japan, which is inconsistent with local studies. Regional calibration over broad areas of mixed oceanic, continental, and island arc structures remains a challenge.

While relative coda measurements can be extremely precise, errors following calibration are estimated to be 0.1 log-ten units or so. Formal error propagation should be an important focus of future work.

We believe that basic research concerning the origin and composition of various types of regional coda, including array analysis, will provide clues as to the temporal evolution of the composition of the coda, and will help make calibration more effective at far regional distances. In addition, investigations into source scaling behavior, including physical mechanisms of stress variation (e.g., Shearer *et al.*, 2006), will lend confidence to the broad area source parameter measurements, which will, in turn, improve our ability to classify event types in routine monitoring work.

Acknowledgments

We acknowledge the leadership and inspiration of the late Keiiti Aki. We also thank Diane Baker, Mike Begnaud, and David Yang for their help organizing waveform data, as well as Bill Walter for advice on source models. Darren Hart and Chris Young facilitated testing that improved our signal processing. R. Herrmann provided moments from Korea and surrounding areas. Two anonymous reviewers as well as editor Haruo Sato provided extensive comments that we found invaluable. This work was supported by the US DOE, contract DE-AC52-06NA25396.

References

Aki, K. (1956). Correlogram analysis of seismograms by means of a simple automatic computer. *J. Phys. Earth* **4**, 71–78.

Aki, K. (1969). Analysis of the seismic coda of local earthquakes as scattered waves. *J. Geophys. Res.* **74**, 615–631.

Aki, K., Chouet, B. (1975). Origin of coda waves: attenuation, and scattering effects. *J. Geophys. Res.* **80**, 3322–3342.

Ammon, C., Randall, G., Julia, J. (2003). Improving estimates of depth, magnitude, and faulting parameters of earthquake in central Asia. *Proceedings of the 25th SRR—Nuclear Explosion Monitoring, Building Knowledge Base, Vol I NNSA*, 24–33, Tucson, Arizona.

Chouet, B., Aki, K., Tsujiura, M. (1978). Regional variation of the scaling law of earthquake source spectra. *Bull. Seismol. Soc. Am.* **68**, 59–70.

de la Torre, T.L., Sheehan, A.F. (2005). Broadband seismic noise analysis of Himalayan Nepal Tibet Seismic Experiment. *Bull. Seismol. Soc. Am.* **95**, 1202–1208, doi: 10.1785/0120040098.
Eken, T., Mayeda, K., Hofstetter, A., Gök, R., Örgülü, G., Turkelli, N. (2004). An application of the coda methodology for moment-rate spectra using broadband stations in Turkey. *Geophys. Res. Lett.* **31**, L11609.
Engdahl, E.R., van der Hilst, R., Buland, R. (1998). Global teleseismic earthquake relocation with improved travel times and procedures for depth determination. *Bull. Seismol. Soc. Am.* **88**, 722–743.
Fehler, M.C., Phillips, W.S. (1991). Simultaneous inversion for Q and source parameters of microearthquakes accompanying hydraulic fracturing in granitic rock. *Bull. Seismol. Soc. Am.* **81**, 553–575.
Ghose, S., Haumburger, M.W., Ammon, C. (1998). Source parameters of moderate-sized earthquakes in the Tien Shan, central Asia from regional moment tensor inversion. *Geophys. Res. Lett.* **25**, 3181–3184.
Hartse, H.E., Phillips, W.S., Fehler, M.C., House, L.S. (1995). Single-station spectral discrimination using coda waves. *Bull. Seismol. Soc. Am.* **85**, 1464–1474.
Kim, W.-Y., Ekstrom, G. (1996). Instruments responses of digital seismographs at Borovoye, Kazakhstan, by inversion of transient calibration pulses. *Bull. Seismol. Soc. Am.* **86**, 191–203.
Kinoshita, S., Ohike, M. (2002). Scaling relations of earthquakes that occurred in the upper part of the Philippine Sea plate beneath the Kanto region, Japan, estimated by means of borehole recordings. *Bull. Seismol. Soc. Am.* **92**, 611–624.
Knopoff, L., Mitchel, R.G., Kausel, E.G., Schwab, E. (1979). A search for the oceanic Lg phase. *Geophys. J. R. Astrron. Soc.* **56**, 211–218.
Lee, W.H.K., Kanamori, H., Jennings, P.C., Kisslinger, C. (Eds.) (2002). *International Handbook of Earthquake and Engineering Seismology, CD Supplement*. Academic Press, Amsterdam.
Madariaga, R. (1979). On the relation between seismic moment and stress drop in the presence of stress and strength heterogeneity. *J. Geophys. Res.* **84**, 2243–2250.
Malagnini, L., Bodin, P., Mayeda, K., Akinci, A. (2006). Unbiased moment-rate spectra and absolute site effects in the Kachachh Basin, India, from the analysis of the aftershocks of the 2001 Mw 7.6 Bhuj earthquake. *Bull. Seismol. Soc. Am.* **96**, 456-466, doi: 10.1785/0120050089.
Mayeda, K. (1993). $m_b(LgCoda)$: A stable single station estimator of magnitude. *Bull. Seismol. Soc. Am.* **83**, 851–861.
Mayeda, K., Walter, W.R. (1996). Moment, energy, stress drop and source spectra of western United States earthquakes from regional coda envelopes. *J. Geophys. Res.* **101**, 11195–11208.
Mayeda, K., Hofstetter, A., O'Boyle, J.L., Walter, W.R. (2003). Stable and transportable regional magnitudes based on coda-derived moment-rate spectra. *Bull. Seismol. Soc. Am.* **93**, 224–239.
Mayeda, K., Gök, R., Walter, W.R., Hofstetter, A. (2005a). Evidence for non-constant energy/moment scaling from coda-derived source spectra. *Geophys. Res. Lett.* **32**, L10306, doi: 10.1029/2005GL022405.
Mayeda, K., Malagnini, L., Phillips, W.S., Walter, W.R., Dreger, D. (2005b). 2-D or not 2-D, that is the question: A northern California test. *Geophys. Res. Lett.* **32**, L12301, doi: 10.1029/2005GL022882.
Mitchell, B.J., Pan, Y., Xie, J.K., Cong, L.L. (1997). Lg coda Q variation across Eurasia and its relation to crustal evolution. *J. Geophys. Res.* **102**, 22767–22779.
Morasca, P., Mayeda, K., Malagnini, L., Walter, W.R. (2005a). Coda derived source spectra, moment magnitudes, and energy-moment scaling in the Western Alps. *Geophys. J. Int.* **160**, 263–275.
Morasca, P., Mayeda, K., Gök, R., Malagnini, L., Eva, C. (2005b). A break in self-similarity in the Lunigiana-Garfagnana region (Northern Apennines). *Geophys. Res. Lett.* **32**, L22301, doi: 10.1029/2005GL024443.
Pasyanos, M.E. (2005). A variable resolution surface wave dispersion study of Eurasia, North Africa, and surrounding regions. *J. Geophys. Res.* **110**, B12301. doi: 10.1029/2005JB003749.
Patton, H.J., Randall, G.E. (2002). On the causes of biased estimates of seismic moment for earthquakes in central Asia. *J. Geophys. Res.* **107**, 2302, doi: 10.1029/2001JB000351.

Phillips, W.S. (1999). Empirical path corrections for regional phase amplitudes. *Bull. Seismol. Soc. Am.* **89**, 384–393.
Phillips, W.S., Stead, R.J. (2006). Regional calibration of peak envelope arrival time. *Seismol. Res. Lett.* **77**, 260–261.
Phillips, W.S., Randall, G.E., Taylor, S.R. (1998). Path correction using interpolated amplitude residuals: An example from central China. *Geophys. Res. Lett.* **25**, 2729–2732.
Phillips, W.S., Hartse, H.E., Taylor, S.R., Randall, G.E. (2000). 1 Hz Lg Q Tomography in central Asia. *Geophys. Res. Lett.* **27**, 3425–3428.
Phillips, W.S., Hartse, H.E., Taylor, S.R., Velasco, A.A., Randall, G.E. (2001). Application of regional phase amplitude tomography to seismic verification. *Pure Appl. Geophys.* **158**, 1189–1206.
Phillips, W.S., Patton, H.J., Aprea, C., Hartse, H.E., Randall, G.E., Taylor, S.R. (2003). Automated broad area calibration for coda based magnitude and yield. *Proceedings of 25th Seismic Research Review—Building the Knowledge Base*, Tucson, Arizona, pp. 437–444.
Phillips, W.S., Patton, H.J., Taylor, S.R., Hartse, H.E., Randall, G.E. (2004). Calibration for coda based magnitude and yield. *Proceedings of 26th Seismic Research Review—Trends in Nuclear Explosion Monitoring*, Orlando, Florida, pp. 449–456.
Phillips, W.S., Hartse, H.E., Rutledge, J.T. (2005). Amplitude ratio tomography for regional phase Q. *Geophys. Res. Lett.* **32**, L21301, doi: 10.1029/2005GL023870.
Randall, G.E., Ammon, C.J., Owens, T.J. (1995). Moment tensor estimation using regional seismograms from a Tibetan Plateau portable network deployment. *Geophys. Res. Lett.* **22**, 1665–1668.
Rautian, T.G., Khalturin, V.I. (1978). The use of the coda for determination of the earthquake source spectrum. *Bull. Seismol. Soc. Am.* **68**, 923–948.
Roecker, S.W., Tucker, B., King, J., Hatzfeld, D. (1982). Estimates of Q in central Asia as a function of frequency and depth using the coda of locally recorded earthquakes. *Bull. Seismol. Soc. Am.* **72**, 129–149.
Saikia, C.K. (2006). Modeling of the 21 May 1997 Jabalpur earthquake in Central India: Source parameters and regional path calibration. *Bull. Seismol. Soc. Am.* **96**, 1448–1473.
Saito, T., Sato, H., Ohtake, M., Obara, K. (2005). Unified explanation of envelope broadening and maximum-amplitude decay of high-frequency seismograms based on the envelope simulation using the Markov approximation: Forearc side of the volcanic front in northeastern Honshu, Japan. *J. Geophys. Res.* **110**, B01304, doi: 10.1029/2004JB003225.
Shearer, P.M., Prieto, G.A., Hauksson, E. (2006). Comprehensive analysis of earthquake source spectra in southern California. *J. Geophys. Res.* **111**, B06303, doi: 10.1029/2005JB003979.
Street, R.L., Herrmann, R., Nuttli, O. (1975). Spectral characteristics of the Lg wave generated by central United States earthquakes. *Geophys. J. R. Astron. Soc.* **41**, 51–63.
Takahashi, T., Sato, H., Ohtake, M., Obara, K. (2005). Scale dependence of apparent stress for earthquakes along the subducting Pacific plate in northeastern Honshu, Japan. *Bull. Seismol. Soc. Am.* **95**, 1334–1345.
Tatham, R.H., Forsyth, D.W., Sykes, L.R. (1976). The occurrence of anomalous seismic events in eastern Tibet. *Geophys. J. R. Astron. Soc.* **45**, 451–481.
Taylor, S.R., Hartse, H.E. (1998). A procedure for estimation of source and propagation amplitude corrections for regional seismic discriminants. *J. Geophys. Res.* **103**, 2781–2789.
Tsujiura, M. (1978). Spectral analysis of the coda waves from local earthquakes. *Bull. Earthquake Res. Inst.* **53**, 1–48.
Walter, W.R., Taylor, S.R. (2001). A revised magnitude and distance amplitude correction (MDAC2) procedure for regional seismic discriminants: Theory and testing at NTS, Lawrence Livermore National Laboratory Report UCRL-ID-146882, http://www.llnl.gov/tid/lof/documents/pdf/240563.pdf.
Wyss, M. (1970). Stress estimates of South American shallow and deep earthquakes. *J. Geophys. Res.* **75**, 1529–1544.
Yang, X. (2002). A numerical investigation of Lg geometrical spreading. *Bull. Seismol. Soc. Am.* **92**, 3067–3079.
Zhu, L., Helmberger, D.V., Saikia, C.K., Woods, B.B. (1997). Regional waveform calibration in the Pamir-Hindu Kush region. *J. Geophys. Res.* **102**, 22,799–33,813.

SEISMIC WAVE SCATTERING IN VOLCANOES

Edoardo Del Pezzo

Abstract

Volcano-tectonic earthquakes produce high-frequency seismograms characterized by impulsive shear mechanism; their seismogram coda reflects the random inhomogeneity of the volcano structure. Consequently, this inhomogeneity can be investigated through the analysis of the coda wave envelopes of the volcano-tectonic events. In this chapter, I will review the main observational results obtained from volcanoes around the World, with the aim of quantifying the scattering and attenuation properties of the volcanic areas. First, I will review the coda-Q observations and their frequency dependence, then I will report on attempts that have been made to separate the intrinsic from the scattering attenuation using multiple scattering and diffusion models, and finally, I will report on the interpretations based on these results. The results show that the coda-Q absolute values characteristic of volcanoes are slightly smaller than those measured in nonvolcanic zones, and that sometimes their frequency dependence is different. It is impossible to deduce by coda-Q observations only whether this difference is controlled more by the intrinsic or the scattering attenuation. The application of multiple scattering models allows separate estimates of the intrinsic and the scattering attenuation coefficients. Results show that volcanoes are highly heterogeneous structures, with a mechanism of seismic wave energy dissipation that tends to be controlled by the scattering phenomena with increasing frequency. For Mt. Vesuvius, Mt. Merapi, and Deception island volcano scattering attenuation prevails at frequencies higher than 2–3 Hz. At Mt. Etna, intrinsic dissipation prevails or is comparable with scattering attenuation for frequencies lower than 8 Hz.

At high frequencies, diffusion approximation is appropriate to describe the energy seismogram envelope. The intrinsic dissipation of shear waves (possibly connected with magma reservoirs, which should decrease the intrinsic Q values for shear waves) only has an important role at low frequencies.

Key Words: Seismic scattering, coda waves, volcanoes. © 2008 Elsevier Inc.

1. Introduction

1.1. Volcanic Earthquakes

Volcanoes are sites of peculiar seismic activity, which is generally classified into four categories: volcano-tectonic (VT) earthquakes, long-period (LP) events, very long-period (VLP) events, and volcanic tremor (Chouet, 2003). In typical observatory practice, this classification is carried out visually, by analyzing the seismogram shapes and/or the spectral contents of the waveforms. Modern research in volcano seismology, aimed at the quantification of the seismic phenomena associated with volcanic eruptions, enlighten the role played by magmatic and hydrothermal fluids in the generation of the seismic waves, making critically important the quantification of the source properties of LP and tremor. As the different events reflect different kinds of source mechanisms, the quantification of their source is crucial to determine the extent and evolution of the magmatic energy. In particular, LP events and tremor generally precede and accompany the eruptive phenomena (Chouet, 2003, and references therein), and are used to assess the eruptive state or to estimate the eruptive potential. This is an obvious reason why monitoring the insurgence of this kind of seismicity is considered to be the most reliable and powerful techniques of volcano monitoring. One of the problems is that the seismogram shape of the volcanic quakes or, similarly, their spectral shape (frequency domain), is greatly dependent on the propagation effects, and hence on the elastic properties of the seismic medium. Consequently, the medium properties need to be studied carefully to correctly adjust for the effects of propagation and to obtain valid event classifications aimed at an understanding of the physics of a volcanic source.

VT events are located inside the volcano structure, generally at shallow depths (down to 10 km). They are the brittle responses of the volcano materials to the magma processes and/or to changes in the thermal state of the rocks (the heating or cooling in the vicinity of a magma body). VT may also reflect far-field stresses acting on heterogeneous materials and changes in the pore pressure. The impulsive shear mechanism of these events produces high-frequency seismograms that reflect the random heterogeneity of the volcano structure in their coda. The heterogeneity of volcanoes can thus be investigated through the analysis of the VT coda wave envelopes. Other volcanic earthquakes, on the contrary, are characterized by non-impulsive and time-persistent sources, which generate complex coda waveforms. In this case, the problem of separating the radiation generated by the scattering phenomena from that related to the source is more difficult.

Because of the impulsive shear mechanism that generates the high-frequency content of the coda, VT earthquakes are the most suitable for investigations into small-scale heterogeneity characterizing the earth medium beneath volcanoes.

1.2. A Brief Review of Coda-Q^{-1} Observation on Volcanoes

The first attempts to quantify the scattering and attenuation parameters for rocks constituting the volcanic structures were carried out using the estimate of coda-Q^{-1}, or Q_C^{-1} of the local VT earthquakes. As is well known, this parameter describes the energy-density decay of the short-period seismogram, $E^{SS}(\mathbf{x}, t)$, recorded at position $\mathbf{x}$ and at the time t due to an impulsive source applied at $\mathbf{x}_0$ in the single-scattering assumption (Sato and Fehler, 1998; Section 3.1.2) as

$$E^{SS}(\mathbf{x},t) = \frac{E_0 g_0}{4\pi|\mathbf{x}-\mathbf{x}_0|^2} K\left(\frac{\beta t}{|\mathbf{x}-\mathbf{x}_0|}\right) H(\beta t - |\mathbf{x}-\mathbf{x}_0|) e^{-\omega t Q_C^{-1}}, \qquad (1)$$

where E_0 is the energy density at the source, g_0 is the scattering coefficient, β is the seismic velocity for the shear waves (taken as a constant), H is the Heaviside step function, K is a term depending on geometrical spreading and distance, and ω is the angular frequency. With short hypocentral distances (source is assumed to be colocated with the receiver) formula (1) can be approximated by

$$E^{SS}(t) \simeq \frac{E_0 g_0}{2\pi\beta^2 t^2} e^{-\omega t Q_C^{-1}}. \qquad (2)$$

Using these assumptions (often respected by VT earthquakes, that are generally recorded by stations close to the source), Q_C^{-1} appears to include both absorption and scattering losses (see Sato and Fehler, 1998, Sections 3.1, 3.3.2, and 7.1.1) although its physical meaning has been controversial for many years. The idea that coda-Q^{-1} accounts for both intrinsic and scattering attenuations prevailed until the appearance in the literature of theoretical (Shang and Gao, 1988), numerical (Frankel and Clayton, 1986), and laboratory (Matsunami, 1991) studies concluding that coda-Q^{-1} is essentially an estimate of intrinsic-Q^{-1}. More recently, Zeng (1991) described the solution of the integral equation for the seismic energy density, $E(\mathbf{x},t)$, recorded at position $\mathbf{x}$ and at the time t due to an impulsive source applied at $\mathbf{x}_0$. The integral equation, that is equivalent to the equation of radiative transfer in case of isotropic scattering, is given by

$$E(\mathbf{x},t) = E_0\left(t - \frac{|\mathbf{x}-\mathbf{x}_0|}{\beta}\right)\frac{e^{-\eta|\mathbf{x}-\mathbf{x}_0|}}{4\pi|\mathbf{x}-\mathbf{x}_0|^2} + \int_V \eta_S E\left(\xi, t - \frac{|\mathbf{x}-\xi|}{\beta}\right)\frac{e^{-\eta|\mathbf{x}-\xi|}}{4\pi|\mathbf{x}-\xi|^2}dV(\xi), \quad (3)$$

where $E_0[t - (|x-x_0|/\beta)]$ represent the impulsive energy radiated by the source, $\eta = \eta_S + \eta_I$ is the total attenuation coefficient, with η_S and η_I, respectively the scattering and the intrinsic attenuation coefficients, and ξ is the variable of integration (describing the spatial position of the scatterers). The quality factors for scattering and intrinsic attenuation can be expressed in terms of attenuation coefficients by

$$Q_{S,I}^{-1} = \eta_{S,I}\beta/\omega, \quad (4)$$

where subscripts S and I stand for scattering and intrinsic, respectively. The assumptions underlying Eq. (3) are more general than those for the Eq. (1), as all multiple scatterings are included. In both Eqs. (1) and (3), scattering is assumed to be isotropic and the medium randomly uniform.

Formulas (1) and (2) have been widely used for fitting the experimental coda envelopes to invert for Q_C^{-1}. Results obtained over time and throughout the World have been used to characterize the average attenuation properties of the zones under study simply by comparing coda-Q^{-1} parameters. Unfortunately, as reported above, the physical interpretation may be confusing. Moreover, there is an additional problem of interpretation of these early results, due to their significant dependence on the time window in which the fit of data with single-scattering model is made. This dependence is often named "lapse time dependence of Q_C." Because of this dependence (Q_C^{-1} decreases with lapse time), the experimental results reported without an explicit specification of the lapse-time window length used for calculations may be not strictly suitable for comparison among the different areas. Figure 1 shows a compilation of Q_C^{-1} observations made in volcanic areas using similar lapse-time window lengths (or I explicitly report the lapse time used). Looking at the plots of Figure 1, it can be noticed that in some cases coda-Q pattern shows a peculiar frequency pattern. This is particularly evident for Kilauea, Mt. S. Helens, and Mt. Vesuvius. A comparison with Q_C^{-1} estimates in stable or tectonically active nonvolcanic areas using data taken from Ibanez (1990) and Del Pezzo *et al.* (1996) is reported in Fig. 2 for the volcanic area of Mt. Etna (Italy) and the nonvolcanic zone of Andalucia (Spain). The plots reproduce estimates obtained at the same lapse-time intervals. Leaving aside the interpretation of the physical meaning, these plots show

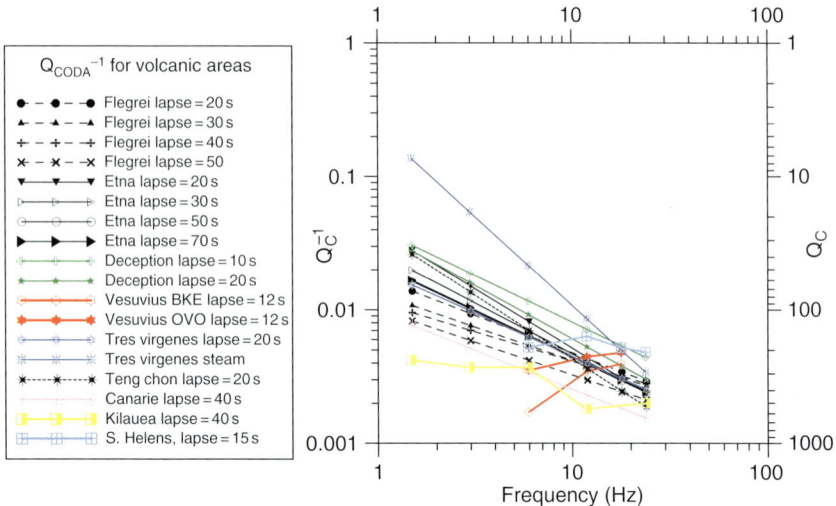

FIG. 1. Q_C^{-1} as a function of frequency, calculated for several volcanic areas around the world. References are: Campi Flegrei (Bianco *et al.*, 1999); Etna (Del Pezzo *et al.*, 1996); Deception (Martinez Arevalo *et al.*, 2003); Tres Virgenes (Wong *et al.*, 2001); Tengchong (Baiji *et al.*, 2000); Canary Islands (Canas *et al.*, 1995); Kilauea (Mayeda *et al.*, 1992); and Mt. S. Helens (Bianco *et al.*, 1999). Lapse-time intervals in which the fit to the single-scattering model has been carried out are also given. Data from the literature with no explicit report of lapse-time interval have been disregarded.

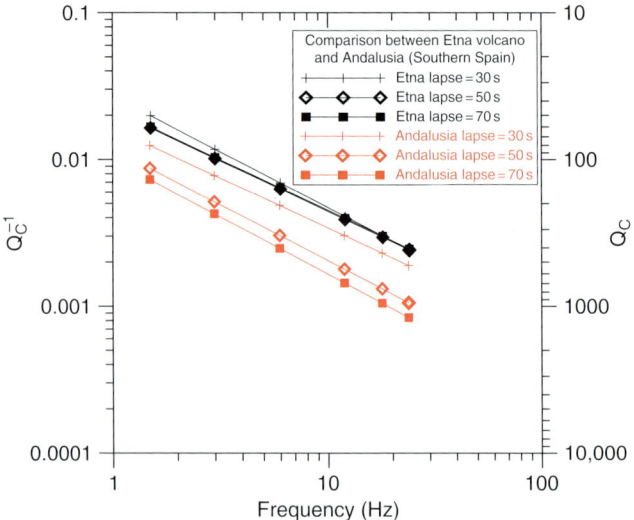

FIG. 2. Comparison of Q_C^{-1} versus frequency, calculated for the same lapse-time interval at Etna volcano and in Andalucia (Granada basin-Southern Spain). Despite the similar frequency patterns, the values of attenuation are different; higher for Etna volcano than for the tectonically active region of Andalucia.

that Etna volcano behaves differently from the tectonically active sedimentary Granada basin, showing an overall higher coda wave attenuation.

The results reported in Fig. 1 show that in some cases (Etna, Kilauea, and Vesuvius), there is a frequency dependence that is weaker than or opposite (Q^{-1} increasing with frequency) to the other regions. Whether these differences are controlled more by intrinsic attenuation or by scattering attenuation is impossible to deduce by examining only Q_C^{-1} observations. A further difficulty in the interpretation of Q_C^{-1} is given by the uniform half-space assumption. Gusev (1995) found that Q_C^{-1} may be closer to the real Q_I^{-1} for earth media characterized by strong velocity gradients. This study demonstrated that coda decay is quantitatively well explained if the scattering coefficient decreases with depth, when the leakage of scattered energy to the bottom cannot be discriminated from intrinsic loss. Margerin et al. (1998) and Wegler (2004) using the diffusion approximation, derived analytical expressions which describe the coda decay due to leakage, in the assumption of a heterogeneous layer superimposed to a transparent half space.

It can be concluded that there is no simple relation between Q_C^{-1} and scattering and intrinsic inverse-Q. Consequently, separate estimates of intrinsic and scattering attenuation coefficients performed by using the realistic assumptions of positive (with depth) seismic velocity gradients and depth-dependent attenuation parameters are necessary to quantify the scattering processes in volcanic regions.

2. Separated Estimates of Intrinsic and Scattering Attenuation

To date, the most complete approach to characterize the earth medium for optimally describing energy propagation and scattering properties has been the radiative transfer theory, which includes multiple scattering of any order (Ryzhik et al., 1996). This theory allows the analytical description of the coda envelope as a function of source–receiver distance and lapse time, in the hypothesis of a uniform medium and isotropic scattering (Zeng, 1991). Numerical simulation is needed to describe more realistic media, characterized by non-isotropic scattering and/or nonuniform velocity and scatterer distributions (Hoshiba, 1991, Gusev and Abubakirov, 1996). This theory can be fitted to the experimental coda envelope to directly invert the scattering and dissipation attenuation coefficients, and is well suited to model the seismic coda on volcanoes.

This inversion for attenuation coefficients has often been carried out with the multiple lapse-time window analysis (MLTWA) technique, as described by Hoshiba (1993). This analysis is based on calculation of the seismogram energy integrals across three successive time windows, as a function of the source–receiver distance and medium parameters. The three energy–distance curves are then fitted to the theoretically determined values, to retrieve the scattering attenuation and the intrinsic dissipation coefficients (for a detailed discussion of this method, see also Sato and Fehler, 1998, pages 190–191). Because of the low velocity of the S waves in the shallowest layers of the volcanic areas (as low as 1.5 km/s), to the short duration of the seismograms of VT earthquakes, and to the limited intervals of source-station distance generally available for these kinds of earthquakes, the integrals in the three successive time windows cannot generally be calculated with sufficient numbers of data points, and the distance range is insufficient to make the fit stable. Consequently, the application of MLTWA to volcanoes may become difficult. For this reason, previous attempts carried out to separately estimate intrinsic and scattering attenuation coefficients for local volcanic earthquakes are based on different and sometimes approximate techniques. In the following sections, I will review the approaches used to experimentally study the scattering properties of volcanic areas and I will discuss the results obtained.

2.1. The Method of Wennerberg

One of the first attempts to separately obtain scattering and intrinsic attenuation parameters in volcanic areas was done for the zones of Etna and Campi Flegrei (Italy) (Del Pezzo et al., 1995) using an approach developed by Wennerberg (1993). This study suggested the possibility of reinterpreting the estimates of single-station Q_C^{-1} values in terms of multiple scattering. It used the approximation given by Abubakirov and Gusev (1990) to the energy-transport theory in the formulation of Zeng (1991) to describe the multiple-scattered wave field in the case of a source colocated with the station. Wennerberg (1993) expresses Q_C^{-1} as a function of Q_I^{-1} and Q_S^{-1} as:

$$Q_C^{-1} = Q_I^{-1} + [1 - 2\delta(\tau)]Q_S^{-1}, \tag{5}$$

where $\delta(\tau) = -1/(4.44 + 0.738\tau)$ and $\tau = 2\pi f t Q_S^{-1}$. Using the definition of $Q_T^{-1} = Q_I^{-1} + Q_S^{-1}$, the total inverse quality factor for S waves, this method allows the estimation of Q_I^{-1} and Q_S^{-1} from (independent) measurements of Q_T^{-1} and Q_C^{-1}. The method is strictly applicable to local earthquake data recorded at stations close to the source for which the direct ray paths share the same volume encompassed by the scattered waves. Results obtained for Etna volcano and Campi Flegrei areas were compared with the tectonically active nonvolcanic zone of Granada Basin, for which the same experimental conditions were encountered (shallow sources located close to the recording stations— 20 s maximum lapse time for Q_C^{-1}). The results for Q_C, Q_T, Q_I, and Q_S are summarized in Table 1 of Del Pezzo et al. (1995). Martinez Arevalo et al. (2003) applied the same method to local VT earthquakes recorded at a small aperture (300 m) array composed of 13 short-period sensors, located in Deception Island, Antarctica. Deception is the most important active volcano of the South Shetland Islands, and is located northeast of the Antarctic Peninsula. Results are reported in Fig. 17 of Martinez Arevalo et al. (2003).

The plot of Fig. 3 shows the results obtained for Etna and Deception [those for Campi Flegrei after being reviewed by Del Pezzo et al. (1996) were shown to be based on a rough estimate of total Q-inverse, and will be discussed later]. For the sake of uniformity with the plots of the present study, I have plotted Q-inverse (instead of Q) with its error

TABLE 1 Separation of inverse Q_I and Q_S for Campi Flegrei

Q_S^{-1}			n
6 Hz	8 Hz	10 Hz	
0.004 ± 0.001	0.005 ± 0.001	0.005 ± 0.001	0.1
0.05 ± 0.01	0.04 ± 0.01	0.031 ± 0.007	0.9
Q_I^{-1}			
6 Hz	8 Hz	10 Hz	
0.00365 ± 0.00007	0.0031±0.0001	0.0028 ± 0.0001	0.1
0.00357 ± 0.00006	0.0031 ± 0.0001	0.0027 ± 0.0001	0.9

Values are re-calculated from Table 2 of Del Pezzo et al. (1996).

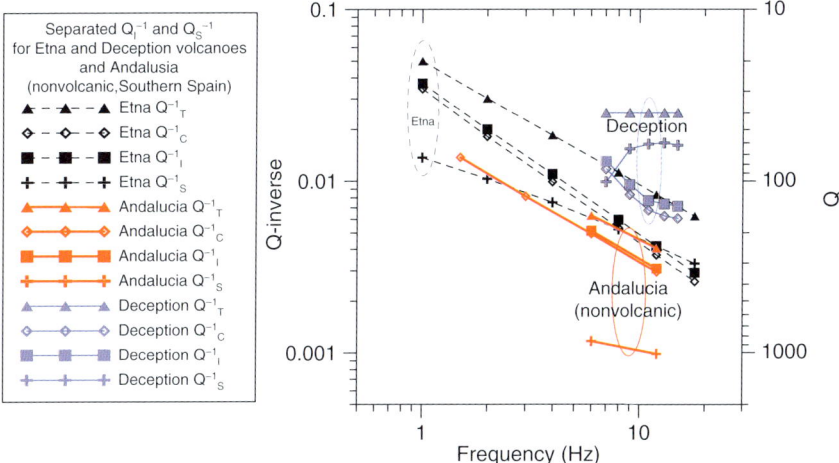

FIG. 3. Q_I^{-1} and Q_S^{-1} obtained using the approximate method of Wennerberg for Etna volcano, Deception Island volcano, and Andalucia.

bars for Etna and Deception Island, and compared the results obtained with those of the tectonically active area of Andalucia. The plot shows that the earth lithosphere beneath volcanoes (Etna and Deception) and the tectonically active area of Andalucia (Granada basin) have different scattering properties: The two volcanic zones are more heterogeneous than the tectonically active zone of the Granada basin. Attenuation at Etna is controlled by intrinsic dissipation at frequencies below 8 Hz, whereas for the higher frequency bands, dissipation and scattering effects are comparable. At Deception, only the frequency bands higher than 7 Hz are available. For high frequencies, scattering effects predominate over the intrinsic dissipation at Deception.

2.2. The Energy-Flux Model

An approximate method, different from that described in the previous section, for obtaining separate estimates of intrinsic and scattering attenuation parameters was applied to Campi Flegrei area, close to Naples, Italy, by Del Pezzo *et al.* (1996). This study used the energy-flux model for coda generation in a uniform medium (Frankel and Wennerberg, 1987)—see formula 3.34 of Sato and Fehler (1998). The energy-flux model is a phenomenological model based on the assumption that coda envelopes, recorded at different distances, approach the same value for increasing lapse time. In contrast to that of Wennerberg (1993), this method, inverts the coda envelope and does not use an independent estimate of Q_T^{-1}. The results are obtained assuming a priori a frequency dependence of Q_S^{-1}. Del Pezzo *et al.* (1996) assume that $Q_S^{-1} = q_0 f^{-n}$ and calculate both Q_S^{-1} and Q_I^{-1} for a suite of n values, spanning the interval between 0.1 and 0.9. Here I report in Table 1 only those for $n = 0.9$ and for $n = 0.1$, corresponding to a strong frequency dependence and an almost constant Q_S^{-1} with frequency, respectively. Results were obtained for three frequency bands (centered at 6, 8, and 10 Hz, respectively).

In the area of Campi Flegrei, scattering phenomena strongly predominate over inelastic dissipation.

2.3. 2-D Transport Theory Applied to Volcanic Tremor

Volcanic tremor is a sustained seismic signal which is often seen in association with magmatic activity. Its importance as a forecasting tool of eruptions has been widely acknowledged since many years (e.g., Chouet, 2003, and references therein). The mechanism of the tremor source is still under debate, although the generation of the sustained ground motion is generally ascribed to the trapping of elastic energy in fluid-filled cavities. Studying the tremor sources is, however, challenging, due to the main reason that tremor signal is quasi-stationary, with no clear onsets and/or clear phases that can be ascribed deterministically to a particular path. Moreover, the signal generally looses coherency with increasing station spacing, making it impossible to adopt classical tools for locating sources and separating path, source, and site effects. An understanding of the scattering properties of the volcanic media for the tremor wave propagation is consequently an important task that should be properly addressed.

The first measurements of the total attenuation for tremor waves were carried out by Del Pezzo et al. (1989) at Etna volcano, who simply calculated the spectral-amplitude decay of the tremor as a function of source-station distance. The results show that total Q-inverse is frequency dependent with data fitting well the relation $Q_T^{-1} = q_0 \left(\frac{f}{f_0}\right)^{-n}$ with $f_0 = 1$ Hz, $n = 0.7$ and $q_0 = \frac{1}{12}$. Lower values of Q_C^{-1} and Q_T^{-1} calculated for VT earthquakes in the same area (see the previous section of the present chapter) were interpreted in terms of a strong depth dependence of attenuation: As the tremor wave propagation is essentially shallow (the tremor is mainly composed of a mixture of surface waves), it samples the highest attenuation layers. A first attempt to separately estimate intrinsic and scattering attenuation coefficients for volcanic tremor has been carried out by Del Pezzo et al. (2001) utilizing data recorded at Etna by Del Pezzo et al. (1989) and at Masaya volcano by Metaxian et al. (1997). The method is based on the energy-transport theory in two dimensions (Sato, 1993). The space and time pattern of the tremor seismic energy is calculated by convolution of the source function with the energy density for an impulsive source radiation. The source time function for tremor is assumed to be constant, in the hypothesis of time- and frequency-stationary emission of seismic energy, and the convolution integral reduces to the following expression

$$E_{\text{Tremor}}(r) \sim \int_{-\infty}^{+\infty} E(r,t) dt = \frac{\frac{1}{2\pi r} \exp\left[-\left(\eta_S + \frac{\eta_I}{v(f)}\right)r\right]}{|v(f)|}$$
$$+ \int_{r/v(f)}^{+\infty} \frac{\eta_S}{2\pi\sqrt{v(f)^2 t^2 - r^2}} \exp\left[\eta_S \sqrt{v(f)^2 t^2 - r^2} - v(f)t\right] \exp(-\eta_I t) dt, \quad (6)$$

where $E_{\text{Tremor}}(r)$ is the tremor energy as a function of distance, r; the frequency, f; the wave speed, $v(f)$; the intrinsic attenuation coefficient, η_I; the scattering attenuation coefficient, η_S; and the lapse time, t. $v(f)$ represents the wave velocity in a dispersive wave field. $E(r,t)$ is the Green's function for an impulsive source (Sato and Fehler, 1998, pages 173–176).

Formula (6) was fit to the experimentally measured tremor energy decay with distance. The best estimates of η_I and η_S were obtained with a grid search method, and the results for Q_S^{-1} and Q_I^{-1} derived from formula (4) are given in Table 2. Despite the trade off between the estimates of Q_I^{-1} and Q_S^{-1}, the results show that the mechanism of

TABLE 2 Separation of inverse Q_I and Q_S for volcanic tremor at Etna and Masaya

Volcano	Frequency (Hz)	Q_I^{-1}	$\sigma_{Q_I^{-1}}$	Q_S^{-1}	$\sigma_{Q_S^{-1}}$
Etna	1	0.2	0.4	0.009	0.005
	2	0.5	2	0.005	0.001
	3	1	10	0.010	0.001
	4	0.3	7	0.009	0.007
	5	0.1	0.2	0.005	0.002
Masaya	2	0.2	0.5	0.04	0.02
	3	0.1	0.5	0.005	>0.001

Values are re-calculated from Table 1 of Del Pezzo et al. (2001).

dissipation is predominant over scattering phenomena in the characterization of the seismic attenuation of tremor for both volcanoes under study. This is in contrast to the earlier results for VT events at Campi Flegrei, Etna, and Deception. This contradictory result may have a geological explanation, as the uppermost layers composing the structure of Etna are spatially homogeneous (smaller amount of scattered energy), all composed by loose and incoherent materials (high intrinsic dissipation).

3. Diffusion Model Applied to Shot Data

3.1. Uniform Half Space

Hereafter I will focus attention on two volcanoes, Vesuvius and Merapi, where most of the studies dealing with the application of the diffusion model (that will be described in the present and in the following sections) have been done.

The transport theory has an important asymptotic approximation in the case of strong scattering: the diffusion theory (Wegler, 2004). This approximation is mathematically much simpler and can be analytically expressed in case of a medium composed of two layers with different characteristics. The analytical expression for a homogeneous earth medium is

$$E(|\mathbf{x}|, t) = E_0 (4\pi D t)^{-p/2} e^{\left(-bt - \frac{|\mathbf{x}|^2}{4Dt}\right)}, \tag{7}$$

where $p = 3$ for body waves and $p = 2$ for surface waves. $p = \beta/\eta_S D$ and $b = \eta_I \beta$. D is named "diffusivity."

The presence of diffusive waves was revealed experimentally in the seismograms of artificial shots fired at Merapi (Indonesia; Wegler and Luhr, 2001) and at Vesuvius (Wegler, 2003) during active tomography experiments. In these two cases, a rapid decrease of direct S-wave energy was observed and detected up to a distance of less than 1 km from the source. Despite this rapid decrease, the coda of the seismogram shots exhibits increasing amplitudes up to a lapse time much greater than the S-wave travel time and a slowly decaying amplitude for longer lapse time. Observations at small aperture arrays set up at Merapi (Wegler and Luhr, 2001) and at Mt. Vesuvius (La Rocca et al., 2001) in the time period of the active experiments showed that the coherence among the array stations is lost at distances smaller than a few tens of meters for frequencies higher than 1 Hz; the polarization properties calculated at the array stations also show that the pattern is chaotic, being the three components of the ground motion almost uncorrelated. This evidence is phenomenologically interpreted by Wegler and Luhr (2001) as the product of a diffusive wave field composing the coda of the seismogram shots.

Fitting the experimentally calculated energy envelopes to the expression of formula (7), it is possible to invert for D and b and to separately obtain the intrinsic and scattering attenuation coefficients.

The pattern of intrinsic and scattering attenuation as a function of frequency (averaged over distance) for Mt. Vesuvius and for Merapi obtained for the shot data is reported in Fig. 4. In both cases, scattering predominates over inelastic dissipation by at least one order of magnitude in the analyzed frequency range. Wegler and Luhr (2001) estimate that the transport mean free path η_S^{-1} is of the order of 100 m for Merapi, whereas Wegler (2003) estimates $\eta_S^{-1} = 200$ m for Mt. Vesuvius [for a wide and exhaustive discussion about the physical meaning of the transport mean free path, see Gusev and Abubakirov (1996)]. In both studies, a dominance of S waves in the coda of the seismogram shots is assumed. This estimate is consistent with the assumption that diffusion is a valid approximation when source-station distance is much greater than the transport mean free path. In their experiments, source-station distance is always greater than 1 km. Almost the same diffusivity value is obtained assuming a 3- or a 2-D propagation.

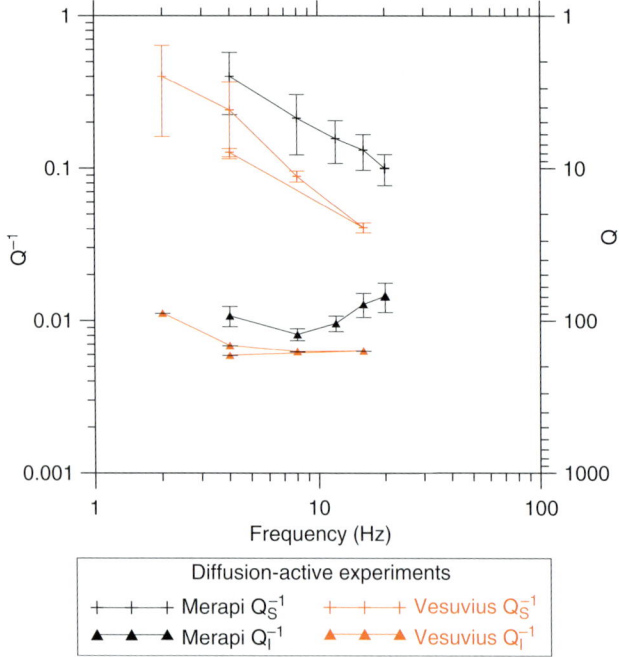

FIG. 4. Diffusion model under the assumption of a half space applied to shot data recorded at Merapi and Mt. Vesuvius. The patterns of both Q_I^{-1} and Q_S^{-1} are similar for the two volcanoes. Scattering attenuation prevails over the intrinsic dissipation of about one order of magnitude.

This is due to the strong influence of the term $e^{\left(-bt-\frac{x^2}{4Dt}\right)}$ in formula (7) with respect to the term accounting for geometrical spreading.

The above results indicate that multiple scattering cannot be neglected in the modeling of seismic wave propagation and in studying seismogram formation in volcanic areas. The comparison of the transport mean free path at Vesuvius and Merapi ($\simeq$ 200m) with the much higher value estimated for the earth's crust ($\simeq$ 200km) leads to the important conclusion that the multiple scattering strongly affects the seismogram shape for sources close to or within volcanoes. An indirect confirm of this important observation comes from high-resolution velocity tomography carried out in volcanoes (see, e.g., Scarpa et al., 2002; Chouet, 2003, and references therein) which shows high-velocity contrasts in small-scale structures. This result cannot be neglected in any modeling of seismic wave propagation in volcanic environments.

3.2. Two-Layer Media

The uniform random medium assumes that heterogeneity is uniformly distributed in the propagation volume. On the other hand, it is quite reasonable and well accepted that the earth properties change with increasing depth. The transport Equation (3) can be solved analytically for a uniform random medium with a uniform scatterer distribution. Numerical simulation is necessary in the case of nonuniform velocity and/or scatterer density (see, e.g., Hoshiba et al., 2001). In the approximation of strong scattering, the diffusion equation can be solved for a two-layer model formed by a shallower diffusive layer over a weakly scattering half space. Margerin et al. (1998) proposed a boundary condition for the layer half space including deterministic reflections. Wegler (2005) presents an improved boundary condition for the diffusion equation connecting a strongly scattering layer to a weakly scattering half space. This condition has a wide range of validity, failing only when the thickness of the upper layer (the strong scattering layer) is smaller than its transport mean free path, and/or when a large contrast in scattering strength between the upper layer and the half space exists. The application of this two-layer model to volcanoes is straightforward, as volcanoes are highly heterogeneous structures in their upper part, due to the presence of lava and ash formations; and generally less heterogeneous in their deeper portion, made up by the last part of the upper crust. An understanding of the diffusive characteristics of the volcanic media using this realistic earth model may explain why the scattering strength apparently decreases with the increasing source–receiver distance when it is estimated assuming a half-space model.

This approach has been used to study Merapi (Wegler, 2005) and Mt. Vesuvius (Wegler, 2004). At Merapi, the application of the theory to the same data set used by Wegler and Luhr (2001) yields new insight in the seismogram interpretation. For shot data (where the source is located at the surface), the coda envelopes well fit to a model with a diffusivity coefficient in the upper layer equal to 0.027 km^2 s^{-1} (corresponding to $Q_S^{-1} = 0.7$ at 6 Hz) and in the lower half space equal to 0.3 km^2 s^{-1} (corresponding to a $Q_S^{-1} = 0.06$). The data inversion yields the position of the half-space boundary at 0.5 km depth below the zero level. This model explains well the differences in the estimates of diffusivity obtained at different source-station distances for the uniform half-space model, fitting the data well at both short and long distances (see, Fig. 5 of Wegler, 2005). For Mt. Vesuvius, Wegler (2004) uses the same data set utilized in the work done assuming a half space (see also Fig. 4 in the present study), and finds nearly the same diffusivity and intrinsic attenuation' as those determined for the 2-D diffusion model.

(a)

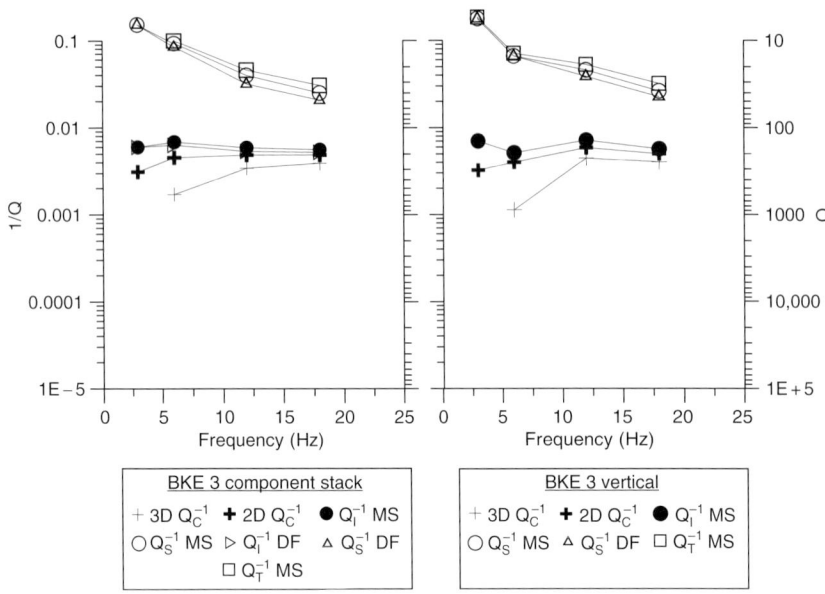

(b)

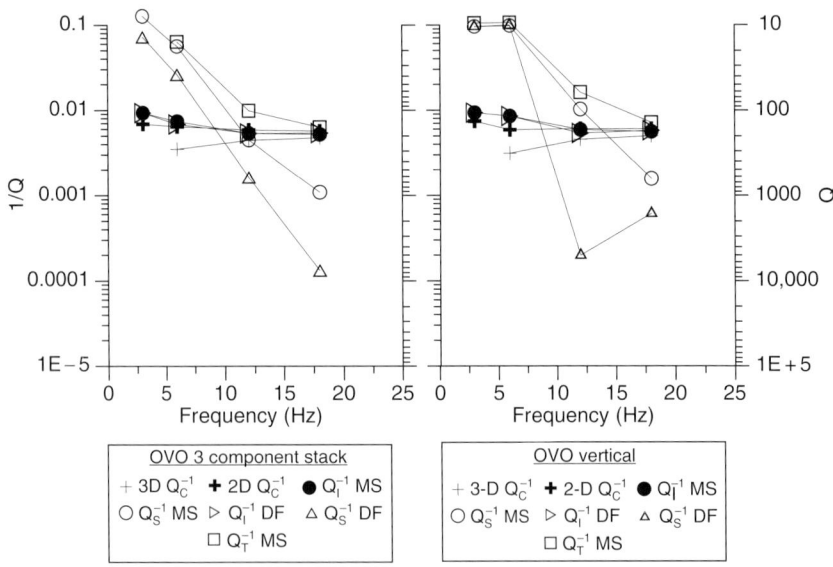

FIG. 5. (a) Left panel shows $Q_C^{-1}, Q_I^{-1}, Q_S^{-1}, Q_T^{-1} (= Q_I^{-1} + Q_S^{-1})$ for BKE station-three component stack (see Del Pezzo *et al.*, 2006, for the station location). The right panel shows the same quantities for BKE vertical component only. (b) The same of for OVO station. 3-D Q_C^{-1} and 2D Q_C^{-1} values were obtained using different geometrical spreading (body waves and surface waves, respectively).

The corresponding Q_S^{-1} spans the interval between 0.6 and 0.07. The values are similar to those obtained at Merapi. At Mt. Vesuvius, the diffusive-layer thickness is found to be of the order of 1 km, and the results for a diffusive layer over a half space are similar to those for a homogeneous model. This is due to the source–receiver distance range used that is always greater than the thickness of the diffusive layer. The results for Mt. Vesuvius appear to be independent of the boundary conditions used.

4. Energy-Transport Theory Applied to Earthquake Data

4.1. Uniform Half Space

Del Pezzo et al. (2006) use earthquakes instead of shot data to measure the scattering properties of the earth materials beneath Mt. Vesuvius. These authors use the energy-transport theory in the approximation of Zeng (1991) given by

$$E_{MS}(\mathbf{x}, t, \eta_i, \eta_S) \simeq E_0 e^{-(\eta_i + \eta_S)\beta t} \left[\delta \frac{\left(t - \frac{|\mathbf{x}|}{v}\right)}{4\pi\beta|\mathbf{x}|^2} + \eta_s \frac{H\left(t - \frac{|\mathbf{x}|}{\beta}\right)}{4\pi\beta|\mathbf{x}|t} \ln \frac{1 + \frac{|\mathbf{x}|}{\beta t}}{1 - \frac{|\mathbf{x}|}{\beta t}} \right]$$

$$+ cH\left(t - \frac{|\mathbf{x}|}{\beta}\right)\left(\frac{3\eta_s}{4\pi\beta t}\right)^{\frac{3}{2}} e^{-\frac{3\eta_s|\mathbf{x}|^2}{4\beta t} - \eta_i \beta t}, \qquad (8)$$

where $c = E_0 \left\{ [1 - (1 + \eta_s \beta t)e^{-\eta_s \beta t}] / \left(4/\sqrt{\pi} \int_0^{\sqrt{3\eta_s \beta t}/2} e^{-\alpha^2} \alpha^2 d\alpha \right) \right\}$; symbols are the same as for Eq. (3). This equation was fitted to the average normalized energy coda envelope calculated for two stations, BKE and OVO, of the local seismic network (see Fig. 1 of Del Pezzo et al., 2006). These two stations are $\sim$1 km distant from each other, and are located respectively eastward and westward from the crater. The assumptions of this theory are constant velocity and scattering coefficient in a uniform random medium.

This study used the stacked S-coda envelope at BKE and OVO obtained by filtering seismograms in four frequency bands, centered respectively at $f_c = 3, 6, 12$, and 18 Hz with a bandwidth of $\pm 0.3 f_c$. The stack is achieved by aligning all the three component envelopes at the P-wave onset time and normalizing to the coda level at 11s lapse time. The average coda is fit to the theoretical model of formula (8) starting at a lapse time of twice the S-wave travel time, after which all the envelopes have a smooth and regular time decay. A further fit of the same data was also done to the diffusion model [Eq. (7)].

The results are given in Fig. 5. Details on the misfit function at 1σ confidence level are given in Del Pezzo et al. (2006). The errors are always of the order of $\pm 25\%$ of the estimated value. From Fig. 5, it appears clear that three-component stack and vertical component average energy envelope share the same pattern and yield the same result. Multiple scattering (MS) and diffusion models (DF) furnish the same results, whereas single-scattering approximation (Q_C^{-1}) gives different results. This last result indirectly indicates that the diffusion regime is also appropriate to describe the seismic energy decay in the coda at Mt. Vesuvius for natural VT earthquakes, confirming the results obtained for shot data. The most important condition for the validity of the diffusion approximation is that the lapse time, t, should be greater than the transport mean free path divided by the wave velocity:

$$t_{\text{lapse}} > L/\beta. \qquad (9)$$

Taking $\beta = 1.5$ km/s I estimate the mean free path L from the estimates of Q_S^{-1} obtained by Del Pezzo et al. (2006) through the formula (Sato and Fehler, 1998)

$$Q_S^{-1} = \frac{\beta}{2\pi f L}. \qquad (10)$$

Results show that for OVO station $L \simeq 0.7$ km for frequencies centered at 3 and 6 Hz and $L > 5$ km for higher frequency; for BKE station $L \simeq 0.5$ km in the whole frequency range investigated. Since the lapse-time window utilized by Del Pezzo et al. (2006) is in the range from 4 to 12 s, inequality (9) is fulfilled except for the highest frequency band at OVO.

4.2. Possible Bias Introduced by Assuming a Uniform Diffusive Layer

On the basis of the results obtained by a velocity tomography study carried out by Scarpa et al. (2002), Del Pezzo et al. (2006) assume a simplified two-layer structure for Mt. Vesuvius: The first layer with $V_p = 2.6$ km/s from the crater top down to the limestone interface overlying a half space with $V_p = 4.5$ km/s. A further assumption is that the S-wave velocity is estimated by the P-wave velocity divided by the V_p/V_s ratio (1.8, averaged from the values reported in Scarpa et al., 2002). Hereafter, I will call MODEL1 the uniform half space and MODEL2 the two-layer model.

Del Pezzo et al. (2006) use the analytical solution for the diffusion equation obtained for a thick layer over a homogeneous half space, described by Wegler (2004) assuming (a) that diffusivity D (see, eq. 7) is constant in the top layer of MODEL2 and (b) that the whole diffusion process takes place in the same layer. They check both the cases of a fully absorbing boundary between the two layers and a fully reflecting interface (Eqs. 16 and 18 of Wegler, 2004, respectively) comparing the energy envelopes obtained for MODEL2 with those calculated for MODEL1. The results are that the coda-energy envelope for MODEL1 calculated with the diffusion parameter, D_{uniform}, is well approximated by the energy envelopes in MODEL2 with $D_{\text{absorbing}} \simeq D_{\text{uniform}}$ and $D_{\text{reflecting}} \simeq 2D_{\text{uniform}}$. This result indicates that an earth model more realistic than the half space may introduce severe biases into estimate of diffusivity. These may be introduced by neglecting the effects of the leakage, as described in Margerin et al. (1998), whereas the diffusion constant may be not severely influenced by the simplified assumptions. The bias introduced by the simplifying assumption of half space has been demonstrated for shot data fired at Mt. Vesuvius by Wegler (2004) and has thus been confirmed also for natural seismicity. This should be taken into account for any comparisons among different volcanic zones.

4.3. Coda-Localization Effects

Aki and Ferrazzini (2000) observed at Piton de la Furnaise volcano (PdF) that the coda-site amplification depends on source position when sources are close to the crater and that most of the scattered energy is produced near the source. This phenomenon has been called coda localization. The same phenomenon was claimed to explain the energy coda shape observed at Merapi volcano by Friederich and Wegler (2005). This study showed that in their area of investigation, the coda-energy envelope had the same systematic decrease with increasing source–receiver distance, different from the general observation of Aki and Chouet (1975) that the coda energy tends to a common level independent of source–receiver distance. The Ioffe Regel criterion $kL < 1$ (k is the wave number and L is the scattering mean free path—Van Tiggelen, 1999) is met in their data. This is the condition for the application of the Anderson localization model (Van Tiggelen, 1999). In this scattering regime, Weaver (1994) found phenomenologically a formula describing the energy-density decay E as a function of distance r and lapse time t, which includes the intrinsic dissipation:

$$E(r,t) = E_0 \exp\left(-\frac{r}{\xi} - \left(\frac{r^{2+n}}{4D_{\text{res}}t\xi^n}\right)^h - bt\right), \qquad (11)$$

where n and h are empirically determined constants (assumed to be respectively equal to 0.46 and 0.76), ξ is the localization length, D_{res} is the residual diffusivity, and b is the intrinsic dissipation coefficient. Friederich and Wegler (2005) compared the Anderson localization model with the half-space diffusion model fitted to their data using both Eqs. (11) and (7). They saw that at Merapi, the energy envelopes as a function of distance and lapse time fit well to formula (11) in the frequency band between 1 and 3 Hz with $D_{\text{res}} \simeq D = 0.12\,\text{km}^2/\text{s}\,;\,b = \frac{2\pi f}{Q_I^{-1}} = 0.2\,\text{s}^{-1}\,;\,\xi = 1.7\,\text{km}$. The condition $kL < 1$ indicates that the Anderson localization regime may be present in their data at low frequency, with a localization length ξ. It is important to note that, however, the spatial localization of coda energy can also be explained, according to Friederich and Wegler (2005) as a result of an inhomogeneous distribution of scattering strength.

For earthquake data recorded at Mt. Vesuvius, the Ioffe-Regel criterion is not met in the investigated frequency bands. Taking $v = 1.5$ km/s and estimating the mean free path L from the estimates of Q_S^{-1} obtained through the formula (10), we have for OVO station $13 < kL < 7 \cdot 10^3$ and for BKE station $6 < kL < 48$, well outside the limits given by $kL < 1$. In conclusion, the difference at OVO should be not a distance effect, as OVO and BKE are almost equally distant from the location centroid of VT earthquakes. The different energy envelope shape between OVO and BKE may be interpreted as being due to different scattering conditions between the N-Western part of Mt. Vesuvius, where OVO is set up, and the Eastern part of the volcano, where most of the other seismic stations are located.

5. Concluding Remarks

In the present chapter, I have described the results obtained from fitting several scattering models to the seismic coda envelopes of volcanic earthquakes and artificial shots fired in volcanic structures. The estimates of coda-Q values on volcanoes (made with the single-scattering model) are only slightly smaller than those measured in nonvolcanic zones, and the frequency dependence of coda-Q in volcanic areas is sometimes different from that in tectonically active regions. However, it is impossible to deduce using only coda-Q observations whether this difference between the coda-Q

estimates in volcanic and nonvolcanic zones is controlled more by intrinsic or by scattering attenuation.

In contrast, the application of multiple scattering models allows a separate estimate of the scattering and intrinsic attenuation parameters. The main results obtained in volcanoes show in general that, at low frequency, intrinsic dissipation is more important, whereas scattering predominates at high frequencies. Volcanoes are consequently heterogeneous structures with a mechanism of seismic wave energy dissipation that tends to be controlled by the scattering phenomena with increasing frequency. For Mt. Vesuvius, Mt. Merapi, and Deception Island, volcano scattering attenuation prevails at frequencies higher than 2–3 Hz, but unfortunately there is no information for lower frequencies. At Mt. Etna, intrinsic dissipation prevails or is comparable with scattering attenuation for frequencies lower than 8 Hz. This result, obtained in the early 1990's with the use of approximate models, has been confirmed by the application of the energy-transport theory to both earthquakes and tremor. At a first sight, it sounds unexpected, as one can image volcanoes as structures where the magma reservoirs, partially filled with melted, high temperature rocks, intrinsically dissipate the seismic energy. This seems to happen only at low frequency. The difference in the pattern of attenuation between low- and high-frequencies can be explained in terms of the scale length of heterogeneity. Volcanoes may represent non self-similar earth medium, characterized by one or more predominant characteristic correlation length.

As a consequence of the high degree of heterogeneity in volcanoes, the seismograms of VT events can visually appear different in shape from the corresponding magnitude seismograms of nonvolcanic earthquakes. In fact, the coda of volcanic earthquakes is longer than that for earthquakes generated with the same magnitude in less heterogeneous environments, including tectonically active nonvolcanic areas. Moreover, the maximum of the energy envelope for volcano earthquakes (VT) is more delayed with respect to the time of S-onset than that for nonvolcanic earthquakes.

Unfortunately, a comparison among the investigated volcanoes is only partial, as seismograms from both shots and VT earthquakes recorded at Mt. Vesuvius have a sufficient signal-to-noise ratio only at high frequency, in contrast to those of Etna and Merapi. Consequently, the best estimate of separated intrinsic- and scattering attenuation is stable for Mt. Vesuvius only at high frequencies, and we have no information in the low frequency band between 1 and 5–6 Hz. Further efforts need be made towards quantifying the observation at low frequencies on this volcano and in general on different volcanoes, to determine if the difference in the attenuation pattern at low and high frequencies could be ascribed to a general phenomenon, peculiar for volcanic environments, or to the geological characteristics of only some volcanoes.

Quantification of the scattering properties is seen also to be very useful in the physical interpretation of the seismic phenomena accompanying volcanic eruptions. Recent results have shown that small changes in the elastic properties of the medium, that have no detectable influence on the first arrivals, are instead amplified by multiple scattering and may thus be readily observed in the coda. This idea is quantified in the concept of coda wave interferometry (see another chapter of this book) that is going to be one of the most promising techniques to monitor the temporal changes of the scattering coefficients of the earth medium (see Gret et al., 2005, and references therein). So, knowledge of the average scattering properties may help in quantifying the effects on the earth medium of the stress changes acting on volcanoes before and during the eruptive periods. This knowledge is also very useful in the interpretation of the attenuation tomographic images obtained in volcanoes. Recent work by Nishigami (1997) and Tramelli *et al.* (2006) among others, show that most of the strong scattering in volcanoes takes place in zones with maximum contrast in their geological characteristics. Tramelli *et al.* (2006) show, for example, that the border between the old caldera rim and the central new caldera zone at Campi Flegrei is characterized by the maximum positive spatial change in the scattering coefficient. As this zone coincides with that of maximum total attenuation, this is an indirect confirmation that the low-total Q zones in volcanoes are often associated with low-scattering Q.

Acknowledgments

This work is financed by INGV-DPC projects V3_4 and V4, MIUR-FIRB project entitled "Analisi del campo d'onda associato al vulcanismo attivo," and EU project "VOLUME." Francesca Bianco, Anna Tramelli, and Luca De Siena are gratefully acknowledged for their suggestions. Mike Fehler and two anonymous reviewers greatly helped in improving the manuscript. Chris Berrie corrected the English style.

References

Abubakirov, I.R., Gusev, A.A. (1990). Estimation of scattering properties of the lithosphere of Kamchatka based on Monte Carlo simulation of record envelope of near earthquake. *Phys. Earth and Planet Inter.* **64**, 52–67.

Aki, K., Chouet, B. (1975). Origin of coda waves: Source, attenuation and scattering effects. *J. Geophys. Res.* **80**, 3322–3342.

Aki, K., Ferrazzini, V. (2000). Seismic monitoring and modeling of an active volcano for prediction. *J. Geophys. Res.* **105**(B7), 16617–16640.

Baiji, L., Jiazheng, Q., Jianqing, Y., Mingong, C., Xuejun, L. (2000). Primary study on attenuation of shear wave and coda in tengchong volcano areas. *J. Seismol. Res. (China)* **23**(2), 136–142.

Bianco, F., Castellano, M., Del Pezzo, E., Ibanez, J. (1999). Attenuation of the short period seismic waves at Mt. Vesuvius, Italy. *Geophys. J. Int.* **138**(1), 67–76.

Canas, J.A., Pujades, L.G., Blanco, M.J., Soler, V., Carracedo, J.C. (1995). Coda-Q distribution in the Canary Islands. *Tectonophysics* **246**, 245–261.

Chouet, B. (2003). Volcano seismology. *Pure Appl. Geophys.* **160**, 739–788.

Del Pezzo, E., Lombardo, G., Spampinato, S. (1989). Attenuation of volcanic tremor at Mt. Etna, Sicily. *Bull. Seismol. Soc. Am.* **79**(6), 1989–1994.

Del Pezzo, E., Ibanez, J., Morales, J., Akinci, A., Maresca, R. (1995). Measurements of intrinsic and scattering attenuation in the crust. *Bull. Seismol. Soc. Am.* **5**, 1373–1380.

Del Pezzo, E., Simini, M., Ibanez, J.M. (1996). Separation of intrinsic and scattering Q for volcanic areas: A comparison between Etna and Campi Flegrei. *J. Volcanol. Geoth. Res.* **70**, 213–219.

Del Pezzo, E., Bianco, F., Saccorotti, G. (2001). Separation of intrinsic and scattering Q for volcanic tremor: An application to Etna and Masaya Volcanoes. *Geophys. Res. Lett.* **28**, 3083–3086.

Del Pezzo, E., Bianco, F., Zaccarelli, L. (2006). Separation of Qi and Qs from passive data at Mt. Vesuvius: A reappraisal of seismic attenuation. *Phys. Earth Planet. Inter.* **159**, 202–212.

Frankel, A., Clayton, R.W. (1986). Finite difference simulations of seismic scattering Implications for the propagation of short-period seismic waves in the crust and models of crustal heterogeneity. *J. Geophys. Res.* **91**, 6465–6489.

Frankel, A., Wennerberg, L. (1987). Energy flux model of seismic coda—separation of scattering and intrinsic attenuation. *Bull. Seismol. Soc. Am.* **77**, 1223–1251.
Friederich, C., Wegler, U. (2005). Localization of the seismic coda at Merapi volcano (Indonesia). *Geophys. Res. Lett.* **32**, L14312, doi:101029.
Gret, A., Snieder, R., Aster, R.C., Kyle, P.R. (2005). Monitoring rapid temporal change in a volcano with coda wave interferometry. *Geophys. Res. Lett.* **32**, L06304, doi:10.1029/2004GL021143.
Gusev, A.A. (1995). Vertical profile of turbidity and coda Q. *Geophys. J. Int.* **123**(3), 665–672.
Gusev, A.A., Abubakirov, I.R. (1996). Simulated envelopes of non isotropically scattered body waves as compared to observed ones: Another manifestation of fractal heterogeneity. *Geophys. J. Int.* **127**, 49–60.
Hoshiba, M. (1991). Simulation of multiple-scattered coda wave excitation based on the energy conservation law. *Phys. Earth Planet. Inter.* **67**, 123–136, doi={10.1016/0031–9201(91)90066-Q}.
Hoshiba, M. (1993). Separation of scattering attenuation and intrinsic absorption in Japan using the multiple lapse time window analysis of full seismogram envelope. *J. Geophys. Res.* **98**(B9), 15809–15824.
Hoshiba, M., Rietbrock, A., Scherbaum, F., Nakahara, H., Haberland, C. (2001). Scattering attenuation and intrinsic absorption using uniform and depth dependent model—Application to full seismogram envelope recorded in Northern Chile. *J. Seismol.* **5**, 157–179, doi: 10.1023/A:1011478202750.
Ibanez, J.M. (1990). Atenuacion de ondas coda y Lg en el sur de Espana y de Italia a partir de sismogramas digitales. Tesis Doctoral, Universidad de Granada, p. 306.
La Rocca, M., Del Pezzo, E., Simini, M., Scarpa, R., De Luca, G. (2001). Array analysis of seismograms from explosive sources: Evidence for surface waves scattered at the main topographical features. *Bull. Seismol. Soc. Am.* **91**(2), 219–231.
Margerin, L., Campillo, M., van Tiggelen, B. (1998). Radiative transfer and diffusion of waves in a layered medium: New insight into coda Q. *Geophys. J. Int.* **134**, 596–612.
Martinez Arevalo, C., Bianco, F., Ibanez, J., Del Pezzo, E. (2003). Shallow seismic attenuation and shear wave splitting in the short period range of Deception Island volcano (Antarctica). *J. Volcanol. Geotherm. Res.* **128**(1–3), 89–113.
Matsunami, K. (1991). Laboratory tests of excitation and attenuation of coda waves using 2-D models of scattering media. *Phys. Earth Planet. Inter.* **67**, 36–47, doi={10.1016/0031–9201 (91)90058-P}.
Mayeda, K., Koyanagi, S., Aki, K. (1992). A comparative study of scattering, intrinsic and coda Q^{-1} for Hawaii, Long Valley and Central California between 1.5 and 15.0 Hz. *J. Geophys. Res.* **97**, 6643–6659.
Metaxian, J.P., Lesage, P., Dorel, J. (1997). Permanent tremor of Masaya volcano, Nicaragua: Wavefield analysis and source location. *J. Geophys. Res.* **102**, 22529–22545.
Nishigami, K. (1997). Spatial distribution of coda scatterers in the crust around two active volcanoes and one active fault system in central Japan: Inversion analysis of the coda envelope. *Phys. Earth Planet. Inter.* **104**, 75–89.
Ryzhik, L.V., Papanicolaou, G.C., Keller, J.B. (1996). Transport equation for elastic and other waves in random media. *Wave Motion* **24**, 327–370.
Sato, H. (1993). Energy transportation in one- and two-dimensional scattering media: Analytic solutions of the multiple isotropic scattering model. *Geophys. J. Int.* **112**(1), 141–146.
Sato, H., Fehler, M.C. (1998). *Seismic Wave Propagation and Scattering in the Heterogeneous Earth*. Springer-Verlag, New York.
Scarpa, R., Tronca, F., Bianco, F., Del Pezzo, E. (2002). High resolution velocity structure beneath Mount Vesuvius from seismic array. *Geophys. Res. Lett.* **29**(21), 2040.
Shang, T., Gao, L. (1988). Transportation theory of multiple scattering and its application to seismic coda waves of impulsive source. *Sci. Sin. (series B, China)* **31**, 1503–1514.

Tramelli, A., Del Pezzo, E., Bianco, F., Boschi, E. (2006). 3-D scattering image of the Campi Flegrei caldera (Southern Italy). New hints on the position of the old caldera rim. *Phys. Earth Planet. Inter.* **155**, 269–280.

Van Tiggelen, B.A. (1999). Localization of waves. In: Fouque, J.P., (Ed.), *Diffuse waves in complex media*, vol. 1, Kluwer Academic Publisher, Dortrecht, The Netherlands, p. 60.

Weaver, R.L. (1994). Anderson localization in time domain: Numerical studies of waves in two dimensional disordered media. *Phys. Rev. B* **49**(9), 5881–5895.

Wegler, U. (2003). Analysis of multiple scattering at Vesuvius volcano, Italy, using data of the TomoVes active experiment. *J. Volcanol. Geotherm. Res.* **128**, 45–63.

Wegler, U. (2004). Diffusion of seismic waves in a thick layer: Theory and application to Vesuvius volcano. *J. Geophys. Res.* **109**, B07303, doi:10.1029/2004JB003048.

Wegler, U. (2005). Diffusion of seimic waves in layered media: Boundary conditions and analytical solutions. *Geophys. J. Int.* **163**, 1123–1135.

Wegler, U., Luhr, B.G. (2001). Scattering behaviour at Merapi Volcano, Java revealed from an active seismic experiment. *Geophys. J. Int.* **145**, 579–592.

Wennerberg, L. (1993). Multiple scattering interpretation of coda-Q measurements. *Bull. Seismol. Soc. Am.* **83**, 279–290.

Wong, W., Rebollar, C.J., Munguia, L. (2001). Attenuation of Coda Waves at the tres Virgenes Volcanic Area, Baja California Sur, Mexico. *Bull. Seismol. Soc. Am.* **91**, 683–693.

Zeng, Y. (1991). Compact solutions for multiple scattered wave energy in time domain. *Bull. Seismol. Soc. Am.* **81**, 1022–1029.

MONITORING TEMPORAL VARIATIONS OF PHYSICAL PROPERTIES IN THE CRUST BY CROSS-CORRELATING THE WAVEFORMS OF SEISMIC DOUBLETS

Georges Poupinet, Jean-Luc Got and Florent Brenguier

Abstract

Doublets or multiplets are earthquakes with nearly identical waveforms. First observed on volcanoes, doublets are found in tectonic environments. Doublets can be relocated relatively with a precision of a few meters. Very good doublets separated by a large time lapse are essential for detecting slow temporal variations of crustal properties. We present basic techniques for selecting and processing doublets. In a seismic database, coherency between all pairs of seismograms is computed and high coherency pairs are candidate doublets. Time delays between waveforms are measured by cross-correlation techniques; a precision of 1 ms is common for good pairs sampled at 100 Hz. Several techniques can relocate one event relatively to the other and P and S delay residuals are obtained. Clock precision remains critical when searching for a few millisecond anomalies. Delays in the coda are analyzed by a cross-spectral moving window (CSMW) or a cross-correlation moving window (CCMW) analysis. Time delay measured in the coda shows variations even when the time elapsed between the two events of a doublet is extremely short. These variations are due to hypocenter separation and to changes in the waves which form the early coda. When seismic velocity is changing homogeneously in the propagation medium, the delay of the coda is proportional to lapse time. Thus, the slope α of the delay in the coda is a very precise (up to 10^{-4}) measurement of the change in S-wave velocity $\Delta V_S/V_S$, $\alpha = -\Delta V_S/V_S$. Relative changes of delay on the horizontal components can detect temporal variations in S-wave splitting and anisotropy. Temporal changes in coda Q may be reflected in the coda amplitude ratio measured in several frequency bands. However, minor changes in sources induce variations in early coda (the coda that just follows the S wave) amplitude ratios, comparable to those due to attenuation changes. Therefore, the interpretation of coda amplitude ratios in terms of coda Q changes should be undertaken in the late coda only, using a statistical approach. Good doublets are seldom, so we present a technique that creates "virtual doublets" from the correlation of seismic noise long sequences. Temporal variations of physical properties in surface layers are recovered by a CCMW analysis of these "virtual doublets." This is an interesting method for measuring strain variations preceding volcanic eruptions. Many other applications should blossom in the near future. At last, we present teleseismic doublets which are a tool for measuring the rate of rotation of the inner core of the Earth. The goal of this chapter is to show that doublet processing is elementary but that the detection of temporal variations of velocity or attenuation remains quite difficult. Excellent doublets should be selected to study temporal variations and such very good natural doublets are few in most seismic regions.

Key Words: Microearthquakes, temporal variation of properties, earthquake forecasting
© 2008 Elsevier Inc.

1. INTRODUCTION

Doublets or multiplets have been noticed in Japan by several seismologists, among them Hamaguchi and Hasegawa (1975) and Tsujiura (1981). They are observed on volcanoes for instance from magmatic type sources (Okada et al., 1981) and on the Moon (Nakamura, 1978). In Northern California, families of similar low-magnitude earthquakes were reported by Geller and Mueller (1980) who suggested that the epicenters are within a radius less than a quarter of the wavelength or about 200–400 m. The waveform similarity implies that the hypocenters are very near and that moment tensor and rupture process are similar; the proximity of hypocenters is not enough. How similar the sources should be is difficult to quantify but coherency between waveforms is an efficient tool to compare events. Two main types of doublets can be distinguished: the short time lapse doublets—spatial doublets—and the long time lapse doublets—temporal doublets. Nakamura (1978), Poupinet et al. (1982, 1984), Ito (1985), Scherbaum and Wendler (1986), Frémont and Malone (1987), Deichmann and Garcia-Fernandez (1992), Got et al. (1994), Gillard et al. (1996), Rubin et al. (1999), Nishimura et al. (2000, 2005), Got and Okubo (2003), Battaglia et al. (2003), and many others used waveform correlation for relative relocation of similar events and mapped very accurately clusters of microearthquakes. Accuracy in relative relocation through the use of cross-correlation time delays motivated very numerous authors to use such differential techniques, mainly after Waldhauser and Ellsworth (2000) double-difference location code. In a search for precursory changes of velocity before large magnitude events, Poupinet et al. (1984) took advantage of the waveform similarity of the entire seismograms to attempt to measure temporal changes with a moving window technique similar to surface wave processing. However, time delay measurement accuracy needed for temporal variation studies is far higher than the one needed for earthquake relocation. The precision in timing is very dependent on the precision of stations clocks. Independent analog stations did not get time precision at the millisecond level. Analog seismic arrays transmitted their signals on radio or phone links and recorded them in a central site with a single reference time base: the unique time base was well suited for comparative time delay studies. The advent of GPS time synchronization improved clock precision but various sources of instrumental delays remain; numerous temporary and permanent stations still have periods with unreliable timing.

The search for temporal changes of crustal velocities and attenuation has been a goal for many years in connection with earthquake precursor studies. Both natural events and artificial sources have been monitored over long period of time. In this chapter, we present basic techniques for detecting temporal variations and discuss their limits and illustrate them with data provided by the Japanese Hi-Net seismic array (Okada et al., 2004).

2. SELECTION OF DOUBLETS

Nowadays seismic arrays store data continuously on hard disks. Seismicity is monitored in real time and origin times and hypocenters catalog are published. Knowing the origin times, event files are extracted from the continuous data. Extraction can also be performed automatically by comparing the amplitude of the signal in several stations to a set threshold. Each event is given an index number. The simplest technique to select doublets is to extract some seconds of record in several good quality stations and to compute the average coherency in a given frequency band for all pairs of events. If coherency is larger than about 90%, we have potential doublets. This technique may not work when noise is correlated from one record to another or during an aftershock sequence because of the short interval between small events. Aster and Scott (1993) used short time P and S windows (~0.5 s) and showed that the median of shear wave three-component cross-correlation maximum gives the best evaluation of the similarity. This method supposes a good knowledge of individual S arrival times. Limited precision in locations usually prevents direct selection from the catalogs, so that selection is usually performed by the direct computation of coherency or cross-correlation from events occurring in kilometer-scale volumes. For studies of temporal variation, the very high coherency large lapse time doublets—temporal doublets—are interesting. For coseismic effects associated with a large magnitude event, pairs with one event before and one event after are selected. Short lapse time doublets—spatial doublets—are interesting for local structural studies and for assessing possible source variations than may create patterns similar to temporal variations (Got et al., 1990; Got and Fréchet, 1993; Got and Coutant, 1997). The procedure to select possible doublets is straightforward and can be implemented as a routine on any database. The probability to find doublets increases with the size of the database, that is, with time.

3. Basic Processing

3.1. Time Delays Measured from Cross-Correlation or Cross-Spectrum

Once a doublet has been selected, travel-time delays between its two events are measured. Figure 1a shows the two P waves for a good doublet recorded by one Hi-net station in Japan. In signal processing, the delay between two signals corresponds to the maximum of the envelope of their cross-correlation in the time domain (Fig. 1b) or to the slope of the phase of their cross-spectrum in the frequency domain (Fig. 1d). Two temporal signals s_1 and s_2 of length $N\Delta$ are sampled at times $i\Delta$, where i is an integer, Δ is the sampling rate, and N is the total number of samples. In the time domain, the cross-correlation $\gamma_{s_1 s_2}$ is

$$\gamma_{s_1 s_2}(n\Delta) = \sum_i \left\{ s_1(i\Delta) s_2(i\Delta - n\Delta) \right\}, \quad (1)$$

where $n\Delta$ is the time delay τ. In the frequency domain, the cross-spectrum $\Gamma_{s_1 s_2}(f)$ is $\Gamma_{s_1 s_2}(f) = s_1(f) s_2^*(f)$, where * is the complex conjugate and f is the frequency. The relation between the two similar signals s_1 and s_2 may be described as a linear filter, the Wiener filter whose transfer function G is such that $G(f) = <\Gamma_{s_1 s_2}(f)>/<\Gamma_{s_2 s_2}(f)>$, where $<>$ denotes smoothing. The Wiener filter (module and phase) characterizes the relationship between the two signals. The delay between the two signals is measured from the Wiener filter phase and amplitude variations are imbedded in its module. The module of the coherency $C(f)$ is defined by

$$C(f) = | <\Gamma_{s_1 s_2}(f)> | / (<\Gamma_{s_1 s_1}(f)><\Gamma_{s_2 s_2}(f)>)^{1/2}, \quad (2)$$

where coherency quantifies the similarity of the recorded signal (Fig. 1c).

The time delay τ is $\tau = \varphi\, (\Gamma_{s_1 s_2}(f))/2\pi f$, where φ is the phase of the cross-spectrum. τ is actually computed from a weighted least squares adjustment of the series of phase samples. Several choices for the weights are possible, the optimal one in the least square sense being inversely proportional to the error in phase measurement. The uncertainty on

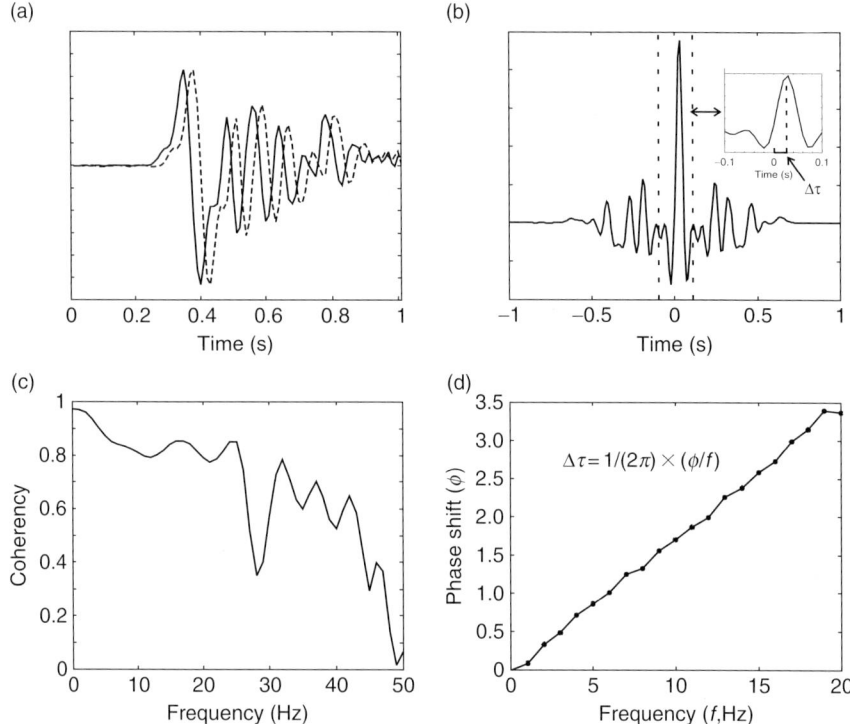

Fig. 1. (a) Example of two P waves recorded by a Hi-net station for doublet 2003/08/25–2005/05/21 (original coordinates of catalog listed in Table 1). (b) Cross-correlation of the two wavelets of Fig. 1a. The delay τ between the two traces is given by the maximum of the cross-correlation. (c) Coherency between the two P-wave traces. (d) Phase of the cross-spectrum of the two wavelets presented in Fig. 1a. The delay is measured from the slope of the phase of the cross-spectrum: $\tau = \varphi$ ($\Gamma_{s1\,s2}(f))/2\pi f$.

the phase is $\sigma^2 = (1/C^2 - 1)/(2BT)$, expressed in radians, where B is the bandwidth of the signal and T is its effective duration. B can be estimated as the width at half-height of the amplitude spectrum and T is the width at half-height of the amplitude of the envelope of the signal window (Mari et al., 1999). The BT product therefore represents the information content of the signal window. After various tests by Frémont (1984), Fréchet (1985) used weights of the form $w(f) = |\Gamma_{s_1 s_2}|(f) \cdot C^2(f)/(1-C^2(f))$. Some clipping has to be applied for high coherency, as this weight diverges for $C = 1$. The uncertainty in time delay measurement is therefore

$$\mathrm{err}(\tau) = \sqrt{\frac{\sum_{i=1}^{n} w_i^2 (\phi_i - af_i)^2}{\sum_{i=1}^{n} w_i^2 f_i^2}}, \qquad (3)$$

where w_i is the weight corresponding to the frequency sample f_i, and a the slope of the cross-spectrum phase. A special attention has to be brought to use a robust phase unwrapping algorithm. Accuracy in the computation of the transfer function $G(f)$ and the coherency $C(f)$ is better when the signals are aligned; standard procedures for computing both $G(f)$ and $C(f)$ therefore involve an iterative alignment of the signals. They avoid the lack in accuracy due to mis-alignment mentioned in Schaff et al. (2004). Longer signal windows provide better accuracy (sample rate being unchanged) in the time delay estimation as far as its physical value remains actually constant over the whole window. Variations of the physical value of the time delay may occur, due to velocity temporal variations or the arrival of seismic waves leaving the source with various directions (see paragraph 7). From USGS Calnet (Northern California seismic array) data, Poupinet et al. (1984) showed the error on the time delays to be as low as ~ 1 ms; computation was made using 1.28-s time windows, the signal (maximum band with 20–25 Hz) being sampled at 100 Hz. Seismic body waves of microearthquakes recorded with a local network may have a small duration and their frequency band may be limited. Anelastic attenuation may therefore make the BT small and imply less precision on the delay.

To compute the delay in the time domain, signals are resampled at a higher rate. The best way is to perform it in the Fourier sense, that is, without modifying its frequency content, by zero-padding the real and imaginary parts of the FFT in the frequency domain and retransform it to the time domain. This step may be applied to the signal FFT or to the cross-spectrum. The delay between $s_1(t)$ and $s_2(t)$ being the delay of the maximum of the envelope of the cross-correlation, it can be measured with a precision better than the original sampling rate. Schaff et al. (2004) showed that correlation measurements were better than travel-time differences, and were usable for relocation of Calaveras fault microearthquakes even for interevent separation distance up to 2 km and correlation coefficient down to 70%.

3.2. Cross-Spectral Moving Window or Cross-Correlation
     Moving Window Technique

A delay can be computed for the entire seismograms, without considering that the record is the sum of many seismic phases following various ray paths and arriving at different times. Phases like P, S, or the coda are physical entities that are independent from each other and their delays may differ. Poupinet et al. (1984) used a moving window technique (CSMW, cross-spectral moving window) to measure the delay of the different seismic phases. Moving window spectrum or multiple filtering are standard techniques in surface waves studies. The analysis can be performed in the frequency domain or in the time domain. We get one delay for a given window and move the window by constant number of samples along the two seismograms. The operation is repeated along the entire length of the seismograms: a curve delay versus lapse time is plotted. In the absence of correlated signals, like before the P arrival time or when the coda becomes smaller than the noise, the delay is erratic. Despite the fact that they are computed by correlating different waveforms, delays remain remarkably stable when the amplitudes are larger than the ambient noise. This may be taken as an independent estimate of the good precision of the measurement (see Fig. 2). Matsumoto et al. (2001), Nishimura et al. (2000, 2005), and Peng and Ben-Zion (2005) have presented excellent applications of this technique on recent data.

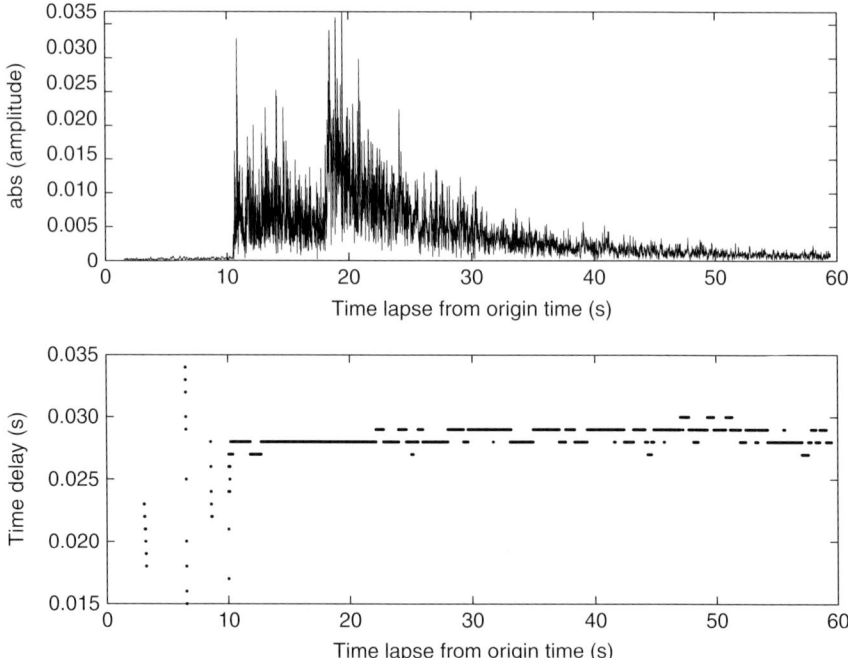

FIG. 2. Application of the cross-correlation moving window technique (CCMW) to the vertical components of Hi-net station N, EDSH, and doublet 2003/08/25–2005/05/21. The envelope and delay are plotted as a function of lapse time. The two traces were aligned using the origin times listed in the Hi-net catalog (Table 1). The delay between the two traces is not zero; there is a small time shift (~0.02–0.03 s) in the difference between the catalog origin times. The stability of the delay as a function of lapse time reflects the fact that there was no significant coda velocity change between 2003 and 2005.

4. RELOCATING DOUBLETS FROM P AND S TRAVEL-TIME DELAYS

4.1. Double-Difference Location

In a standard earthquake location, we measure the arrival times of P waves (and S). Then, we compute the sum of the square of the differences between the observed travel times and the computed travel times for different inputs. The hypocenter and origin time minimize this misfit function: we obtain an absolute location. In a doublet study, the measurements are P delays (or S delays), that is, differences in arrival times between the two events. The location is found by minimizing the square of the differences between measured delays and computed delays. Therefore, differences of differences—double difference—are computed. The location is relative: one event is positioned with respect to the other. There are various techniques for relocating doublets, but they usually apply as far as the interevent distance is small regarding the hypocentral distance. The simplest method is to perform a grid search of the minimum of the misfit function (Poupinet et al., 1984). This technique is not the most computationally efficient; linear inversion is a better tool. However, grid search is still valuable in heterogeneous media, when nonlinearity could pose numerical problems. Fréchet (1985) proposed a simple linear approach. He considered a location difference vector $\mathbf{r} = (\partial x, \partial y, \partial z)$ and the slowness vector $\mathbf{s}_k$ (whose module is supposed to be constant around the hypocenters). The delay is the dot product of the slowness vector by the relative position vector

$$\tau_k = \mathbf{s}_k \cdot \mathbf{r}, \tag{4}$$

For a doublet, the later equation leads to a set of n equations (n = number of stations) with four unknowns, $\partial x, \partial y, \partial z$, and Δh_o, the origin time difference minus the arbitrary time shift introduced for delay computation,

$$\tau_k = \Delta h_o + (\partial x \sin A_k \sin I_k + \partial y \cos A_k \sin I_k + \partial z \cos I_k)/V, \tag{5}$$

where A_k and I_k are respectively the azimuth and the take-off angle of the wave vector and V is the velocity around the hypocenters. Fréchet (1985) solved these equations by a singular value decomposition technique. Accuracy in the computation of double-difference location increases as the square root of the number of similar events contained in a multiplet, as far as the probability density function of the time delay is gaussian (Got et al., 1994). To avoid the inversion of huge matrices, Got et al. (1994) directly built the Hessian matrix without storing the derivative matrix and used Cholesky decomposition for solving the weighted normal equations. This approach is linear and limited to the case where interevent distance is small compared to the hypocentral distance: all rays leaving the sources and reaching one station have a similar wave vector. It is, however, fast and accurate and convenient. In a hydraulic injection experiment with seismograms sampled at 20,000 Hz, Fréchet et al. (1989) achieved a precision of a few centimeters in relative distance between tiny magnitude events. The algorithm has been generalized to the case of extended multiplets by relocating them progressively (Got et al., 1994). Jordan and Sverdrup (1981) wrote a more general algorithm and used it to relocate teleseismic clusters. Waldhauser and Ellsworth (2000) generalized the linear approach to distant events, each ray being characterized by a wave vector. This allows the computation of the geometric center of the cluster if event separation is sufficient and if the modeling of P (and S) velocities is correct. Wolfe (2002), Menke and Schaff (2004), Michelini and Lomax (2004), and Monteiller et al. (2005) discussed some properties of double-difference locations. Zhang and Thurber (2003) and Monteiller et al. (2005) applied double-difference location using 3D tomographic models. Unfortunately, doublets used for monitoring temporal variations are made from very similar events which are extremely close; they do not allow the accurate determination of their absolute position.

4.2. Two Synthetic Examples with IASP91 Travel Times

Let us present the delays computed for two examples of synthetic doublets. Travel times of body waves are listed in propagation tables like IASP91 or are computed. The output formats of the software "ttimes" (Kennett et al., 1995) were modified to get a precision of 0.1 ms. We used the distribution of stations of the Japanese Hi-net array (Obara et al., 2005), and the hypocenter is that of the first event listed in Table 1. The first example is a shallow doublet (Fig. 3a) at a depth of 5 km and the second example is a deep doublet (Fig. 3b) at a depth of 65 km. Both pairs of events are 15 m apart on the horizontal plane and 15 m apart in depth. The circles in Fig. 3a and b are the theoretical

TABLE 1. Hypocenters of an excellent doublet recorded by Hi-net array the in Japan

Date	Origin time	Latitude	Longitude	Depth	Magnitude
2003/08/25	06:08:31.78 (JST)	35.686 N	140.127 E	65.94	2.2
2005/05/21	04:20:31.39 (JST)	35.683 N	140.127 E	65.10	2.4

delays. Figure 3a exhibits two distinct sinusoidal curves: one sinusoid corresponds to stations at short distance ($<0.8°$) and the other to stations at large distance ($>0.8°$). Figure 3b shows more dispersed delays but a single sinusoidal pattern remains. How can we explain these simple patterns? In a distance interval where apparent velocity is constant, the delay τ_k is a sinusoidal function of azimuth; the amplitude of the sinusoid is related to the distance between the two hypocenters, r_{12}, and its phase to the azimuth of the vector joining the two events, az_{12}:

$$\tau_k = r_{12} * (\partial \mathbf{TP}(\Delta_k, h)/\partial \Delta) * \cos(az_k - az_{12}) + \partial z * (\partial \mathbf{TP}(\Delta_k, h)/\partial h) + \Delta h_0. \quad (6)$$

Δ_k and az_k are the distance and azimuth to station k. $r_{12} * \cos(az_k - az_{12})$ is the projection of r on the unit horizontal vector toward station k. ∂z is a depth change and Δh_0 the difference in origin times. $\partial \mathbf{TP}(\Delta_k, h)/\partial \Delta$ and $\partial \mathbf{TP}(\Delta_k, h)/\partial h$, the distance and depth derivatives of the hodochron $\mathbf{TP}(\Delta, h)$ are both standard outputs of travel-time computations. For a simple velocity model, $\partial \mathbf{TP}(\Delta, h)/\partial \Delta$ or its inverse c_k, the apparent velocity, remains nearly constant on large distance intervals. For crustal events (Fig. 3a), the P_g has an apparent velocity of $\sim$6 km/s and P_n $\sim$8 km/s. The sinusoid at short distance corresponds to P_g and the sinusoid at longer distance to P_n. The average offset from the zero base line is due to the change in depth ∂z. Also an error in origin times causes an offset in the base line of delays. For a deep event, the apparent velocity is varying continuously and becomes constant at large distance, so the sinusoidal pattern appears better for long distance stations; the scatter of delays is larger at short distances. These synthetic examples show that we can get the location difference vector from the delays versus azimuth graph. Difference in origin times or depth change can be retrieved from the offset of the base line with respect to zero: we subtract the mean (or the median) delay. The amplitude and phase of the sinusoid at large distances give the amplitude and azimuth of the epicenter difference vector.

4.3. Possible Technical and Intrinsic Difficulties

Any instrumental error in the database affects the precision of delays measurements. Clock precision should be $\sim$20–100 times better than the sampling rate, because the cross-correlation achieves a precision 10 times better than the sampling rate. Large clock errors, for instance, 0.01 s, are easy to pinpoint, but errors which are of the same size as the variance of travel times (induced by 3D structure fluctuations) are nearly impossible to detect. Mistakes like the labeling of components or a change in the transfer function of a station can also be detected (see, e.g., Rubin, 2002). In a study of temporal change, we compute residuals of P and S delays, that is, the difference between delays and theoretical delays; we cannot use raw measurements of P and S delays. The precision of residuals depends on the precision of the velocity model; our frequent limitation in model

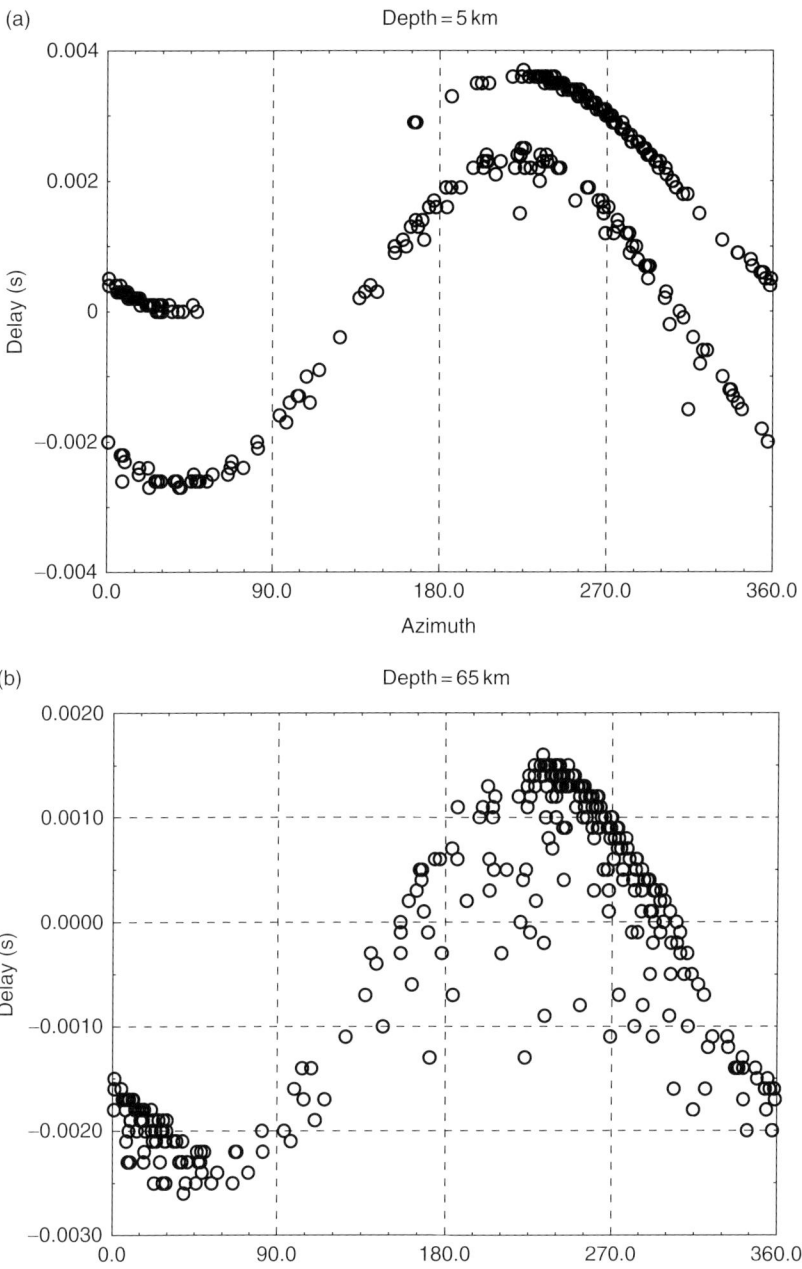

Fig. 3. Theoretical delays for two doublets. (a) A crustal doublet at a depth of 5 km. (b) A deep doublet at a depth of 65 km. The locations are the first epicenter listed in Table 1. Both pairs of events are 15 m apart on the horizontal plane and 15 m apart in depth. The station have the geographical distribution of the Hi-net Japanese array. The travel times are computed with IASP91 tables.

description limits our capacity to detect temporal changes from non perfect doublets. In case of a very good doublet, the mislocation vector is nearly null and the observed delays are the basic data containing directly the information on temporal travel-time changes. The search for very good doublets is a must for tracking temporal variations.

5. AN EXAMPLE OF OBSERVED DELAYS: AN EXCELLENT DOUBLET IN JAPAN

Figure 4 is a map of S–P delays for doublet 2003/08/25–2005/05/21 (hypocenters from the catalog listed in Table 1) recorded by the Hi-net Japanese array. The delays are in millisecond and are extremely small. The distance between the two events is of the order of a few meters. We should keep in mind that these two events originate at a depth of 65 km.

Figure 5 is an azimuthal plot of P and S raw delays for the same doublet. Residuals of delays are similar to raw measured delays when the two events are at the same location. The variance of P and S delays is of the order of 3–4 ms.

Let us consider some typical cases of travel-time changes. If velocity is changing in a small area beneath contiguous stations, and if the relocation process and the instrumental changes do not completely erase the anomaly, we should detect the change in the residuals. Like in any hypocenter search, the spatial distribution of stations is a key ingredient. Temporal changes may be "mapped" into location changes. For instance, an isotropic change in velocity in the entire region will be essentially mapped in the origin

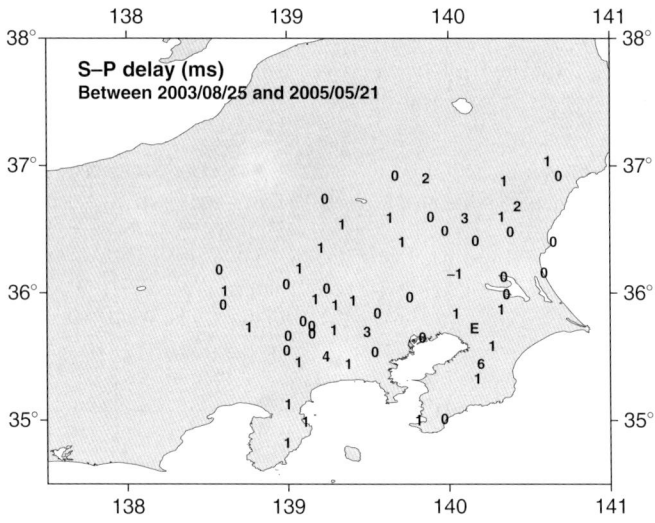

FIG. 4. Map of S–P delays in millisecond for doublet 2003/08/25–2005/05/21 recorded by the Hi-net Japanese network. The location of the doublet is marked by E on the map. The Hi-net timing precision is given as much smaller than 1 ms. The small size and the spatial distribution of delays show that the two events have nearly the same hypocenter location within a few meters. Notice that both events originate at a depth of 65 km.

MONITORING TEMPORAL VARIATIONS OF PHYSICAL PROPERTIES 383

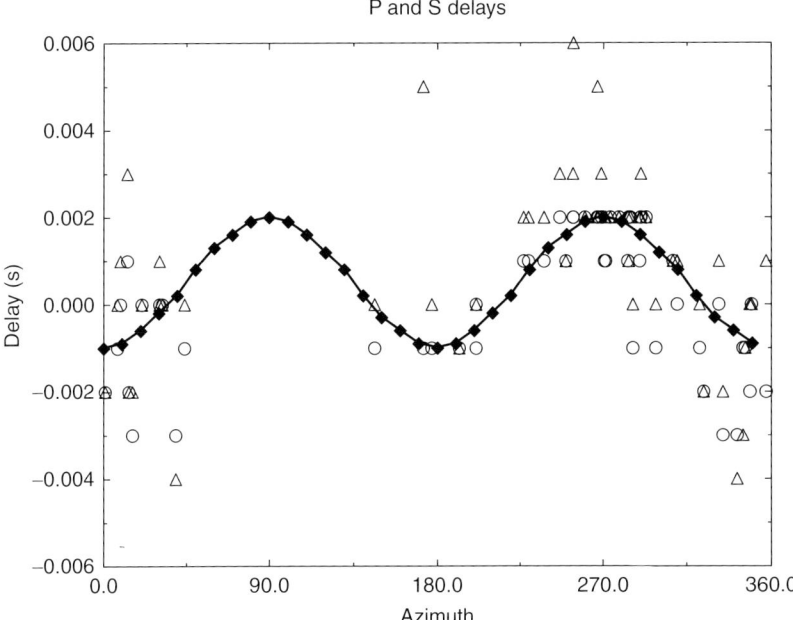

FIG. 5. P–S delays in millisecond as a function of azimuth for doublet 2003/08/25–2005/05/21 recorded by the Hi-net array. The open circles correspond to P and the triangles to S delays. The two events are distant by few meters and the sinusoidal hypocentral correction is very low amplitude compared to the variance of the data. An azimuthal pattern is observable. The squared curve shows the theoretical pattern expected from a temporal variation in anisotropy: it shows a sin (2×Azimuth) term and not the sin (Azimuth) term related to mislocation.

time and depth difference. A large azimuthal gap in stations coverage will prevent detection of some "one-sided" temporal changes: they would be equivalent to a location difference vector.

Despite an uneven coverage in azimuth, Fig. 5 exhibits a sin (2 azimuth) pattern. Clock errors cannot cause the observed pattern because, according to the manufacturer, the time precision of Hi-net GPS clocks is significantly better than 1 ms. What would be the effect of a stress change in the region? Velocity is dependent on stress (Nur and Simmons, 1969); a change in stress induces a small directional variation in velocity. A temporal change in anisotropy gives a sin (2 azimuth) pattern in opposition with the sin(azimuth) pattern related to location difference. The squares in Fig. 5 show the delay expected when anisotropy has changed between the dates of occurrence of the two events of a doublet. Notice that this is a very small effect that can be hidden by the mislocation pattern when the two events of a doublet are not exactly similar. Also, seasonal variations in the weathered layer can be of the order of a few milliseconds and should be checked before any firm conclusion.

More complex simulations of a scenario in which only a small region is affected by such a change in anisotropy can be computed. The detection of anisotropy related changes is important because strain rate at depth could be measured from precise seismic observations. Several active seismic experiments have a similar goal; shots or a continuous vibration sources are activated at regular time interval, and a transit time is monitored as a function of time. One limiting factor for this type of experiments is that the weathered surface layers may be sensitive to periodic seasonal effects.

Furumoto *et al.* (2001) repeated explosions in the Kanto-Tokai region and found travel-time changes as large as 10 ms over a 10-year period. Nishimura *et al.* (2005) repeated shots in the vicinity of Iwate volcano and reported a 1% velocity decrease associated with a M6.1 event in 1998 followed by a partial recovery of velocity until 2002.

6. Slope of the Delay in the Coda and the Measurement of S-Velocity Temporal Variation

According to Aki (1969), coda waves are essentially S-wave energy backscattered from randomly distributed inhomogeneities in the volume surrounding the source and the station. The first remarkable observation is that the two records of a good doublet are nearly identical for a very long duration; randomness generates exactly the same signal. Poupinet *et al.* (1984) found experimentally that delay versus lapse time curves are often straight lines but exhibit a slope, null, positive, or negative: the delay increases or decreases proportionally to lapse time following the arrival of S. One possible explanation is that average S velocity changes in the volume between the two dates of a doublet. In the

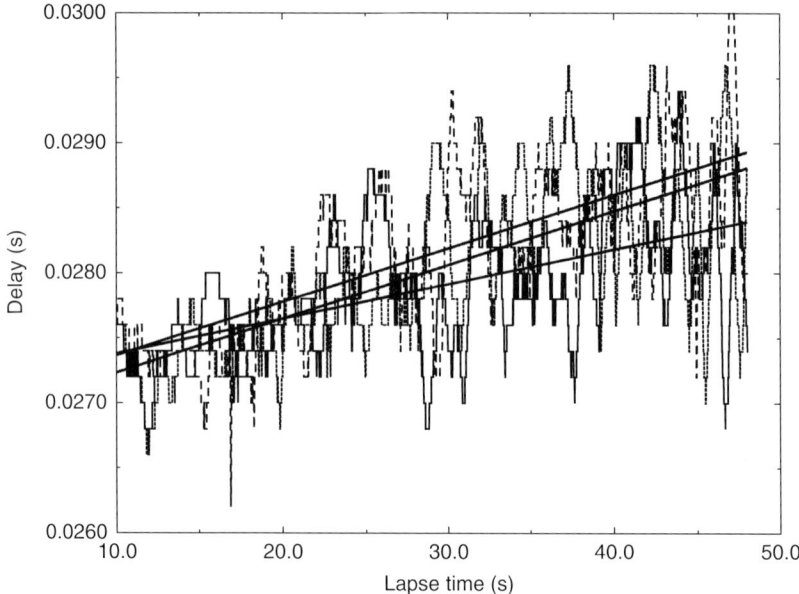

Fig. 6. Delay versus lapse time curves for Hi-net station N, NRTH, and for doublet 2003/08/25–2005/05/21. In a simple model, α, the slope of these curves can be related to a temporal change in S velocity by the relationship: $\alpha = -\Delta V_S/V_S$.

hypothesis coda waves are backscattered S waves, a homogeneous V_S change will induce a delay in the coda at lapse time t, $\tau(t) = -t\, \Delta V_S/V_S$. The slope α of the delay versus time is proportional to the S velocity change: $\alpha = -\Delta V_S/V_S$. The lapse time t on which $\tau(t)$ is measurable being as large as one or several minutes, the precision on α can reach $\sim 10^{-4}$ or a few 10^{-5}. Therefore, we should be able to detect an average change in S velocity in a large region with a precision better than 10^{-4}. This is an elementary type of coda interferometry (Snieder et al., 2002). The measurement of the slope is not dependent on the precision in timing; it would be erroneous if the clock was drifting over a few 10 s. One important question is to know if the coda is generated in the deep crust; Phillips and Aki (1986), Dainty and Toksöz (1990), Mayeda et al. (1991, 1992), and Koyanagi et al. (1992) argue that the coda is essentially generated in the near vicinity of stations so that any temporal change may be very superficial. Figure 6 shows an example of a delay versus time lapse curve for doublet 2003/08/25–2005/05/21. The average slope of the three lines measured on the three components of station N, NRTH, is $\sim 3 \times 10^{-5}$ and it could correspond to a temporal S velocity change $-\Delta V_S/V_S$.

From simple synthetic seismogram computations, Poupinet et al. (1996) illustrated more complex patterns of coda delay configurations due to simple changes in the deep structure of volcanoes.

7. Possible Artifacts in $\Delta V_S/V_S$ Measurement: Arguments from the Coda of Spatial Doublets

Studying temporal variations in the coda implies to know whether variations of the delay as a function of lapse time can occur in the absence of temporal variations of crustal properties. Got and Coutant (1997) studied a 71-event cluster recorded in south flank of Kilauea volcano (Hawaii) by the HVO (Hawaiian Volcano Observatory) seismic network from 1979 to 1983 and selected from this cluster 27 subclusters forming 83 doublets occurring during 1-week periods or less (see Fig. 7a). Such short periods do not contain any eruption or intrusion, in such a way the medium properties may be assumed as constant during this time. Cross-spectral time delays were computed along the entire seismograms, using 1.28 s signal windows sampled at 100 Hz. Relative relocations were performed from the P-wave time delays with an average accuracy of 50 m horizontally and vertically. Events were found to be located on a subhorizontal decollement plane beneath the Kilauea south flank (see, e.g., Got et al., 1994), and cover a 1 km × 1 km wide, 300–500 m high, volume. They may be considered as an (earthquake) array, playing at depth the same role as seismic station networks at the earth surface. Given the short average interevent distance in the cluster, having regard to the hypocentral distance range (10–100 km), waves were assumed to leave the hypocenters with the same slowness vector for each station. The complete set of 83 relocated event pairs was used to infer the slowness vector direction for each station as a function of lapse time from the time delay computed for each signal window and the event relative positions. This computation was performed from the onset of the P wave up to four times the S travel time. For each station and each signal window, a linear system was solved:

$$r_{ij} \cdot s^k = \Delta T_{ij}^k, \tag{7}$$

where $\mathbf{r}_{ij}$ is the relative position vector for events i and j, $\mathbf{s}^k$ is the slowness vector for station k and $\Delta T_{ij}^k = \delta t_{ij}^k - \delta t_{orij}$, where is the time delay computed between each signal for a given time window for the station k and the events i and j, and is the error in origin time difference.

The previous equation may be written, in matrix form

$$\mathbf{R}\mathbf{s}^k = d^k, \tag{8}$$

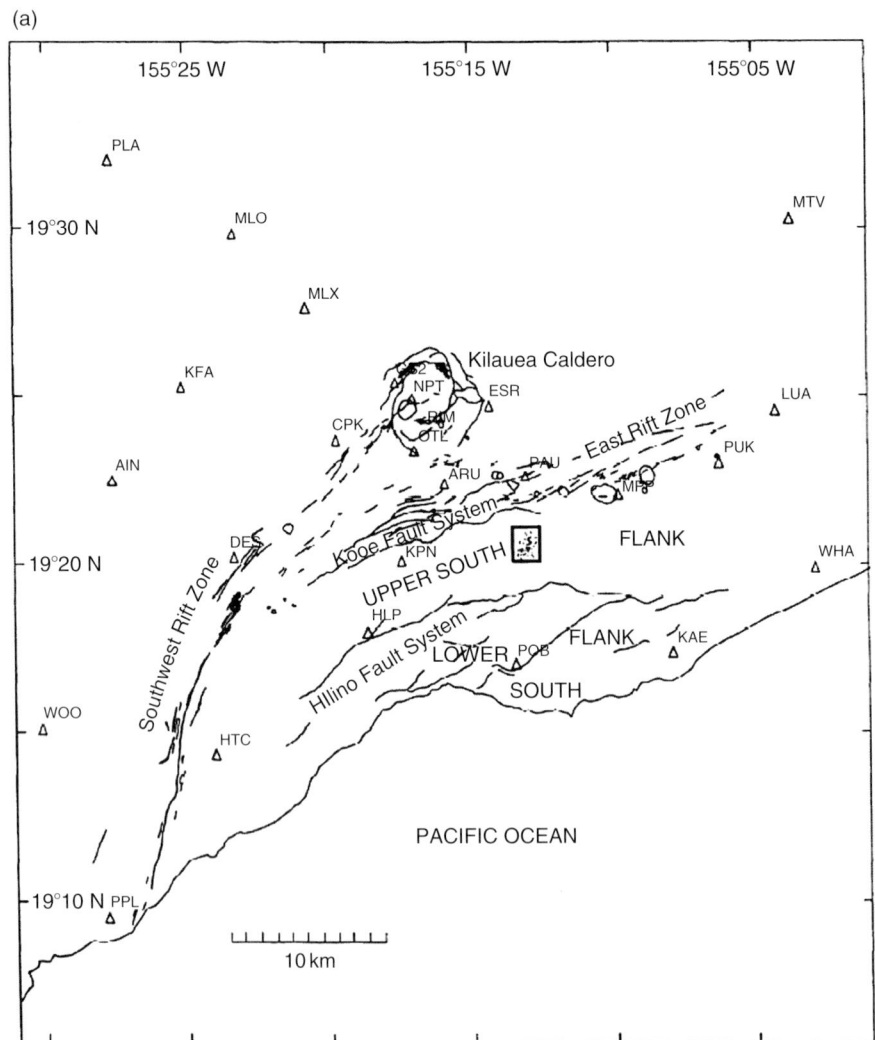

FIG. 7. (a) Map of the USGS-HVO (Hawaii Volcano Observatory) stations (open triangles) on Kilauea volcano, Hawaii, and main geological features.

(Continued)

MONITORING TEMPORAL VARIATIONS OF PHYSICAL PROPERTIES 387

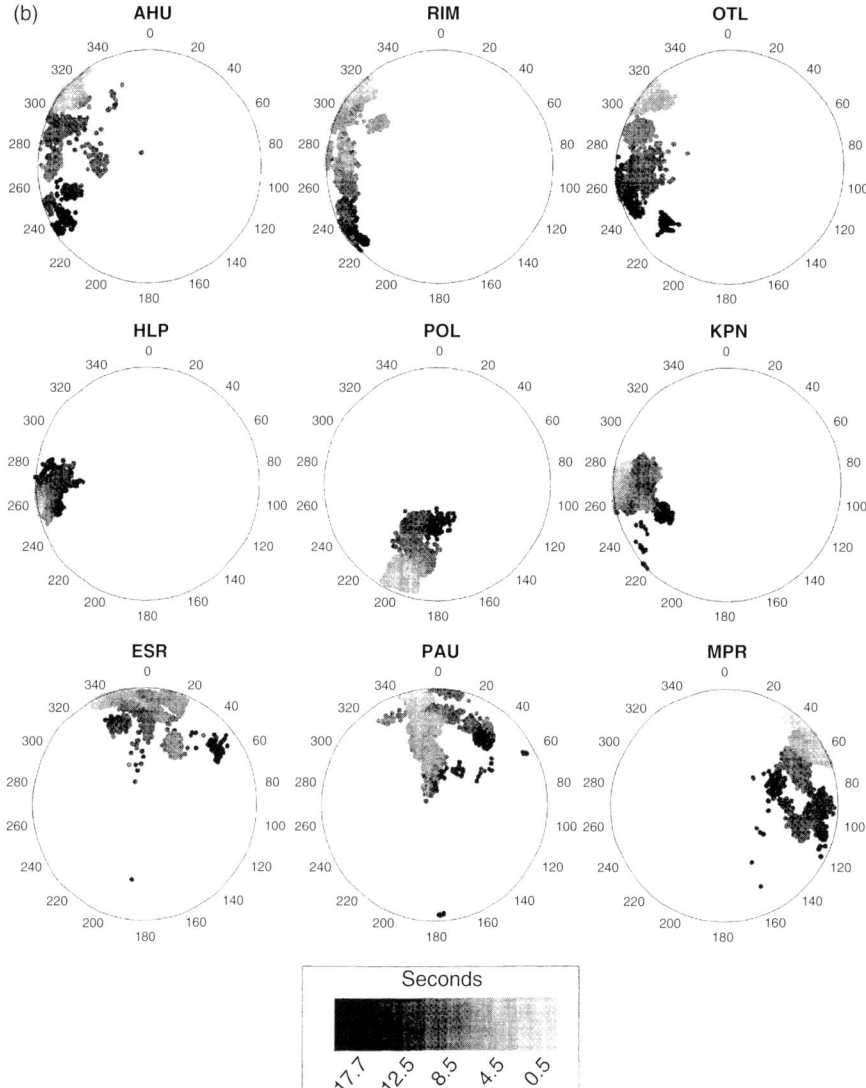

FIG. 7. (b) Stereogram showing the direction of arrival of energy in the early coda of several seismic stations, from Got and Coutant (1997).

where **R** is the matrix containing the relative positions r_{ij}, $\mathbf{s}^k$ is the slowness vector for the station k and $\mathbf{d}^k$ is the time delay vector containing, whose solution is

$$s^k = \left(R^T C_d^{-1} R\right)^{-1} R^T C_d^{-1} d^k, \tag{9}$$

where $\mathbf{C}_d$ is the data covariance matrix.

Time delay computation was performed using the moving window technique from the P-wave onset to the end of the coda. Results of this computation show variations of the delay as a function of the lapse time, even for event pairs that are very close in time, for which there is a few probability to have a temporal variation of the medium properties. To explore the origin of such variations, the direction of the slowness vector was computed for each signal window. The computation converges easily, showing that for each signal window, the whole set of time delays computed for the 83 event pairs is explained by only one slowness vector. It proves that purely geometrical propagation effects may explain the variation of the time delay in the coda. The order of magnitude of this time delay variation is controlled by the interevent distance and the seismic velocity around the hypocenters. Results of the slowness vector computation (Fig. 7b) show that the seismic energy dominating each window leaves the hypocenter in directions close to the station one, for lapse times up to $4t_s$. For some stations (e.g., stations AHU, RIM, OTL in Fig. 7), the emission remains coherent and shows a rotation of the azimuthal plane with the propagation time. For others (e.g., stations HLP, POL, KPN), the energy remains very concentrated in the direction of the station and reveals reverberation close to the station: most of the travel time is spent in the most superficial and slow layers, rather than at depth. For other stations (e.g., ESR, PAU, MPR), the energy is more randomly distributed through wave packets, though it does not cover the whole space around the source. Similar conclusions have been reached for comparable geological settings by numerous authors and various methods [see, e.g., Phillips and Aki (1986), Dainty and Toksöz (1990), Mayeda et al. (1991, 1992), Koyanagi et al. (1992), Dodge and Beroza (1997), Rubinstein and Beroza (2004)]. Notice that Dodge and Beroza (1997) used beam forming from an array of similar earthquakes to compute the slowness power spectrum of coda waves.

This study shows that for small-magnitude earthquakes, seismic energy following S waves may never reach the random scattering regime, which may be reached at longer propagation times. Slow superficial layers exhibiting strong impedance contrasts with surrounding or underlaying formations may contribute strongly to build the early coda. Seismic energy coming from these slow superficial layers may mask scattered seismic energy coming from depth. A suitable way to avoid contamination of the deep scattered wave field by the superfical reverberation is to use data recorded by buried seismic station networks. Hi-net seismic network offers from this point of view extremely favorable characteristics, as most of the stations are buried at depth of some (a few) hundred meters and may be located below the more unconsolidated sediment layers responsible for the superficial reverberation. Another important conclusion is that the early coda bears the signature of the actual source radiation, recorded at the surface; this information clearly appears by studying the time delays computed in the coda of spatial earthquake doublets. Such a conclusion was already reached by Got et al. (1990) and Got and Fréchet (1993), by studying amplitude ratios of earthquake doublets in Central California. Got and Coutant (1997) have shown that typical relative variations of the time delay for spatial doublets were in the 0–1% interval. An interevent distance of 100 m and S velocity $\sim$3 km/s give a time delay of $\sim$0.035 s in the direction of the relative position vector. A station located in the azimuth of the doublet may show a time delay decrease in the coda in the range 0.3–0.7%, that is, typically in the range of the variations that may be measured by using temporal doublets. A practical consequence is that the study of temporal variations using time delays or amplitude of microearthquake doublets, which has been undertaken again recently (e.g., Li et al., 2007), should use extremely close events (some meters—less than a few tens of meters for the usual S-wave velocities encountered in the upper crust), to avoid a significant time delay contribution due to purely geometrical propagation effects.

8. Search for Temporal Variation of S-Wave Splitting

S-wave anisotropy is widespread. Anisotropy is related to layering and to the distribution of cracks and fluids inside rocks; it is dependent on stress (see, e.g., Crampin, 2001). The S-wave train is composed of two S waves, a fast one and a slow one, separated by a small time delay. The usual procedure to detect S-wave splitting is to plot the horizontal S-particle motion: the particle motion displays a linear segment at the beginning of the S-wave train and then becomes chaotic. The fast axis is the direction of the initial S linear segment and the slow axis is perpendicular to it. The delay $\tau_{\text{splitting}}$ between the fast wave and the slow wave and φ the polarization angle define S-wave splitting. Let us suppose that the S-wave train is recorded on the radial and transverse components S_{NS} and S_{EW}. Considering rays close to the vertical, we have the relationships

$$S_{\text{fast}}(t) = S_{\text{NS}}(t)\cos\varphi - S_{\text{EW}}(t)\sin\varphi,$$

$$S_{\text{slow}}(t) = S_{\text{NS}}(t - \tau_{\text{splitting}})\sin\varphi + S_{\text{EW}}(t - \tau_{\text{splitting}})\cos\varphi, \qquad (10)$$

In the case of vertical S waves, the horizontal components are rotated and their cross-correlation is computed for different polarization angles and delays. The maximum of the cross-correlation gives φ and $\tau_{\text{splitting}}$. If a change in anisotropy occurs on the S ray path between the two dates of a doublet, the new slow wave is

$$S'_{\text{slow}}(t) = S_{\text{NS}}(t - \tau_{\text{splitting}} - \partial\tau_{\text{splitting}})\sin\varphi + S_{\text{EW}}(t - \tau_{\text{splitting}} - \partial\tau_{\text{splitting}})\cos\varphi, \qquad (11)$$

where $\partial\tau_{\text{splitting}}$ is the temporal change in splitting. This delay can also be defined as

$$\partial\tau_{\text{splitting}} = \tau_{\text{slow}} - \tau_{\text{fast}}, \qquad (12)$$

where τ_{fast} is the delay between the two fast component S records, and τ_{slow} the delay between the two slow component S records. Both τ_{fast} and τ_{slow} are measured by cross-correlating very similar waveforms and are very precise measurements. The phases between the N–S and E–W components will also vary with time.

Aster et al. (1990) studied shear wave splitting on the records of the Anza array in southern California. They measured S-wave splitting on individual events with an automated processing and checked the temporal variations of splitting on changes in multiplets waveforms. Their very precise analysis shows the complexity of S-wave splitting measurements and the ambiguity between sources, local effects and large scale temporal changes. Bockelman and Harjes (2000) used a doublet technique to observe velocity changes after the injection of water in the KTB deep borehole in Germany. Saiga et al. (2003) found evidence for temporal changes in anisotropy in Japan related to an earthquake in the Tokai region. Peng and Ben-Zion (2005) studied spatiotemporal changes in anisotropy in the aftershocks sequences of the 1999 M7.4

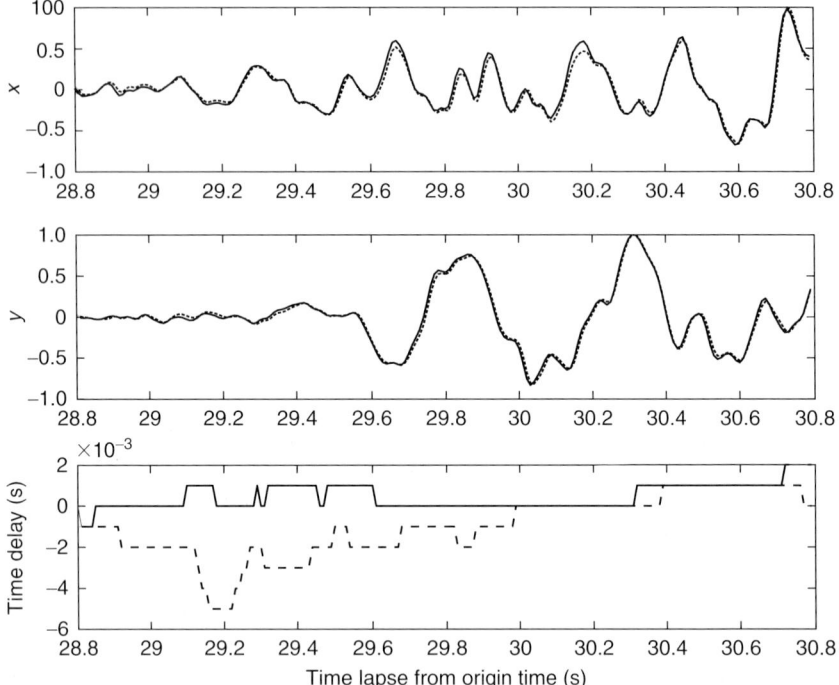

FIG. 8. Zoom of the delay versus lapse time for the two horizontal components of the S waves for Hi-net station N, HGAH, and doublet 2003/08/25–2005/05/21. The upper traces are the N–S and E–W components. In the bottom diagram, the continuous trace is the delay between the two N–S components and the dashed trace the delay between the two E–W components. A delay of ~1–2 ms is observable on one component at the beginning of the S-wave train but not on the other component. This may correspond to a variation of S-anisotropy between the two dates of the doublet.

Izmit and M7.1 Düzce earthquakes in Turkey; temporal change in splitting delay associated with the Düzce earthquake are smaller than 2%. In a controlled source experiment using the ACROSS vibrator, Ikuta and Yamaoka (2004) found that the S velocity between the surface and a depth of 800 m slowed down by 0.4% and 0.1% after two M>6.4 distant earthquakes and then recovered its initial value.

Figure 8 zooms on the horizontal components of one Hi-net station for doublet 2003/08/25–2005/05/21 and shows a possible variation in anisotropy between 2003 and 2005: the delay is different by 2 ms at the beginning of the S wave than at the end; this could correspond to a variation of anisotropy of about 3% in one station of Hi-net array.

9. Search for Temporal Variation of Coda Attenuation

After Chouet's (1979) pioneered temporal Q-coda studies, Q coda has been considered as a key parameter for monitoring physical changes in the crust [see Sato (1988) for an early review; Hiramatsu et al. (2000) for an application to Kobe earthquake]. Jin and Aki (1993) presented evidence for Q-coda temporal variations that could be correlated with changes in the seismicity rate. More recently, Aki (2004) strongly advocated Q-coda decay and seismicity rate in a certain magnitude interval as essential seismic precursors to large earthquakes. Most Q-coda studies are statistical analysis of the rate of decay of the amplitude of coda waves using a large number of earthquakes, and temporal comparisons involve sets of earthquakes having different source parameters. Another approach using doublets is tempting, but it is limited by the smaller number of data. Starting from the expression given by Aki (1969) and Chouet (1979), the logarithm of the spectral ratio of the amplitudes of the coda of two events in station i is

$$\text{Log}(SR_i(f, t)) = -\pi f t \Delta Q / Q^2 + SR_{i0}, \tag{13}$$

where f is frequency, t lapse time, Q coda Q, and ΔQ the change in coda Q. For one station, SR_{i0} is a constant. So for a given frequency and ΔQ, the log of the spectral ratio is proportional to lapse time. Got et al. (1990) processed doublet data to monitor Q^{-1} temporal changes in Coyote Lake region in California with this model; they did not find any significant Q-coda temporal change. Moreover, they showed that there are many possible variations in source radiation (source size and rupture velocity) that may produce amplitude changes which are not related to temporal changes, even when using coda waves from highly similar and close events (located within some meters). Coda waves showing such coherent source effects belong to the early coda, which constitutes most of the coda of the small-magnitude earthquakes studied: in that case it seems that scattering regime is not reached. Therefore, small-magnitude earthquake doublet amplitude ratios should be more surely used to infer the absence of temporal variations rather than to find temporal variations. Beroza et al. (1995) performed a detailed analysis of 21 doublets recorded between 1978 and 1991 in the Loma Prieta earthquake region. They placed an upper bound of 5% on preseismic, coseismic, and postseismic change in coda Q in the epicentral region of Loma Prieta.

10. "Virtual Doublets" Computed by Cross-Correlating Seismic Noise

As mentioned previously, very good doublets (i.e., having similar source processes occurring within a few meters or tens of meters) are seldom and restricted to very active tectonic or volcanic environments. This limitation can be partially overcome by using explosions as repetitive seismic sources (Li et al., 2006). However, this *active* monitoring technique suffers from the difficulty to repeat explosions in remote and hardly accessible terrains (volcanoes). Here, we present a new technique of real-time passive monitoring using cross-correlations of ambient seismic noise as repetitive seismic sources.

The basic idea is that a cross-correlation of random seismic wave fields such as coda or noise recorded at two receivers yields the Green function, that is, the impulse response of the medium at one receiver as if there was a source at the other (Weaver and Lobkis, 2001; Campillo and Paul, 2003; Campillo, 2006; Wapenaar, 2006). This property has been used for imaging the crust at regional scales (Shapiro et al., 2005; Sabra et al., 2005; Yang et al., 2007) and, more recently, has been applied to infer the internal structure of the Piton de la Fournaise volcano at La Réunion island (Brenguier et al., 2007a). By computing noise cross-correlations between different receiver pairs for consecutive time periods, we make each receiver to act as a virtual highly repetitive seismic source. Green functions computed for consecutive time periods may therefore be considered as virtual doublets. They can then be used to detect temporal perturbations associated to small velocity changes (<1 %, Stehly et al., 2007).

We applied this method to study the Piton de la Fournaise volcano on La Réunion island (Fig. 9a). We used the continuous seismic noise recorded during year 2006 by 21 vertical short period receivers operated by IPG Paris to compute 210 cross-correlation

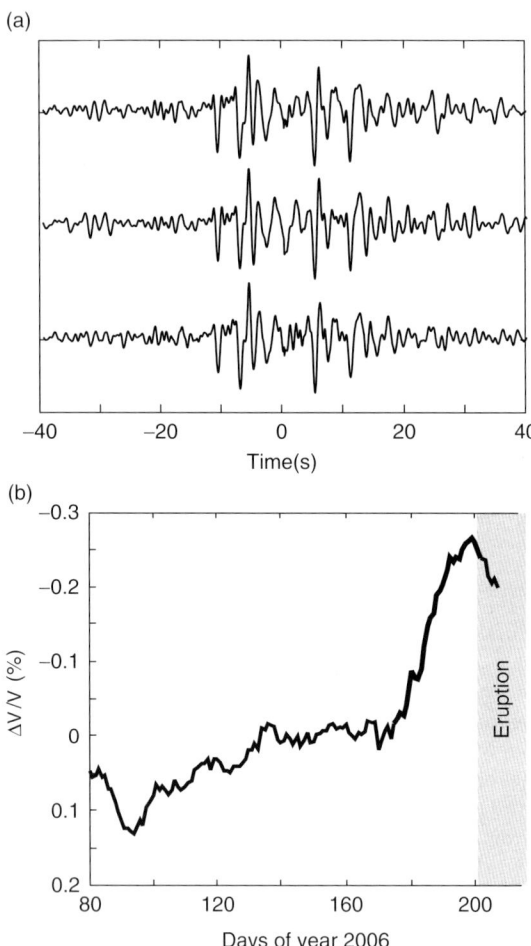

FIG. 9. (a) "Virtual doublets" computed by cross-correlating the seismic noise in two stations on the Piton de la Fournaise volcano (Réunion Island) on different dates (40, 14, and 1 day before an eruption). Notice the excellent similarity of the waveforms. (b) Relative velocity changes before the eruption of July 2006 computed by processing virtual doublets. The separation distance between the two stations is 8 km. The velocity change is equal to minus the slope of the delay versus lapse time in the coda, $\alpha = -\Delta V_S/V_S$.

functions corresponding to all possible receiver pairs. We used the spectral band between 0.1 and 0.9 Hz where the recovered Green functions have been demonstrated to consist of Rayleigh waves that are sensitive to the structure at depths down to 2 km below the edifice surface (Brenguier *et al.*, 2007a). The cross-correlation functions obtained by correlating 18 months of seismic noise are called the reference Green functions. The temporal evolution was then tracked by comparing the reference Green functions with current Green functions computed by correlating the noise from a 10-day-long moving window. We applied the technique of CSMW to measure time shifts between the current and reference Green functions and estimate the relative velocity changes ($\Delta v/v$) by using the technique described in Section 6.

Figure 9b shows the temporal evolution of the relative velocity changes before the eruption of July 2006. The figure shows a clear precursor to the volcanic eruption characterized by a decrease of relative seismic velocities measured with an unprecedent accuracy (0.05%). This precursor starts about 20 days before the eruption and reaches -0.3% of relative velocity change few days before the eruption. This type of precursor was also observed for five other eruptions of the Piton de la Fournaise volcano (Brenguier *et al.*, 2007b). We interpret the observed decreases in seismic velocities as an effect of the dilatation of a part of the edifice resulting from the magma pressurization within the volcano plumbing system similar to observations at Mount Etna (Patanè *et al.*, 2003). This new direct observation of the dilatation of volcanic edifices should thus improve our ability not only to forecast eruptions but also to a priori assess their intensity and environmental impact. Finally, this new technique of passive monitoring using virtual doublets may also be useful in other geophysical, engineering, and geotechnical applications that require nondestructive monitoring of the media.

11. PKP FROM TELESEISMIC DOUBLETS AND THE ROTATION OF THE INNER CORE

A few large magnitude earthquakes have similar waveforms and are recorded at large distance; they are called teleseismic doublets. They are extremely useful for tracking temporal variations deep inside the Earth. We illustrate this with an application on the rotation of the inner core relatively to the mantle of the Earth. Song and Richards (1996) studied the differential travel times of PKP(DF) and PKP(BC) for South Sandwich Islands earthquakes recorded in COL, Alaska. They noticed a change in PKP(DF) travel time as a function of date. Their interpretation is that this change in PKP(DF) travel time is caused by superrotation of the inner core of the Earth relative to the mantle: the inner core would rotate $1°$ per year faster than the daily rotation of the mantle and crust. The very small size of the residual anomaly, 0.3 s for 30 years, led to questions about the robustness of this observation. Poupinet *et al.* (2000) showed that teleseismic doublets should contribute to this debate by improving on the time delay precision. Song and Richards's data were reprocessed by pairs and the best doublets were selected from their data. After alignment of the two records, the relative delays between PKP(DF), PKP (BC), and PKP(AB) are measured by correlation as presented in Chapter 3, this volume. Observed PKP(BC) delay–PKP(DF) delay is plotted versus observed PKP(BC) delay–PKP(DF) delay. Theoretical PKP(BC)–PKP(DF) and PKP(AB)–PKP(BC) are computed with "ttimes" (Kennett *et al.*, 1995) for hypocenters positioned on a 3D grid around the doublet. PKP(AB)–PKP(BC) is proportional to distance but depends also on depth which is not precisely known. These theoretical points form a narrow band—the mislocation band. If PKP(DF) changes significantly with date, a measurement lies outside the mislocation band. Figure 10 is a mislocation diagram for PKP phases from South Sandwich events recorded in COL, Alaska. Full lines show the mislocation band in which the data should be in the absence of inner core rotation. The dotted lines correspond to different PKP(DF) travel-time changes, as expected in the case of inner core rotation. There is no need to relocate the events and to compute observed travel-time residuals. Some controversies followed because phase reversals of core phases may occur and are sometimes difficult to assess. The precision of the delay depends on the similarity of the waveforms and when doublets are not perfect, the polarity of PKP can reverse. Zhang *et al.* (2005) found a very good South Sandwich Islands doublet which shows a temporal change in PKP(DF). Doublets have only been recently used in global seismology and despite their rare occurrence, they have a great potential for detecting temporal variation anywhere inside the Earth. The study of differential travel times of clusters of events could also contribute to an improved mapping of structure at depth.

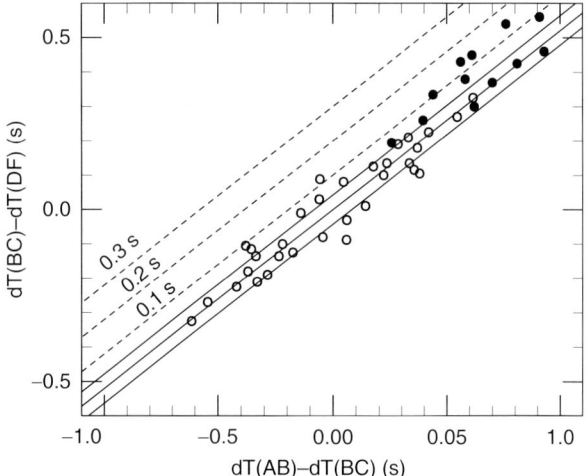

FIG. 10. Mislocation diagram for South Sandwich Islands events recorded in COL, Alaska. The two seismograms of a teleseismic doublet are aligned; the three PKP phase delays, dT(DF), dT(BC), and dT(AB) are measured by correlation. Open and full circles are raw measurements from a pair of events: dT(AB)–dT(BC) on the x axis and dT(BC)–dT(DF) on the y axis. Open circles are short lapse time pairs and full circles are large time lapse pairs (>15 years). In the absence of PKP(DF) temporal travel-time change, the circles should remain close to the central lines (full lines), the mislocation band. We get a mislocation band because locations and depths of the events of a pair are not exactly similar. It is computed from theoretical travel times using IASP91. In case of inner core rotation, the travel time of DF is changing; the measurements should be offset from the mislocation band (dashed lines correspond to a 0.1, 0.2, 0.3 s temporal travel-time change). The distance from the center of the dT(AB)–dT(BC) axis is proportional to the distance between the two events. For instance, our black circles are far apart: they are not very good doublets. These data illustrate the precision necessary for measuring temporal variations of PKP(DF) travel time. Zhang *et al.* (2005) have found a few excellent doublets showing a temporal variation of PKP(DF) travel time.

12. Conclusion

Finding doublets in a digital seismic data bank can be automated by performing a systematic computation of the coherency between events. This computation does not need the determination of accurate arrival times. The basic tools for doublet processing are cross-correlation or cross-spectrum moving window techniques. Delays between P and S waves are input for relative relocation; they are measured with a precision of a few milliseconds on a large array and better on a smaller scale. After relocation, P and S delay residuals are obtained; they may contain information on temporal variations of P and S travel times in the crust. To achieve precision, events must be recorded by a dense seismic array with good azimuthal coverage. With such constraints, in most seismic regions, there are a few high quality doublets. We illustrate a possible change in crustal anisotropy with time in Central Japan. The small size of the residual change supposes that timing is perfect. Clock precision is not presently verifiable; it would be extremely helpful to be able to a posteriori check clocks. To avoid time delay variation in the near coda due to simple geometrical effects, temporal change studies should be restricted to the late coda of very close and highly similar doublets. The late coda contains information on the stability of the medium or on an average temporal change in S velocity: in a simple model where the coda is made from backscattered S energy, the slope of the delay versus lapse time is equal to $-\Delta V_S/V_S$. The rate of decrease of the amplitude of the late coda of highly similar earthquakes allows a precise measurement of temporal changes in Q coda. Natural doublets studies can complement active artificial source experiments like ACROSS; the weathered zone is the site of large changes in velocity so that the presence of both the source and the receiver near the surface makes measurements less sensitive to deep layer properties. The S-wave phase can also be processed to find extremely small temporal changes in S-wave splitting. A systematic and routine processing of doublets and multiplets on the best seismic arrays would certainly improve investigations of temporal changes in crustal properties. The Japanese Hi-net array and its database are very adapted for this kind of study. The use of "virtual doublets" computed by cross-correlating seismic noise is a promising technique: it should provide precise measurements of strain variations in the upper crustal layer, on a large scale, and can provide one of the best tool for forecasting eruption and contribute to earthquake risk assessment.

Acknowledgements

Part of the processing was performed during GP stay in Prof. Haruo Sato's laboratory in Tohoku University, Sendai. GP thanks Professor Sato for his hospitality and for his constant help. We thank Dr. Katsuhiko Shiomi and the Japanese National Research Institute for Earth Science and Disaster Prevention (NIED) for providing the Hi-net data and difficult to access technical information on the array. Dr. Mare Yamamoto and Prof. Takeshi Nishimura introduced GP to Hi-net data processing, and GP thanks them for their unselfish cooperation. Reviews by Dr. William Ellsworth, Prof. Takeshi Nishimura, and Prof. Haruo Sato were essential for improving the manuscript.

References

Aki, K. (1969). Analysis of the seismic coda of local earthquakes as scattered waves. *J. Geophys. Res.* **74**, 615–631.

Aki, K. (2004). A new view of earthquake and volcano precursors. *Earth Planets Space* **56**, 689–713.

Aster, R.C., Scott, J. (1993). Comprehensive characterization of waveform similarity in microearthquake data sets. *Bull. Seismol. Soc. Am.* **83**, 1307–1314.
Aster, R.C., Shearer, P.M., Berger, J. (1990). Quantitative measurements of shear wave polarizations at the Anza seismic network, southern California: Implications for shear wave splitting and earthquake prediction. *J. Geophys. Res.* **95**, 12449–12473.
Battaglia, J., Got, J.-L., Okubo, P. (2003). Location of Long Period events below Kilauea using seismic amplitudes and precise relative relocation. *J. Geophys. Res.* **108**, 2553, doi:10.1029/2003JB002517.
Beroza, G.C., Cole, A.T., Ellsworth, W.L. (1995). Stability of coda wave attenuation during the Loma Prieta, California, earthquake sequence. *J. Geophys. Res.* **100**, 3977–3988.
Bockelman, G.H.R., Harjes, H.-P. (2000). Evidence for temporal variation of seismic velocity within the upper crust. *J. Geophys. Res.* **105**, 23879–23894.
Brenguier, F., Shapiro, N.M., Campillo, M., Nercessian, A., Ferrazzini, V. (2007a). 3D surface wave tomography of the Piton de la Fournaise volcano using seismic noise correlations. *Geophys. Res. Lett.* **34**, L02305, doi:10.1029/2006GL028586.
Brenguier, F., Shapiro, N.M., Campillo, M., Ferrazzini, V., Duputel, Z., Coutant, O., Nercessian, A. (2008). Towards forecasting volcanic eruptions using seismic noise. *Nature Geoscience*, doi:10.1038/ngeo104.
Campillo, M. (2006). Phase and correlation in random seismic fields and the reconstruction of the Green function. *Pure Appl. Geophys.* **163**, 475–502.
Campillo, M., Paul, A. (2003). Long-range correlations in the diffuse seismic coda. *Science* **299**, 547–549.
Chouet, B. (1979). Temporal variation in the attenuation of earthquake coda near Stone Canyon, California. *Geophys. Res. Lett.* **6**, 143–146.
Crampin, S. (2001). Developing stress-monitoring sites using cross-hole seismology to stress-forecast the times and magnitudes of future earthquakes. *Tectonophysics* **338**, 233–245.
Dainty, A.M., Toksöz, N. (1990). Array analysis of seismic scattering. *Bull. Seismol. Soc. Am.* **80**, 2242–2260.
Deichmann, N., Garcia-Fernandez, M. (1992). Rupture geometry from high-precision relative hypocenter locations of microearthquake clusters. *Geophys. J. Int.* **110**, 501–517.
Dodge, D., Beroza, G.C. (1997). Source array analysis of coda waves near the 1989 Loma Prieta, California mainshock: Implications for the mechanism of coseismic velocity changes. *J. Geophys. Res.* **102**, 24437–24458.
Fréchet, J. (1985). *Sismogénèse et doublets sismiques.* University Grenoble, Thèse d'Etat, 207 pp.
Fréchet, J., Martel, L., Nikolla, L., Poupinet, G. (1989). Application of the cross-spectral moving window technique (CSMWT) to the seismic monitoring of forced fluid migration in a rock mass. *Int. J. Rock Mech. Min. Sci. & Geomech. Abstr.* **26**(3–4), 221–233.
Frémont, M.-J. (1984). *Mesure des variations temporelles des paramètres de la croûte terrestre et d'effets de sources par traitement de doublets de séismes.* University Grenoble, Thèse de 3ème cycle, 224 pp.
Frémont, M.-J., Malone, S.D. (1987). High precision relative locations of earthquakes at Mount St Helens—Washington. *J. Geophys. Res.* **92**, 10223–10236.
Furumoto, M., Ichimori, Y., Hayashi, N., Hiramatsu, Y. (2001). Seismic wave velocity changes and stress build-up in the crust of the Kanto-Tokai region. *Geophys. Res. Lett.* **28**, 3737–3740.
Geller, R.J., Mueller, C.S. (1980). Four similar earthquakes in Central California. *Geophys. Res. Lett.* **7**, 821–824.
Gillard, D., Rubin, A., Okubo, P. (1996). Highly concentrated seismicity caused by deformation of Kilauea's deep magma system. *Nature* **384**, 343–346.
Got, J.-L., Coutant, O. (1997). Anisotropic scattering and travel time delay analysis in Kilauea volcano, Hawaii, earthquake coda waves. *J. Geophys. Res.* **102**, 8397–8410.
Got, J.-L., Fréchet, J. (1993). Origins of amplitude variations in seismic doublets: Source or attenuation process. *Geophys. J. Int.* **114**, 325–340.

Got, J.-L., Okubo, P. (2003). New insights into Kilauea's volcano dynamics brought by large-scale relative relocation of microearthquakes. *J. Geophys. Res.* **108**, 2337, doi:10.1029/2002JB002060.
Got, J.-L., Poupinet, G., Fréchet, J. (1990). Changes in source and site effects compared to coda Q^{-1} temporal variations using microearthquakes doublets in California. *Pageophys.* **134**, 195–228.
Got, J.-L., Fréchet, J., Klein, F.W. (1994). Deep fault geometry inferred from multiplet relative location beneath the south flank of the Kilauea. *J. Geophys. Res.* **99**, 15375–15386.
Hamaguchi, H., Hasegawa, A. (1975). Recurrent occurrence of the earthquakes with similar wave forms and its related problems. *J. Seism. Soc. Jpn.* **28**, 135–169 (in Japanese).
Hiramatsu, Y., Hayashi, N., Furumoto, M. (2000). Temporal changes in coda Q-1 and b value due to the static stress change associated with the 1995 Hyogo-ken Nambu earthquake. *J. Geophys. Res.* **105**, 6141–6151.
Ikuta, R., Yamaoka, K. (2004). Temporal variation in the shear wave anisotropy detected using the Acurately Controlled Routinely Operated Signal System (ACROSS). *J. Geophys. Res.* **109**, B09305, doi:10.1029/2003JB002901.
Ito, A. (1985). High resolution relative hypocenters of similar earthquakes by cross-spectral analysis method. *J. Phys. Earth* **33**, 279–294.
Jin, A., Aki, K. (1993). Spatial and temporal correlation between coda Q^{-1} and seismicity and its physical mechanism. *J. Geophys. Res.* **94**, 14041–14059.
Jordan, T.H., Sverdrup, K.A. (1981). Teleseismic location techniques and their application to earthquake clusters in the South-Central Pacific. *Bull. Seismol. Soc. Am.* **71**, 1105–1130.
Kennett, B.L.N., Engdahl, E.R., Buland, R. (1995). Constraints on seismic velocities in the Earth from travel times. *Geophys. J. Int.* **122**, 403–416.
Koyanagi, S., Mayeda, K., Aki, K. (1992). Frequency-dependent site amplification factors using the S-wave coda for the island of Hawaii. *Bull. Seismol. Soc. Am.* **82**, 1151–1185.
Li, Y.-G., Chen, P., Cohran, E.S., Vidale, J.E., Burdette, T. (2006). Seismic Evidence for Rock Damage and healing on the San Andreas Fault Associated with the 2004 M 6.0 Parkfield Earthquake. *Bull. Seismol. Soc. Am.* **96**(4B), S349–S363.
Li, Y.-G., Chen, P., Cochran, E.S., Vidale, J.E. (2007). Seismic velocity variations on the San Andreas fault caused by the 2004 M6 Parkfield earthquake and their applications. *Earth Planets Space* **59**, 21–31.
Mari, J.-L., Glangeaud, F., Coppens, F. (1999). *Signal Processing for Geologists and Geophysicists*. Editions Technip, Paris, 458 pp.
Matsumoto, S., Obara, K., Yoshimoto, K., Saito, T., Ito, A., Hasegawa, A. (2001). Temporal change in P-wave scatterer distribution associated with the M6.1 earthquake near Iwate volcano, northeastern Japan. *Geophys. J. Int.* **145**, 48–58.
Mayeda, K., Koyanagi, S., Aki, K. (1991). Site amplification from S-wave coda in the Long Valley caldera region, California. *Bull. Seismol. Soc. Am.* **81**, 2194–2213.
Mayeda, K., Koyanagi, S., Hoshiba, M., Aki, K., Zeng, Y. (1992). A comparative study of scattering, intrinsic, and coda Q-1 for Hawaii, Long Valley, and Central California between 1.5 and 15.0 Hz. *J. Geophys. Res.* **97**, 6643–6659.
Menke, W., Schaff, D. (2004). Absolute earthquake locations with differential data. *Bull. Seismol. Soc. Am.* **94**, 2254–2264.
Michelini, A., Lomax, A. (2004). The effect of velocity structure errors on double-difference earthquake locations. *Geophys. Res. Lett.* **31**, doi:10.1029/2004GL019682.
Monteiller, V., Got, J.L., Virieux, J., Okubo, P. (2005). An efficient algorithm for double-difference tomography and location in heterogeneous media, with an application to the Kilauea volcano. *J. Geophys. Res.* **110**, B12306, doi:10.1029/2004JB003466.
Nakamura, Y. (1978). A1 Moonquakes—Source distribution and mechanism. Proceedings Lunar Planetary Science Conference, 9th, Houston, March 13–17, 3589–3607.
Nishimura, T., Uchida, N., Sato, H., Ohtake, M., Tanaka, S., Hamaguchi, H. (2000). Temporal changes of the crustal structure associated with the M6.1 earthquake on September 3, 1998, and the volcanic activity of Mount Iwate, Japan. *Geophys. Res. Lett.* **27**, 269–272.

Nishimura, T., Tanaka, S., Yamawaki, T., Yamamoto, H., Sano, T., Sato, M., Nakahara, H., Uchida, N., Hori, S., Sato, H. (2005). Temporal changes in seismic velocity of the crust around Iwate volcano, Japan, as inferred from analyses of repeated active seismic experiment data from 1998 to 2003. *Earth Planets Space* **57**, 491–505.

Nur, A., Simmons, G. (1969). Stress-induced velocity anisotropy in rocks: An experimental study. *J. Geophys. Res.* **74**, 6667–6674.

Obara, K., Kasahara, K., Hori, S., Okada, Y. (2005). A densely distributed high-sensitivity seismograph network in Japan: Hi-net by National Research Institute for Earth science and Disaster Prevention. *Rev. Sci. Instrum.* **76**, 021301–021312.

Okada, H., Watanabe, H., Yamashita, H., Yokohama, I. (1981). Seismological significance of the 1977–78 eruptions and magma intrusion process of Usu volcano. *J. Volcanol. Geotherm. Res.* **9**, 311–334.

Okada, Y., Kasahara, K., Hori, S., Obara, K., Sekiguchi, S., Fujiwara, H., Yamamoto, A. (2004). Recent progress of seismic observation networks in Japan—Hi-net, F-net, K-NET and KiK-net. *Earth Planets Space* **56**, xv–xxviii.

Patanè, D., De Gori, P., Chiarabba, C., Bonaccorso, A. (2003). Magma ascent and the pressurization of Mount Etna's volcanic system. *Science* **299**, 2061–2063.

Peng, Z., Ben-Zion, Y. (2005). Spatiotemporal variations of crustal anisotropy from similar events in aftershocks of the 1999 M7.4 Izmit and M7.1 Düzce, Turkey, earthquake sequences. *Geophys. J. Int.* **160**, 1027–1043.

Phillips, W.S., Aki, K. (1986). Amplification of coda waves from local earthquakes in Central California. *Bull. Seismol. Soc. Am.* **76**, 627–648.

Poupinet, G., Glangeaud, F., Côte, P. (1982). P-time delay measurement of a doublet of microearthquake Proceedings IEEE ICASSPP82, Paris, 3–5 May, 1516–1519.

Poupinet, G., Ellsworth, W., Fréchet, J. (1984). Monitoring velocity variations in the crust using earthquake doublets: An application to the Calaveras fault, California. *J. Geophys. Res.* **89**, 5719–5731.

Poupinet, G., Ratdomopurbo, A., Coutant, O. (1996). On the use of earthquake multiplets to study fractures and the temporal evolution of an active volcano. *Annali di Geofisica* **39**, 253–264.

Poupinet, G., Souriau, A., Coutant, O. (2000). The existence of an inner core super-rotation questioned by teleseismic doublets. *Phys. Earth Planet. Inter.* **118**, 77–88.

Rubin, A. (2002). Using repeating earthquakes to correct high-precision catalogs for time dependent stations delays. *Bull. Seism. Soc. Am.* **92**, 1647–1659.

Rubin, A., Gillard, D., Got, J.-L. (1999). Streaks of microearthquakes along creeping faults. *Nature* **401**, 635–641.

Rubinstein, J.L., Beroza, G.C. (2004). Evidence for widespread nonlinear strong ground motion in the Mw 6.9 Loma Prieta earthquake. *Bull. Seismol. Soc. Am.* **94**, 1595–1608.

Sabra, K.G., Gerstoft, P., Roux, P., Kuperman, W.A., Fehler, M.C. (2005). Surface wave tomography from microseisms in Southern California. *Geophys. Res. Lett.* **32**, 14311–14314.

Saiga, A., Hiramatsu, Y., Ooida, T., Yamaoka, K. (2003). Spatial variation in the crustal anisotropy and its temporal variation associated with a moderate-sized earthquake in the Tokai region. *Geophys. J. Int.* **154**, 695–705.

Sato, H. (1988). Temporal change in scattering and attenuation associated with the earthquake occurrence—A review of recent studies on coda waves. *Pageophysics* **128**, 465–497.

Schaff, D.P., Bokelmann, G.H., Ellsworth, W.L., Zanzerkia, E., Waldhauser, F., Beroza, G.C. (2004). Optimizing correlation techniques for improved earthquake location. *Bull. Seismol. Soc. Am.* **94**, 705–721.

Scherbaum, F., Wendler, J. (1986). Cross spectral analysis of Swabian Jura (WS Germany) three-component microearthquake recordings. *J. Geophys.* **60**, 157–166.

Shapiro, N., Campillo, M., Stehly, L., Ritzwoller, M.H. (2005). High-resolution surface-wave tomography from ambient seismic noise. *Science* **307**, 1615–1618.

Snieder, R., Grêt, A., Douma, H., Scales, J. (2002). Coda wave interferometry for estimating non linear behaviour in seismic velocity. *Science* **295**, 2253–2255.

Song, X., Richards, P.G. (1996). Seismological evidence for differential rotation of the Earth's inner core. *Nature* **382**, 221–224.

Stehly, L., Campillo, M., Shapiro, N.M. (2007). Travel time measurements from noise correlation: Stability and detection of instrumental errors. *Geophys. J. Int.* **171**, 223–230.

Tsujiura, M. (1981). Activity mode of the 1980 earthquake swarm off the coast of the Izu Peninsula. *Bull. Earthqu. Res. Inst.* **56**, 1–24.

Waldhauser, F., Ellsworth, W.L. (2000). A double-difference earthquake location algorithm: Method and application to the Northern Hayward fault. *Bull. Seismol. Soc. Am.* **90**, 1353–1368.

Wapenaar, K. (2006). Green's function retrieval by cross-correlation in case of one-sided illumination. *Geophys. Res. Lett.* **33**, L19304, 1–6.

Weaver, R.L., Lobkis, O.I. (2001). Ultrasonics without a Source: Thermal fluctuation correlations at MHz frequencies. *Phys. Rev. Lett.* **87**, 1–4134301.

Wolfe, C.J. (2002). On the mathematics of using difference operators to relocate earthquakes. *Bull. Seismol. Soc. Am.* **92**, 2879–2892.

Yang, Y., Ritzwoller, M.H., Levshin, A.L., Shapiro, N.M. (2007). Ambient noise Rayleigh wave tomography across Europe. *Geophys. J. Int.* **168**, 259–274.

Zhang, H., Thurber, C.H. (2003). Double difference tomography: The method and its application to the Hayward fault, California. *Bull. Seismol. Soc. Am.* **93**, 1875–1889.

Zhang, J., Song, X., Li, Y., Richards, P.G., Sun, X., Waldhauser, F. (2005). Inner core differential motion confirmed by earthquake waveform doublets. *Science* **309**, 1357–1360.

SEISMOGRAM ENVELOPE INVERSION FOR HIGH-FREQUENCY SEISMIC ENERGY RADIATION FROM MODERATE-TO-LARGE EARTHQUAKES

HISASHI NAKAHARA[1]

ABSTRACT

Studies of high-frequency (above 1 Hz) earthquake source processes are important not only to clarify the earthquake source process on smaller length scales but also to quantitatively predict strong ground motion. However, the application of conventional waveform inversion methods is not straightforward for high frequencies, because random heterogeneities in the Earth cause incoherent scattered waves and the source process is also hard to treat deterministically. To obviate these difficulties, seismogram envelope inversion methods have been developed since the 1990s for clarifying high-frequency earthquake source processes. In this chapter, we first give a broad discussion of the methods in terms of data types, Green's function, source parameters, inversion methods, and so on. We developed an envelope inversion method in 1998, in which we used theoretical envelope Green's functions based on the radiative transfer theory as a propagator from a source to a receiver, and estimated the spatial distribution of high-frequency seismic energy radiation from an earthquake fault plane. We have applied the envelope inversion method to nine moderate-to-large earthquakes. Here, we compile the results and clarify some characteristics of high-frequency seismic energy radiation from moderate-to-large earthquakes. Concerning a scaling of high-frequency radiated energy, logarithm of the high-frequency seismic energy is found to be proportional to the moment magnitude with a coefficient of proportionality of 1, which is different from 1.5 for whole-band seismic energy. Moreover, a regional difference in high-frequency seismic energy radiation is detected for the earthquakes analyzed: Earthquakes in offshore regions of northeastern Japan are found to be more energetic by about an order of magnitude than inland earthquakes in Japan and Taiwan. Regarding the spatial relations, we find four earthquakes in which high-frequency radiation occurs dominantly at the edges of asperities (areas of large fault slip); in four cases there is no correlation between locations of high-frequency radiation and asperities. For one earthquake, we have no fault slip model. So far, reasons for the variation are not known yet, heterogeneous distribution of stress, strength, and material properties may control the variability. These characteristics will provide important information for the study of high-frequency earthquake source process and improvements for predicting strong ground motion.

KEY WORDS: Envelope inversion, high-frequency seismic energy, prediction of strong ground motion. © 2008 Elsevier Inc.

---

[1] Author thanks e-mail: naka@zisin.geophys.tohoku.ac.jp

1. INTRODUCTION

Studies on high-frequency (usually higher than 1 Hz) earthquake source processes of moderate-to-large earthquakes are important not only for clarifying the earthquake source processes in detail but also for quantitatively predicting strong ground motion. However, it is not easy to apply conventional waveform inversion methods for high frequencies. One of the reasons is that random heterogeneities in the Earth cause scattering and produce incoherent wave trains in seismograms (Sato and Fehler, 1998). Another reason is that the source process of moderate-to-large earthquakes becomes complex at higher frequencies so it is hard to treat it deterministically (e.g., Koyama, 1994).

Historically, various approaches have been developed to clarify high-frequency earthquake source process of moderate-to-large earthquakes. Hypocenter determination of subevents may be the most primitive method, in which arrival times of large-amplitude phases are picked on strong-motion records and their source locations are determined. If the large-amplitude phases have not originated from structures (e.g., reflected or scattered or refracted waves), the phases are attributed to earthquake sources. Waveform data other than these large-amplitude phases are neglected in this kind of analysis. Moreover, information on the amplitude of the phases is not considered. Measurements of earthquake source spectra had been conducted and led to heterogeneous fault rupture models (e.g., Gusev, 1983; Papageorgiou and Aki, 1983; Koyama, 1985). This kind of study uses amplitude information of all seismograms but lacks spatial resolution, because the amplitude source spectrum is calculated from seismograms in a long time window. To improve the spatial resolution, it is necessary to investigate the temporal change in observed signals. So, time series of seismogram envelopes have been used since the 1990s to invert for high-frequency earthquake source processes. (e.g., Gusev and Pavlov, 1991; Cocco and Boatwright, 1993; Zeng et al., 1993; Kakehi and Irikura, 1996).

In particular, Zeng et al. (1993) and Kakehi and Irikura (1996) have succeeded in clarifying the spatial distribution of the intensity of high-frequency wave radiation on earthquake fault planes. They have also enabled the comparison of the results to the spatial distribution of fault slip obtained by conventional waveform inversions in lower frequency ranges. The relationship between the locations of high-frequency wave radiation and the locations of asperities (regions with large fault slip) identified by studying low-frequency data is an important characteristic to be explained by the theory of dynamic earthquake rupture. At the same time, this relation is recently one of the central issues in strong-motion seismology from a perspective of the prediction of broadband strong ground motion (e.g., Irikura and Miyake, 2001). Regarding the lower frequencies, fault slip models found by waveform inversion methods have been accumulating since the middle of 1980s. Based on the slip models, some statistical characteristics in slip distributions on earthquake faults have been successfully extracted (e.g., Somerville et al., 1999; Mai and Beroza, 2000). On the contrary, the number of high-frequency envelope inversion analyses is much smaller than that of waveform inversion analyses in lower frequencies. Accordingly, our knowledge about the high-frequency seismic wave radiation is much smaller than that about fault slip models. However, the number of high-frequency envelope inversion analyses has been gradually increasing since the 1990s. Therefore, we here make the first trial to extract statistical characteristics in high-frequency wave radiation based on the previous results obtained by envelope inversions. Moreover, we discuss the relationship between the locations of high-frequency wave radiation and those of low-frequency wave radiation. The compilation will provide us with important information which should be incorporated into the simulation of broadband strong ground motion.

2. ENVELOPE INVERSION METHODS

2.1. General Framework

As far as the Earth can be assumed to be a linear system, an observed seismogram $u(t)$ can be expressed by the convolution between a source time function $S(t)$, an impulse response for propagation paths $P(t)$, that is the Green's function, and a receiver (site) response function $R(t)$ as

$$u(t) = S(t)^* P(t)^* R(t), \qquad (1)$$

where * means convolution. If each of the time functions is a quasi-stationary, mutually uncorrelated, and narrowband random signal with zero mean, the similar convolution relation may be valid for the instantaneous power or the envelope of $u(t)$ as

$$<u(t)^2> = <S(t)^2>^* <P(t)^2>^* <R(t)^2>, \qquad (2)$$

where <> means an ensemble average. For more detailed explanation, please refer to Chapter 5 in Ishimaru (1978). A schematic illustration is shown in Fig. 1. Practically, the

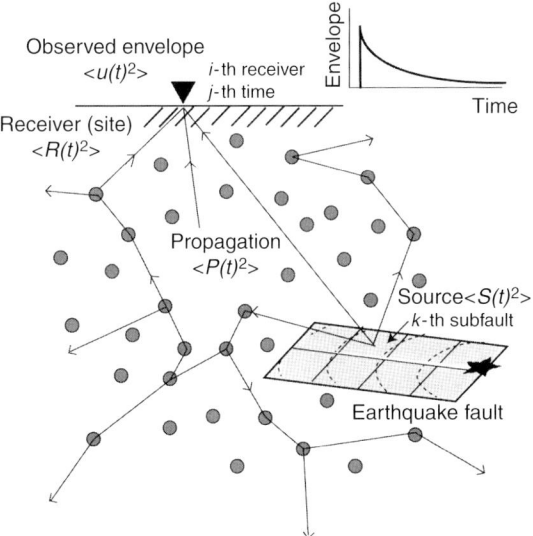

FIG. 1. Schematic illustration of the envelope inversion method of Nakahara et al. (1998). Seismic energy radiated from a double-couple source located at the center of the kth subfault on an earthquake fault (gray parallelogram) is multiply scattered, amplified beneath the ith receiver (solid triangle) and reaches the receiver at the jth time. Gray circles represent point-like isotropic scatterers for S waves randomly distributed in a three-dimensional space. The background S-wave velocity structure is assumed to be homogeneous.

ensemble average is substituted by a moving time average. Therefore, $<u(t)^2>$ can be calculated as a mean-squared (MS) envelope or a squared envelope of the analytic signal of the observed seismogram $u(t)$. Assuming that high-frequency seismic waves are incoherent, we can calculate the observed envelope by the convolution of each envelope for the source, the propagation path, and the site effects, respectively. This characteristic facilitates a direct calculation of an envelope of Green's function, which we call the envelope Green's function.

We note that spatial coordinates do not appear explicitly in Eqs. (1) and (2). So, the equations are for a point source and a receiver in a strict sense. For multiple sources and finite-sized faults, an additional convolution with respect to the spatial coordinates is necessary. Such an extension is straightforward for Eq. (1) because of the superposition principle for linear systems. However, the extension of Eq. (2) has to rely on an assumption that energy radiated from different subsources is additive. This assumption is equivalent to the incoherence of high-frequency seismic waves.

Most envelope inversion methods are based on the finite-fault version of Eq. (2). However, some others make a forward calculation of synthetic seismograms based on the finite-fault version of Eq. (1) and the envelope is used just in fitting for inversion. However, differences between the approaches are small because signals are assumed to be random with zero mean.

2.2. A Classification of Current Envelope Inversion Methods

It may be helpful to find similarities and differences in the envelope inversion methods which have been proposed to date. Here, we make a classification of the methods in terms of data types, frequency ranges, source parameters, Green's functions, inversion methods, and so on as shown in Table 1. The source is modeled as a point source, multiple point sources, or finite-sized faults. Frequency ranges are higher than 0.45 Hz in all the methods. Types of the Green's function are empirical or theoretical, and the data are seismograms or envelopes. Estimation of source parameters is conducted by trial-and-error methods, inversion methods, or deconvolution methods.

Iida and Hakuno (1984) is a pioneering paper in using temporal change in absolute amplitude of acceleration seismograms for estimating intensity of source radiation on earthquake fault planes. Because the amount of available data was small, trail-and-error modeling was conducted for the 1968 Tokachi-Oki, Japan, earthquake and 1978 Miyagi-Ken-Oki, Japan, earthquake. Gusev and Pavlov (1991) performed deconvolution of MS envelopes of far-field P-wave velocity seismograms of the 1978 off Miyagi, Japan, earthquake (M 7.6), and estimated the location of a "short-period radiator," which corresponds to a centroid for high-frequency wave radiation. Cocco and Boatwright (1993) deconvolved MS envelopes of acceleration records and estimated the power rate function for an aftershock (M_L 5.9) of the 1976 Friuli earthquake. Kakehi and Irikura (1996) estimated high-frequency wave radiation areas on the fault of the 1993 Kushiro-Oki earthquake (M_W 7.6) by using root MS envelopes of acceleration seismograms. These four studies used seismograms of small earthquakes as empirical Green's functions. When there are records of small events available which have the same location and focal mechanism as a target large event (mainshock), realistic propagation effects can be naturally included in the empirical Green's function. However, that is not always the case. Petukhin et al. (2004) relaxed this constraint a little and used average envelopes of small events located in the same area as a mainshock as the envelope Green's function, and inverted MS envelopes of squared velocity records of

TABLE 1. Envelope inversion methods for earthquake source studies

References	Data type	Frequency (Hz)	Source model	Source parameters	Green's function	Estimation methods
Iida and Hakuno (1984)	Near-field acc. 2 horiz. cmp.	No description	Finite fault	Acceleration radiation	Empirical seismograms	Forward
Gusev and Pavlov (1991)	Teleseismic sqr. vel. P-wave vert. cmp.	0.45–1.75	Point source	Seismic energy	Empirical envelope	Deconvolution and inversion
Zeng et al. (1993)	Near-field sqr. disp. 2-cmp. sum	>5	Finite fault	Displacement radiation	Ray theory envelope	Inversion
Cocco and Boatwright (1993)	Near-field sqr. acc. 2-cmp. sum	No description	Multi point source	Acceleration radiation	Empirical envelope	Deconvolution and inversion
Kakehi and Irikura (1996)	Near-field rms. acc. 3-cmp.	2–10	Finite fault	Acceleration radiation	Empirical seismograms	Inversion
Nakahara et al. (1998)	Near-field S-wave sqr. vel. 3-cmp. sum.	1–16	Finite fault	Seismic energy	Radiative transfer theory envelope	Inversion
Petukhin et al. (2004)	Near-field S-wave sqr. vel. 3-cmp. sum.	1–16	Finite fault	Seismic energy	Empirical envelope	Inversion

the 1992 Avachinsky Gulf earthquake (Mw 6.8). Regarding the empirical Green's function method, estimated source parameters are relative to those of a small event for which the empirical Green's function is used.

The use of theoretical Green's function is superior in terms of estimating absolute values of source parameters as far as a reference station is correctly selected. However, the number of the studies using theoretical envelope Green's functions is very small. Zeng et al. (1993) inverted MS envelopes of displacement seismograms from the 1989 Loma Prieta earthquake by using ray-theoretically calculated Green's functions, and mapped the high-frequency source radiation intensity on the earthquake fault. Nakahara et al. (1998) inverted MS envelopes of squared velocity seismograms using the radiative transfer-based theoretical envelope Green's function, and estimated the spatial distribution of high-frequency seismic energy radiation on the fault plane of the 1994 far east off Sanriku earthquake (Mw 7.7).

Finally, it is worth while to refer to two pioneering studies dealing with envelopes though they are not inversion studies in a strict sense. Midorikawa and Kobayashi (1979) proposed a method to calculate a velocity response spectrum on seismic bedrock due to the rupture of a finite-sized fault. The method estimates an average empirical envelope of a velocity motion of an oscillator with a certain natural period on the seismic bedrock due to the rupture of a subfault. The envelope of the oscillator due to the entire fault is obtained by summing up contributions from all subfaults. The maximum amplitude of the envelope corresponds to the velocity response at the period on the seismic bedrock. The method serves for a forward modeling of strong ground motion based on a fault model. Koyama and Zheng (1985) proposed a technique to estimate spectral amplitude at a frequency of about 1 Hz from envelopes of teleseismic P-wave displacement records obtained by short-period sensors of the World Wide Standardized Seismic Network. After the correction of propagation effects and instrumental responses, they estimated the short-period seismic moment for 79 large earthquakes, and verified that the radiation of high-frequency seismic waves is incoherent.

2.3. The Method of Nakahara et al. (1998)

Here, we make a brief explanation of the envelope inversion method of Nakahara et al. (1998), because we will be mainly concerned with the results based on the method. The method uses the theoretical envelope Green's function developed by Sato et al. (1997) based on the radiative transfer theory (e.g., Chandrasekhar, 1960). The radiative transfer theory, sometimes called the energy transport theory or the multiple scattering theory, was first introduced to seismology by Wu (1985) for stationary cases. Later, it was extended to time-dependent cases numerically by Gusev and Abubakirov (1987) and theoretically by Zeng et al. (1991), because seismic energy is usually radiated from a transient (approximately impulsive) earthquake source. Because the radiative transfer theory was initially used to explain coda envelopes of local earthquakes, the source radiation pattern, which has a large effect on early parts but a small effect on later coda parts, had not been included in the modeling. Sato et al. (1997) succeeded in introducing the radiation pattern for a point shear dislocation (double couple) source and enabled the synthesis of S-wave seismogram envelopes from the direct waves through coda. The study paved a way to the application of the radiative transfer theory to detailed earthquake source studies.

Here, we briefly explain the envelope Green's function used in our envelope inversion method. As shown in Fig 2(a), point-like isotropic scatterers are assumed to be distributed randomly and homogeneously in an infinite medium, where the background S-wave velocity V is constant. Only S waves are considered in the modeling. A double-couple source is located at the origin from which seismic energy of unit amplitude is impulsively radiated. In the framework of the radiative transfer theory, the energy density $E_G(\mathbf{x}, t)$ at a location $\mathbf{x} = (r, \theta, \phi)$ in a spherical coordinate system and time t can be expressed by the following integral equation:

$$E_G(\mathbf{x},t) = R(\theta,\phi)G(\mathbf{x},t) + g_0 V \int\!\!\int\!\!\int d\mathbf{x}' \int_{-\infty}^{\infty} dt'\, G(\mathbf{x} - \mathbf{x}', t - t') E_G(\mathbf{x}', t'), \tag{3}$$

where $r = |\mathbf{x}|$, θ is the zenith angle and ϕ is the azimuth angle, and g_0 is the total scattering coefficient characterizing the scattering power per unit volume. The function E_G is the envelope Green's function in the scattering medium. $R(\theta, \phi)$ is the radiation pattern of S-wave energy which is normalized as $\oint R(\theta, \phi) d\Omega = 4\pi$. The first term in the right-hand side means the coherent part corresponding to the direct wave. The second term means the scattered energy, which is given by integrating the contributions from the last scattering point $\mathbf{x}'$ and at lapse time t'. The propagator function $G(\mathbf{x}, t)$ is expressed as

$$G(\mathbf{x},t) = \frac{\exp(-(g_0 V + \eta)t)}{4\pi V r^2} \delta\!\left(t - \frac{r}{V}\right) \quad \text{for} \quad t \geq 0. \tag{4}$$

This is characterized by geometrical spreading, time lag due to propagation, and exponential decay due to intrinsic absorption and scattering attenuation of seismic energy. Intrinsic absorption η is related to the intrinsic Q value as $Q_i^{-1} = \eta/\omega$ for an angular frequency ω.

For a double-couple source with the fault normal vector in the first axis and slip vector in the second axis as shown in Fig. 2(a), we can explicitly express $R(\theta, \phi)$ in terms of the spherical harmonics with the order n of up to four as

$$R(\theta,\phi) = \sum_{n=0,2,4} \sum_{m=-n}^{n} a_{nm} Y_{nm}(\theta,\phi) = \sqrt{4\pi} Y_{0,0}(\theta,\phi) + \frac{5}{7}\sqrt{\frac{4\pi}{5}} Y_{2,0}(\theta,\phi)$$
$$- \frac{2}{7}\sqrt{\frac{4\pi}{9}} Y_{4,0}(\theta,\phi) + \frac{\sqrt{280\pi}}{21}\left(Y_{4,4}(\theta,\phi) + Y_{4,-4}(\theta,\phi)\right), \tag{5}$$

where θ is measured from the null axis (the third axis), and ϕ is measured from the fault normal (the first axis).

We solve Eqs. (3–5) for the envelope Green's function $E_G(\mathbf{x}, t)$ by using the Fourier transform in space, the Laplace transform in time, and the spherical harmonics expansion with respect to radiation angles. The energy density can be written in a spherical harmonics expansion with the expansion coefficients of the radiation pattern:

$$E_G(\mathbf{x},t) = \sum_{n=0}^{\infty} E_{G,n}(r,t) \sum_{m=-n}^{n} a_{nm} Y_{nm}(\theta,\phi), \tag{6}$$

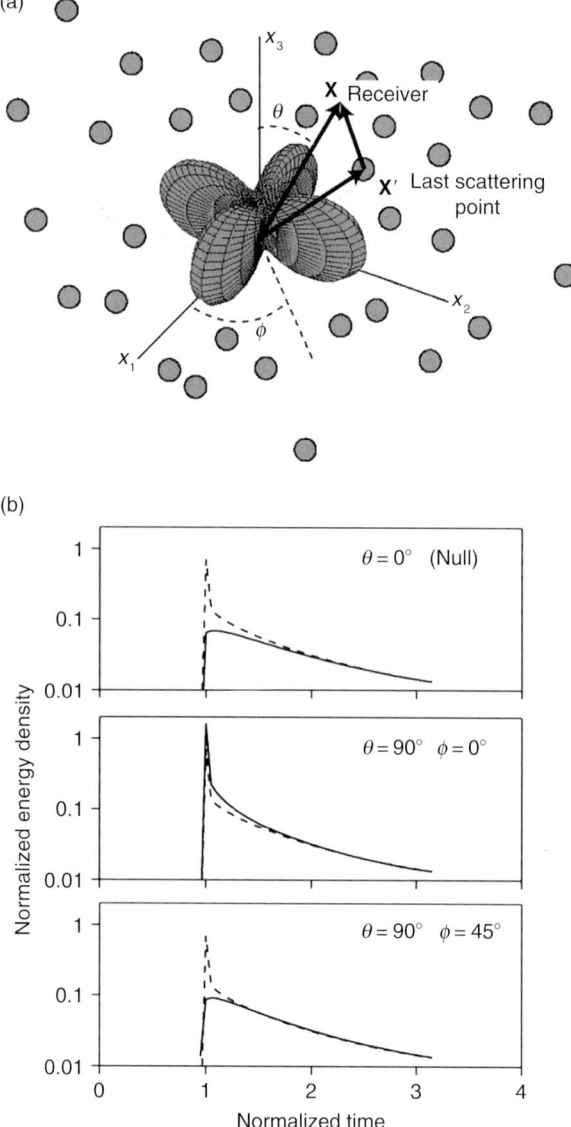

FIG. 2. (a) Configuration of a double-couple source with the fault normal vector in x_1 axis and the slip vector in x_2 axis. (b) Theoretical envelopes calculated for the double-couple source shown in (a) (solid curves) and those for an isotropic source (broken curves). Horizontal axis is normalized time and vertical axis is normalized energy density. Envelopes at three different receivers are shown from top to bottom. Although early parts of envelopes clearly exhibit a difference in energy density due to the radiation pattern, the difference becomes smaller and smaller as time elapses.

where

$$E_{G,n}(r,t) = \frac{e^{-(g_0 V+\eta)t}}{4\pi r^2 V}\delta\left(t-\frac{r}{V}\right)H(t) + \frac{g_0 e^{-(g_0 V+\eta)t}}{4\pi r V t}Q_n\left(\frac{(Vt/r)^2+1}{2(Vt/r)}\right)H\left(t-\frac{r}{V}\right)$$
$$+ \frac{g_0^2 V^2}{2\pi r i}\int_{-\infty}^{\infty}d\omega\frac{e^{i\omega t}}{2\pi}\int_{-\infty}^{\infty}dk\frac{e^{ikr}}{2\pi}ku_n(kr)\frac{\overline{\overline{G}}_n(k,i\omega)\overline{\overline{G}}_0^2(k,i\omega)}{1-g_0 V\overline{\overline{G}}_0(k,i\omega)}. \quad (7)$$

The first term is the coherent term, the second is the single scattering term, and the third is multiple scattering term with the order higher than or equal to 2. Q_n is the Legendre polynomial of the second kind. Function $u_n(\mathbf{x})$ originates from the spherical Bessel function and is defined as

$$u_n(x) \equiv \sum_{s=0}^{n}\frac{i^{s-n}(n+s)!}{s!(n-s)!(2x)^s}. \quad (8)$$

Function $\overline{\overline{G}}_n(k,s)$ corresponds to the Laplace transform of the spherical Bessel function:

$$\overline{\overline{G}}_n(k,s) = \frac{1}{kV}\left(\frac{kV}{2(s+gV+\eta)}\right)^{n+1}\frac{\sqrt{\pi}\Gamma(n+1)}{\Gamma(n+(3/2))}$$
$$_2F_1\left(\frac{n+1}{2},\frac{n+2}{2},n+\frac{3}{2},-\left(\frac{kV}{s+gV+\eta}\right)^2\right), \quad (9)$$

where Γ is the Gamma function, and $_2F_1$ is the Gauss's hypergeometric function.

We can numerically calculate $E_{G,n}(r,t)$ using the fast Fourier transform (FFT) algorithm over frequency and wave number for given three parameters of g_0, V, and Q_i^{-1}. In Fig. 2(b), we give examples of calculated theoretical envelope Green's functions for a double-couple source in solid curves and those for an isotropic source in broken curves. Time on the horizontal axis and energy density on the vertical axis are both normalized. Envelopes at three different receivers are shown. A prominent character of the envelope Green's function $E_G(r,t)$ is a long tail which follows the direct wave and decays slowly due to the scattering. Moreover, early parts of envelopes clearly exhibit a difference in energy density due to the radiation pattern. However, the difference becomes smaller as time elapses. Envelopes for the double-couple source are found to converge to those for the isotropic source after twice the direct S-wave travel time. The energy density of the higher-order modes diminishes faster than that of the lower-order modes because of multiple isotropic scattering. Therefore, only the lowest 0th mode corresponding to spherical source radiation dominates at large lapse times. From the examples, it is found that detailed information on the focal mechanism can be extracted from early part of the coda whose lapse time is smaller than twice the direct S-wave travel time.

Using this envelope Green's function, we formulated an envelope inversion method to estimate the spatial distribution of energy radiation from an earthquake fault and site amplification factors. The method is schematically illustrated in Fig. 1. A rupture is assumed to propagate with a constant rupture velocity of V_r from the initial rupture point.

The fault plane is composed of subfaults. When a rupture front passes through the kth subfault, the energy W_k is radiated from a double-couple source on the subfault with a time history of $f_k(t)$. The integral of $f_k(t)$ over the transit time of the rupture front is normalized as 1. The radiated energy is multiply scattered in the course of propagation through the scattering medium, and reaches the ith receiver at the jth time, and is modified by a subsurface structure in the vicinity of the receiver (Fig. 1). We further assume that seismic energies radiated from different subfaults are additive (waves are incoherent), so that the energy density for the ith receiver at the jth time is the sum of the radiated energy from the subfaults. Then, we can formulate the theoretical energy density C_{ij} as

$$C_{ij} = R_i^2 \sum_k W_k F_{ijk}, \qquad (10)$$

where

$$F_{ijk} = \int f_k(t') E_G\left(\mathbf{x}_i - \mathbf{x}_k, t_j - t'\right) dt'. \qquad (11)$$

F_{ijk} is the convolution of the envelope Green's function and the energy radiation time history. R_i is the receiver (site) amplification factor for velocity amplitude at the ith receiver. Under the framework of Eq. (2), this corresponds to the assumptions that $<S(t)^2> = W_k f_k(t)$, $<P(t)^2> = E_G(\mathbf{x},t)$, and $<R(t)^2> = R_i^2 \delta(t)$. The values of W_k and R_i are estimated so as to minimize the residual between observed envelopes and synthesized ones in the following least squares sense:

$$\sum_i \sum_j \left(\frac{1}{\max_j O_{ij}}\right)^2 |O_{ij} - C_{ij}|^2 \to \text{Min.}, \qquad (12)$$

where O_{ij} is the observed energy density at the ith receiver and the jth time. We normalize both the observed envelopes and the synthesized ones by the observed maximum value at each receiver to set the weight of all receivers equal. To simplify the inversion, we further assume that $f_k(t)$ is a box-car function with the same duration time of Δt for all subfaults. Because Eq. (10) is nonlinear for the radiated energy W_k and the site amplification factor R_i, the equation is iteratively solved by the following procedures: (i) Assuming values of V_r and Δt, (ii) Setting the initial value of R_i for all receivers. The value is assumed to be 2 for a reference hard rock site on the surface, and 1 for a reference hard rock site in the subsurface. (iii) Solving Eq. (10) for the radiated energy W_k by the linear least squares method. (iv) Estimating the site amplification factors by fixing the radiated energy calculated in step (iii). (v) Iterating steps (iii) and (iv) until the residual between the observed envelopes and synthesized ones does not change with increasing number of iterations. We thus estimate the best-fit values of W_k and R_i for various sets of V_r and Δt. The final result is obtained by choosing the solution having the minimum residual among them.

Finally, we mention about a few points which are necessary for practical applications of the inversion method. First, we need to make corrections of travel times and takeoff angles by using a horizontally layered structure, because theoretical envelopes are calculated for a medium with homogeneous background S-wave velocity. Second, we

3. DATA ANALYSIS AND THE RESULTS

3.1. An Example of Practical Data Analysis

need to select a reference receiver carefully to estimate absolute values of radiated seismic energy. Velocity logging data and site amplification factors estimated by other previous studies are useful for the purpose. Third, strong nonlinear site effects may affect our inversion results, because our modeling is based on the linear elastic theory. For an accurate estimation of radiated seismic energy, a reference receiver is required to never experience the nonlinear effect.

We explain analysis procedures of our inversion method by taking the 2003 Miaygi-ken Oki, Japan, earthquake (Mw 7.0) as an example. This is an intraslab earthquake which took place at a depth of about 70 km in the subducted Pacific plate beneath northeastern Japan. We use strong-motion seismograms recorded at 18 stations of the K-NET and Kik-net within epicentral distance of 50 km. A reference station is set at a subsurface (depth of about 100 m) of MYGH12 station, denoted as MYGB12 in this study, because logging data show high seismic velocity at the site. Three component acceleration records are numerically integrated to velocity records, and are band-pass filtered in four octave-width frequency bands of 1–2, 2–4, 4–8, and 8–16 Hz. We square band-passed velocity records, take the sum of three components, and then smooth them by taking a moving average using a time window of 2 s. Multiplying them by a density of the crust (2.5×10^3 kg/m^3), we obtain seismogram envelopes having the unit of energy density (J/m^3). A time window from the S-wave onset to the lapse time of 51.2 s is used for the inversion analysis. The end of the time window is set to be smaller than twice the direct S-wave travel time.

The envelope Green's function is calculated using scattering parameters (g_0 and Q_i^{-1}) estimated at Onagawa (ONG) station (a solid square in Fig. 3) by Sakurai (1995) using envelopes of smaller events in the region. The background S-wave velocity of the scattering medium is estimated to be 4.09 km/s. Travel times and takeoff angles of S waves are corrected using a structure with four horizontal layers. For the inversion analysis, we set a fault plane with a length of 30 km and a width of 25 km dipping to the west (see Fig. 3), and divide it into 30 subfaults each of which is a 5×5 km^2. The geometry of the fault plane is assumed as strike = 193°, dip = 69°, and rake = 87° with reference to a focal mechanism obtained from far-field body waves by Yagi (2003). Rupture velocity and duration time of the box-car source time function are estimated by grid search.

A contour map in Fig. 4 shows residuals between observed envelopes and synthesized ones for all the four octave-width frequency bands plotted for various rupture velocities and source duration times. A solid star marks parameters for which the minimum residual is obtained. The residuals are normalized by the minimum one. The duration time was estimated to be 1.6 s. Rupture velocity is 3.8 km/s. In Fig. 5, we show the spatial distribution of seismic energy radiation on the fault plane in a gray shade, in which a darker color corresponds to larger energy radiation. A solid star shows the initial rupture point. High-frequency seismic energy was mainly radiated from two regions on the fault. The first one is around the initial rupture point and the other is a northern deeper part of the fault. This spatial pattern looks common irrespective of frequency band analyzed. Total amount of seismic energy radiation is 8.3×10^{15} J in 1–16 Hz. Observed envelopes

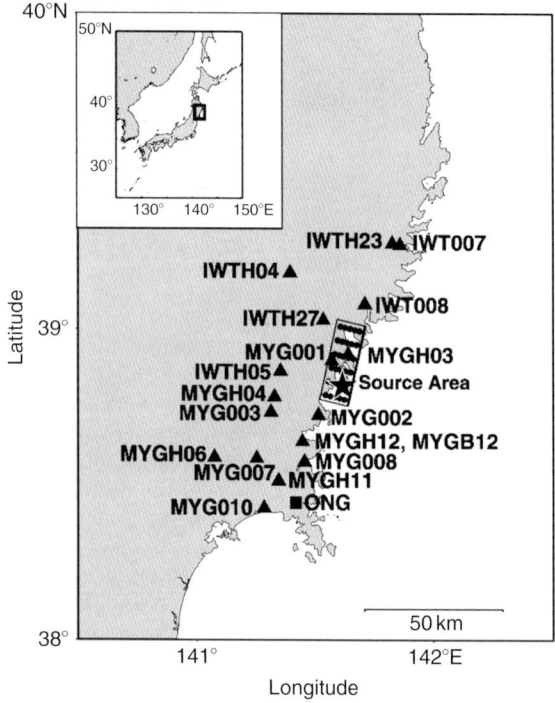

FIG. 3. Map of stations and assumed fault plane for the 2003 Miyagi-ken Oki, Japan, earthquake. An enclosed rectangle in an inserted map is enlarged. The hypocenter is shown by a solid star. Rectangle is the assumed fault plane, which dips to the west. Locations of subfaults are shown by solid circles. Each subfault is a 5×5 km^2. Strong-motion records at 18 stations (solid triangles) are used for the inversion analysis. Epicentral distance of these stations is smaller than 50 km. MYGB12 station is a reference for the site amplification factor. A solid square is ONG station where scattering parameters were estimated by Sakurai (1995).

(solid) and synthesized ones (broken) for 4–8 Hz are shown in Fig. 6. Two peaks are clearly found in observed envelopes at most of the stations. Generally, the peaks are well explained by the synthesized envelopes. The first and the second peaks are attributed to the strong energy radiation from the initial rupture point and the northern deeper part of the fault, respectively.

For the estimation of errors in our inversion results, we perform the following procedure. First, we produce synthetic envelopes for the best-fit distribution of energy radiation in Fig. 5 and best-fit site amplification factors and by adding random noise to the synthetic envelopes. The random noise is assumed to obey an exponential distribution. To the amplitude of the noise, a root MS residual between an observed envelope and a synthesized envelope from the best-fit solutions in Fig. 5 is assigned at each station. The amplitude of the noise is up to 15% of the maximum amplitude of the signal. Repeating the envelope inversion by changing random noise 100 times, we estimate the spatial distribution of energy radiation. In Fig. 7, we show standard deviation of

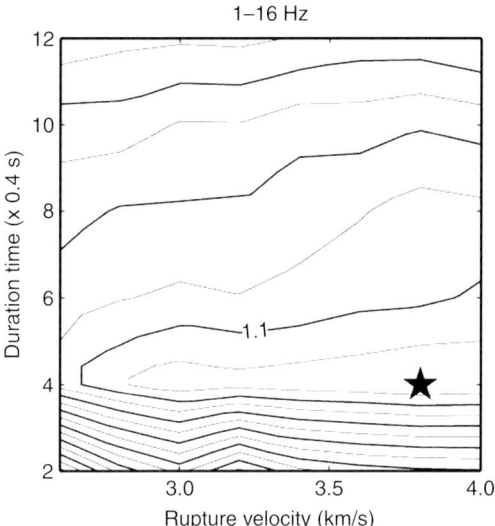

FIG. 4. Contour map of residuals between observed envelopes and synthesized ones with respect to rupture velocity and duration time of energy radiation for each subfault of the 2003 Miyagi-ken Oki, Japan, earthquake. Residuals are normalized by the minimum value which is shown by a solid star. Best-fit parameters are obtained for a rupture velocity of 3.8 km/s and duration time of 1.6 s.

estimated seismic energy normalized by the true solution in Fig. 5, that is, the coefficient of variation (CV). Errors are considered to be small in the parts where CV is small. Two parts of strong high-frequency energy radiation shown in Fig. 5 are found to be located in the region shaded by black, which confirms that estimation errors in seismic energy are less than 20% for the two parts.

From a waveform inversion analysis of both teleseismic and near-field seismograms in a frequency range between 0.05 Hz and 0.5 Hz, Yagi (2003) estimated the spatial distribution of slip on the fault plane. A contour of the slip is shown in Fig. 8. The maximum slip amount reaches about 1.7 m. From the comparison between the slip distribution and high-frequency seismic energy radiation (shaded map in Fig. 8), both the high-frequency and the low-frequency waves are radiated around the initial rupture point. But the other region of high-frequency radiation in the northwestern part does not overlap an asperity and rather corresponds to the edge of the asperity. Therefore, the spatial relationship between the location of high-frequency radiation and that of low-frequency radiation is not simple for this event.

4. COMPILATION OF THE RESULTS

We have applied the envelope inversion method of Nakahara *et al.* (1998) to nine earthquakes around Japan with moment magnitude (Mw) of from 5.9 to 8.3. Among the earthquakes, three are interplate earthquakes, one is an intraslab earthquake, and five are

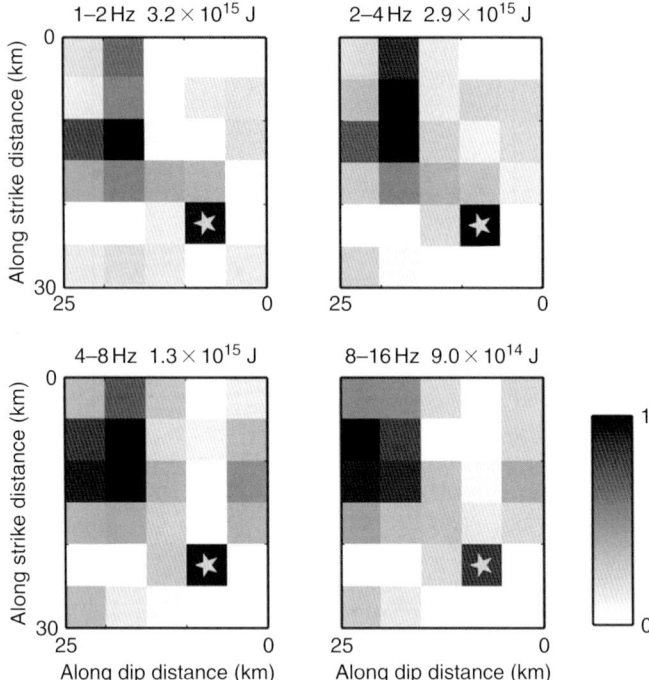

FIG. 5. Spatial distribution of high-frequency seismic energy radiation from the fault plane. Results for 1–2, 2–4, 4–8, and 8–16 Hz, respectively, are shown in a grayscale. Darker shade means stronger energy radiation. A star is the initial rupture point. Numerals are total energy radiation from the entire fault plane. High-frequency energy is mainly radiated from two regions: around the initial rupture point and a northern deeper part.

inland earthquakes. Focal mechanisms differ among the earthquakes. Although the number of nine cases is small, we compile the results in Table 2, and make one of the first trials to extract statistical characteristics of high-frequency seismic energy radiation from moderate-to-large earthquakes. We put our focus on the following three subjects: (1) Frequency dependence of high-frequency radiated energy. (2) A scaling relationship between high-frequency radiated energy and earthquake magnitude. (3) A spatial relationship between locations of asperities (areas of large fault slip) and locations of high-frequency energy radiation. The second subject can be studied only by our envelope inversion analysis, because our method is capable of dealing with absolute values of seismic energy.

4.1. Frequency Dependence of High-Frequency Seismic Energy

First, we examine the theoretical frequency dependence of high-frequency seismic energy. If the source spectrum obeys the omega-squared model (e.g., Aki, 1967;

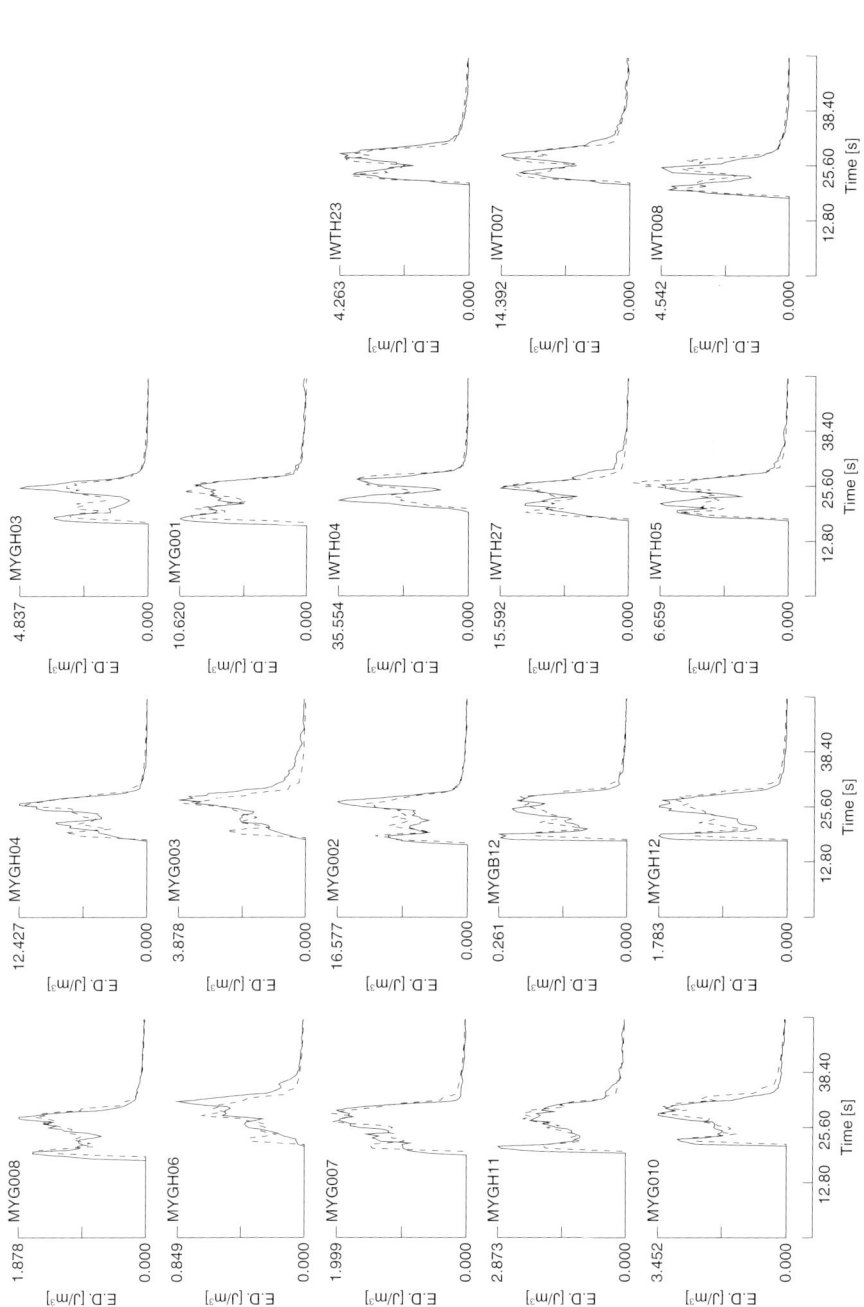

FIG. 6. Comparison between observed envelopes (solid curve) and synthesized ones (broken curve) for the 4–8 Hz band. Synthesized envelopes explain observed ones well.

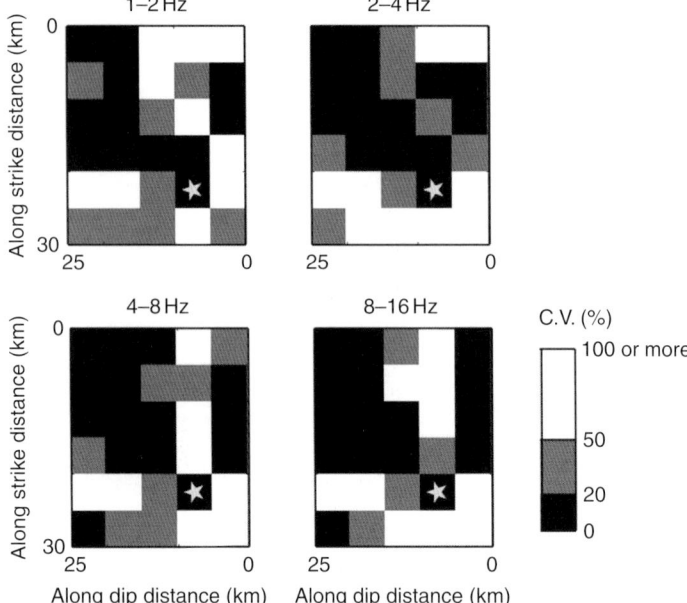

FIG. 7. Standard deviation of high-frequency seismic energy radiation for 1–2, 2–4, 4–8, and 8–16 Hz, respectively, estimated for 100 synthetic data. Standard deviation normalized by the mean value, the coefficient of variation (CV), is shown in a grayscale. Darker shade means smaller CV, which shows a smaller error in the estimation. Two parts of strong high-frequency radiation are located in the regions with darker shade.

Brune, 1970), the acceleration source spectrum $a(\omega)$ becomes flat at frequencies higher than the corner frequency, and the amplitude level, often denoted as A, can be expressed as

$$a(\omega) \equiv A \propto \sigma L, \qquad (13)$$

where ω is the angular frequency, σ is a stress parameter, and L is a characteristic length scale of a fault. If the stress parameter σ is constant, A is proportional to L. A velocity source spectrum $v(\omega)$ in the frequency range is

$$v(\omega) = \frac{A}{\omega}. \qquad (14)$$

S-wave energy spectrum in an octave-width frequency band $E_{\text{HF}}(\omega)$, which is directly obtained from our envelope inversion method, is expressed as

$$E_{\text{HF}}(\omega) = \frac{1}{10\pi\rho\beta^5} \int_{\omega}^{2\omega} \left|v(\omega')\right|^2 d\omega' = \frac{1}{20\pi\rho\beta^5} \frac{A^2}{\omega}, \qquad (15)$$

where ρ is the density and β is the S-wave velocity. Therefore, the seismic energy is expected to decrease with frequency with a power of -1.

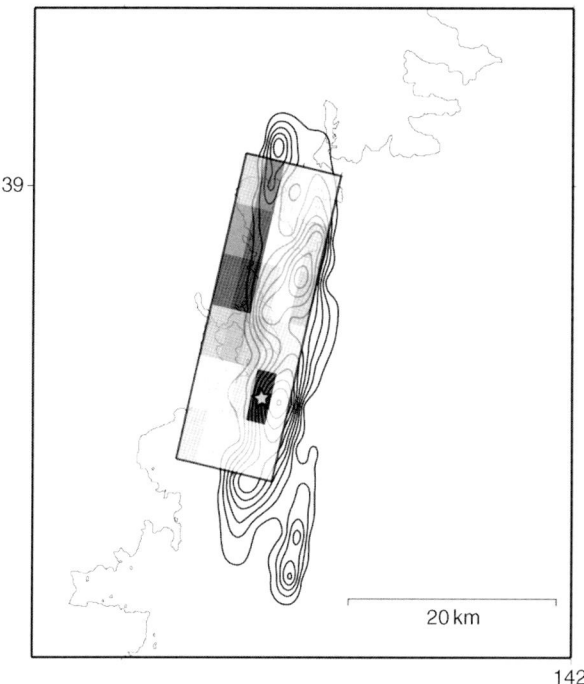

FIG. 8. Comparison between the regions of high-frequency (HF) radiation and those of low-frequency (LF) radiation for the 2003 Miyagi-ken Oki, Japan, earthquake. Sum of HF energy radiation over the 1–16 Hz band is shown by a grayscale. Darker shade means stronger energy radiation. Slip distribution is shown in a contour, which was estimated by Yagi (2003) by inversion analysis of both of near-field and teleseismic waves in the frequency ranges between 0.05 Hz and 0.5 Hz. Comparing both results, LF- and HF- waves are radiated from nearly the same region around the initial rupture point (solid star) and they are complementary at the northern deeper part of the fault.

In Fig. 9, we plot E_{HF} versus frequency relations for nine earthquakes shown in Table 2. Seismic energy in an octave-width band is found to decrease with increasing frequency. Here, we fit regression lines to the data to estimate the power of the decay of seismic energy with frequency. The results are tabulated in Table 3. The mean value of the power for all the events is estimated to be −0.99 which is close to the theoretical expectation of −1, though it shows variation from −0.38 for the Sanriku event to −1.85 for the Kobe event. The median value is −0.70. To have a more closer look at the frequency dependence, we find a slight increase of energy in 1–8 Hz band for the Sanriku event. We also note that the energy in the 8–16Hz band falls off rapidly for the Sanriku event and the Kobe event. Although someone might think that this is caused by f_{max}, the two datasets are probably contaminated by instrumental response which we could not correct. For events other than Sanriku and Kobe, we can not detect a rapid fall off for higher frequency which might be associated with f_{max}.

TABLE 2. Summary of our envelope inversion analyses

Event (Mw)	Source location	Focal mechanism type	HF location and LF location	Reference
1994 Off-Sanriku, JAPAN (7.6)	Plate boundary	Thrust	Complementary	Nakahara et al. (1998)
1995 Kobe, JAPAN (6.9)	Inland	Right-lateral strike slip	Otherwise	Nakahara et al. (1999)
1998 Northern Iwate, JAPAN (5.9)	Inland	Reverse	Complementary	Nakahara et al. (2002)
1999 Chi-Chi, Taiwan (7.6)	Inland	Thrust	Complementary	Nakahara et al. (2006)
2000 Western Tottori, JAPAN (6.7)	Inland	Left-lateral strike slip	Complementary	Nakahara (2003)
2003 Off-Miyagi, JAPAN (7.0)	Intraslab	Reverse	Otherwise	Nakahara (2005a)
2003 Off-Tokachi, JAPAN (8.3)	Plate boundary	Thrust	Otherwise	Nakahara (2004)
Largest aftershock of the 2003 Off-Tokachi, JAPAN (7.3)	Plate boundary	Thrust	Indeterminate	Nakahara (2004)
2004 Niigata Chuetsu, JAPAN (6.6)	Inland	Reverse	Otherwise	Nakahara (2005b)

HF, high frequency; LF, low frequency.

4.2. Scaling of High-Frequency Seismic Energy

Here, we discuss a scaling relationship between high-frequency seismic energy and earthquake magnitude. As shown in the previous subsection, our observation of high-frequency seismic energy in an octave-width frequency band for individual events shows variation from the theoretical expectation based on Eq. (15). However, it may be acceptable to use the Eq. (15) as a reference since it fits the average of the observed data. For a fixed frequency band, we obtain the following relation by taking the logarithm of both sides of Eq. (15)

$$\log E_{\mathrm{HF}}(\omega) \propto 2 \log A \sim \log L^2 \sim M. \tag{16}$$

The final proportionality comes from a well-known empirical relationship between earthquake magnitude and fault area (e.g., Kanamori and Anderson, 1975). Therefore, the logarithm of high-frequency seismic energy in an octave-width band is predicted to be proportional to earthquake magnitude. Or, high-frequency seismic energy is proportional to fault area. This is an important relation which only holds for high-frequency seismic energy. It should be noted that the coefficient of proportionality is 1, which is different from 1.5 in the Gutenberg–Richter's relation (Gutenberg and Richter, 1956) for whole-band seismic energy. We also find that the high-frequency seismic energy becomes half when a frequency is doubled.

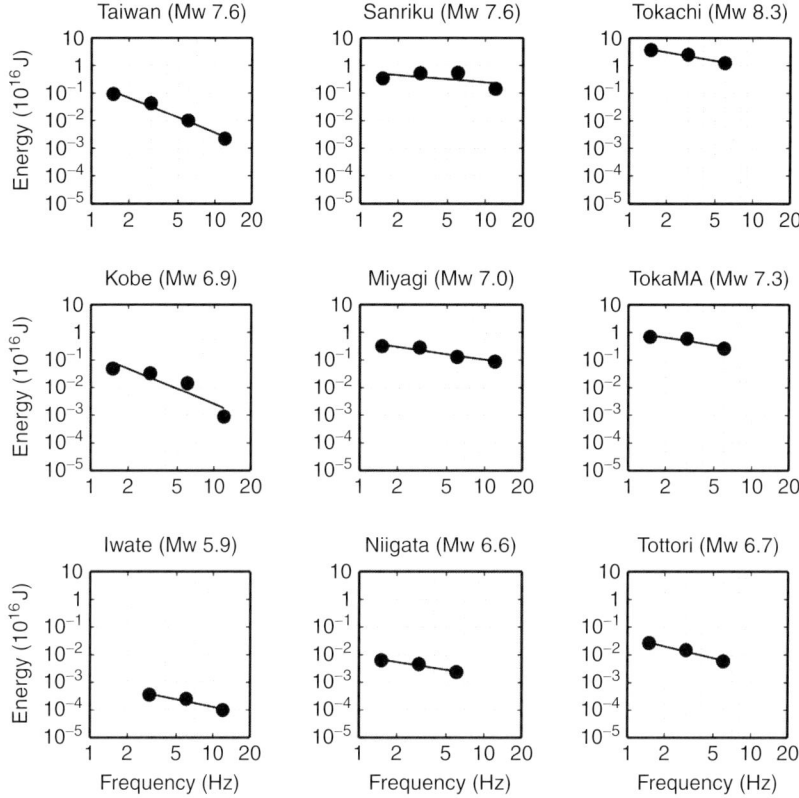

FIG. 9. Frequency dependence of high-frequency seismic energy for nine earthquakes. Solid circles are data and solid lines are regression lines. High-frequency seismic energy in an octave-width frequency band decreases with increasing frequency.

TABLE 3. Power of the decay of high-frequency seismic energy with frequencies

Event (Mw)	Power
1994 Off-Sanriku, JAPAN (7.6)	-0.38 ± 0.41
1995 Kobe, JAPAN (6.9)	-1.85 ± 0.58
1998 Northern Iwate, JAPAN (5.9)	-0.91 ± 0.26
1999 Chi-Chi, Taiwan (7.6)	-1.83 ± 0.17
2000 Western Tottori, JAPAN (6.7)	-1.08 ± 0.15
2003 Off-Miyagi, JAPAN (7.0)	-0.66 ± 0.13
2003 Off-Tokachi, JAPAN (8.3)	-0.79 ± 0.15
Largest aftershock of the 2003 Off-Tokachi, JAPAN (7.3)	-0.70 ± 0.28
2004 Niigata Chuetsu, JAPAN (6.6)	-0.70 ± 0.14

Then, we compile the results for nine earthquakes shown in Table 2. Here, we divide the data into two categories: one is inland earthquakes and the other is offshore earthquakes. For offshore earthquakes, there is no discrimination between interplate earthquakes and intraplate ones. Observed high-frequency seismic energy in 1–2, 2–4, 4–8, and 8–16 Hz, respectively, is shown against Mw in Fig. 10. Observed data are shown by open symbols. For the calculation of theoretically expected values from the Eq. (15), we assume that $\rho = 2.5 \times 10^3$ [kg/m^3] and $\beta = 3.5$ [km/s]. In addition, we adopt a relation $A = 5.3 \times 10^{12} M_0^{1/3}$ [Nm/s^2] in which M_0 is measured in [Nm], corresponding to Brune's stress drop of 9.7 [MPa] (97 [bars]). The value was obtained from measurements for 12 inland earthquakes by Dan et al. (2001). Theoretically expected values thus calculated are shown by solid, long-broken, short-broken, and dotted lines for 1–2, 2–4, 4–8, and 8–16Hz, respectively. Our observation matches with the expectation for inland earthquakes as shown in Fig. 10(a). This suggests that our estimates for high-frequency seismic energy are consistent with independent estimates by Dan et al. (2001) which used a different method. However, we can not explain levels of high-frequency seismic energy for offshore earthquakes in Fig. 10(b) by using A of Dan et al. (2001). Our observation seems larger than the expectation by about 10 times. Because the energy is proportional to A^2 as shown in Eq. (15), we multiply the A value by 3.16 (square root of 10), corresponding to Brune's stress drop is about 54.6 [MPa] (546 [bars]), and compare again with our observation for offshore earthquakes in Fig. 10(c). The new expected values can roughly explain our observation for the offshore earthquakes. This is not a strict fit to the data but a rough estimate. However, this implies that the offshore earthquakes in northeastern Japan radiate about 10 times more high-frequency seismic energy than inland earthquakes in Japan and Taiwan with the same magnitude. This tendency was also reported for the same region by Satoh (2004) and for regions to the south by Takemura et al. (1989) and Kato et al. (1998), though it seems to contradict an empirical rule that the static stress drop is higher for intraplate events than for interplate events (e.g., Kanamori and Anderson, 1975). Kato et al. (1998) referred to the depth dependence of the stress drop as a possible cause.

Two points have been clarified from our studies. First, the high-frequency seismic energy is proportional to fault area. This relation is a manifestation that the high-frequency seismic waves are incoherent. Because this point was first pointed out by Koyama and Zheng (1985), our result is a confirmation of their results. Second, there exists a regional difference in the excitation level of high-frequency seismic energy. Although the reason is not clear, this result is practically important for quantitative prediction of strong ground motion.

4.3. Spatial Relationship Between Asperities and High-Frequency Sources

For broadband simulations of strong ground motion due to an earthquake fault, it is necessary to specify regions of high-frequency radiation on the fault as well as those of low-frequency wave radiation (asperities). So, a spatial relationship between these two kinds of regions is of particular interests in strong-motion seismology.

Here, we classify the relation into 3 cases: (1) Complementary, (2) Matching, and (3) Otherwise. The complementary case means that locations of high-frequency radiation are at peripheries of asperities. The matching case means the both locations are the same. The otherwise case includes any others but the two. For example, if there are several

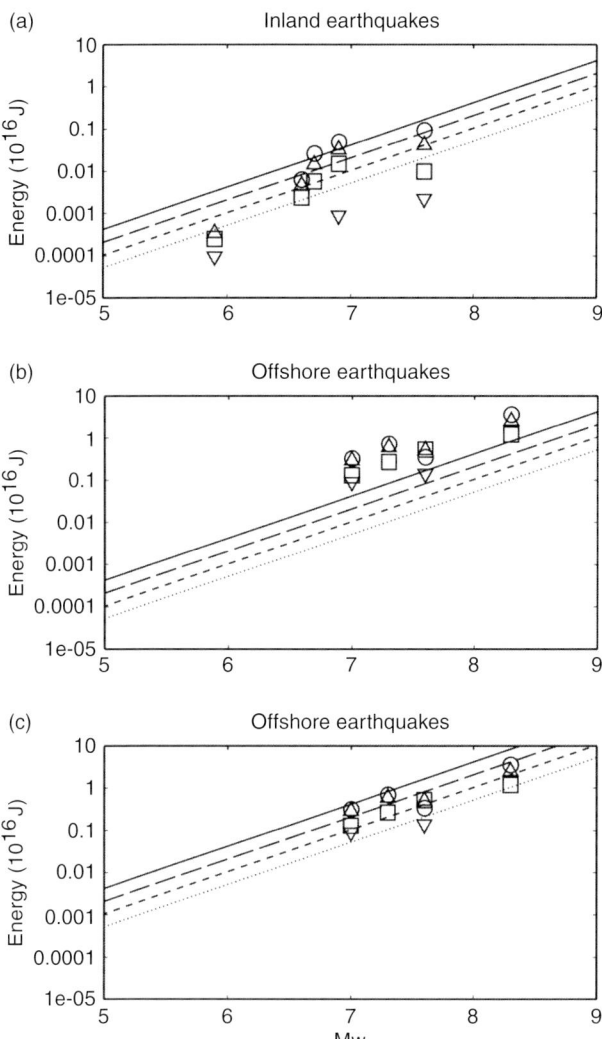

FIG. 10. Scaling relation of high-frequency seismic energy radiation in four octave frequency bands with moment magnitude. Open circles, open triangles, open squares, and open inverted triangles are observed energy for 1–2, 2–4, 4–8, and 8–16 Hz, respectively. Solid, long-broken, short-broken, and dotted straight lines with a proportionality factor of 1 show theoretically expected values for 1–2, 2–4, 4–8, and 8–16 Hz, respectively. Figure (a) shows the results for 5 inland earthquakes with the expected values from the A value of Dan et al. (2001). Figure (b) is for offshore earthquakes composed of 3 interplate earthquakes and 1 intraslab earthquake with the expected values from the A value of Dan et al. (2001). Figure (c) is for the offshore earthquakes composed of 3 interplate earthquakes and 1 intraslab earthquake with the expected values from 3.16 (square root of 10) times the A value of Dan et al. (2001).

asperities and high-frequency sources on a fault plane, some are complementary but some are matching. Such a case is identified as the otherwise case. In terms of this classification, four events are identified as complementary, none is matching, and four are otherwise. Unfortunately, the comparison is impossible for one event because no fault slip models have been estimated for lower frequencies.

According to envelope inversion studies by other groups (shown in Table 4), the complementary relation is reported for the 1989 Loma Prieta earthquake (Ms 7.1) (Zeng et al., 1993) and for the 1993 Kushiro-Oki, Japan, earthquake (Mw 7.6) (Kakehi and Irikura, 1996). The otherwise relation is reported for the 1993 Hokkaido-Nansei-Oki, Japan, earthquake (Mw 7.7) (Kakehi and Irikura, 1997), for the 1994 Northridge earthquake (Ms 6.7) (Hartzell et al., 1996) and the 1995 Kobe, Japan, earthquake (Mw 6.9) (Kakehi et al., 1996).

On the contrary, from the simultaneous fitting of displacement waveforms and acceleration envelopes in a 0.2–10 Hz band for 12 crustal earthquakes (Mw 4.8 – 6.0) in Japan, Miyake et al. (2003) estimated source areas, which they call the strong-motion generation areas, on each fault plane. They found that the strong-motion generation areas coincide with those of asperities.

The classification conducted in this subsection is still qualitative. We hope to introduce more quantitative measures to characterize the relationship between locations of low-frequency radiation and those of high-frequency radiation in the near future. In terms of this viewpoint, Gusev et al. (2006) conducted a quantitative analysis. They estimated slip rate time functions in lower frequencies and seismic luminosity (source) time functions in high frequencies (0.5–2.5 Hz) at the same time for 23 intermediate-depth earthquakes with magnitude of 6.8 and larger. Calculating correlation coefficients between the two kinds of time functions after corrections of propagation effects, they estimated the mean correlation coefficient of 0.52, and concluded that this mean value indicates the genuine difference in distributions of low-frequency radiation and high-frequency radiation.

Finally, we discuss this spatial relation based on kinematic heterogeneous fault-rupture models (e.g., Papageorgiou and Aki, 1983; Koyama, 1985). In the models,

TABLE 4. Summary of envelope inversion analyses by other authors

Event (Mw)	Source location	Focal mechanism type	HF Location and LF location	Reference
1989 Loma Prieta (6.9)	Inland	Reverse	Complementary	Zeng et al. (1993)
1993 Kushiro-Oki, JAPAN (7.6)	Intraslab	Down-dip extension	Complementary	Kakehi and Irikura (1996)
1993 Hokkaido-Nansei–Oki, JAPAN (7.7)	Plate boundary	Thrust	Otherwise	Kakehi and Irikura (1997)
1994 Northridge, California (6.6)	Inland	Reverse	Otherwise	Hartzell et al., (1996)
1995 Kobe, JAPAN (6.9)	Inland	Right-lateral strike slip	Otherwise	Kakehi et al. (1996)

HF, high frequency; LF, low frequency.

many small-scale asperities or patches are assumed to be randomly distributed on the fault. Generation of high-frequency seismic waves is due to the rupture of the small-scale asperities. On the contrary, low-frequency waves are radiated by the coherent rupture of the entire fault. Relative contribution of the coherent rupture and the incoherent rupture may be a key. If the coherent rupture is dominant, we expect that the rupture propagates smoothly and stops at the edge of the fault. For example, Madariaga (1976) investigated the dynamic rupture of a circular crack. He showed that slip (moment release) is large at the center of the fault and high-frequency seismic waves are strongly radiated from edges of the fault. This is an example of a complementary relation. On the contrary, if the contribution of the incoherent rupture becomes large, we expect that high-frequency waves are strongly radiated from throughout the fault plane due to the rupture of small-scale asperities, whereas slip is large at the center of the fault due to the coherent rupture. This is an explanation for the matching case. Based on the consideration, the relationship between the locations of high- and low-frequency radiation may be understood by a relative weight of the coherent rupture and the incoherent rupture, which may be controlled by the heterogeneity of stress and/or strength on the earthquake fault: Complementary for homogeneous faults and matching and/or otherwise for heterogeneous faults. But the consideration here is under a frame of kinematic rupture models. Dynamic rupture simulations should be conducted for asperities with various degrees of heterogeneities in stress and/or strength for more quantitative considerations. Kato (2007) may be the first step forward to this direction.

5. Conclusions

We have made a brief review of envelope inversion studies for high-frequency seismic wave radiation from moderate-to-large earthquakes. Several methods have been proposed since the 1990s. An assumption on the incoherency of high-frequency seismic waves facilitates direct convolution of each envelope for source, path, and site effects. Thanks to the methods, it became possible to image earthquake source process at high frequencies and to compare the results to those from lower frequencies. On the basis of results for 9 earthquakes so far analyzed by us, we have clarified the following two characteristics in high-frequency seismic energy radiation. First, logarithm of the high-frequency seismic energy is proportional to the moment magnitude with a coefficient of proportionality of 1 as is theoretically expected. Moreover, a regional difference in the high-frequency seismic energy radiation has been detected: Earthquakes in offshore regions in northeastern Japan are found to be more energetic by about an order of magnitude than inland earthquakes in Japan and Taiwan. Second, spatial relationships between the locations of asperities and the locations of high-frequency radiation have been summarized. Among 9 earthquakes analyzed by us, 4 are complementary, none is matching, 4 are otherwise, and 1 is indeterminate. According to analyses of 5 earthquakes by other groups, 2 are reported to be complementary, none is matching, 3 are otherwise. In total, 6 are complementary, none is matching, 7 are otherwise, and 1 is indeterminate among 14 earthquakes. Reasons for the variation are not yet known. However, heterogeneities in the distribution of stress, strength, and material properties on and around earthquake faults may control the variation. The two characteristics found for high-frequency seismic energy radiation will give important information for the study of high-frequency earthquake source processes.

And they may also contribute to improving the accuracy of predicting strong ground motion, because the locations of high-frequency radiation on an earthquake fault greatly affect ground motion at nearby stations.

ACKNOWLEDGMENTS

Most of the results were obtained through collaborations with H. Sato, M. Ohtake, T. Nishimura, and R. Watanabe. We thank the National Research Institute for Earth Science and Disaster Prevention, Japan for providing us with strong-motion data recorded by the K-NET and the Kiknet. A figure of the slip distribution for the 2003 Miyagi-Oki earthquake was provided by Y. Yagi of Tsukuba University. We greatly appreciate thoughtful comments from an associate editor, M. Fehler, and two reviewers, A. Gusev and Y. Zeng.

REFERENCES

Aki, K. (1967). Scaling law of seismic spectrum. *J. Geophys. Res.* **72**, 1217–1231.
Brune, J.N. (1970). Tectonic stress and the spectra of seismic shear waves from earthquakes. *J. Geophys. Res.* **75**, 4997–5009.
Chandrasekhar, S. (1960). *Radiative Transfer*. Dover, New York, pp. 393.
Cocco, M., Boatwright, J. (1993). The envelopes of acceleration time histories. *Bull. Seismol. Soc. Am.* **83**, 1095–1114.
Dan, K., Watanabe, M., Sato, T., Ishii, T. (2001). Short-period source spectra inferred from variable-slip rupture models and modeling of earthquake faults for strong motion prediction by semi-empirical method (in Japanese with English abstract). *J. Struct. Constr. Eng. AIJ* **545**, 51–62.
Gusev, A.A. (1983). Descriptive statistical model of earthquake source radiation and its application to an estimation of short period strong ground motion. *Geophys. J. R. Astron. Soc.* **74**, 787–808.
Gusev, A.A., Abubakirov, I.R. (1987). Monte-Carlo simulation of record envelope of a near earthquake. *Phys. Earth Planet. Inter.* **49**, 30–36.
Gusev, A.A., Pavlov, V.M. (1991). Deconvolution of squared velocity waveform as applied to the study of a noncoherent short-period radiator in the earthquake source. *Pure Appl. Geophys.* **136**, 235–244.
Gusev, A.A., Guseva, E.M., Panza, G.F. (2006). Correlation between local slip rate and local high-frequency seismic radiation in an earthquake fault. *Pure Appl. Geophys.* **163**, 1305–1325.
Gutenberg, B., Richter, C.F. (1956). Magnitude and energy of earthquakes. *Ann. Geofis.* **9**, 1–15.
Hartzell, S., P. Liu, and C. Mendoza (1996). The 1994 Northridge, California, earthquake: Investigation of rupture velocity, risetime, and high-frequency radiation, *J. Geophys. Res.*, **101**, 20091-20108.
Iida, M., Hakuno, M. (1984). The difference in the complexities between the 1978 Miyagiken-oki earthquake and the 1968 Tokachi-oki earthquake from a viewpoint of the short-period range. *Nat. Disaster Sci.* **6**, 1–26.
Irikura, K., Miyake, H. (2001). Prediction of strong ground motions for scenario earthquakes (in Japanese with English abstract). *J. Geogr.* **110**, 849–875.
Ishimaru, A. (1978). *Wave Propagation and Scattering in Random Media*, Academic Press, San Diego, Volume 1, pp. 270.
Kakehi, Y., Irikura, K. (1996). Estimation of high-frequency wave radiation areas on the fault plane by the envelope inversion of acceleration seismograms. *Geophys. J. Int.* **125**, 892–900.
Kakehi, Y., Irikura, K. (1997). High-frequency radiation process during earthquake faulting—Envelope inversion of acceleration seismograms from the 1993 Hokkaido-Nansei-Oki, Japan, earthquake. *Bull. Seismol. Soc. Am.* **87**, 904–917.

Kakehi, Y., Irikura, K., Hoshiba, M. (1996). Estimation of high-frequency wave radiation areas on the fault of the 1995 Hyogo-ken Nanbu earthquake by the envelope inversion of acceleration seismograms. *J. Phys. Earth* **44**, 505–517.

Kanamori, H., Anderson, D.L. (1975). Theoretical basis of some empirical relations in seismology. *Bull. Seismol. Soc. Am.* **65**, 1073–1095.

Kato, K., Takemura, M., Yashiro, K. (1998). Regional variation of source spectra in high-frequency range determined from strong motion records (in Japanese with English abstract). *Zisin* **2**(51), 123–138.

Kato, N. (2007). How frictional properties lead to either rupture-front focusing or cracklike behavior. *Bull. Seismol. Soc. Am.* **97**, 2182–2189.

Koyama, J. (1985). Earthquake source time function from coherent and incoherent rupture. *Tectonophysics* **118**, 227–242.

Koyama, J. (1994). General description of the complex faulting process and some empirical relations in seismology. *J. Phys. Earth* **42**, 103–148.

Koyama, J., Zheng, S.H. (1985). Excitation of short period body waves by great earthquakes. *Phys. Earth Planet. Inter.* **37**, 108–123.

Madariaga, R. (1976). Dynamics of an expanding circular fault. *Bull. Seismol. Soc. Am.* **66**, 639–666.

Mai, M., Beroza, G. (2000). Source scaling properties from finite-fault-rupture models. *Bull. Seismol. Soc. Am.* **90**, 604–615.

Midorikawa, S., Kobayashi, H. (1979). On estimation of strong earthquake motions with regard to fault rupture (in Japanese with English abstract). *Trans. Arch. Inst. Jpn.* **282**, 71–81.

Miyake, H., Iwata, T., Irikura, K. (2003). Source characterization for broadband ground-motion simulation: Kinematic heterogeneous source model and strong motion generation area. *Bull. Seismol. Soc. Am.* **93**, 2531–2545.

Nakahara, H. (2003). Envelope inversion analysis for the high-frequency seismic energy radiation by using Green's functions in a depth dependent velocity structure: The 2000 Western Tottori earthquake, IUGG 2003 meeting SS02/04A/D-024, Sapporo, Japan.

Nakahara, H. (2004). High-frequency envelope inversion analysis of the 2003 Tokachi-Oki, JAPAN, earthquake (Mw8.0), AGU fall meeting S13D-1082, San Francisco, California.

Nakahara, H. (2005a). High-frequency envelope inversion analysis of the May 26, 2003 Miyagi-Ken-Oki, JAPAN, earthquake (Mj 7.0) (in Japanese). *Chikyu Monthly* **27**, 39–43.

Nakahara, H. (2005b). High-frequency envelope inversion analysis of the 2004 Niigata-Ken Chuetsu earthquake (Mw 6.6) (in Japanese), Japan earth and planetary science joint meeting, S079–006.

Nakahara, H., Nishimura, T., Sato, H., Ohtake, M. (1998). Seismogram envelope inversion for the spatial distribution of high-frequency energy radiation from the earthquake fault: Application to the 1994 far east off Sanriku earthquake, Japan. *J. Geophys. Res.* **103**, 855–867.

Nakahara, H., Sato, H., Ohtake, M., Nishimura, T. (1999). Spatial distribution of high-frequency energy radiation on the fault of the 1995 Hyogo-Ken Nanbu earthquake (Mw 6.9) on the basis of the seismogram envelope inversion. *Bull. Seismol. Soc. Am.* **89**, 22–35.

Nakahara, H., Nishimura, T., Sato, H., Ohtake, M., Kinoshita, S., Hamaguchi, H. (2002). Broadband source process of the 1998 Iwate Prefecture, Japan, earthquake as revealed from inversion analyses of seismic waveforms and envelopes. *Bull. Seismol. Soc. Am.* **92**, 1708–1720.

Nakahara, H., Watanabe, R., Sato, H., Ohtake, M. (2006). Spatial distribution of high-frequency seismic energy radiation on the fault plane of the 1999 Chi-Chi, Taiwan, earthquake (Mw 7.6) as revealed from an envelope inversion analysis, Submitted to PAGEOPH.

Papageorgiou, A.S., Aki, K. (1983). A specific barrier model for quantitative description of inhomogeneous faulting and the prediction of strong ground motion Part I. Description of the model. *Bull. Seismol. Soc. Am.* **73**, 693–722.

Petukhin, A.G., Nakahara, H., Gusev, A.A. (2004). Inversion of high-frequency source radiation of M6.8 Avachinsky Gulf, Kamchatka, earthquake using empirical and theoretical envelope Green functions. *Earth Planets Space* **56**, 921–925.

Sakurai, K. (1995). Separation of scattering loss and intrinsic absorption based on a multiple non-isotropic scattering model (in Japanese), Master thesis Tohoku University, Sendai, Japan.

Sato, H., Fehler, M. (1998). *Seismic Wave Propagation and Scattering in the Heterogeneous Earth*. Springer-Verlak, New York.

Sato, H., Nakahara, H., Ohtake, M. (1997). Synthesis of scattered energy density for the non-spherical radiation from a point shear dislocation source based on the radiative transfer theory. *Phys. Earth Planet. Inter.* **104**, 1–13.

Satoh, T. (2004). Short-period spectral level of intraplate and interpolate earthquakes occurring off Miyagi Prefecture (in Japanese with English abstract). *J. JAEE* **4**, 1–4.

Somerville, P., Irikura, K., Graves, R., Sawada, S., Wald, D., Abrahamson, N., Iwasaki, Y., Kagawa, T., Smith, N., Kowada, A. (1999). Characterizing crustal earthquake slip models for the prediction of strong ground motion. *Seismol. Res. Lett.* **70**, 59–80.

Takemura, M., Hiehata, S., Ikeura, T., Uetake, T. (1989). Regional variation of source properties for middle earthquakes in a subduction region (in Japanese with English abstract). *Zisin* 2(42), 349–359.

Wu, R.S. (1985). Multiple scattering and energy transfer of seismic waves -seperation of scattering effect from intrinsic attenuatuon- I. Theoretical modeling. *Geophys. J. R. Astron. Soc.* **82**, 57–80.

Yagi, Y. (2003). Source rupture process of the 2003 Miyagi-ken-oki earthquake determined by joint inversion of teleseismic body waves and strong ground motion data, http://iisee.kenken.go.jp/staff/yagi/eq/east_honshu20030526.

Zeng, Y., Su, F., Aki, K. (1991). Scattering wave energy propagation in a random isotropic scattering medium 1. Theory *J. Geophys. Res.* **96**, 607–619.

Zeng, Y., Aki, K., Teng, T.L. (1993). Mapping of the high-frequency source radiation for the Loma Prieta Earthquake, California. *J. Geophys. Res.* **98**, 11981–11993.

ON THE RANDOM NATURE OF EARTHQUAKE SOURCES AND GROUND MOTIONS: A UNIFIED THEORY

Daniel Lavallée[1]

Abstract

The synthesis of fundamental principles of physics and of the theory of probability provides a coherent and unified picture of earthquake variability from its recording in the ground motions to its inference in source models. This theory, based on the representation theorem and the (generalized) Central Limit Theorem, stipulates that the random properties of the ground motions and the source for a single earthquake should both be (approximately) distributed according to a Lévy law. The Lévy law is a special class of probability law. According to the (generalized) Central Limit Theorem, a sum of Lévy random variables is simply a Lévy random variable. The Gauss and the Cauchy laws are special cases of the Lévy law.

Random models are best suited to describe the spatial heterogeneity embedded in the earthquake source model of slip (or stress). For this purpose, we have developed a random model that can reproduce the variability in slip amplitude and the long-range correlation of the slip spatial distribution. Analysis of slip spatial distribution shows that a non-Gaussian law, that is, the Lévy law, is better suited to describe the distribution of slip amplitude values over the fault. Furthermore, a comparison of the random properties of the source and of the ground motions for the 1999 Chi-Chi and 2004 Parkfield earthquakes demonstrates that the slip distribution and the peak ground acceleration (PGA) can be described by the Lévy law. Additionally, the tails of the probability density functions (PDFs) characterizing the slip and the |PGA| are controlled by a parameter, the Lévy index, with almost the same values as predicted by the (generalized) Central Limit Theorem. Thus, from the source to the ground motion, the Lévy index provides a universal law describing the tail of the PDF.

The PDF tail controls the frequency at which extreme large events occur. These large events correspond to the large stress drops—or asperities—distributed on the fault surface and to the large PGA observed in the ground motion. The theory and the results suggest that the frequency of these events is coupled: the PDF of the |PGA| is a direct consequence of the PDF of the asperities.

Key Words: Earthquake, source model, ground motion, random model, Lévy law.
© 2008 Elsevier Inc.

1. Introduction

In a discussion about his motivation to study turbulence, Kolmogorov indicated the following: "I took an interest in the study of turbulent flows of liquids and gases in the late 1930s. It was clear to me from the very beginning that the main mathematical

---

[1] Author thanks e-mail: daniel@crustal.ucsb.edu

instrument in this study must be the theory of random functions of several variables (random fields) which had only then originated" (Kolmogorov, 1991). More recently, Kagan observed that "seismicity is the turbulence of solid" (Kagan, 1992; cited in Andrews, 1980, p. 3869). From this, one can infer that a proper "mathematical instrument" in the study of the "turbulence of solid" "must be the theory of random functions." In this chapter, we intend to illustrate the full consequences of this approach.

The investigation of the random properties of source models of several earthquakes suggests that the probability law that best describes the slip variability, or heterogeneity, is either a Cauchy law or the more general Lévy law (Lavallée and Archuleta, 2003, 2005; Lavallée et al. 2006a). The Gauss (or Normal) law is less appropriate. The tail of the Gauss probability density function (PDF) decreases too quickly to predict with accuracy the frequency of large fluctuations observed in the slip inversions. These large fluctuations correspond to the large stress drops—or asperities—distributed over the fault surface. The list of earthquake source models investigated includes the 1979 Imperial Valley, the 1989 Loma Prieta, the 1994 Northridge, the 1995 Hyogo-ken Nanbu (Kobe), the 1999 Chi-Chi and the 2004 Parkfield earthquakes.

Kostrov and Das (1988, p. 234) observed that heterogeneity in the source parameters "would be manifested in the complexity of the pulse shapes." Lavallée and Archuleta (2005) take a step further and assume that observation of a non-Gaussian law of the Lévy type in the ground motion metrics (Gusev, 1989, 1996; Tumarkin and Archuleta, 1997) could have its origin in the spatial variability of the slip over the fault surface. The basic idea supporting this hypothesis rests on the fundamental properties of linear wave propagation and the Lévy random variables. The principle of superposition stipulates that the sum of linear waves is also a linear wave. Consequently, during an earthquake the ground motion recorded at a given distance from the fault is essentially the sum of seismic waves emitted by point sources distributed over the fault surface. Similarly, the (generalized) Central Limit Theorem postulates that the sum of Lévy random variables is also a Lévy random variable (Zolotarev, 1986; Uchaikin and Zolotarev, 1999). These two fundamental principles can be used to infer the random properties of the radiated field and *at fortiori* of ground motion metrics. During an earthquake, the rupture front propagates over the fault surface; as the rupture front reaches different points on the fault, each point source will emit a wave with an amplitude proportional to the slip (or stress released, Andrews, 1980). Since the slips are distributed according to a Lévy law, the point source wave amplitudes will also be. Because the point source wave amplitudes are distributed according to the Lévy law, the sum of these amplitudes observed at a given distance from the sources will also be distributed according to the Lévy law (see Fig. 1 for a schematic illustration).

In this chapter, we will discuss the theoretical framework and present empirical evidence that earthquake source models, and the ground motions generated during the rupture process, are coupled through their random properties. A fundamental consequence of this coupling is to relate the PDF tails of the source and of the ground motion through an invariant measure: the Lévy index of the Lévy law. Not only heterogeneity in the source parameters will be mirrored in the complexity of the ground motions (as suggested by Kostrov and Das, 1988, p. 234), but their random properties will be also!

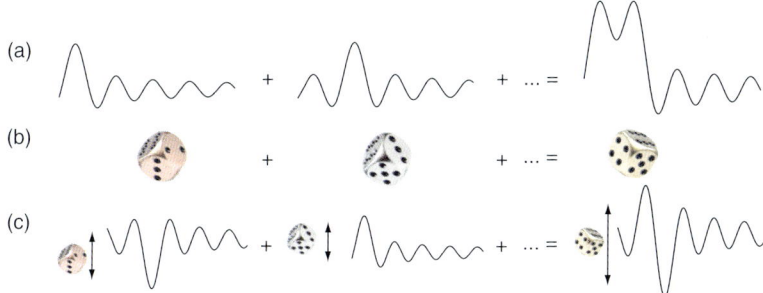

FIG. 1. Two of the most fundamental principles in science are founded on the similitude in properties between a single event and a sum of these events: the principle of superposition of linear waves (a), and the Central Limit Theorem (b). In (c), the two principles are combined to predict that the sum of linear waves with amplitude distributed according to the Lévy law will result in a wave signal with amplitude also distributed according to the Lévy law.

2. RANDOM MODEL OF EARTHQUAKES SLIP SPATIAL DISTRIBUTION AND CONSEQUENCES FOR THE GROUND MOTIONS

2.1. From the Source ...

Up to now, the spatial heterogeneity observed in the slip (or stress drop) has been interpreted and simulated with the help of random models. According to Andrews (1980), both the slip and the stress drop are related by a linear relation—see also the Appendix in Mai and Beroza (2002). (In a uniform medium, the Fourier transform of the stress drop is proportional to the magnitude of the wave number times the Fourier transform of the slip.) Because of this relation, we assume that we can derive the random properties of the stress from the slip or vice versa. (Thus, we will refer to slip random property and stress drop random property interchangeably.)

A random model can be understood (at least approximately) as a set of random variables characterized by a given probability law, and a set of rules and operations to combine the random variables. The set of rules and operations can be chosen in such a way to control the functional behavior of the correlation function and spectrum (two-point statistics), the three-point statistics, and so on. Different rules and probability laws have been reported to model the spatial heterogeneity of the slip or stress drop distribution.

In Boore and Joyner (1978), the rule consists of a simple normalization of the random variables, while it consists of a normalization and smoothing of the random variables in Oglesby and Day (2002). In the past, the rule of choice implemented in almost every random model of source heterogeneity consisted of operations that insured that the Fourier amplitude (or spectrum) of the source decreased asymptotically with a power law behavior. In a two-dimensional random model, the spectrum is proportional to $k^{-\nu_{2D}}$ where $k = |\mathbf{k}|$ is the 2D radial wavelength number, while in a one-dimensional random model the spectrum is proportional to $k_1^{-\nu_{1D}}$, where k_1 is the wave number component along the length of the fault. (For an isotropic random function, the exponent of the 2D spectrum is related to the exponent for the spectra of 1D profiles by the following relation $\nu_{2D} = \nu_{1D} + 1$—see Peitgen and Saupe, 1988.) This rule essentially constrains the functional behavior of the autocorrelation function of the random model. Several values have been proposed for the parameter v_{2D} (or v_{1D}). They are based on theoretical considerations (for a review, see Herrero and Bernard, 1994, and references therein; see Lavallée et al., 2006a), numerical simulations (Liu-Zeng et al. 2005), or empirical studies of slip inversions (Mai and Beroza, 2002; Lavallée et al., 2006a). A proper choice for the value for the parameter v_{1D} (or v_{1D}) and the method used to generate the power law behavior $k^{-v_{2D}}$ is still the object of investigation and debate. In recent modeling of source heterogeneity, the power law behavior is generated using either a von Karman model (Guatteri et al., 2003; Hartzell et al., 2005; Liu et al., 2006), or a fractional Brownian motion (fBm) (Lavallée and Archuleta, 2003; Liu-Zeng et al., 2005).

In the literature, there is a large focus on the power spectrum of the random model of sources. However, much less attention has been paid to the fundamental question of the probability law that will govern the distribution of the random variables used to generate the random model of slip. The distribution of the random variables is the basic feature in the random model that controls the slip variability over the fault surface and the expectation to observe large slip values. The distribution is also the foundation of the random model, in the sense that the rules are applied to the random variables to obtain, for instance, the proper power spectrum. Furthermore, application of the rules assigned to the random variables may drastically change the law governing their distribution. For instance, a sum of random variables distributed according to a uniform distribution gives a random variable distributed according to the Gauss law. It is thus unfortunate to neglect this basic feature of random modeling. There is no discussion of the probability law in the random model derived in Mai and Beroza (2002) and subsequent synthetic slip spatial distribution based on this model (Guatteri et al., 2003; Hartzell et al., 2005). In Boore and Joyner (1978), the slip values are distributed according to a Uniform law, while Oglesby and Day (2002) choose a Uniform law to generate synthetic stress spatial distribution. It should be noted that under the relationship between the slip and the stress drop given by Andrews (1980), it is not possible to have both the slip and the stress drop distributed according to a Uniform law (see below for details on sum of random variables). Liu-Zeng et al. (2005) use a random model based on Gauss random variables. Liu et al. (2006) developed a random model of slip distributed according to the Cauchy law (Lavallée and Archuleta, 2003) to compute broadband ground motions that compared rather well to near-source data recorded during the 1994 Northridge earthquake.

There are more than 80 probability laws (just including those for continuous random variables) available to generate random variables that can be used to model the slip spatial variability (for a listing of the probability laws, see http://mathworld.wolfram.com/topics/ContinuousDistributions.html). In theory, all these laws can be used and tested (providing the rather remarkable opportunity of publishing many papers—a sure career move). However, early investigation of numerical modeling of fault ruptures (Mikumo and Miyatake, 1978; Fakao and Furumoto, 1985) and the pioneering works of Gusev (1989, 1992) suggest that the choice can be restrained to probability laws with a PDF characterized by "heavy tails" such as the Pareto law (Gusev, 1989) or the Lévy law (Gusev, 1992).

In general, the PDF of the Lévy law is characterized by a "heavy tail" (except for the Gauss law, a special case of the Lévy law). Four parameters (α, β, γ, and μ) are needed to characterize the Lévy law. The parameter $\alpha (0 < \alpha \leq 2)$ controls the rate of fall off of the "heavy tail" of the PDF (except for $\alpha = 2$ corresponding to the Gauss law). The parameter

$\beta(-1 \leq \beta \leq 1)$ controls the departure from symmetry of the PDF curve. The parameter $\gamma (\gamma > 0)$ is mainly responsible for the width of the PDF, while the parameter $\mu(-\infty < \mu < \infty)$ controls the location of the PDF. A fundamental property of the Lévy law is that a sum of Lévy random variables $X(\alpha, \beta_j, \gamma_j, \mu_j)$ maps onto a Lévy random variable

$$X(\alpha, \beta, \gamma, \mu) = \sum_j A_j X(\alpha, \beta_j, \gamma_j, \mu_j) + B, \tag{1}$$

where A_j and B are real numbers. Furthermore, the Lévy index α characterizing the random variable $X(\alpha, \beta, \gamma, \mu)$ is identical to the Lévy index of the summation of the Lévy random variables $X(\alpha, \beta_j, \gamma_j, \mu_j)$. Additional details and the relationship between the quadruple $(\alpha, \beta, \gamma, \mu)$ and the parameters $(\beta_j, \gamma_j, \text{and } \mu_j)$, A_j and B are discussed in Zolotarev (1986) and in Uchaikin and Zolotarev (1999). The expression in Eq. (1) can be understood as the general formulation of the Central Limit Theorem. The origin of the term *stable*, also used to name the Lévy law, is attributed to the property in Eq. (1)—Janicki and Weron (1994).

Lavallée and Archuleta (2003) introduced a random model of slip spatial heterogeneity for the 1979 Imperial Valley earthquake (Archuleta, 1984) with random variables distributed according to the Lévy law. The random model is similar to an fBm, except that in the fBm, the random variables are usually distributed according to the Gauss law. The wave number spectrum of the slip spatial distribution decreases as a power law function of the wave number. So, fBm provides the proper rule to reproduce the power spectrum of the slip, but the Gauss law turns out to be a poor proxy to reproduce the spatial variability observed in the slip spatial distribution. Subsequent investigations of the random properties of source models of several earthquakes, the 1989 Loma Prieta, 1994 Northridge, 1995 Hyogo-ken Nanbu (Kobe), and the 1999 Chi-Chi earthquakes, established that the probability law that best describes the slip variability is either the Cauchy law (a special case of the Lévy law) or the more general Lévy law (Lavallée and Archuleta, 2005; Lavallée et al. 2006a).

The (discrete) random model of the slip Δu_x is given by the following expression in one dimension:

$$\Delta u_x \propto \sum_{s=2-N/2}^{1+N/2} \left| \frac{k_1}{2\pi} \right|^{-v_{1D}/2} F_s[X_x] \exp\left[\frac{-2\pi i (x-1)(s-1)}{N}\right], \tag{2}$$

where the random variables X_x are distributed over a 1D lattice (or grid) of length N. The index x is the integer spatial component along the 1D lattice. The discrete variable s is related to wave number k_1 by $k_1 = 2\pi(s-1)/N$; $F_s[X_x]$ is the discrete Fourier transform of the random variables (for $s \leq 0$ in Eq. (2), the index $s = N + s$ in $F_s[X_x]$). The exponent v_{1D} measures the deviation of the wave number spectrum from flat, that is, white noise spectrum with $v_{1D} = 0$. We assume that $k_1^{-v_{1D}/2} F_s[X_x] \to 0$ at $s = 1$. (On the question of the scaling property of the slip spatial distribution, see also Section 2 in Lavallée et al., 2006a.)

Given the relation (Eq. 2), it is possible in principle to invert Δu_x to recover the set of random variables X_x and then to compute the probability law that fits the PDF associated with X_x. For this, it is sufficient to recognize that the 1D power spectrum $P(k_1)$ for Δu_x is given by the following relation:

$$P(k_1) = |F_s[\Delta u_x]|^2 \propto k_1^{-v_{1D}}. \tag{3}$$

Equation (3) can be used to compute the values of the parameter v_{1D}. Using the value of v_{1D}, the random variables X_x can be computed using

$$X_x \propto F_x^{-1}\left[F_s(\Delta u_x) \times k_1^{v_{1D}/2}\right], \tag{4}$$

where F_x^{-1} is the Fourier inverse. (For a generalization of the model to 2D, see Lavallée et al., 2006a.)

Ji (2006; personal communication) draws a parallel between kinematic source inversion, understood as a mapping of the radiated seismic energy onto the source, and watching a star with a telescope. When the telescope is poorly focused, the image of the star is blurred. Focusing the telescope concentrates light rays over smaller subregions, and allows us to picture the star image with additional details that were absent at lower resolution. Computing source inversion, refining the fault geometry, and selecting ground motions and Green functions is analogous to refocusing the telescope. It provides additional details in the slip spatial distribution. Formulating the random property of the slip with the help of the random model in Eq. (2) is in good agreement with this analogy. In Eq. (7), spatial variability at higher resolution is obtained by laying additional randomness at smaller scales (or higher wave numbers). However, the random variables are independent of the length scale or resolution. In other words, at every length scale (or wavelength), the random variables are governed by the same probability law.

Finally, it is important to review (although briefly) the limitations of the random model discussed in this section (for a detailed discussion of the limitations, see Lavallée et al., 2006a). First, the power law behavior of the slip spatial distribution [see Eq. (3)] is limited to a finite range of k given (approximately) by $10^{-4}\,\text{m}^{-1} \leq k \leq 10^2\,\text{m}^{-1}$ (see Andrews, 1980). Second, the power spectrum, and the exponent in Eq. (3) can only be computed in average. This raises questions about the accuracy of the v_{1D} (or v_{2D}) that cannot be easily resolved in view of the quality, the resolution, and the number of events (or subfaults) available in current slip models. Third, the range of values for the random variables X_x is also bounded, which implies that the Lévy PDF is truncated. The rationale for the truncation is that physical or geophysical parameters are usually bounded between finite values. For instance, the maximum slip value that can be observed is limited by the finite size of the fault. Peak ground velocity (PGV) and peak ground acceleration (PGA) are also limited to a finite range of values. Thus, modeling these parameters with the Lévy law—or other probability laws (McGuire, 2004)—requires a modification to the asymptotic behavior of the probability law. An instance where the effect of the truncation may have significant implications is in the computation of the random properties of the ground motions under limit conditions. This question is not only theoretical, for it has fundamental consequences in computing the probability associated with seismic hazard (PSHA –see McGuire, 2004) for the nuclear waste repository at Yucca Mountain and in predicting the observation of large PGA over a very long period of time (Bommer et al., 2004; Andrews et al., 2007).

In principle, it is possible to modify the asymptotic behavior of the PDF of the Lévy law while preserving—at least approximately—the properties of replication of the Lévy random variables given by the (generalized) Central Limit Theorem. For instance, one can assume that the power law behavior of the PDF tail holds up to a very large cutoff random variable. The sum of random variables with values (sufficiently) smaller than the cutoff will converge to a Lévy random variable. Alternatively, it is also possible to consider an explicit functional form for the truncated tail (for details, see Paul and Baschnagel, 1999 and references therein;

Voit, 2001; Sornette, 2004). In this chapter, we assume that the property of the Lévy random variables given by Eq. (1) is not affected by the truncation. Thus, in addition to the five parameters (α, β, γ, μ, and v_{1D}) needed to specify the random model of slip given in Eq. (2), the two cutoff values bounding the range of random variables X_x will have to be provided (see also Lavallée and Archuleta, 2003).

2.2. ... to the Ground Motion

The formulation of the slip spatial distribution in terms of Lévy random variables has fundamental consequences on the radiated field generated by the rupture motions as well as on the ground motions recorded at the surface during an earthquake. First, consider the consequences of this formulation on the slip. In Eq. (1), the random model of the Δu_x is essentially obtained through a linear sum of random variables distributed according to the Lévy law. Indeed, consider the expression for the real part of the Fourier transform $F_s[X_x]$ in Eq. (2), which can be written as follows:

$$\text{Re}[F_s[X_x]] \propto \sum_{x=1}^{N} X_x \cos\left[\frac{2\pi(x-1)(s-1)}{N}\right]. \tag{5}$$

For any fixed value of s, the left-hand side of Eq. (5) is a sum of Lévy random variables modulated by real numbers given by the cosine functions estimated at s. Thus, according to Eq. (1), $\text{Re}[F_s[X_x]]$ is a random variable distributed according to the Lévy law. Note, however, that the parameters (β, γ, and μ) of the random variable $\text{Re}[F_s[X_x]]$ vary with s. The Lévy index α remains invariant, and is thus the same for $\text{Re}[F_s[X_x]]$ and the random variable X_x. Again we emphasize this fundamental property of the summation of Lévy random variables. The same considerations apply to the imaginary part of $F_s[X_x]$ which is thus also distributed according to the Lévy law. In a similar way, when expanding $\exp[-2\pi i(x-1)(s-1)/N]$ in term of sines and cosines, the sum in the left-hand side of Eq. (6) is performed over Lévy random variables $\text{Re}[F_s[X_x]]$ or $\text{Im}[F_s[X_x]]$, modulated by the product of a cosines (or sines) with the function $|k_1/2\pi|^{-\nu/2}$. For instance, the real part of Δu_x in Eq. (2) is given by

$$\text{Re}[\Delta u_x] \propto \sum_{s=2-N/2}^{1+N/2} \left|\frac{k_1}{2\pi}\right|^{-\nu/2} \left\{ \text{Re}[F_s[X_x]]\cos\left[\frac{2\pi(x-1)(s-1)}{N}\right] \right.$$
$$\left. + \text{Im}[F_s[X_x]]\sin\left[\frac{2\pi(x-1)(s-1)}{N}\right] \right\}. \tag{6}$$

For any fixed value of the position x and fixed value of s, the product $\{|k_1/2\pi|^{-\nu/2}\cos[2\pi(x-1)(s-1)/N]\}$, or $\{|k_1/2\pi|^{-\nu/2}\sin[2\pi(x-1)(s-1)/N]\}$, is reduced to a real number. These products are factors of the random variables $\text{Re}[F_s[X_x]]$ and $\text{Im}[F_s[X_x]]$, respectively, and correspond to the real numbers A_j in Eq. (1). As far as the random properties are concerned, the sum on the left-hand side of Eq. (6) is equivalent to the sum on the left-hand side of Eq. (5), thus the slip obtained through Eq. (6) is a random variable distributed according to the Lévy law. Again, the random variable Δu_x is characterized by a set of parameters (β, γ, and μ) which are functions of the position x, but α is invariant under spatial translation and fixed by the Lévy parameter of the random variable X_x. By the same order of approximation, assuming that the relationship between the slip and the stress drop given in Andrews (1980), the stress drop is also distributed according to the Lévy law with the same parameter α.

Now, consider the expression for the ith component of the far-field seismic $u_i(\mathbf{x}, t)$ observed at point $\mathbf{x}$ for a finite dislocation source buried in a homogeneous, isotropic, unbounded medium discussed in Aki and Richards (2002; Eq. 10.4). The derivation of this result is based on the representation theorem for seismic sources. For the sake of simplicity, we assume that the slip variability is mainly along a given direction, for instance, the strike slip as for the 1979 Imperial Valley earthquake (Archuleta, 1984). The expression for the far field consists of an integration over the fault surface that involves essentially a linear transformation (derivative with respect to time) of the slip function $\Delta u(\boldsymbol{\xi}, t)$, and other parameters such as the density, the distance between the receiver and the point source $r = |\mathbf{x} - \boldsymbol{\xi}|$, the medium velocity, and so on. The variable $\boldsymbol{\xi}$ corresponds to the position of the source point over the fault surface. Let us assume furthermore that the integration over the fault surface can be reduced to a summation (see Appendix B in Heaton and Hartzell, 1989; Tumarkin and Archuleta, 1994). The expression for the far field can be approximated by the following sum:

$$u_i(\mathbf{x}, t) \propto \sum_{j=1}^{N_{\text{Subfault}}} \Delta \dot{u}_j(t') f_j(r), \tag{7}$$

where $\Delta \dot{u}_j(t')$ is the slip velocity, t' a lagged time, and $f_j(r)$ a function that involves all the other parameters and variables. The fault is divided in N_{Subfault} with index j. (A similar formulation between the pulse shape of P waves and a sum of tractions is discussed in Kostrov and Das, 1988. For the purpose of this demonstration and to keep it simple, the contribution of the P waves and S waves are regrouped in a single term—see Aki and Richards, 2002. Eq. 10.4). Now, if we assume that the only source of randomness in Eq. (7) is the slip spatial distribution, that is, if we assume that $\Delta u_j(t) \propto X_j g_j(t)$ where X_j is a Lévy random variable, and $g_j(t)$ a function of time and of the position j on the fault then Eq. (6) can be rewritten as

$$u(\mathbf{x}, t) \propto \sum_{j=1}^{N_{\text{Subfault}}} X_j \dot{g}_j(t') f_j(r). \tag{8}$$

For a fixed time t' and a fixed position $\mathbf{x}$, the sum in Eq. (8) is a sum of random variables distributed according to the Lévy law multiplied by a real number. In Eq. (8), the real number is given by the product of $\dot{g}_j(t')$ by $f_j(r)$ at the specific time t' and position r. According to Eq. (1), the far-field $u_i(\mathbf{x}, t)$ is also distributed according to the Lévy law. As for the random slip, each of the parameters (β, γ, μ) of the Lévy law characterizing the randomness of $u_i(\mathbf{x}, t)$ are a function of the position and time. However, the Lévy index α characterizing the tail fall off of the PDF of $u_i(\mathbf{x}, t)$ is identical to the parameter α that controls the tail fall off of the PDF of the slip. The same conclusion applies to the ground motion variables such as the acceleration and velocity. The formulation of the far-field displacement based on the representation theorem and the assumption that the probability law controlling the slip spatial distribution is the Lévy law implies that the power law controlling the PDF tails is invariant from the source to the ground motion. Another (but equivalent) formulation of the random property of the ground motions is discussed in Lavallée and Archuleta (2005)—see also Fig. 1. In the Appendix, we discuss a generalization of Eq. (7).

Finally, it should be noted that in the formulation discussed in Appendix B of Heaton and Hartzell (1989), the slip value at coarser resolution for the jth subfault is itself an integral performed over a slip spatial distribution that varies at higher resolution. In that sense, the slip value at coarser resolution corresponds to some sort of average. Assuming that the Central Limit Theorem can be applied to the slip spatial variability at higher resolution, this will constrain the probability law of the slip values observed at coarser resolution to converge to the Lévy law. Thus, observations of slip values distributed according to the Lévy law at coarser resolution can be understood as a consequence of the formulation given in the Appendix B Heaton and Hartzell (1989).

3. THE 2004 PARKFIELD EARTHQUAKE

3.1. Random Model of the Source

In view of the density of near-source data, the 2004 Parkfield earthquake is arguably one of the best-recorded earthquakes in history. It provides an ideal candidate for evaluating and validating the coupling of the random properties between the slip and the radiated field discussed above. Furthermore, abundance of near-source records (see Fig. 2) provides an exceptional opportunity to test inversion methods and the random properties of the source parameters. Using different sets of ground motion data, Custódio *et al*. (2005) computed several source models of the Parkfield earthquake. The sets of ground motion data differ by the number and the location of the stations used in the inversion. All the source models used in Custódio *et al*. (2005) are based on a method to invert the kinematic source parameters developed by Liu and Archuleta (2004). The sets selected for our investigation of the random property of slip spatial distribution are listed in Table 1.

For each dip and strike slip spatial distribution, we compile the parameters of the random model discussed in Section 2.1 (see Fig. 3A and B). The procedure used to compile the parameters can be summarized in three steps (for additional details, see Lavallée *et al*., 2006a).

1. The power spectrum is computed for each of the horizontal (along the strike) layers (layers are values at constant depth) of the slip component (dip or strike). The mean power spectrum of the horizontal layers is computed (see Fig. 4). The values of the exponents v_{1D} are reported in Tables 2 and 3.
2. Using Eq. (4), each layer of the slip spatial distribution is filtered in the Fourier space. We assume that the resulting field corresponds to a field of (uncorrelated) random variables X_i and we compute the PDF of X_i (see Figs. 5–8).
3. The PDF of X_i is fitted with the PDF of theoretical probability laws. Three candidates are considered: the Gauss law, the Cauchy law, and the more general Lévy law. The parameters of the probability laws are obtained by minimizing the following expressions:

$$\sum_{i=1}^{N} |\mathrm{PDF}(X_i) - p(X_i; \alpha, \beta, \gamma, \mu)|, \tag{9}$$

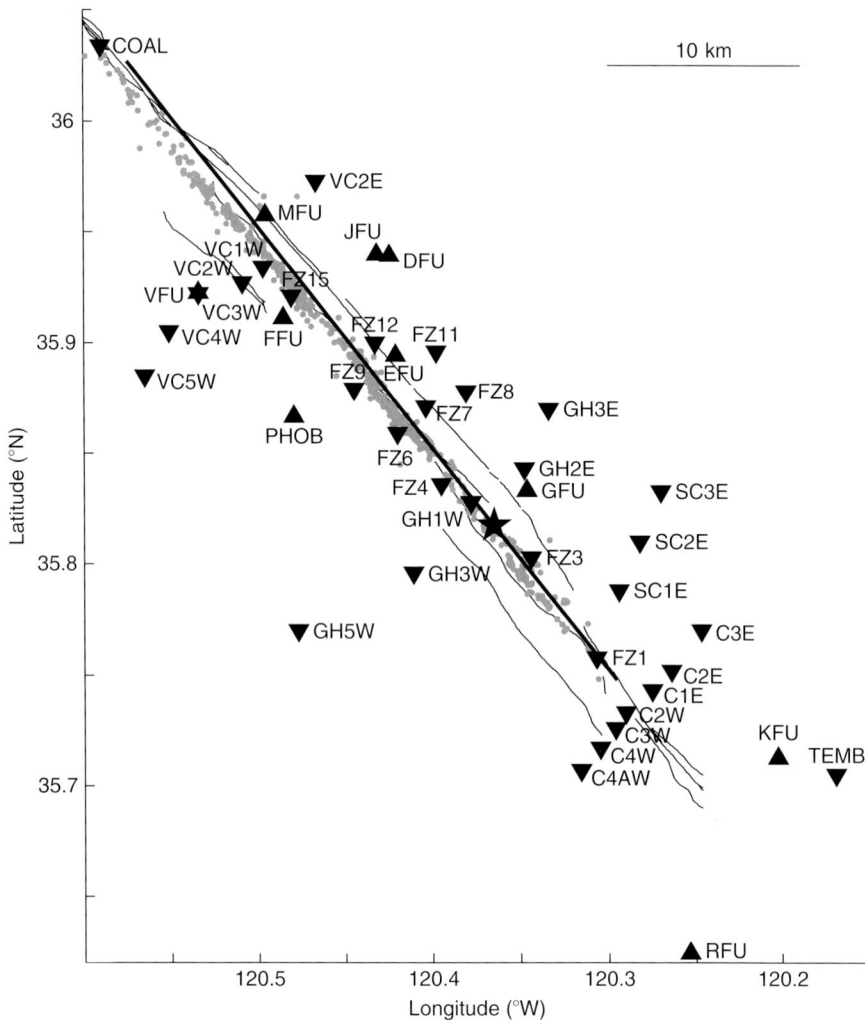

FIG. 2. Map of the Parkfield segment of the San Andreas fault is illustrated with the 43 three-component strong-motion instruments used in Custodio *et al.* (2005). The stations used in the inversions are the 9 USGS stations (▲) with digital recorders and 34 CGS stations (▼) with analog recorders. The template also includes the 2004 Parkfield epicenter (black star), modeled fault profile (black line); and aftershocks as located by Hardebeck and Michael (2004) (gray dots).

where N is the number of random variables, and $p(X_i; \alpha, \beta, \gamma, \mu)$ is the theoretical PDF of X_i for either the Gauss ($\alpha = 2$, no parameter β), Cauchy ($\alpha = 1, \beta = 0$), or Lévy law. In parallel with fitting the parameters of the Lévy law by optimizing Eq. (9), we also fitted the parameters by optimizing an expression similar to Eq. (9) but given for the characteristic function and its (corresponding) absolute value. (The characteristic function is defined as the Fourier transform of the PDF.) The parameters obtained by fitting

TABLE 1. The four slip models used in this section are based on the following set of ground motion data

Set	Number of stations	Stations
All	All 43 stations	All stations illustrated in Fig. 2.
1	17	C1E C2W C3E C4AW EFU FFU FZ12 FZ4 FZ8 GH2E JFU KFU MFU PHOB SC1E VC1W VC3W
2	25	C2E C3W C4W COAL EFU FZ1 FZ12 FZ15 FZ3 FZ6 FZ8 FZ9 GFU GH3E GH3W GH5W JFU RFU SC2E SC3E TEMB VC2W VC4W VC5W VFU
5	24	C1E C2W C4AW COAL EFU FZ12 FZ15 FZ3 FZ6 FZ8 FZ9 GFU GH3E GH5W JFU KFU SC2E VC2W VC4W VFU

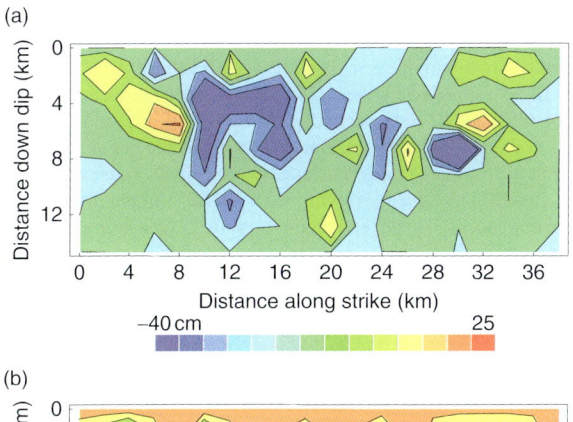

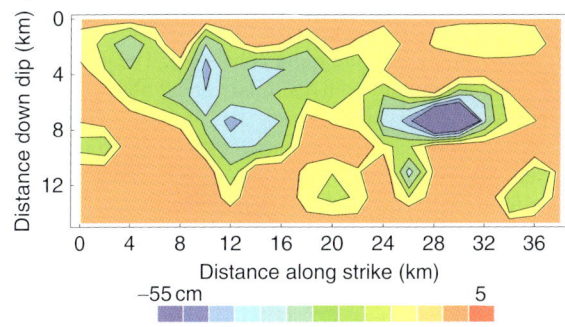

FIG. 3. Source inversion of the 2005 Parkfield earthquake (Custodio *et al.*, 2005) based on strong ground records (set All). The model fault plane is divided in subfaults located at 2 km intervals in the strike direction and 1.84 km in the dipping direction. The spatial distributions of the dip slip is illustrated in the top panel (a), while the strike slip is illustrated in the bottom panel (b).

the characteristic function and its absolute value are in good agreement with those reported in this chapter. The procedure used to estimate the parameters of the Lévy law is discussed in details in the Appendix in Lavallée *et al.* (2006a). Note also, that in Janicki and Weron (1994, Section 3.7), histograms of generated Lévy random variables

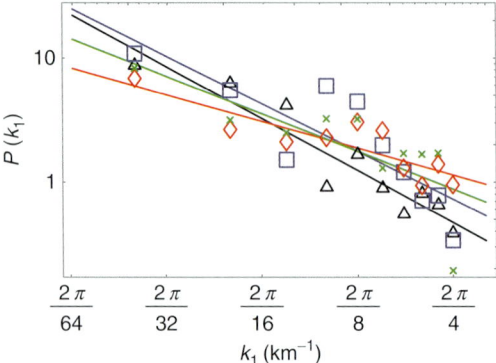

FIG. 4. The mean power spectrum $P(k_1)$ and the best straight lines that fit the log–log curve are reported for the strike slip distributions in Table 1: set All (△), set 1 (□), set 2 (◇), and set 5 (×). The quality of the fit, as estimated by the values of the linear correlation coefficient (in absolute values) is 0.94, 0.84, 0.87, and 0.76, respectively, for the set All, 1, 2, and 5. These values are within the range of values computed for other earthquakes (see Lavallée et al., 2006a). Variations in the slopes of $P(k_1)$ and in the estimated linear correlation coefficients suggest that the convergence to a single exponent v_{1D} is difficult to achieve with such a small sample of values. These results also suggest that the convergence is also dependent on the number of stations used in computing the slip distribution.

are compared to the theoretical curves of the PDFs for several values of the Lévy index. The theoretical curves and the histograms are also in good agreement.

The curves of the Gaussian, Cauchy and Lévy laws that best fit the PDF are illustrated in Figs. 5–8. The parameters of the Gaussian, Cauchy and Lévy laws are reported in Table 2 and 3. The values of the parameter α estimated for both the strike and the dip components of the four slip models are quite consistent. The spread of ± 0.3 around the value $\alpha \cong 1.1$ is in good agreement with a similar spread observed when fitting the PDF of 200 synthetic Lévy random variables (see Appendix in Lavallée et al. 2006a for details). These results suggest that estimation of the parameter α is relatively independent of the number of stations (provided that this number is large enough). The estimation is also independent of the stations selected for computing the slip spatial distribution. Since α controls the relative decreasing expectation to observe large slip values, the result suggests that the decreasing power law of the PDF tail is well captured by source inversions computed with the different sets reported in Table 1. This is quite an achievement in view of the number of subfaults used to compute the PDF ($N=180$). This result is in good agreement with the theory discussed in Section 2.2. A value for the Lévy index α close to 1 is also comparable to the value computed for the slip inversion of the Imperial Valley earthquake (Lavallée and Archuleta, 2003), the 1989 Loma Prieta and the 1994 Northridge earthquake (Lavallée et al. 2006a), and the 1999 Chi-Chi earthquake (Lavallée and Archuleta, 2005; and next section).

The values of the parameter β reported in Tables 2 and 3 are close to 0 for (almost) every slip inversion. This suggests that positive and negative fluctuations in the slip spatial distribution are equally probable. As discussed in Lavallée et al. (2006a), the value of the parameter μ is an artifact of the operations [see Eq. (4)] used to compute the random variables X_i, thus its value is not really relevant.

TABLE 2. Parameters of the random model for the dip slip of four settings presented in Table 1

Set	Power law exponent v_{1D}	Gauss law		Cauchy law		Lévy law			
		μ	σ	γ	μ	α	β	γ	μ
All	0.92	0.08	1.62	0.95	0.03	1.2	0.14	1.02	0.5
1	0.69	−0.18	1.71	1.14	−0.3	0.88	0.0	1.13	−0.3
2	0.6	0.1	1.71	1.	0.04	1.28	0.11	1.15	0.4
5	0.82	−0.19	1.60	1.07	−0.22	0.93	0.08	1.05	−1.

TABLE 3. Parameters of the random model for the strike slip of four settings presented in Table 1

Set	Power law exponent v_{1D}	Gauss law		Cauchy law		Lévy law			
		μ	σ	γ	μ	α	β	γ	μ
All	1.40	0.25	0.86	0.63	0.23	1.13	−0.18	0.55	−0.34
1	1.28	0.31	1.09	0.79	0.3	1.13	−.09	0.71	−0.05
2	0.72	0.42	2.12	1.37	0.26	0.87	0.16	1.35	−0.74
5	1.01	0.42	1.49	1.02	0.28	1.01	0.	0.97	0.34

Of the five parameters needed to describe the random model of the slip, only the parameters v_{1D} and γ vary significantly (i.e., relative to the other parameters) from one slip inversion to another. The fact that both parameters are affected in a similar way is essentially due to the filtering process used to generate the random variables X_i. The amount of filtering in Eq. (4) essentially controls the spreading of the X_i and thus the width of the PDF, which in turn is (mainly) characterized by the parameter γ. Using different slip distributions with a similar range of values, and filtering the slip distributions with Eq. (4) but with an exponent v_{1D} that varies significantly from one slip inversion to another, will generate white noise X_i characterized by different ranges of values. For a fixed value of α (e.g., in Tables 2 and 3, for $\alpha = 2$ and $\alpha = 1$), the interdependence between v_{1D} and γ is clearly illustrated and shows that when v_{1D} is growing, γ is decreasing. The results summarized in Tables 2 and 3 suggest that an accurate estimation of the parameter v_{1D}, and thus the amount of correlation in the slip variability, is difficult to achieve. The small number of subfaults used in computing v_{1D} may be responsible for a slow convergence to the real value of v_{1D} and may explain the variation reported in Tables 2 and 3 (see also discussion in Lavallée et al., 2006a). Alternatively, the variation in v_{1D} may be due to a dependency in the number of stations and/or the location of the stations used in computing the slip inversion. Additional investigations are needed to understand the origin of variations in the estimated parameter v_{1D} reported in Tables 2 and 3. Fortunately, according to Eq. (5) and numerical simulations with generated Lévy random variables (see discussion on *pseudo-white noise*, in Lavallée et al., 2006a), estimation of the parameter α is not too dependent on an accurate estimate of the parameter v_{1D}.

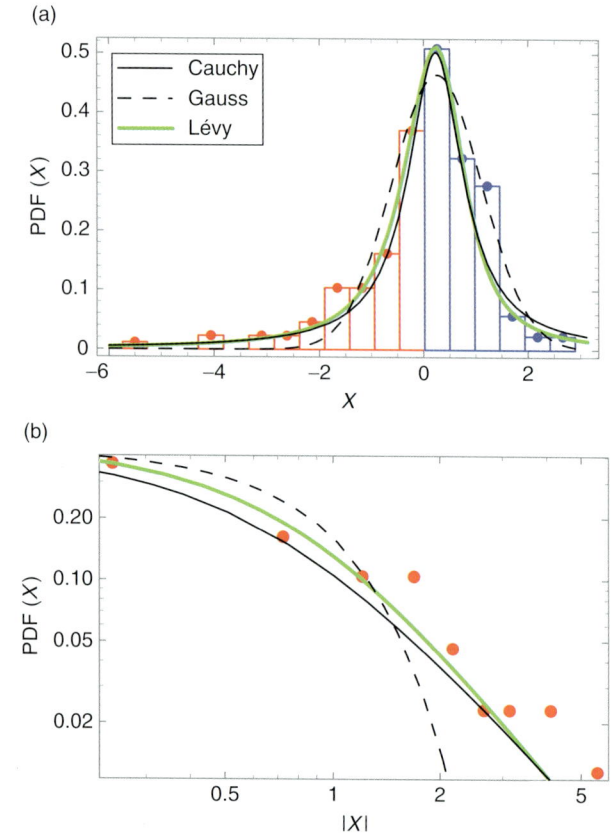

FIG. 5. (a) The (discrete) probability density function (PDF) (red and blue dots and bars) associated with the filtered strike slip X of the set All is compared to the curves of the three probability laws that best fit the PDF: the Cauchy law (black curve), the Gauss law (dashed curve) and the Lévy law (green curve). The left side of the PDF ($X < 0$) is colored in red while the right ($X > 0$) side is in blue. The magnitude of the random variables, that is, the filtered slip, is given by X. (b) The left tails of the curves on a log–log plot to emphasize the fit for values far into the tail. The PDF tail decreases according to a power law $X^{-\alpha-1}$, with $\alpha \approx 1$. Note that according to the Gauss law, the large events—last points on the left-hand side of the graphics—have almost a zero probability of being observed. The parameters of the Gauss, the Cauchy and the Lévy laws are reported in Table 2.

3.2. Random Model of the Ground Motion PGA

Under the assumption that the slip spatial heterogeneity can be interpreted in terms of a random model and the representation theorem (approximated by Eq. (7)), we have found that the random properties of the ground motion are constrained by a sum of random variables. According to the (generalized) Central Limit Theorem, a sum of Lévy random variables with identical α value, but with different values for β, γ, and μ, is a Lévy random variable characterized by the same α values [see Eq. (1)]. With respect to its random property, the ground motion recorded at a given station can be understood as a

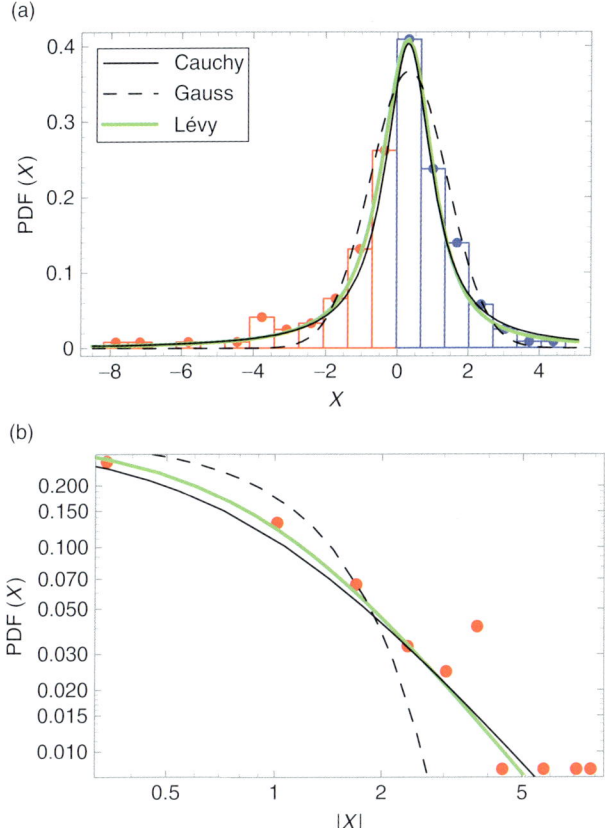

FIG. 6. Same as Fig. 5 but for the random variables associated with the strike slip of the set 1. The parameters of the Gauss, the Cauchy, and the Lévy laws are reported in Table 3.

Lévy random variable modulated by a function of time. At any time, the ground motion is thus a Lévy random variable characterized by the same α, but the parameters β, γ, and μ are functions of time [see Eq. (8)].

The PGAs recorded during the 2004 Parkfield earthquake are used to test this hypothesis. The signal recorded at a station is divided into three components: N-S, E-W, and U-D. In this analysis, we are using the PGA of the three components. Furthermore, we consider the PDF of the absolute value of the PGA as it is traditionally done in seismology. We assume that the functional behavior of the "heavy tails" of the PDF of |PGA| is essentially similar to the functional behavior of the "heavy tails" of the PDF of PGA. The |PGA| is one of the most important ground motion metrics used in computing the PGA probability associated with seismic hazard (PSHA)—see McGuire (2004). By definition, the PGA is an extreme "event" in the ground motion signal. The parameter α controls the attenuation of the PDF tail, and thus the frequency of extreme events. It is thus particularly relevant to compute the PDF of the |PGA| to test the hypothesis

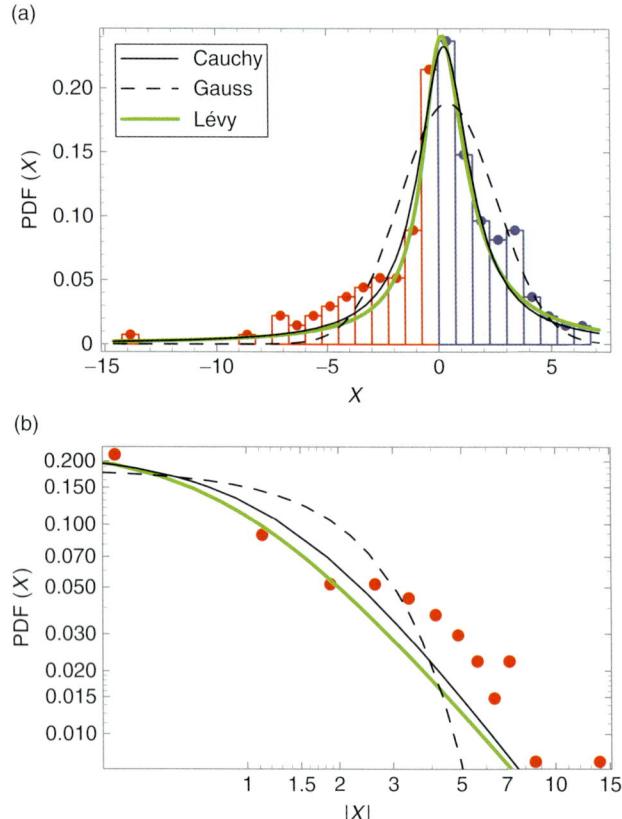

FIG. 7. Same as Fig. 5 but for the random variables associated with the strike slip of the set 2.

discussed above. More specifically, we want to compare the functional behavior of the tail of the PDF for the |PGA| to the "heavy tails" of the PDF of the random variable X associated to the slip (see Figs. 5–8).

The stations selected for this analysis are located within a closest distance to the rupture surface that varies from 0 km to 11 km (Fig. 9). To test the effect of the distance on the computed random properties, these stations are divided into three subsets. The first set includes the stations located between 0 km and 5 km, the second set includes the stations located between 0 km and 7.5 km, and the third set includes all the stations located between 0 km and 11 km. For each set, the PDF of the |PGA| is computed. Assuming that the PDF of the |PGA| can be approximated by the Lévy law, we compute the parameters of the Lévy law that fit the PDF curves. The procedure used to compute the parameters is identical to the one discussed in Section 3.1 for the random variable X. The results are reported in Table 4 and illustrated in Figs. 10 and 11.

Note that the PGA values recorded at different stations are surely characterized by PDFs with different values of β, γ, and μ. The PGA values do not correspond to a white noise. The variation in β, γ, and μ surely affects the shape of the PDF computed for the PGA,

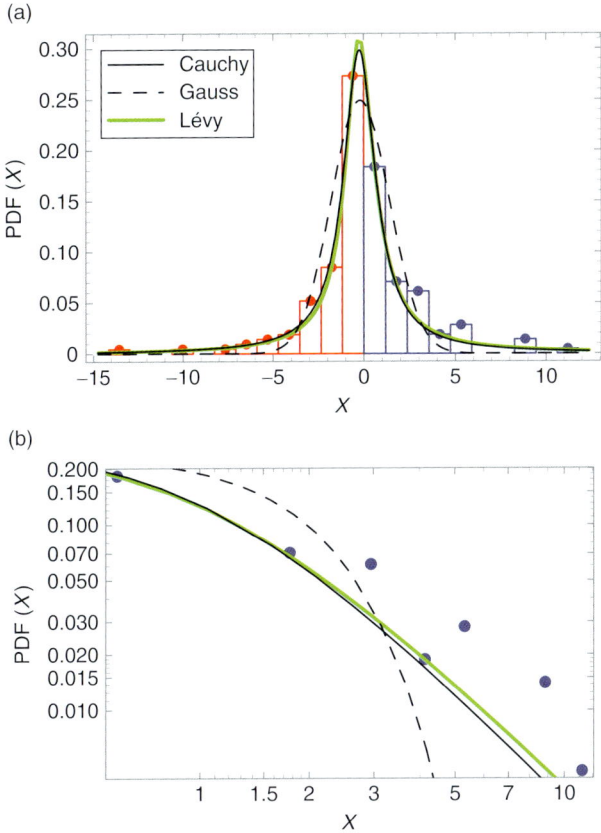

FIG. 8. Same as Fig. 5 but for the random variables associated with the dip slip of the set 5. In (b), the positive tails of the curves are illustrated on a log–log plot.

in particular the width of the PDF. We assume that these variations in the parameters values do not significantly affect the computation of α. Consistency in the values of α estimated for the three different sets supports this assumption (see also Gusev, 1996).

3.3. Random Model of the Ground Motion PGV

The ground motion velocity is the time integral (or sum) of the ground motion acceleration. If the ground motion acceleration is distributed according to the Lévy law at any time, a time integral (or sum) over the recorded ground motion will also be distributed accorded to the Lévy law. According to Eq. (8), the Lévy index α should be invariant under this linear transformation.

To the same order of approximation and for the same stations used to estimate the PDF of the |PGA|, we compute the PDF of the |PGV|. Again, we assume that the PDF of the |PGV| can be approximated by the Lévy law, and we compute the parameters of the Lévy law that best fit the PDF curves. The results are reported in Table 5 and illustrated in Fig. 12.

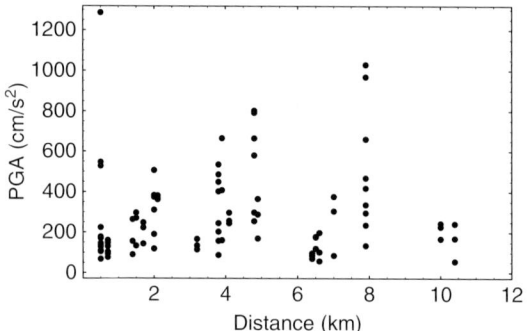

Fig. 9. The peak ground acceleration (PGA) amplitude estimated at different stations as a function of the closest distance between the station and the fault rupture.

Table 4. Parameters of the Lévy law that best fit the probability density function (PDF) of the peak ground acceleration (PGA) for stations within distance intervals with size of 10, 7.5, and 5 km

Location of the stations (km)	Number of events	Lévy law			
		α	β	γ	μ
0–5	65	0.95	1.	51.5	−613.
0–7.5	77	1.04	1.	75.0	1236.
0–11	92	1.11	0.93	125.	561.

The number of PGA events used to compute the PDF for every interval is given in the second column.

These results show that the PDF of the |PGA| and the PDF of the |PGV| can be approximated by the Lévy law. The estimated parameter α is (almost) invariant for stations located between 0 km and 11 km, and (almost) independent of the size of the distance intervals used to compute the PDF. The values of α are also in good agreement with the values reported in Tables 2 and 3. This suggests that the same parameter α controls the rate of decrease in the PDF tails of the |PGA|, of the |PGV|, and of X.

4. The 1999 Chi-Chi Earthquake

4.1. Random Model of the Source

The results discussed in Section 3 are in good agreement with the results obtained for the 1999 Chi-Chi earthquake (Lavallée and Archuleta, 2005). The same procedure discussed in Section 3.1 was applied to the dip and the strike slip components computed by Zhang *et al.* (2003). The power spectrum of the dip and the strike slips are illustrated in Fig. 13. As for other earthquake slip inversions (see Lavallée *et al.* 2006a), the power spectrum is attenuated as $k_1^{-v_{1D}}$. The values of the parameters v_{1D} are listed in Table 6.

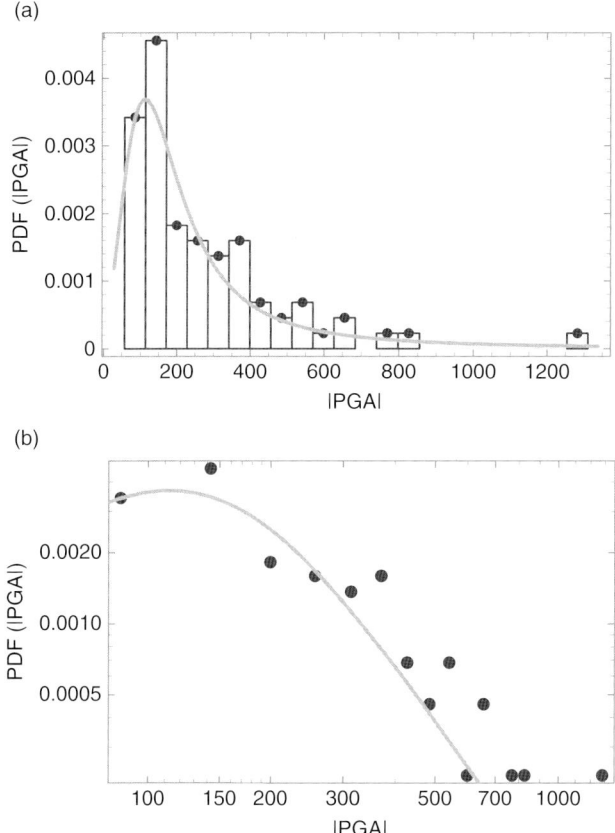

FIG. 10. (a) Using ground motions recorded for the 2004 Parkfield earthquake, the probability density function (PDF) of the peak ground acceleration |PGA|, for the stations located within a distance of 0–7.5 km of the rupture surface, is compared to the curves of the Lévy law that best fit the PDF. (b) The positive tails of the curves are plotted on a log–log scale. The PDF tails decrease according to a power law $|PGA|^{-\alpha-1}$ with α close to 1.

Using the parameter v_{1D}, the white noise X is obtained by filtering the dip (or strike) slip (see Eq. (4)). The parameters of the Gauss, the Cauchy, and the Lévy laws that best fit the PDF of X are computed and listed in Table 6 (see Figs. 14 and 15). As for the Parkfield earthquake, the Lévy law (with α close to 1) and the Cauchy law ($\alpha = 1$) give a much better representation of the decrease in the PDF tail of X.

4.2. Random Model of the Ground Motion PGA

The model presented in Section 2.2 was first tested for the PGA recorded at the surface during the 1999 Chi-Chi earthquake. The stations are located with a closest distance to the rupture surface that varies from 0 km to 30 km (Fig. 16). Contrary to the Parkfield

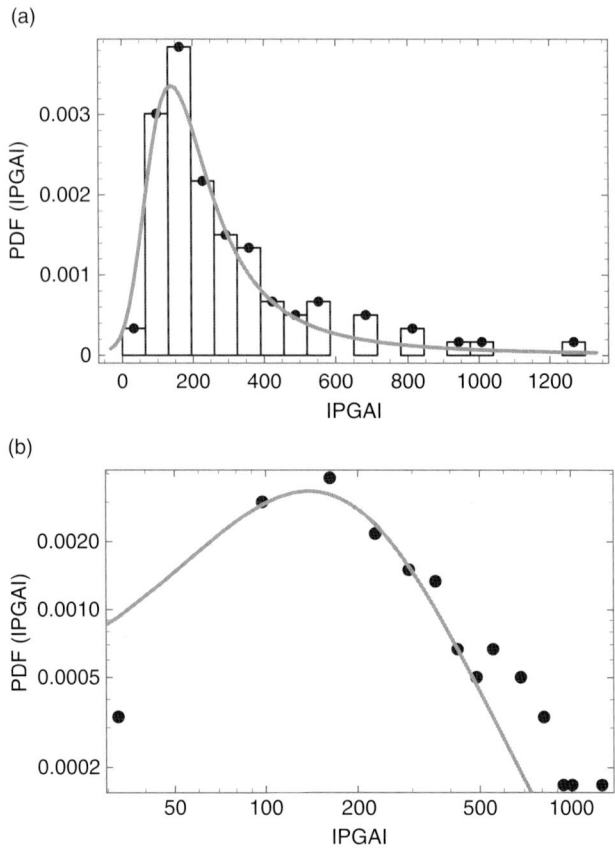

FIG. 11. Same as Fig. 10 but for the stations located between a distance of 0–11 km.

TABLE 5. Parameters of the Lévy law that best fit the probability density function (PDF) of the peak ground velocity (PGV) for stations within distance intervals with size of 10, 7.5, and 5 km

Location of the stations (km)	Number of events	Lévy law			
		α	β	γ	μ
0–5	65	1.23	1.	6.15	21.7
0–7.5	77	1.22	1.	8.39	25.6
0–11	92	1.21	0.94	7.97	25.5

The number of PGV events used to compute the PDF for every interval is given in the second column.

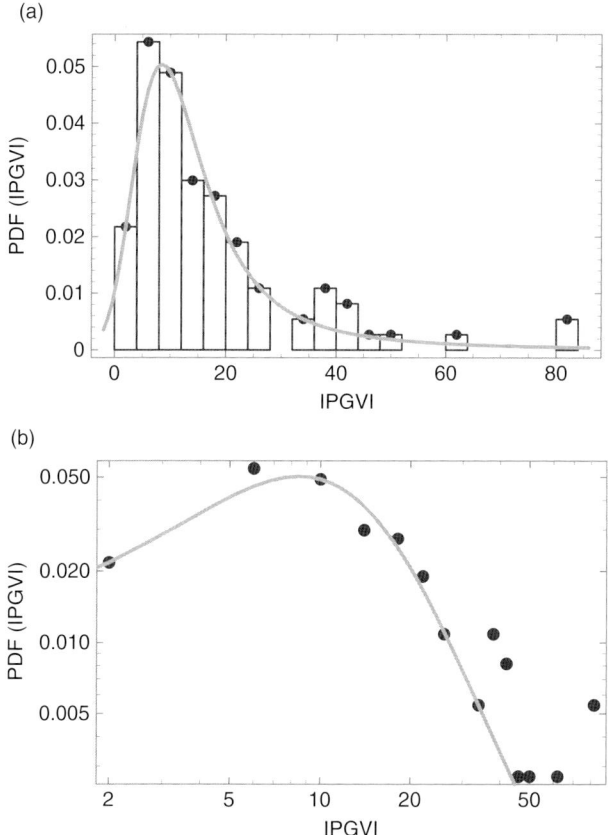

FIG. 12. Same as Fig. 10 but for the peak ground velocity recorded for station between a distance of 0–11 km.

earthquake (see Fig. 9), the spatial distribution of the PGA is clearly a function of the distance. However, we can assume that variations in the random property are (approximately) independent of the distance to the fault if we only consider the stations located within a window of a narrow size (along the horizontal axis). There is a trade-off to be made between the size of the window and the number of PGA values. The larger the size of the window, the larger the number of PGA values within the window will be. From a statistical point of view, a larger number of events provides a more reliable estimate. To test the effect of the location of the stations on the computed PDF, Lavallée and Archuleta (2005) computed the PDF of the PGA for stations located within windows of different sizes and with the window centers located at different positions. Several settings were investigated in Lavallée and Archuleta (2005). For each setting, the parameters of the Lévy law were computed. Except for 2 of the 13 settings, the values of α are close to 1. There is a very small deviation from $\alpha = 1$ for the three PDFs computed within windows with a size of 10 km (see Table 7 and Figs. 17 and 18). In these settings, the largest window size used was 10 km. The number of stations included within

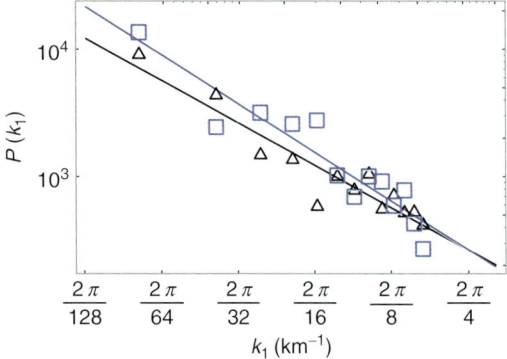

FIG. 13. The mean power spectrum $P(k_1)$ computed for the dip slip ($\triangle$) and the strike slip ($\square$) is illustrated as a function of the wave number k_1. The straight line corresponds to the best fits. The quality of the fit, as estimated by the values of the linear correlation coefficient (in absolute values), is 0.89 for the dip slip and 0.88 for the strike slip. These results suggest that the power law behavior is observed for scale length that ranges from 3 km to 72 km.

TABLE 6. Parameters of the random model for the dip and strike slips of the 1999 Chi-Chi earthquake

	Power law exponent	Gauss law		Cauchy law		Lévy law			
	v_{1D}	μ	σ	γ	μ	α	β	γ	μ
Dip slip	1.11	9.2	25.9	19.1	9.8	.95	−0.3	14.6	79.
Strike slip	1.27	−7.5	17.3	13.	−7.7	1.	0.3	12.3	9.7

a broader distance interval is larger than the number of stations within a narrower distance interval. Accordingly, the PDF of the former are computed with more accuracy—especially the PDF tail.

The results for the Chi-Chi earthquake corroborate the results obtained for the Parkfield earthquake (see Section 3) and validate the model discussed in Section 2. The results discussed in this section and in Sections 3.2 and 3.3 also confirm that Lévy type laws with "heavy tails" are a better proxy than Gaussian or log-normal laws to reproduce the asymptotic behavior of the PDF of ground motion metrics (Gusev, 1989; Tumarkin and Archuleta, 1997). For a discussion on fitting the PDF of the |PGA| with log-normal laws see Abrahamson (1988) and references therein.

5. Limitations of the Model

The random properties of the ground motions, and *a fortiori* of the ground motion metrics, are not completely determined by the random properties of the slip. Only the law that governs the fall off of the PDF of the ground motion metrics is essentially

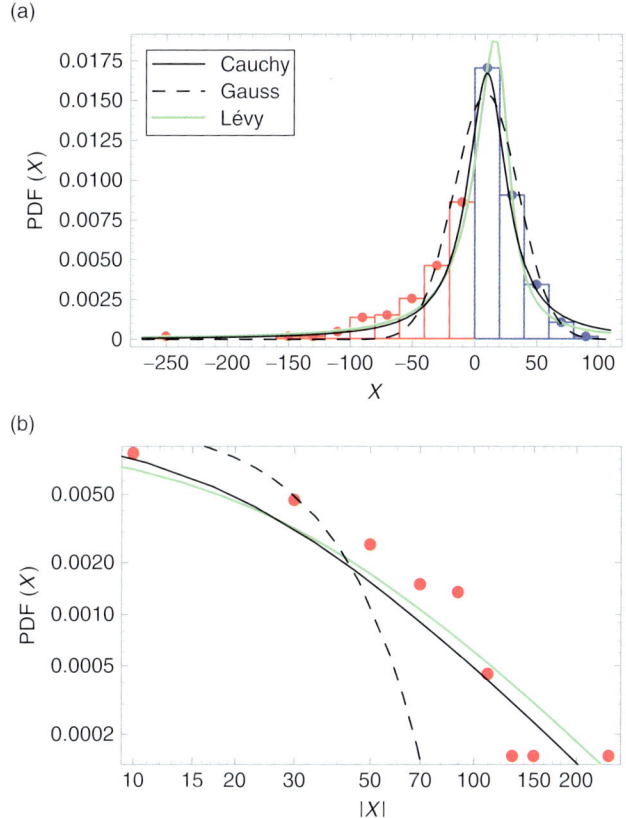

FIG. 14. (a) The (discrete) probability density function (PDF) (red and blue dots and bars) associated with the filtered dip slip X of the Chi-Chi earthquake is compared to the curves of the three probability laws that best fit the PDF: the Cauchy law (black curve), the Gauss law (dashed curve) and the Lévy law (green curve). The left side of the PDF ($X < 0$) is colored in red while the right ($X > 0$) side is in blue. The magnitude of the random variables, that is, the filtered slip, is given by X. (b) The left tails of the same curves are illustrated on a log–log plot to emphasize the fit for values well into the tail. The misfit of the Gaussian PDF is more obvious in this plot. The parameters of the Gauss, the Cauchy, and the Lévy laws are reported in Table 6.

constrained by the random properties of the slip. The interplay between the rupture dynamics and the heterogeneity in the fault's surrounding medium will also affect the distribution of ground motion values. However, according to the theory discussed above, these effects will mainly affect the width and the location (position) of the PDF of the ground motions, for example it will affect the mean value of the |PGA| and its variance (see also the Appendix).

Strictly speaking, according to the theory given above, it is the displacement, velocity, and acceleration that can be approximated by Lévy random variables, not the absolute value of such variables. However, we were mainly interested in investigating the functional behavior of the tail of the PDF and to verify that the Lévy law can reproduce

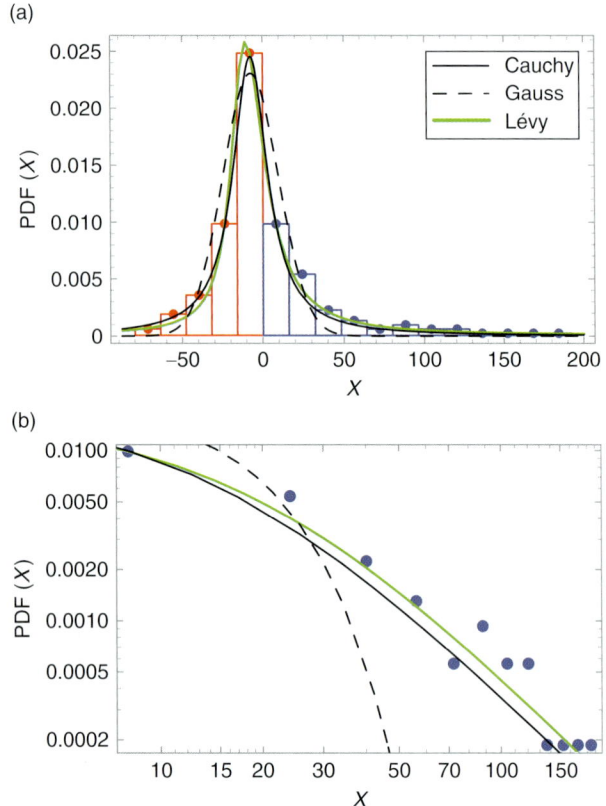

FIG. 15. Same as Fig. 14 but for the random variables associated with the strike slip of the Chi-Chi earthquake.

the PDF functional behavior with a parameter α predicted by the (generalized) Central Limit Theorem. Because the values of the PGA are distributed symmetrically with respect to the value 0, it is also possible to investigate the behavior of the PDF tails of the PGA. We do not expect that the conclusion reached in this study will be affected significantly by using this procedure.

Furthermore, it should be noted that the PDF values computed for the small values of the |PGA| is an artifact of the Lévy law used to fit the PDF. A more accurate description of the probability of the |PGA| will require fitting the small |PGA| values with a different probability law. Only a small number of values are affected by this limitation. The Lévy law is a better approximation of the probability associated with the larger values of |PGA|, for instance, for those values that exceed the |PGA| value that characterizes the maximum of the PDF curve. The same limitations apply to the |PGV|.

Recorded ground motions under different soil conditions may or may not include nonlinear effects (see Archuleta et al., 2003; Bonilla et al., 2005 for illustrations of modeling wave propagation including the nonlinear soil dynamics). Taking advantage of the strong-motion data recorded during the Northridge earthquake, Field et al. (1997)

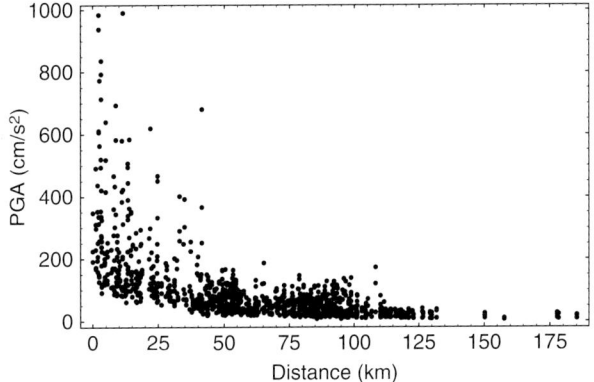

FIG. 16. The peak ground acceleration |PGA| amplitude of the three components of recordings estimated at different stations as a function of the closest distance between the station and the rupture surface.

TABLE 7. Parameters of the Lévy law that best fit the probability density function (PDF) of the peak ground acceleration (PGA) for stations within different distance intervals

		Lévy law			
Location of the stations (km)	Number of events	α	β	γ	μ
0–10	102	0.95	1.	46.2	−492.
10–20	99	0.96	0.58	33.9	−259.
20–30	60	1.03	1.	28.4	754.

The number of PGA events used to compute the PDF for every interval is given in the second column.

went on to infer pervasive nonlinear soil response. However, O'Connell (1999) explained much of the same data used by Field *et al.* (1997) with a linear site response and scattering of waves in the upper kilometers of the Earth's crust. In this study, we ignore site amplification effects for the following reasons. First, taking into account shallow soil properties would require subdividing the limited samples accordingly, thus reducing the number of stations used to compute the PDF. Second, it is usually understood that nonlinear effects deamplify the signal recorded at a given site (though in some cases nonlinearity may create PGA, Archuleta, 1998; and Bonilla *et al.* 2005). In terms of the random properties of the |PGA|, this implies rescaling (or narrowing) and translating the PDF curve. These transformations are mainly controlled by the scale parameter γ and the location parameters μ. In Section 3.2, we ignored variability in these parameters from one station to another, pointing out that the relevant parameter for our analysis is the Lévy index α. Finally, using ground motions recorded at the surface and at the bottom of the boreholes during the 2003 Tokachi-oki earthquake, we found that the tails of PDF of PGA decrease with power law behaviors controlled by Lévy indexes with values close to 1 (Lavallée *et al.*, 2006b). These results suggest that nonlinear shallow soil properties may not influence the rate of decrease in the PDF tails of PGA recorded at the surface.

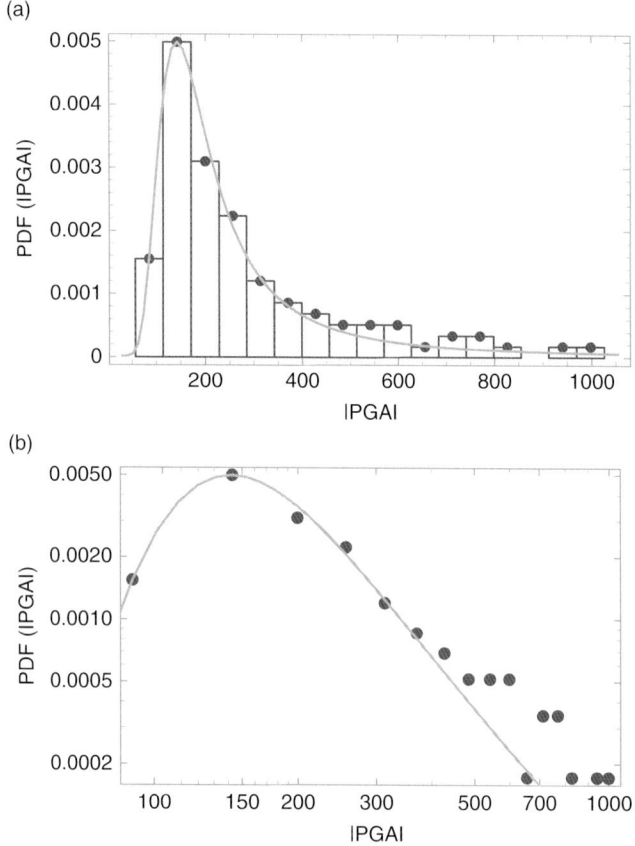

FIG. 17. (a) The probability density function (PDF) of the peak ground acceleration |PGA| for the stations located between a distance of 0–7.5 km is compared to the curves of the Lévy law that best fit the PDF (gray curve). (b) The positive tails of the curves are plotted on a log–log scale.

To the same order of approximation, we can also ignore directivity effects. The consistency in the estimate of the parameters α for different windows (see Tables 4–6) suggests that this approximation is correct (see also Lavallée et al., 2006a).

Besides the Lévy law, other probability laws can also be used to reproduce the typical shape observed for the PDF of the |PGA| (see also the Appendix). For instance, the Gamma and the Weibull laws have similar functional behaviors for certain parameter values (for other examples, see Abrahamson, 1988). Relying only on statistical tests to identify the law that best reproduces a functional shape of the PDF curve will not provide insight into understanding the physical origin of the random nature embedded in the ground motions. Furthermore, the determination of the best probability law for a given set of a data will essentially be grounded on statistical tests, with the consequence that different data sets may be best approximated by different probability laws.

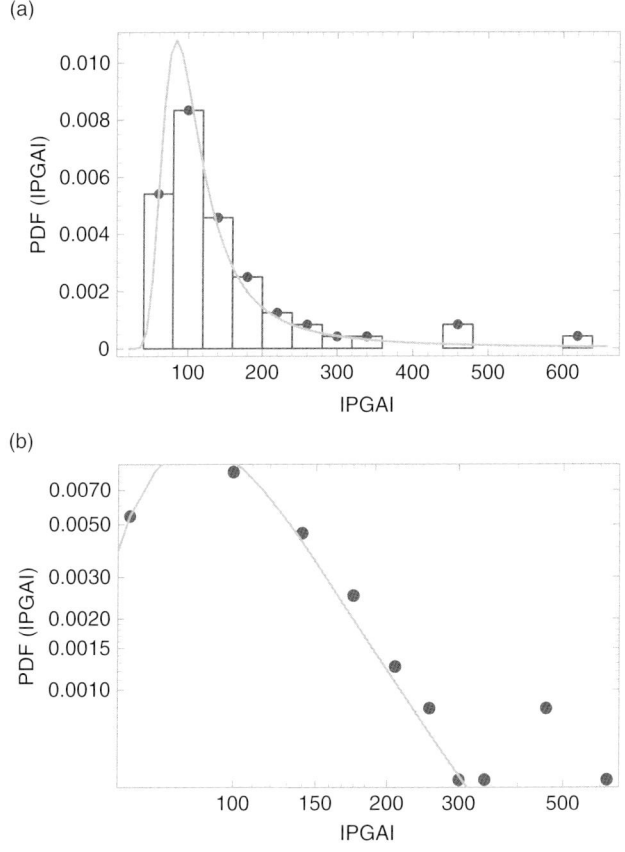

FIG. 18. Same as Fig. 17 but for the stations located between a distance of 20–30 km.

Although not a model limitation *per se*, the question of the quantification of the accuracy of—or uncertainty on—the parameters estimated in Sections 3 and 4 deserve a brief discussion. There is no simple procedure to compute the accuracy of the estimated parameters of the random model. However, it is possible to compute the range of "plausible" values for the parameters of the random model. For the sake of simplicity, we focus on the Lévy index α. In the Appendix of Lavallée *et al*. (2006a), the computed PDF (and characteristic functions) of generated Lévy random variables were fitted with the Gauss, Cauchy, and Lévy laws. The fitted parameters were compared to the parameters used to compute the Lévy random variables. In these numerical experiments, the number of generated random variables was chosen to match the number of events (or subfaults) in a source inversion—see also Janicki and Weron (1994), for numerical experiments that involve a larger number of random variables. These numerical experiments allow us to quantify, although approximately, the degree of accuracy that can be expected under "ideal conditions" when estimating the parameters of the Gauss, Cauchy, and Lévy laws. The numerical experiments discussed in Lavallée et al. (2006a) suggest that the parameter α can be determined within a spread of ± 0.3. Another procedure to infer the interval of "plausible" values consists of using the computed values of the parameters for empirical data. For instance, the range of values computed for the parameter α for the four dip-slip models of the Parkfield earthquake varies from 0.88 to 1.28 (see Table 2). The range for the strike slip goes from 0.87 to 1.13 (see Table 3). For different earthquake source models, the range of values estimated for the parameter α goes from 0.87 (2004 Parkfield earthquake, this chapter) to 1.56 (1995 Hyogo-ken Nanbu—see Lavallée et al., 2006a). For the slip, the range magnitude for the parameter α is thus ~ 0.7. For the ground motion metrics, the range of values can be estimated by comparing the values of α obtained for different subsets of data discussed in Sections 3.2 and 4.2. The range of α values for PGA recorded during the 2004 Parkfield earthquake goes from 0.95 to 1.11 (see Table 4) while the range goes from 0.95 to 1.03 (see Table 7) for the events recorded during the 1999 Chi-Chi earthquake (see Lavallée and Archuleta, 2005 for additional results). The values obtained for the ground motion metrics are thus within the range of values estimated for the different earthquake source models. The range of values for α computed for empirical data is thus consistent with the range for numerical experiment discussed in the Appendix of Lavallée et al. (2006a).

6. CONCLUSION: FROM RANDOMNESS TO INVARIANCE

The concept of conservation laws plays a fundamental role in science. This concept specifies that a physical quantity is invariant under time, space, or scale translation, regardless of the complexity of the interactions or mechanisms involved in the process under consideration. Furthermore, application of the concept does not require a detailed knowledge of these interactions and mechanisms. In the second section of this chapter, we argue that, based on the synthesis of the representation theorem and the (generalized) Central Limit Theorem, the random properties of the slip and of the ground motion should be distributed according to the Lévy law. Our investigation of the random properties of the source model and of the 1999 Chi-Chi earthquake confirmed this hypothesis (Lavallée and Archuleta, 2005).

In Section 3, we investigated the random properties of several source models computed for the Parkfield earthquake (Custódio et al. 2005). For each source, we compiled the parameters of the random model and compared them to the parameters of the PDF of the PGA and the PDF of the PGV. We found that the tails of the PDF characterizing the slip, the |PGA|, and the |PGV| are governed by a parameter α with almost the same values as predicted by the (generalized) Central Limit Theorem. This parameter, the Lévy index α that controls the rate of decrease in the frequency of large fluctuating slip values, |PGA| values and |PGV| values, can be considered an invariant measure.

In an earlier work (Lavallée and Archuleta, 2003), we stressed the fact that a description of the slip spatial distribution in terms of a Cauchy law has a significant consequence on the distribution of asperities over the fault surface. (Asperities are usually defined as regions with large slip or stress drop values on the fault.) The results presented in this chapter show that the PDF of the |PGA| (and the PDF of the |PGV|) is a direct consequence of the PDF of the asperities as first suggested by Gusev (1989).

An important application of the theory outlined in this chapter will be the validation and the calibration of source models. Often for a given earthquake, several candidates are available with spatial configurations that differ significantly from one source model to another (http://www.seismo.ethz.ch/srcmod; Mai *et al.*, 2005). The optimal inversion can be found by comparing the random properties of the source to the random properties of the ground motion, assuming that enough data are available. Numerical simulations of dynamic rupture require realistic complexity in fault geometry, material properties, and stress state. Again, intercomparison of the random properties of the stress drop and generated synthetic ground motions can be used to insure that these properties are in agreement with each other as predicted by the theory outlined in this chapter.

In devising a random model of slip inversions, we have stressed the importance of properly describing the probability law that governed the slip spatial distribution. In Section 2, we show that the representation theorem implies that the randomness in the radiated field is a direct consequence of the randomness embedded in the slip. Then, using the (generalized) Central Limit Theorem, we are able (at least approximately) to constrain the probability law governing the radiated field to the Lévy law. This suggests that the basic principles governing the combination of random variables are to random modeling what calculus is to physics!

Acknowledgments

The author thanks S. Custodio for kindly providing the source model for the 2004 Parkfield earthquake, and W. Zhang for the source model for the 1999 Chi-Chi earthquake. I gratefully acknowledge discussions with R.J. Archuleta, B. Erickson, M. Fehler, D.M. Higdon, C. Ji, K. Koketsu, P. Liu, H. Miyake, D. Sornette, and J. Schmedes. Thank you to L. Malenfant for editing a draft version of this chapter. The research discussed in this chapter was supported by KECK Grant No. 19990997, and LANL/IGPP Grant No. 04-08-16L-1532. This research was supported by the Southern California Earthquake Center. SCEC is funded by NSF Cooperative Agreement EAR-0106924 and USGS Cooperative Agreement 02HQAG0008. The SCEC contribution number for this chapter is 1136. This is ICS contribution No. 0815.

Appendix. Generalization of the Random Model of the Radiated Field

A.1. Near-Field Displacement

The random properties of the far-field displacement $u(\mathbf{x}, t)$ can be extended to the near-field seismograms if we adopt the formulation given by Aki and Richards (2002, Eq. 10.39). Here, we assume that the slip spatial distribution is the only source of randomness. In Aki and Richards (2002, Eq. 10.39), the near field is essentially related to the slip through relationships that involve the slip or linear transformations of the slip such as the integral or derivative. To the same order of approximation, the results discussed in Section 2.2 can be extended to the near-field displacement.

A.2. Product of Random Variables

For the sake of simplicity, we consider again the expression for the far-field displacement $u(\mathbf{x}, t)$ given by expression (Eq. 8), with X_j distributed according to the Lévy law with a Lévy index α_X.

$$u(\mathbf{x}, t) \propto \sum_{j=1}^{N_{\text{Subfault}}} X_j \dot{g}_j(t') f_j(r). \tag{A.1}$$

We assume that besides the slip heterogeneity, there is another source of randomness in the formulation of the far field. For instance, additional random effects can be due to heterogeneity in the medium surrounding the fault. Heterogeneity in the medium can be approximated by assuming that the medium velocity is related to a random variable Y (Fouque et al., 2007). First, we assume that this effect is independent of the sum over the N_{Subfault} subfault. Thus, the expression for $u(\mathbf{x},t)$ can be rewritten as

$$u(x, t) \propto Y \sum_{j=1}^{N_{\text{Subfault}}} X_j \dot{g}_j(t') f_j(r). \tag{A.2}$$

For a fixed time t' and position r, and using Eq. (1), the expression for $u(\mathbf{x}, t)$ is proportional to the product of two random variables

$$u(\mathbf{x}, t) \propto YX. \tag{A.3}$$

If we assume that Y is also distributed according to the Lévy law with Lévy index α_Y—a rather general assumption that also includes the Gauss law as a special case—then $u(\mathbf{x}, t)$ is given by the product of two Lévy random variables. In general, the product of two Lévy random variables with Lévy indexes α_X and α_Y is not a Lévy random variable. However, the product is in the basin of attraction (for addition) of the Lévy law with a Lévy index given by $\alpha = \text{Min}[\alpha_X, \alpha_Y]$ (see Breiman, 1965; Sornette, 2006, personal communication; and Marsan, 2005 for an application). Let us assume that the "attractive" property is (nearly) valid for truncated Lévy random variables and that heterogeneities in the crust are distributed according to the Lévy law. The results obtained in Sections 3 and 4 suggest that the Lévy index of the medium heterogeneities will be bounded between the value obtained for the slip ($\alpha_Y \approx \alpha_X$ close to 1) and a Gauss law ($\alpha_Y = 2$).

If we assume that the slip is distributed according to a Cauchy law and that the medium heterogeneity is distributed according to a Gauss law, we can compute the analytical PDF associated with $u(\mathbf{x}, t)$. (In general, other nonrandom parameters in Eq. A.3 can be absorbed in the definition of the probability laws for Y or X. To keep it simple, we ignore the nonrandom parameters.)

The analytical expression for the product of two random variables w_1 and w_2, with respective PDF $p_1(w_1)$ and $p_2(w_2)$, can be obtained by using either this relationship

$$p(u) = \int_{-\infty}^{\infty} dw_1 \int_{-\infty}^{\infty} dw_2 \delta(u - w_1 w_2) p_1(w_1) p_2(w_2) \tag{A.4}$$

or by computing the Jacobian of the inverse transformation (for details, see Rohatgi and Ehsanes Saleh, 1976) and producing the following expression

$$p(u) = \int_{-\infty}^{\infty} dw_1 \frac{p_1(w_1)}{|w_1|} p_2\left(\frac{u}{w_1}\right). \tag{A.5}$$

Consider the hypothesis where p_1 corresponds to the Gauss law ($\alpha_1 = 2$) with mean $\mu_1 = 0$ and standard deviation $\sigma_1^2 = 1$, and where p_2 corresponds to the Cauchy law ($\alpha_2 = 1$) with $\mu_2 = 0$ and $\gamma_2 = 1$. Then, $p(u)$ takes the following expression

$$p(u) = \frac{Exp(u^2/2)\Gamma(0, u^2/2]}{\sqrt{2\pi^3}}, \quad u \neq 0, \tag{A.6}$$

where $\Gamma(0, u^2/2)$ is the incomplete gamma function. The asymptotic expansion of Eq. (A.6) gives

$$p(u) \sim \sqrt{\frac{2}{\pi^3}} \frac{1}{u^2} + O\left(\frac{1}{u^4}\right), u \gg 1. \tag{A.7}$$

In the expression in Eq. (A.7), the PDF tail is decreasing as $u^{-\alpha-1}$ with $\alpha = 1$, the Lévy index of the Cauchy law. Consider now the hypothesis where both p_1 and p_2 are distributed according to the Cauchy law ($\alpha = 1$) with $\mu_1 = \mu_2 = 0$ and $\gamma_1 = \gamma_2 = 1$. Computing $p(u)$ using Eq. (A.5) gives,

$$p(u) = \frac{Log(u^2)}{\pi^2(u^2 - 1)}, \quad u \neq 0 \tag{A.8}$$

and the asymptotic expansion

$$p(u) \sim \frac{2Log(u)}{\pi^2} \frac{1}{u^2} + O\left(\frac{1}{u^4}\right) \sim \frac{2}{\pi^2} \frac{1}{u^2} + O\left(\frac{1}{u^3}\right), \quad u \gg 1, \tag{A.9}$$

when using the following asymptotic expansion for $Log(z) = (1 - 1/z) + (1/2)(1 - 1/z)^2 + (1/3)(1 - 1/z)^3 + \ldots$ (see Abramowitz and Stegun, 1972, p. 68, Eq. 4.1.25). Again, the PDF tail in Eq. (A.9) is decreasing as $u^{-\alpha-1}$ with $\alpha = 1$. The results in Eqs. (A.7) and (A.9) show that predicting the frequency of large values of u is principally constrained by the probability law that governs the random variables with the smaller α value. If the ground motions can be understood as a product of random variables, it is thus fundamental to correctly describe the probability law that governs each random variable involved in the product (see Lavallée and Archuleta, 2005). The results in Eqs. (A.6) and (A.8) show that, in general, the product of random variables is not distributed according to a log-normal law (see Galambos and Simonelli, 2004).

In theory, the expression for $p(u)$ can be computed for any values of the parameters μ_1, μ_2, γ_1 and γ_2. However, the expression for $p(u)$ is rather complicated and cumbersome. For this reason, we only provide analytical solutions of $p(u)$ for the parameter values given above.

In principle, the analytical expression for $p(u)$ in Eq. (A.5) can be computed with the help of either *Mathematica* or *Maple*. However, for the cases discussed above, the solutions computed by *Mathematica* or *Maple* lead sometimes to awkward analytical expressions. For instance, the expression for the PDF may include an imaginary term like $I\pi^{-1}$ (see also Glen *et al.*, 2004). Computation of $p(u)$ by traditional methods (that usually requires the indispensable Abramowitz and Stegun, 1972; Gradshteyn and Ryzhik, 1994) is usually free of such incongruities!

Finally, we consider the more general case where the additional random effect is also a function of the position on the faults (e.g., due to variations in the fault geometry or using the more general formulation of the far-field displacement, see Aki and Richards, 2002, Eq. 10.4). The expression for $u(\mathbf{x}, t)$ can be approximated by the following expression

$$u(\mathbf{x}, t) \propto \sum_{j=1}^{N_{\text{Subfault}}} X_j Y_j \dot{g}_j(t') f_j(r), \tag{A.10}$$

where Y_i is a random variable with a Lévy index α_Y. If we assume that Y_i is also distributed according to the Lévy law, then the random variable $U_i = X_i Y_i$ is in the basin of attraction (for addition) of the Lévy law with a Lévy index $\alpha_U = \text{Min}[\alpha_X, \alpha_Y]$. The expression for $u(\mathbf{x}, t)$

$$u(\mathbf{x}, t) \propto \sum_{j=1}^{N_{\text{Subfault}}} U_j \dot{g}_j(t') f_j(r) \tag{A.11}$$

is thus a sum of random variables U_j with a PDF tail that decreases as $U^{-\alpha_U - 1}$. According to a theorem due to Gnedenko and Kolmogorov (1954), a sum of many independent random variables with PDFs that have a power law tail $U^{-\alpha_U - 1}$ with an index $0 < \alpha_U < 2$ is also distributed according to the Lévy law (here, we are borrowing a formulation of the theorem given in Voit, 2001, p. 101). If $\alpha_U \geq 2$, the sum converges to the Gauss law. For a fixed time t' and position r, the sum in Eq. (A.11) is equivalent to a sum of random variables that meet the requirement of the theorem formulated by Gnedenko and Kolmogorov (1954) and $u(\mathbf{x}, t)$ is thus distributed according to the Lévy law with a Lévy index α_U. Here, we are making the assumptions that the theorem remains valid for truncated Lévy random variables, and that U_J can be approximated by an independent random variable (or conversely that the theorem applies to nonindependent variables).

In Section 2 and in this Appendix, we assumed that the slip took place along the strike direction. The random model of ground motion displacement can be generalized to include the effect of slip along the dip direction. Additional "linear" random contributions can also be included to the random model of the ground motion displacement. In principle, this can be achieved by following the procedure discussed in this Appendix and in Section 2.

This *exposé* on the random nature of earthquakes started with a quotation of Kolmogovov and concludes with a theorem that he coauthored!

References

Abrahamson, N.A. (1988). Statistical properties of peak ground accelerations recorded by the SMART 1 array. *Bull. Seismol. Soc. Am.* **78**, 26–41.

Abramowitz, M., Stegun, I.E. (1972). Handbook of Mathematical Functions, U.S. Department of Commerce, National Bureau of Standards, p. 1046.

Aki, K., Richards, P.G. (2002). Quantitative Seismology. 2nd University Science Books, Sausalito, California, p. 700.
Andrews, D.J. (1980). A stochastic fault model 1. Static case. *J. Geophys. Res.* **78**, 3867–3877.
Andrews, D.J., Hanks, T.C., Whitney, J. (2007). Physical limits on ground motion at Yucca Mountain. *Bull. Seism. Soc. Am.* ftp://ehzftp.wr.usgs.gov/jandrews.
Archuleta, R.J. (1984). A faulting model for the 1979 Imperial Valley earthquake. *J. Geophys. Res.* **89**, 4559–4585.
Archuleta, R.J. (1998). Direct observation of nonlinear soil response in acceleration time histories. *Seism. Res. Lett.* **69**, 149.
Archuleta, R.J., Liu, P., Steidl, J.H., Bonilla, L.F., Lavallée, D., Heuze, F. (2003). Finite-fault site-specific acceleration time histories that include nonlinear soil response. *Phys. Earth Planet. Inter.* **137**, 153–181.
Bommer, J.J., Abrahamson, N.A., Strasser, F.O., Pecker, A., Bard, P.-Y., Bungum, H., Cotton, F., Fäh, D., Sabetta, F., Scherbaun, F., Studer, J. (2004). The challenge of defining upper bounds on earthquake ground motions. *Seismol. Res. Lett.* **75**(1), 82–95.
Bonilla, L.F., Archuleta, R.J., Lavallée, D. (2005). Hysteretic and dilatant behavior of cohesionless soils and their effects on nonlinear site response: Field data observations and modeling. *Bull. Seism Soc. Am.* **95**, 2373–2395, doi: 10.1785/0120040128.
Boore, D.M., Joyner, W.J. (1978). The influence of rupture incoherence on seismic directivity. *Bull. Seismol. Soc. Am.* **68**, 283–300.
Breiman, L. (1965). On some limit theorems similar to the arc-sin law. *Theor. Prob. Appl.* **10**(2), 323–331.
Custódio, S., Liu, P., Archuleta, R.J. (2005). The 2004 M_w6.0 Parkfield, California, earthquake: Inversion of near-source ground motion using multiple datasets. *Geophys. Res. Lett.* **32**, L23312, doi: 10.1029/2005GL024417.
Fakao, Y., Furumoto, M. (1985). Hierarchy in earthquake size distribution. *Phys. Earth Planet. Inter.* **37**, 149–168.
Field, H.E., Johnson, P.A., Beresnev, I.A., Zeng, Y. (1997). Nonlinear ground-motion amplification by sediments during the 1994 Nothridge earthquake. *Nature* **390**, 599–602.
Fouque, J.-P., Garnier, J., Papanicolaou, G., Sølna, K. (2007). Wave Propagation and Time Reversal in Randomly Layered Media Springer, Berlin.
Galambos, J., Simonelli, I. (2004). Products of Random Variables. Pure and Applied Mathematics, Marcel Dekker, New York, p. 323.
Glen, A.G., Leemisb, L.M., Drew, J.H. (2004). Computing the distribution of the product of two continuous random variables. *Comput. Stat. Data Anal.* **44**(3), 451–464.
Gnedenko, B.V., Kolmogorov, A.N. (1954). Limit Distributions for Sums of Independent Random Variables Addison-Wesley, Reading, Massachusetts, p. 264.
Gradshteyn, I.S., Ryzhik, I.M. (1994). Table of Integrals, Series, and Products 5th edn, Academic Press, Inc., Boston, p. 1204.
Guatteri, M., Mai, P.M., Beroza, G.C., Boatwright, J. (2003). Strong ground motion prediction from stochastic-dynamic source models. *Bull. Seismol. Soc. Am.* **93**, 301–313.
Gusev, A.A. (1989). Multiasperity model fault model and the nature of short-periods subsources. *Pure Appl. Geophys.* **136**, 515–527.
Gusev, A.A. (1992). On relation between earthquake population and asperity population on the fault. *Tectonophysiscs* **211**, 85–98.
Gusev, A.A. (1996). Peak factors of Mexican accelerograms: Evidence of a non-Gaussian amplitude distribution. *J. Geophys. Res.* **101**, 20,083–20,090.
Hartzell, S., Mariagiovanna, G., Mai, P.M., Liu, P.C., Fisk, M. (2005). Calculation of broadband time histories of ground motion, Part II: Kinematic and dynamic modeling using theoretical Green's functions and comparison with the 1994 Northridge earthquake. *Bull. Seismol. Soc. Am.* **95**, 614–645.

Heaton, T.H., Hartzell, S.H. (1989). Estimation of strong ground motions from hypothetical earthquakes on the Cascadia subduction zone, Pacific Northwest, 1989. *Pure Appl. Geophys.* **129**, 131–201.
Herrero, A., Bernard, P. (1994). A kinematic self-similar rupture process for earthquake. *Bull. Seismol. Soc. Am.* **84**, 1216–1228.
Janicki, A., Weron, A. (1994). Simulation and Chaotic Behavior of Stable Stochastic Processes M. Dekker, New York, p. 355.
Kagan, Y.Y. (1992). Seismicity: Turbulence of solids. *Nonlin. Sci. Today* **2**, 1–13.
Kolmogorov, A.N. (1991). *In:* Tikhomirov, V.M., (ed.), Selected Works of A.N. Kolmogorov. Vol. I, *Mathematics and Mechanics*, Kluwer, Dordrecht, p. 552.
Kostrov, B.V., Das, S. (1988). Principles of Earthquake Source Mechanics Cambridge University Press, Cambridge, p. 286.
Lavallée, D., Archuleta, R.J. (2003). Stochastic modeling of slip spatial complexities for the 1979 Imperial Valley, California, earthquake. *Geophys. Res. Lett.* **30**(5), 1245, doi: 10.1029/2002GL015839.
Lavallée, D., Archuleta, R.J. (2005). Coupling of the random properties of the source and the ground motion for the 1999 Chi Chi earthquake. *Geophys. Res. Lett.* **32**, L08311, doi: 10.1029/2004GL022202.
Lavallée, D., Liu, P., Archuleta, R.J. (2006a). Stochastic model of heterogeneity in earthquake slip spatial distributions. *Geophys. J. Int.* **165**, 622–640.
Lavallée, D., Miyake, H., Koketsu, K. (2006b). On the random nature of earthquake ground motion recorded at the Surface and at the bottom of the boreholes: The 2003 Tokachi-Oki earthquake. *EOS Trans. Am. Geophys. Union* **87**(52), Fall Meet. Suppl., Abstract S43B-1393.
Liu, P.C., Archuleta, R. (2004). A new nonlinear finite fault inversion with 3D Green's functions: Application to 1989 Loma Prieta, California, earthquake. *J. Geophys. Res.* **109**, B02318doi: 10.1029/2003JB002625.
Liu, P., Archuleta, R., Hartzell, S. (2006). Prediction of broadband ground-motion time histories: Hybrid low/high-frequency method with correlated random source parameters. *Bull. Seismol. Soc.* **96**(6), doi: 10.1785/0120060036. To be printed December, 2006.
Liu-Zeng, J., Heaton, T., DiCaprio, C. (2005). The effect of slip variability on earthquake slip-length scaling. *Geophys. J. Int.* **162**(3), 841–849, doi: 10.1111/j.1365–246X.2005.02679.x.
Mai, P.M., Beroza, G.C. (2002). A spatial random-field model to characterize complexity in earthquake slip. *J. Geophys. Res.* **107**, 2308, doi: 10.1029/2001JB000588.
Mai, P.M., Spudich, P., Boatwright, J. (2005). Hypocenter locations in finite-source rupture models. *Bull. Seismol. Soc. Am.* **95**(3), 965–980.
Marsan, D. (2005). The role of small earthquakes in redistributing crustal stress. *Geophys. J. Int.* **141**, doi:10.1111/j.1365–246X.2005.02700.x.
McGuire, R.K. (2004). *In* Seismic Hazard and Risk Analysis Monograph by Eathquake Engineering Research Institute, p. 221.
Mikumo, K.T., Miyatake, T. (1978). Dynamic rupture process on a 3D fault with nonuniform friction and near-field seismic waves. *Geophys. J. R. Astron. Soc.* **54**, 417–438.
O'Connell, D.R.H. (1999). Replication of apparent nonlinear seismic response with linear wave propagation models. *Science* **283**, 2045–2050.
Oglesby, D.D., Day, S.M. (2002). Stochastic fault stress: Implications for fault dynamics and ground motion. *Bull. Seismol. Soc. Am.* **92**, 3006–3021.
Paul, W., Baschnagel, J. (1999). Stochastic Processes, Springer, Berlin, p. 231.
Peitgen, H.O., Saupe, D. (1988). The Science of Fractal Images, Springer-Verlag, New York, p. 312.
Rohatgi, V.K., Ehsanes Saleh, A.K. (1976). *In* An Introduction to Probability Theory Mathematical Statistics, 2nd edn. Wiley, New York, p. 716.
Sornette, D. (2004). Critical Phenomena in Natural Sciences. Springer, Berlin, p. 528.

Tumarkin, A.G., Archuleta, R.J. (1994). Empirical ground motion prediction. *Ann. Geofis.* **37**(6), 1691–1720.
Tumarkin, A.G., Archuleta, R.J. (1997). Stochastic ground motion modeling revisited. *Seismol. Res. Lett.* **68,** 312.
Uchaikin, V.V., Zolotarev, V.M. (1999). Chance and Stability, VSP, Utrecht, The Netherlands, p. 570.
Voit, J. (2001). The Statistical Mechanics of Financial Market Springer, Berlin.
Zhang, W., Iwata, T., Irikura, K., Sekiguchi, H., Bouchon, M. (2003). Heterogeneous distribution of the dynamic source parameters of the 1999 Chi-Chi, Taiwan, earthquake. *J. Geophys. Res.* **108** (B5), 2232, doi:10.1029/2002JB001889.
Zolotarev, V.M. (1986). *In* One-Dimensional Stable Distributions, American Mathematical Society, Providence, Rhode Island, p. 284.

GLOSSARY

Ludovic Margerin, Haruo Sato, Michael C. Fehler
and Yingcai Zheng

Anisotropic (nonisotropic) scattering/Transport mean free path
In a random medium, when the scatterers have a tendency to diffract energy preferentially in some directions, one introduces the notion of anisotropic scattering. In such a situation, it usually takes several scattering events before a beam propagating initially in a well-defined direction has distributed a significant amount of energy in all space directions. The typical length scale over which a beam "loses memory" of its initial direction is the transport mean free path. It can be significantly larger than the scattering mean free path.

Anisotropic (nonisotropic, anisometric) random media
Random media statistically characterized by different characteristic distances in different directions. For example, the horizontal characteristic distance is larger than the vertical characteristic distance in sedimentary layers.

Born approximation
A perturbation method used for the calculation of scattering waves due to a specific obstacle or a localized inhomogeneity, where the wave is written as a sum of the primary wave and scattered wave. This approximation is valid only when the scattered wave amplitude is smaller than the incident wave amplitude.

Coda attenuation
The coda wave envelopes of local earthquakes smoothly decay with increasing lapse time. The decay can be empirically described as a product of a geometrical factor and an exponential decay factor, where the latter is called coda attenuation. Generally, coda attenuation depends on frequency and is characteristic of a given region. The relation between coda attenuation and direct-wave attenuation by either scattering or intrinsic mechanisms is not well established.

Coda normalization method
A method used for measurements of earthquake magnitudes, wave attenuation per distance, and site amplification factors from the coda wave amplitude at a fixed lapse time within some frequency band. This method is based on the assumption that the distribution of coda energy of a local earthquake is spatially uniform in a given region at a sufficiently large lapse time.

Coda waves
Wave trains in the tail portion of a seismogram. They are interpreted as waves that are multiply or singly scattered by distributed heterogeneities in the Earth. This term is often used as names for wave trains following the direct phase: P coda, S coda, and Lg coda.

Differential scattering cross-section/Total scattering cross-section

A measure of scattering power of an obstacle. The differential scattering cross-section, which is proportional to the square of scattered wave amplitude, means the scattered-wave energy generation per time in a solid angle for a unit incident energy flux density. Its integral over all the solid angle gives the total scattering cross-section.

Diffusion constant/Transport mean free path

In a strongly scattering medium it is possible to describe the energy propagation of multiply scattered waves with a simple scalar diffusion equation for the total energy. This equation is identical to the heat conduction equation in thermal processes. The parameter that determines the strength of the scattering process is the diffusion constant. It is the product of the energy propagation velocity and transport mean free path divided by the space dimension. The strength of the scattering increases as the diffusion constant decreases.

Earthquake doublet

A pair of earthquakes that ruptured at nearly the same location with nearly identical focal mechanisms (or identical source waveforms) but at different occurrence times. Doublets are often used to monitor the temporal changes of Earth medium properties.

Envelope

Time trace connecting the peak values of oscillating waves. It is often used to characterize the shape of a short-period seismogram. The smoothed curve of the squared waves is called the mean square (MS) envelope, which represents the time trace of wave energy density.

Fresnel zone

In ray theory (infinite-frequency approximation), the propagation path of waves between two points A and B in an inhomogeneous medium is represented by a line or ray connecting A and B along which the travel time is minimal (or extremal to be exact). The wavefield at B due to a source at A solely depends on medium properties along that ray, however this is not correct for finite-frequency wave propagation for which one can show, in the light of diffraction and interference, that the wave field at B usually depends on medium properties at points within a volume around that ray and this volume is called the Fresnel zone. The wavefield at B may have different sensitivity due to changes at different points within the Fresnel zone. For a propagation distance L, the size of the Fresnel zone in the transverse dimension scales like $\sqrt{L\lambda}$, where λ is the wavelength.

Intrinsic attenuation

A mechanism of wave amplitude decay caused by the transfer of wave oscillation energy into heat.

Lapse time

The time measured from the origin time of an earthquake.

GLOSSARY

Markov approximation

Consider the propagation of harmonic plane waves through media with random velocity fluctuation, where the wavelength is shorter than the correlation distance. The master equation that governs the statistical moments of the wave field on a plane perpendicular to the global ray direction is derived from the parabolic equation as the average over an ensemble of random media. The parabolic equation describes one-way propagation without backscattering. Using this approximation, the wavefield at some distance from the source depends only on the wavefield at a slightly smaller distance. This is called the Markov approximation since it has roots in the concept of Markov process in which the probability of future events is dependent only on most recent events.

Mean wave-field/Mean intensity/Dyson and Bethe-Salpeter equations

Consider the following thought experiment. Launch a point source at $\mathbf{r}_0$ in a random medium and record the wave field at $\mathbf{r}$. Repeat the experiment for all possible realizations of the random medium and compute the average of all the wave records. This defines the mean wave field, or mean Green's function, at $\mathbf{r}$ due to a point source at $\mathbf{r}_0$. In statistical wave theory, the equation that governs mean Green's function is known as the Dyson equation. It is usually formulated in the frequency domain for a monochromatic source. If instead of the wave field one measures the average energy, one obtains the mean intensity. The equation that governs the mean intensity in a random medium is known as the Bethe-Salpeter equation. It forms the rigorous theoretical basis for the radiative transfer equation.

Parabolic approximation

Consider the propagation of harmonic plane waves through a medium with velocity inhomogeneity whose characteristic scale is longer than the wavelength. The wave equation can be approximated by a parabolic-type equation since the second derivative with respect to the global ray direction can be neglected. The parabolic wave equation describes a one-way wavefield with no turning waves and it is widely used in seismic imaging or underwater acoustics.

Radiative transfer theory

A theory that describes the energy propagation in scattering media, where the key parameters are the scattering coefficient and the background velocity. Originally it was a phenomenological theory based on energy conservation and causality; however, it can be theoretically derived from the Bethe-Salpeter equation based on the stochastic wave theory in random media.

Random media/Ensemble of random media

A statistical concept of inhomogeneous elastic media, where the fractional fluctuations of elastic coefficients are random functions of locations. An ensemble of random media is a collection of random media characterized by the same statistical moments. Autocorrelation functions of the fractional fluctuations of elastic coefficients are often used for the characterization. Physical quantities calculated by their average over the ensemble can be compared with observed quantities. "Discrete" random media contain randomly distributed discrete obstacles (cracks, cavities, etc.) whose spatial distribution can also be characterized by statistical means.

Reciprocity

Reciprocity is a general property of the acoustic and elastic wave equations. It can be summarized with this simple sentence: "If you can see me, I can see you." It means that the response measured at r_2 due to a source at r_1 is rigorously identical to the response measured at r_1 due to a source at r_2. Reciprocity can be broken in the presence of an external field. For instance the rigid Earth rotation breaks the reciprocity for long period seismic waves.

Rytov approximation

A perturbation method to solve the parabolic wave equation in weakly inhomogeneous media. The wave amplitude is expressed as a product of the unperturbed wave and the exponential of a surrogate function (the complex phase function), which must be iteratively solved. This approximation is valid when the transverse variations of the perturbation term of the phase fluctuation over distances of about the wavelength should be smaller than the square root of the velocity fractional fluctuation. This method is known as the method of smooth perturbations because of this smoothness condition. The Rytov approximation can account for some diffraction effects and it is still valid when the geometrical ray theory breaks down, and even better than the Born approximation for long-range wave scattering. The Rytov approximation is not valid for backscattering. It is often used for the line of sight propagation problems.

Scattering amplitudes

The scattering amplitude is the amplitude of spherically outgoing scattered wave due to a specific obstacle or a localized inhomogeneity for the incidence of a plane wave with unit amplitude. In general, the scattering amplitude is nonisotropic and is a function of angle, size, and shape of scatterer, and frequency.

Scattering attenuation

A mechanism of attenuation that results in amplitude decay of the pulse due to scattering from distributed heterogeneities. This mechanism causes the redistribution of wave oscillation energy in space and time without energy loss.

Scattering coefficient/Total scattering coefficient

A measure of scattering power of a unit volume of heterogeneous media. It is a product of 4π, the differential scattering cross-section and the number density of scattering obstacles. Total scattering coefficient is the average of the scattering coefficient over all the solid angles.

Scattering mean free path

In a random medium, the scattering mean free path is the typical length scale beyond which a plane wave has been significantly attenuated by scattering. It also represents the average distance between two scattering events in the multiple-scattering theory. It is the reciprocal of the total scattering coefficient.

Transmission fluctuations
Statistical fluctuations of phase and amplitude of transmitted waves through random velocity inhomogeneities. These fluctuations can be used to determine the spectrum of the velocity inhomogeneities.

Wave parameter D
In a random medium with large-scale inhomogeneities compared with the wavelength, an initially planar wavefront will be distorted by diffraction within the Fresnel volume. This generates fluctuations of arrival times and amplitudes compared to the undistorted plane wave. The nondimensional wave parameter D, defined as four times the ratio of the squared Fresnel zone width to the squared average size of inhomogeneities is a key parameter to quantify the fluctuations of phase and amplitude along the wavefront.

Weak/Anderson localization
Weak localization is an interference effect that occurs in a multiple scattering random medium. It results in an increase or enhancement of the mean intensity in a zone of linear dimension about one wavelength around the source. A consequence of weak localization is the reduction of the amount of energy transported away from the source. In the case of extremely strong scattering, the transport of energy can be completely blocked by interference effects, a phenomenon known as Anderson localization.

WKBJ (Wentzel-Kramers-Brillouin-Jeffreys) approximation
A high-frequency approximation method originally developed for the semiclassical calculation of the Schrödinger equation in quantum mechanics. Consider the plane wave propagation through a medium with smooth velocity inhomogeneity, of which the characteristic scale is longer than the wavelength. The solution to the wave equation assumes a form called the WKBJ ansatz (or trial solution) in which the phase and amplitude separates. With this ansatz, the original wave equation reduces to two equations, one about the phase and the other the amplitude and they can be solved approximately. The WKBJ method solves only transmitted waves and it assumes no reflections, thus preserving the transmitted energy flux.

Index

A

Acoustic logging
  array processing, 250–252
  dipole and monopole, 250
  guided and head waves
    investigations, 253–255
  measurement, 249
  tools used, 249, 250
  waves generation, 250
Akaike's information criterion
  (AIC), 231
Angular coherence
  function (ACF), 22
Anisotropic heterogeneities,
  211–213
Apparent attenuation
  mechanism, 125
Asymptotic scaling, seismic waves
  attenuation, 162, 163
Autoregressive (AR) reflection sheet
  definition, 223
  phase spectra, 234
  travel-time fluctuation, 231

B

Bethe-Salpeter equation, 85
Biot's slow wave, 124
Biot's wave-induced flow equation,
  124, 129, 130
Birch's law, 66
Borehole seismic measurement
  acoustic logging
    array processing, 250–252
    dipole and monopole, 250
    guided and head waves
      investigations, 253–255
    measurement, 249
    tools used, 249, 250
    waves generation, 250
  crosswell seismic survey
    products and sources, 255
    reflection image, 257–259
    velocity tomogram, 256, 257
  vertical seismic profiling
    reflection image, 260, 261
    walkaway and 3D, 260
    zero-offset, 259, 260

Born approximation scattering
  coefficients, 83–85
Brittle-ductile interaction
  hypothesis (BDIH),
  291, 292

C

Calibration methodology
  band-dependent processing and
    parameters, 323
  coda spectral results
    level comparison, ENH and
      XAN, 344
    network coda vs.
      ground-truth moment
      magnitudes, 343
    source parameters, 342
  coda start time, 331
  2-D path and interstation site
    coda spreading model, 336
    ESH model and
      curves, 336, 337
    path calibration steps,
      337–340
  flowchart representation, 325
  intrastation site
    mean relative site, 335
    methods, 333
  master station constraint groups,
    326–330
  measured coda amplitude, 326
  shape function and amplitude
    measurement, 333
  source to coda transfer function
    Brune source model, 340
    catalog magnitudes, 341
    reweighted transfer
      function, 342
CDP transform algorithm, 257
Centimeter-scale heterogeneities,
  126, 127
Chernov theory, 21, 22
Chi-Chi earthquake
  ground motion PGA model
    probability density function
      (PDF), 447, 448
    random property
      variation, 447

source random model, 444, 445
Coda energy density
  definition, 265
  diffusion–absorption model,
    273–275
  predicted assumptions, 266
  single isotropic scattering
    approximation, 266, 267
Coda energy distribution
  nonuniform distribution
    coda normalization method,
      273, 276, 277
    diffusion–absorption model,
      273–275, 277
    in higher regions, 275, 276
    radiative transfer theory, 269
    spatio-temporal correlation,
      289–292
  uniform distribution
    coda normalization
      method, 270
  intrinsic absorption
    and scattering
    attenuation, 271
  isotropic/nonisotropic
    scattering models, 269
  S-coda waves, 268
Coda envelope
  inversion analysis and scattering
    coefficient, 302
  energy density and residuals,
    302, 303
  fault system, 304, 305
  focal mechanism, 305, 306
Coda normalization method
  nonuniform distribution, 273,
    276, 277
  uniform distribution, 270
Coda Q_C^{-1}
  2D finite difference method, 266
  temporal decay rate
    frequency dependence,
      279–281
    geographic variation, 281–284
    lapse time dependence,
      278, 279
    temporal variation, 284–289
Coda waves
  characteristics

Coda waves (*cont.*)
 back-scattering, 266
 coherence distance, 267
 scattering strength, 312
 seismogram composition, 311, 312
 conventional seismogram-stacking methods, 168
 definition, 265
 envelope and inversion analysis, 302, 306
 Kirchhoff coda migration, 306
 seismic signal detection, 307–309
 single back-scattering model, 265, 266
 slant-stacked waveforms, 309–311
 teleseismic P, 170–174, 187, 188
Coherence function
 constant background medium, 35
 delta-correlated assumption, 34, 35
 plane waves
  bessel function, 32
  correlation functions, 29, 30
  fourier transform pairs, 31
Coherence tensor, 7
Covariance function
 anisometric fluctuations, 100, 101
 Gaussian, 115
 longitudinal, 113
 quasi-homogeneous fluctuations, 99
 small offsets, 104–106
 travel-time variance, 106
Cross-correlating seismic noise
 definition, 393
 impulse response, 391, 392
 relative velocity changes, 392, 393
Cross-correlation moving window technique (CCMW), 377, 378
Crosswell seismic survey
 products and sources, 255
 reflection image, 257–259
 velocity tomogram, 256, 257
Crustal inhomogeneity, 305

D

Disk-shaped transducer, 224–226
Double passage effect (DPE)
 geometry, 103
 zero offset, 106, 107
Dynamic poroelasticity, 129–133

E

Earthquake doublet waveforms
 P and S phase time delay relocation
  crustal and deep doublet, 379–381
  double-difference location, 378, 379
  instrumental error, 380
  temporal change, 380, 382
  selection, 374, 375
 spatial doublet coda
  cross-spectral time delays computation, 385, 386
  slowness vector computation, 387, 388
 S-P observed delays
  anisotropy detection, 383, 384
  azimuthal function, 382, 383
  S-velocity temporal variation, 384
 teleseismic doublets
  definition, 394
  PKP phase delays, 393, 394
 temporal variation search
  coda attenuation, 390, 391
  S-wave splitting, 389, 390
 time delay measurement
  cross-correlation and coherency module, 375, 376
  fast Fourier transform (FFT), 377
  phase uncertainty, 376, 377
  virtual doublets, 391–393
Earthquake, random model
 Chi-Chi earthquake
  ground motion PGA model, 447, 448
  source random model, 444, 445
 Parkfield earthquake
  ground motion PGA model, 440–443
  ground motion PGV model, 443, 444
  source random model, 435–440
 slip spatial distribution, ground motions
  distributed random variables, 434, 435
  far-field seismic source, 434
  Lévy randomvariable, 433
  slip/stress distribution spatial heterogeneity
   Lévy randomvariable, 432, 433
   limitations, 432
   probability laws, 430, 431
   variable normalization and smoothing, 429, 430
Earthquake source process. *See* High-frequency seismic energy radiation
Effective wave number
 1-D random media, 138
 3-D random media
  coherent phase and statistical smoothing, 151, 152
  correlation function and fluctuation spectrum, 153
  weak-wavefield-fluctuation regime, 151, 154
Elastic wave propagation, random media
 laboratory experiments
  heterogeneity statistical description, 221–223
  scale-invariant expression, 227–230
  wave fields, 223–227
 wave fluctuations
  EHM/SRM boundaries, 241, 242
  scattered waves diffraction, 242, 243
  small-scale heterogeneities, 240, 241
 waveform analysis
  common approaches, 230, 231
  cross spectrum, 231–236
  shear-wave particle velocities, 236–238
  travel-time fluctuation, 231
  waveform envelope, 238–240
Elastic waves, weak localization effect
 Bethe-Salpeter equation, 10
 2-D chaotic cavity, 7
 inhomogeneous medium, 1
 reciprocity properties, 4
Energy-flux model, 359
Energy transport theory. *See* Radiative transfer theory
Envelope inversion
 classification
  frequency range, 404
  Green's function, 404, 406
  velocity response envelope, 406
  double-couple source configuration, 407, 408

Index 471

finite-fault version, 404
high-frequency seismic energy
   radiation
   frequency dependence,
      414–418
   scaling, 418–420
   vs. asperities, 420–423
practical applications, 410, 411
practical data analysis
   envelope residuals, 411, 413
   frequency radiation spatial
      relationship, 413, 417
   observed and synthetic
      envelopes, 412, 415
   spatial distribution of energy
      radiation, 412, 414
   standard deviation, 413, 416
   S-wave velocity, 411
radiative transfer theory, 406, 407
seismogram envelope, 403, 404
spherical Bessel function,
   407, 409
theoretical energy density, 410
Equivalent homogeneous material
   (EHM). See Scattering random
   medium (SRM)
Extended Street-Herrmann model
   path curves, 337
   spreading function, 336

F

Flexural wave, 250
Fourier transform
   angular spectrum, 56
   wandering effect, 55
Fresnel filters, 151, 155, 156

G

Gassmann's equation, 130
Gaussian autocorrelation function
   three-dimensional random elastic
      media, plane wavelet
      incidence
      plane S-wavelet, 63, 64
      P-wave envelope
         characteristics, 60–62
      P-wavelet ISDs, 57–60
      teleseismic P-wave envelope
         analysis, 62
   three-dimensional random elastic
      media, radiation from point
      source
      P-wavelet ISDs, 72–74
      spherically outgoing P-wave
         envelopes, 74–76
      spherically outgoing
         S-wavelet, 77, 78

two-dimensional random elastic
   media
   plane wavelet incidence,
      64–66
   radiation from point source,
      78–81
Gaussian correlation model, 128,
   140, 145
Geometrical optics (GO)
   basic elements
   applicability, 101, 102
   equations, 97–99
   quasi-homogeneous
      fluctuations (QHF)
      model, 99
   travel-time covariance
      function, 100, 101
   numerical simulation
   longitudinal correlation scale
      vs. distance, 118
   refractive index fluctuations,
      115
   travel-time fluctuations, 115,
      116
   travel-time variance, 115, 117
   statistical parameters, 96, 97, 114,
      117
   travel-time fluctuations
   reflection geometry, 102–107
   refraction geometry, 107–115
Green's function, 129, 154
Green's functions
   high-frequency seismic energy
      radiation
   double-couple source
      configuration, 407, 408
   energy density, 410
   practical applications, 410,
      411
   propagator function and
      energy density, 407
   spherical Bessel function, 409
   spherical harmonics
      expansion, 407, 409
Green's tensors
   first-order statistical smoothing,
      135
   homogeneous and isotropic
      media, 132
   scattered wavefields, 133, 134
   spatial Fourier transform, 136

H

Heterogeneity scale, 126–128
Heterogeneity statistical description
   averaged auto-correlation, 222
   layered structure, 223

Oshima granite distribution, 221
Heterogeneous plate model, 210, 211
High-frequency seismic energy
   radiation. See also Envelope
   inversion
   envelope Green's functions
   double-couple source
      configuration, 407, 408
   energy density, 410
   practical applications, 410,
      411
   propagator function and
      energy density, 407
   spherical Bessel function, 409
   spherical harmonics
      expansion, 407, 409
frequency dependence
   acceleration source spectrum,
      414, 416
   decay power, 417, 419
   octave-width frequency band,
      416, 417
scaling
   fixed frequency band, 418
   inland and offshore
      earthquakes, 420, 421
vs. asperities
   complementary and matching
      spatial relations, 420,
      422
   kinematic heterogeneous
      fault-rupture models,
      422, 423
   slip rate and luminosity time
      functions, 422
High-Q and high-V models, 208–210

I

Inland earthquakes, 420, 421
Intensity spectral density (ISD)
   plane wavelet incidence, 54, 55
   radiation from point source,
      71, 72
Interlayer flow attenuation, 1-D
   random media, 158
Intrinsic attenuation, 123. See also
   Seismic waves attenuation

J

Joint transverse and angular
   coherence function (JTACF),
   22

K

Kirchhoff coda migration, 306
Kirchhoff depth migration, 257

Kramers–Kronig equations, 152, 153
Kramers–Krönig relation, 84

L

Laboratory scale-model experiments, 219, 220
Large aperture seismic array (LASA), 21
Laser Doppler vibrometer (LDV), 220, 221
Logarithmic amplitude, 21

M

Macroscopic heterogeneities, 126
Mesoscopic heterogeneities, 126
Monte Carlo algorithms
  applications
    acoustic wave scattering, 175
    different approaches, 176
    isotropic S-wave scattering, 175
  computer-generated random numbers, 174
  implementation, 176
  interfaces
    Snell's law, 180
    standard Earth models, 179
  interinsic attenuation
    definition, 185, 186
    PREM Earth model, 186, 187
  particle trajectories
    free surface, 179
    ray theory, 178
    S-wave polarizations, 178, 179
  powerful tool, 188
  scattering angles
    Birch's law, 185
    Born scattering coefficients, 182
    computational advantages, 185
    Gaussian function, 182
    ray-centered coordinate system, 184
    scattering at high frequencies, 174
  shallow-and deep-earthquake coda amplitudes, 170
  Shearer and Earle heterogeneity model, 189
  sources
    double-couple, 178
    event biasing, 177
    isotropic radiation, 177
    teleseismic P coda, 187, 188
Monte Carlo solution, 84, 85

Mt. Merapi
  coda-localization effects, 367
  diffusion model, 361–365
  two-layer media, 363, 364
  uniform half space, 361–363
  volcano scattering attenuation, 368
Mt. Vesuvius
  coda-localization effects, 367
  coda-Q^{-1} observation, 355
  diffusion model, 361–365
  two-layer media, 363–365
  uniform diffusive layer, 366
  uniform half space, 361–363, 365, 366
  volcano scattering attenuation, 368
Multiple lapse-time window analysis (MLTWA) technique, 357
Multiple scattering theory. *See* Radiative transfer theory
Multiple signal classification (MUSIC) spectrum, 308

N

Non-isotropic heterogeneity structure, 215
Numerically simulated envelopes
  finite difference simulations
    mean square envelopes, 68, 69
    pulse shape, plane wavelet, 66
    P-wavelet and S-wavelet incidence, 67, 68
  Markov envelopes *vs.* FD envelopes, 68, 69
Numerical simulations
  Gaussian correlation function, 34, 35
  random velocity model, 35
  WKBJ Green's function, 36

O

O'Doherty–Anstey (ODA) formalism
  locally layered media, 148
  wave fields self-averaging properties, 149, 150
Offshore earthquakes, 420, 421
One-dimensional poroelastic random media, seismic waves
  interlayer flow, 158, 159
  numerical validation, 161
  reciprocal quality factor, 158
  and wave-induced flow attenuation, 159, 160
Oshima granite
  distribution, 221–223
  quartz grain distribution, 243
  three-component waveforms, 238
  wave form fluctuation, 225
  wave particle velocities, 236
Oyo Geospace, orbit vibrator, 255

P

Pacific plate events
  anelastic attenuation properties, 203, 204
  frequency selective propagation, 201, 202
  low-frequency precursors, 201
Parabolic wave equation, 52, 79
Parkfield earthquake
  ground motion PGA model
    Lévy random variable, 440, 441
    probability density function (PDF), 441–443
  ground motion PGV model, 443, 444
  source random model
    mean power spectrum $P(k1)$, 438
    parameter compilation, 435, 436
    parameter variation, 439
    probability density function (PDF), 440
    source inversion, 437
    strike and dip components, 438, 439
Peak ground acceleration (PGA)
  ground motion random model
    Lévy random variable, 440, 441
    probability density function (PDF), 441–443, 447, 448
    random property variation, 447
Peak ground acceleration (PGA) pattern
PHS events, 198, 199
  radial-component records, 200, 201
Perturbation theory, 1-D random media wavefields, 124, 163
Plane P-wavelet, 51, 52
Plane S-wavelet
  Gaussian ACF, 63, 64
  intensity spectral density, 63
  parabolic wave equation, 62
Poroelastic scattering equation, seismic waves, 133, 134
Pseudo-Rayleigh wave, 250

Index

Q

Quasi-homogeneous fluctuations (QHF) model, 99

R

Radiative transfer theory
  envelope Green's functions
    double-couple source configuration, 407, 408
    energy density, 410
    practical applications, 410, 411
    propagator function and energy density, 407
    spherical Bessel function, 409
    spherical harmonics expansion, 407, 409
Radiative transfer theory (RTT)
  conversion and large-angle scattering, 83, 84
  Monte Carlo solution, 84, 85
Random porous media, seismic waves attenuation
  2-D and 3-D random media, 152–154
  effective wave number, 149–154
  ODA formalism, 148, 149
Ray theory
  sample-scale heterogeneity, 240
  sample surfaces, 241
  travel-time fluctuation, 231
  wave propagation, 230
Reciprocity, 4
Reflection geometry
  covariance function
    small offsets, 104–106
    travel-time variance, 106
  double passage effect (DPE)
    geometry, 103
    zero offset, 106, 107
Refraction geometry
  constant velocity gradient
    auxiliary dimensionless random field, 107
    refractive index, 108
    inverse problem solution
      inhomogeneity scale lengths and standard deviation, 112
      longitudinal and transverse correlation scales, 112–114
      transverse length, 113
    offset dependence, 110–112
    variance, 108–110, 115, 117
Relative fluid displacement, 130
Rytov
  heterogeneous medium, 25, 26
  monochromatic wave, 25
Rytov approximation, 148, 151–154

S

Scale-invariant expression
  scattering random medium, 229, 230
  subsurface heterogeneities, 229
  wave frequencies, 227, 228
Scattered waves diffraction, 242, 243
Scattering random medium (SRM)
  EHM boundaries, 241, 242
  wave propagation method, 229, 230
S-coda waves, 268
SCSN. *See* Southern California Seismic Network
Seismic analysis code software, 322
Seismic arrays
  scatterer distribution, 312–315
  signal detection
    ray direction, 308
    slant stacking, 307, 308
    subsurface structure, 307
  single-scattering model and slant-stacked waveforms
    energy density, 309
    energy level of envelopes, 310, 311
  wave characteristics
    scattering strength, 312
    seismogram composition, 311, 312
Seismic coda waves
  apparent stress distribution, 345
  band-dependent processing and calibration parameters, 323
  calibration methodology
    coda start time, 331
    intrastation site, 333–335
    master station constraint groups, 326–330
    measured coda amplitude, 326
    shape and amplitude measurement, 331–333
    source to coda transfer function, 340–342
  coda techniques, 320
  data Analysis
    event merging, 321
    flowchart representation, 325
    seismic analysis code software, 322
    waveform data and instrument response information, 321
  2-D transfer function, 347
  flowchart representation, 325
  Green's function technique, 320
  regional coda techniques, 319
  spectral analyzing seismometers, 320
  spectral results
    level comparison, ENH and XAN, 344
    network coda *vs.* ground-truth moment magnitudes, 343
    source parameters, 342
  stability, 319, 320
  static stress drops, 348
  types, 346
Seismic doublets
  observed delays, S–P delays
    anisotropy detection, 383, 384
    azimuthal function, 382, 383
  P and S phase time delay relocation
    crustal and deep doublet, 379–381
    double-difference location, 378, 379
    instrumental error, 380
    temporal change, 380, 382
  selection, 374, 375
  spatial doublet coda
    cross-spectral time delays computation, 385, 386
    slowness vector computation, 387, 388
  S-velocity temporal variation, 384
  teleseismic doublets
    definition, 394
    PKP phase delays, 393, 394
  temporal variation search
    coda attenuation, 390, 391
    S-wave splitting, 389, 390
  time delay measurement
    cross-correlation and coherency module, 375, 376
    cross-correlation moving window technique (CCMW), 377, 378
    fast Fourier transform (FFT), 377
    phase uncertainty, 376, 377
  virtual doublets, 391–393
Seismic network data
  coda envelope
    energy residuals, 302, 303
    fault system, 304, 305
    focal mechanism, 305, 306

474　Index

Seismic network data (*cont.*)
　Kirchhoff coda migration, 306
Seismic scattering
　coda envelope
　　energy density, 302
　　energy residuals, 302, 303
　　fault system, 304, 305
　　focal mechanism, 305, 306
　in deep earth
　　envelope-stacking technique, 169, 170
　　PKP precursors, 167
　　small-scale arrays, 169
　　teleseismic P coda, 170–174
　　waveform stacking advantages, 168
　density and strength, 312
　distribution, 312–315
　envelope-stacking methods, 188
　Monte Carlo methods
　　applications, 175, 176
　　computer-generated random numbers, 174
　　implementation, 176
　　interfaces, 179, 180
　　intrinsic attenuation, 185–187
　　particle trajectories, 178, 179
　　powerful tool, 188, 189
　　scattering angles, 181–185
　　scattering at high frequencies, 174
　　scattering events, 180, 181
　　sources, 177, 178
　　teleseimic P coda, 187, 188
　seismic array data, 309–311
Seismic waveform analysis
　common approaches, 230, 231
　cross spectrum
　　amplitide spectra, 233
　　lapse time frequency bands, 235, 236
　　phase spectra, 233, 234
　　statistical analysis, 231, 232
　shear-wave particle velocities
　　polarization and propagations, 237, 238
　　S-wave particle, 236, 237
　travel-time fluctuation, 231
　waveform envelope
　　shear-wave sources, 238, 239
　　S-wave envelope, 239, 240
Seismic wave-induced flow
　attenuation and dispersion
　　Biot's equations, 129, 130
　　correlation function, 140, 141
　　1-D effective wave number, 138

effective fast wave number, 135–138
first-order statistical smoothing approximation, 134, 135
Green's tensors, 132–134
homogeneous poroelastic composite, 130, 131
point source response, 132
poroelastic scattering equation, 133, 134
P-waves, 139, 142–145
spectral filter function and fluctuation spectrum, 140, 141
mechanism, 124
Seismic waves
　elastic scattering medium, 1
　logarithm of energy, 2
　source mechanism and wavefield polarization
　　acoustic and elastic waves, 10–13
　　Bessel function, 11
　　Bethe-Salpeter equation, 8, 10
　　diffuson and cooperon, 8
　　elastic waves, 2-D chaotic cavity, 7
　　Feynman diagrams, 9
　　Green function, 11
　　Heisenberg time, 6
　　multiple scattering formalism review, 7–9
　　transport mean free path, 2
　weak localization effect
　　application, 13–17
　　coherent back scattering, 5
　　enhancement zone, 6
　　Lamb mode, 15
　　reciprocity property, 4
　　scalar partial wave, 3
　　scattering media measurement, 16, 17
　　speckle pattern, 15
　　surface waves, measurement of, 13–15
Seismic waves attenuation
　asymptotic scaling, 162, 163
　attenuation coefficient, 148, 149, 152, 154
　Biot's slow wave, 124
　conversion scattering, 135, 138, 140
　1-D poroelastic random media
　　interlayer flow, 158, 159
　　numerical validation, 161
　　reciprocal quality factor, 158

　　and wave-induced flow attenuation, 159, 160
　elastic scattering, 125, 138, 159
　exponential correlation function, 129, 142, 146
　mesoscopic and macroscopic heterogeneity, 126–129
　phase velocity, 138, 142, 147
　P-waves
　　correlation function, 145, 146
　　differently correlated fluctuations, 144
　　reciprocal quality factor, 142, 143
　　solid and fluid phase parameters, 142
　　variances, 142
　　velocity for, 144–146
　in random porous media
　　2-D and 3-D random media, 152–154
　　effective wave number, 149–154
　　ODA formalism, 148, 149
　scattering attenuation and asymptotic behavior, 154–158
　statistical smoothing method, 148
　spatial autocorrelation function, 126, 136, 140, 141
　wave-induced flow
　　asymptotic behavior, 145–148
　　and dispersion, 138–145
　　dynamic poroelasticity, 129–133
　　and effective wave number, 135–138
　　first-order statistical smoothing approximation, 134, 135
　　mechanism, 124
Seismogram envelope analysis
　regional and local earthquake seismograms, 45, 46
　teleseismic P-waves
　　energy-flux model, 46–48
　　scattering attenuation, 48
　　statistical parameters, lithospheric inhomogeneity, 48, 50
　　travel-time fluctuations, 45
　　wave field fluctuations, 46
　　vector-wave envelope, Markov approximation, 48, 49, 51
Seismograms, 195
　Kirchhoff migration, 306

wave composition, 311, 312
Semblance cross-correlation method
   head wave sensitivity, 253, 255
   wave velocity formation, 251, 252
Semblance-weighted slant stacking, 313, 314
Single back-scattering model, 265, 266
Single isotropic scattering approximation, 266, 267
Slant stacking, 307
Southern California Seismic Network (SCSN), 306
Spatio-temporal correlation
   brittle-ductile interaction hypothesis (BDIH), 291, 292
   coda energy and seismicity, 289–292
Spherical Bessel function, 409
Spherically outgoing P-wavelet, 70, 71
Spherically outgoing S-wavelet
   Gaussian ACF, 77, 78
   intensity spectral density, 76, 77
Stoneley wave, 250
Street-Herrmann model, 336
Subduction waveguide
   2D FDM models
      anisotropic heterogeneities, 211–213
      heterogeneity scale effect, 214, 215
      heterogeneous plate model, 210, 211
      high-Q and high-V models, 208–210
      P-and S-wave velocities, 204, 205
      plate thickness, 213, 214
      quasi-lamina structure, 206, 207
      slab guided waves, 208
   large ground motions, 197
   Pacific plate events
      anelastic attenuation properties, 203, 204
      frequency selective propagation, 201, 202
      low-frequency precursors, 201
      peak ground acceleration (PGA) pattern, 198–201
      zone plates, 197, 198
S-wave attenuation
   attenuation parameter, 281
   coda normalization method, 270
   frequency dependence, 271

T

Teleseismic doublets
   definition, 393
   PKP phase delays, 393, 394
Teleseismic P-waves
   energy-flux model, 46–48
   lithospheric inhomogeneity, statistical parameters, 48, 50
   scattering attenuation, 48
   travel-time fluctuations, 45
   wave field fluctuations, 46
Temporal decay rate, coda energy
   frequency dependence, 279–281
   geographic variation, 281–284
   lapse time dependence, 278, 279
   temporal variation, 284–289
TFMCF. *See* Two-frequency mutual coherence function
Three-dimensional random elastic media
   plane wavelet incidence
      angular spectrum, 56
      Gaussian ACF, 56–62
      intensity spectral density, 54, 55
      plane P-wavelet, 51, 52
      plane S-wavelet, 62–64
      random media ensemble, 52
      TFMCF, stochastic master equation, 52–54
      wandering effect, 55
      wave equations, inhomogeneous media, 51
   radiation from point source
      Gaussian ACF, 72–76
      intensity spectral density (ISD), 71, 72
      spherically outgoing P-wavelet, 70, 71
      spherically outgoing S-wavelet, 76–78
      TFMCF, stochastic master equation, 71
Tikhonov regularization, 256
Tramsmission fluctuations
   acoustic waves
      stratified medium, 23
      WKBJ green function, 23–25
   Chernov theory, 21, 22
   coherence function
      constant background medium, 34
      delta-correlated assumption, 33, 34
      plane waves, 29–32

      heterogeneous medium, 25, 26
   numerical simulations
      Gaussian correlation function, 34, 35
      random velocity model, 35
      WKBJ Green's function, 36
   phase
      delta-correlated assumption, 36, 37
   plane wave incidence, 26–28
Transverse coherence function (TCF), 21
Travel-time fluctuations
   reflection geometry
      covariance function, 104–106
      double passage effect (DPE), 106, 107
   refraction geometry
      constant velocity gradient, 107, 108
      inverse problem solution, 112–115
      offset dependence, 110–112
      variance, 108–110, 115, 117
Two-dimensional finite-difference method (FDM) models
   anisotropic heterogeneities, 211–213
   heterogeneity scale effect, 214, 215
   heterogeneous plate model, 210, 211
   high-Q and high-V models, 208–210
   P-and S-wave velocities, 204, 205
   plate thickness, 213, 214
   quasi-lamina structure, 206, 207
   slab guided waves, 208
Two-dimensional random elastic media
   plane wavelet incidence
      Gaussian ACF, 64–66
      vs. numerically simulated envelopes, 66–69
   radiation from point source, 78–83
      Gaussian ACF, 78–81
      Markov envelopes *vs.* FD envelopes, 82–83
Two-frequency mutual coherence function (TFMCF)
   cylindrical wave analytic solutions, 90
   plane wave analytic solutions
      three dimensions, 87, 88
      two dimensions, 88, 89
   plane wavelet incidence
      angular spectrum, 56

Two-frequency mutual
    coherence function
    (TFMCF) (cont.)
  definition, 52–53
  Gaussian ACF, 56, 57
  Markov approximation, 53
  stochastic master equation,
    53, 54
  radiation from point source, 71
  spherical wave analytic solutions,
    89, 90

U

Uniform diffusive layer, 366

V

Vector-wave envelope synthesis,
    Markov approximation
  Born approximation scattering
    coefficients
      conversion and large-angle
        scattering, 83, 84
      Monte Carlo solution, 84, 85
  plane wavelet incidence
    angular spectrum, 56
    Gaussian ACF, 56–62, 64–66
    intensity spectral density, 54,
      55
    plane P-wavelet, 51, 52
    plane S-wavelet, 62–64
    random media ensemble, 52
    TFMCF, stochastic master
      equation, 52–54
  two-dimensional random
    elastic media, 64–69
    vs. numerically simulated
      envelopes, 66–69
    wandering effect, 55
  wave equations,
    inhomogeneous
    media, 51
  radiation from point source
    Gaussian ACF, 72–76, 78–81

intensity spectral density
    (ISD), 71, 72
  Markov envelopes vs. FD
    envelopes, 82, 83
  spherically outgoing
    P-wavelet, 70, 71
  spherically outgoing
    S-wavelet, 76–78
  TFMCF, stochastic master
    equation, 71
  realistic ACFs, 85, 86
Vertical seismic profiling (VSP)
  reflection image, 260, 261
  walkaway and 3D, 260
  zero-offset, 259, 260
Virtual doublets, seismic noise
  definition, 393
  impulse response, 391, 392
  relative velocity changes, 392,
    393
Volcanic tremor, 360, 361
Volcanoes
  classification, 353, 354
  coda-Q^{-1} observation, 354–357
  diffusion model
    two-layer media, 363–365
    uniform half space, 361–363
  energy-transport theory
    coda-localization effects, 366,
      367
    uniform diffusive layer, 366
    uniform half space, 365, 366
  intrinsic and scattering
    attenuation
    2-D transport theory, 360, 361
    energy-flux model, 359
    MLTWA technique, 357
    volcanic tremor, 360, 361
    Wennerberg method, 358, 359
Volcano-tectonic (VT) earthquakes
  coda-localization effects, 366,
    367
  coda-Q^{-1} observation, 354–357
  2-D transport theory, 360, 361

energy envelop, 368
  shape, 368
  uniform half space, 365, 366
  Wennerberg method, 358, 359
von Kármán correlation model, 128,
    129, 141, 162
VSP. See Vertical seismic profiling

W

Walkaway VSP, 260
Wandering effect, 55
Wave envelope synthesis, Markov
    approximation, 44
Wave fields
  autoregressive (AR) reflection
    sheet, 223
  fluctuations, 225–227
  generation, 223–225
WaveSonic™, 250
Weak localization effect
  applications
    dispersion relation of surface
      waves, 13–15
    Lamb mode, 15
    scattering media, 16, 17
  Bessel and Green functions, 11
  Bethe-Salpeter equation, 10
  coherence tensor, elastic
    wavefield, 7–9
  coherent backscattering, 5
  elastic waves, 6
  multiple scattering paths, 4
  scalar partial wave, 3
  seismology, 1
  transport theory, 7
Wennerberg method, 358, 359
WKBJ Green's function, 23–25

Z

Zero-offset VSP, 259, 260
Z-Seis, piezoelectric source, 255,
    256